The Physics of Interstellar Travel

What is the physics behind getting a spacecraft to the nearest stars? What science can it do when it gets there? How can it send back data over enormous distances? Drawing on established physics, Coryn Bailer-Jones explores the various challenges of getting an uncrewed spacecraft to a nearby star within a human lifetime. In addition to propulsion methods such as nuclear rockets and laser sails, this book examines critical issues such as navigation, communication, and the interstellar medium. Starting from fundamental concepts, readers will learn how a broad spectrum of physics – ranging from relativity to optics, and thermodynamics to astronomy – can be applied to address this demanding problem. Assuming some familiarity with basic physics, this volume is a comprehensive and self-contained introduction to interstellar travel, and an indispensable guide for studying the literature on deep space exploration. This title is also available as open access on Cambridge Core.

Coryn Bailer-Jones is a staff member at the Max Planck Institute for Astronomy in Heidelberg where he does research on stellar and Galactic astrophysics. He teaches physics at Heidelberg University and in 2017 published the textbook *Practical Bayesian Inference*. For many years he played a leading role in the European Space Agency's space astrometry mission, Gaia.

"*The Physics of Interstellar Travel* delivers an impressively rigorous and clear exposition of the scientific and engineering challenges and remarkable promise of future interstellar missions. I found this work to bridge theoretical depth with practical vision, making it indispensable reading for aerospace professionals and science enthusiasts alike. With thorough analysis, technical clarity, and inspiring scope, Bailer-Jones sets a new benchmark for scientific literature on interstellar exploration."

Les Johnson, Chief Technology Officer, NASA MSFC
(retired), author of *A Traveler's Guide to the Stars*

"This is a comprehensive extension of current textbooks in astronautics into the realm of missions once thought to be impossible. The book's mathematical rigour should clarify for rising students the realization that steps we take today can result in practical outcomes, with the goal of reaching another star conceivable by the end of this century. Where science and engineering have not yet taken us, this textbook illustrates the direction of steps forward, aiding the community in the construction of the needed roadmap."

Paul Gilster, author of *Centauri Dreams*

"This book beams you up – literally! In this engaging and rigorous textbook, Coryn Bailer-Jones lays out the physics foundations behind interstellar travel. From nuclear rockets to laser sails, from the hazards of the interstellar medium to the challenge of beaming data home across light years, this book shows how physics can be applied to one of humanity's boldest ambitions. Clear, comprehensive, and thought-provoking, *The Physics of Interstellar Travel* is essential reading for students, researchers, and anyone curious about how to reach the stars."

Professor Andreas Hein, Université du Luxembourg

"An authoritative and pedagogical guide for anyone captivated by the prospect of deep-space exploration, *The Physics of Interstellar Travel* is grounded in existing proven physics while providing the insight to envision spacecraft reaching the nearest stars within a human lifetime."

Professor Sara Seager, Massachusetts Institute of Technology,
author of *Exoplanet Atmospheres*

"Coryn Bailer-Jones' book is aimed primarily at "undergraduate science students who have completed courses in basic physics", although it certainly merits a much wider audience. Based on his BSc and MSc course delivered at Heidelberg University over several years, it is authoritative, beautifully written, and a compulsive read. The book's broad scope serves as a valuable guide to the field's diverse literature, with applications extending beyond that of interstellar travel alone. It is, above all, highly stimulating. This book will surely help to focus serious interest in long-term plans for one of the important next steps in our exploration of the universe."

Professor Michael Perryman, ESA (retired) and University
College Dublin, author of *The Exoplanet Handbook*

The Physics of Interstellar Travel

CORYN A. L. BAILER-JONES

Max Planck Institute for Astronomy, Heidelberg

CAMBRIDGE
UNIVERSITY PRESS

Shaftesbury Road, Cambridge CB2 8EA, United Kingdom

One Liberty Plaza, 20th Floor, New York, NY 10006, USA

477 Williamstown Road, Port Melbourne, VIC 3207, Australia

314–321, 3rd Floor, Plot 3, Splendor Forum, Jasola District Centre, New Delhi – 110025, India

Cambridge University Press is part of Cambridge University Press & Assessment,
a department of the University of Cambridge.

We share the University's mission to contribute to society through the pursuit of
education, learning and research at the highest international levels of excellence.

www.cambridge.org
Information on this title: www.cambridge.org/9781009689328
DOI: 10.1017/9781009700429

First published 2026

A catalogue record for this publication is available from the British Library

A Cataloging-in-Publication data record for this book is available from the Library of Congress

ISBN 978-1-009-68932-8 Hardback

Cambridge University Press & Assessment has no responsibility for the persistence
or accuracy of URLs for external or third-party internet websites referred to in this
publication and does not guarantee that any content on such websites is, or will
remain, accurate or appropriate.

For EU product safety concerns, contact us at Calle de José Abascal, 56, 1°, 28003 Madrid, Spain, or email
eugpsr@cambridge.org.

Contents

Preface

What is the physics behind getting a spacecraft to the nearest stars? How fast could it go? What would it encounter along the way? How can a spacecraft navigate autonomously? What science would it do at the target star? And how can we send back the data over enormous distances?

This book is about the physics of interstellar spaceflight. If a spacecraft is to get to even the nearest star within 50 years, it needs to travel at almost 10% the speed of light. This is well over a thousand times faster than the fastest spacecraft that has already left the solar system. For fundamental physical reasons, no chemical rocket can ever achieve anything close to this. We therefore need to look at alternative propulsion methods, such as nuclear fusion rockets and laser-propelled sails. These don't yet exist, but they are physically possible, and important steps are being been taken towards achieving them. Sending a robotic spacecraft is doable, and not so very far beyond our current technological reach.

The physics of interstellar travel is about much more than propulsion, however. In this book we will examine a broad range of topics, including how to navigate in deep space, the effect of high velocity particle impacts, the instruments needed to study exoplanets close up, how to communicate using individual photons, and much more besides. On this journey we will delve into many areas of physics, including classical mechanics, special relativity, optics, electromagnetism, thermodynamics, materials, nuclear and particle physics, astronomy, and of course rocket science. We will stick to established physics – I don't consider outlandish ideas such as warp drives and wormholes – but we won't be overly constrained by current technology or budgets.

Many articles and a number of books have been written about specific aspects of interstellar travel. Some of these are rather technical or assume a high level of background knowledge, so can be hard to connect to the basic physics. By looking at the fundamental concepts, and taking a broad view, I aim to give the reader a sense of the big picture and an understanding of the physical principles at work.

This book is targeted primarily at those who have completed introductory university courses in physics. By bringing together a wide range of topics and drawing on recent literature, this book should also be useful for researchers. I show how physics that has been learned in sometimes abstract ways can be applied in a practical setting with a clear goal. Many topics that I cover have a wider relevance too, such as the principles of rockets, Keplerian orbits, satellite navigation systems, relativistic optical effects, energy from nuclear reactions, and the basics of communication.

As the scope is broad, I cannot go into great detail in all topics, and I omit some potential technologies. My goal is to identify key issues and some proposed solutions, but without discussing every subtlety or giving a complete road map for how to achieve interstellar travel. Having read this book, the reader should be well equipped to dive deeper into topics covered by other books and articles. The appendix provides both a discussion of the literature sources I have used and suggestions for further reading.

In addition to being suitable for independent study, this book can serve as the basis for a full semester lecture course on interstellar travel, or parts could be selected for a shorter course. The content is modular, so chapters on specific topics could also be used to support other courses, for example on propulsion methods or relativity applications. Section 1.6 gives an overview of the chapters and their dependencies. Even if one is not particularly keen on interstellar travel, it is an excellent vehicle for learning about many topics in physics (excuse the pun). Learning is consolidated by doing, so I provide some exercises at the end of each chapter.

This project has grown out of a seminar course and a full semester lecture course I have taught to physics BSc and MSc students at Heidelberg University over a number of years. Their feedback and suggestions have been very useful in developing this work. For fruitful discussions and helpful comments on the manuscript I would like to thank Nora Bachmann, Werner Becker, Ian Crawford, Christian Fendt, Elisa Haas, David Hägele, Andreas Hein, René Heller, Tom Herbst, Ramon Khanna, Angelique Kahle, Pekka Janhunen, Les Johnson, Anne Jones, Lisa Kaltenegger, Laura Kreidberg, Stan Letchev, Richard London, Kelvin Long, Thomas Müller, David Messerschmitt, Carl Mungan, Kevin Parkin, Sabina Pauen, Markus Pössel, Bhavesh Rajpoot, Nick Sleep, Paul Steimle, Martin Tajmar, Ted von Hippel, Fabian Walter, and the anonymous reviewers. I thank colleagues and friends for their interest in, and encouragement of, this project. I am grateful to Roxanne Daruwalla, Vince Higgs, Zoë Lewin, and Aleksandra Serocka at Cambridge University Press (CUP) for their assistance in preparing the manuscript, and to Simone Kronenwett, the librarian at my institute, the Max Planck Institute of Astronomy (MPIA), for acquiring some of the literature. Special thanks go to the MPIA and Laura Kreidberg as well as to the Max Planck Society via the Max Planck Digital Library for supporting the publication of this book as open access.

Plots and sketches not credited to other sources were made by the author using R and inkscape. Artificial intelligence tools were not used to help write this book (other than to check a few commas). Despite my best efforts, there are no doubt some errors in this book. Notification of possible mistakes is welcome. Corrections, as well as additional information and extra exercises, can be found at www.mpia.de/homes/bailer-jones/pit.html or via the CUP website.

Introduction

1.1 The interstellar journey starts

On 5 September 1977, just before 09:00 local time, a Titan rocket lifted off from Cape Canaveral in North America. After just two minutes, the rocket was 40 km above the Earth and moving at many times the speed of sound. The launch almost went disastrously wrong at this point because the first stage engines cut off too early. Fortunately, the Centaur upper stage had just enough fuel to compensate. This gave the rocket's payload enough kick to escape Earth's gravity and to send it on its interplanetary mission.

The payload was the Voyager 1 spacecraft; its mission, to explore the outer planets. For the next year and a half, Voyager 1 sped outwards through the solar system, reaching Jupiter in March 1979. Passing within just four Jupiter radii of the planet's surface, it used its cameras, spectrometers, and magnetometers to gather unrivalled information about the planet and its moons as it sped by (figure 1.1).

A carefully planned trajectory allowed the spacecraft to swing around Jupiter and continue on towards Saturn. After another year and a half, in November 1980, it reached the ringed planet and scrutinized it as well as possible in the short time available during the flyby. Moving too fast and having insufficient fuel to slow down, it performed a gravity assist around Saturn to take it past the planet's largest moon, Titan, on its way out.

Moving now at 22 km s^{-1} (4.6 au yr^{-1}) relative to the Sun, Voyager 1 headed off into deep space. Travelling more or less in a straight line, its velocity slowly dropped as it climbed out of the Sun's gravitational well. But it already exceeded the escape velocity of the Sun, and short of an extremely unlikely collision, it was now guaranteed to escape the solar system forever.

For three decades Voyager 1 continued its lonely journey outward, periodically returning information about the gas and dust it encountered, and even taking a snapshot of the distant Earth against the empty blackness of space. Then, in August 2012, when it was 121 au from the Sun, Voyager 1 measured an unremarkable, yet significant, decrease in the ambient particle density. It had crossed the heliopause, what is often taken as the otherwise invisible boundary between the solar system and interstellar space. Voyager 1 had become the first human-made object to leave the solar system.

As of mid 2025, nearly 50 years after its launch, Voyager 1 is 167 au from the Sun and moving at 17 km s^{-1} (3.6 au yr^{-1}). It has just enough energy to operate a couple of instruments and communicate feebly with the Earth. If it were travelling directly towards our nearest star, Proxima Centauri, Voyager 1 would take 77 thousand years to get there. It will instead be hundreds of thousands of years before it gets much closer to any other star. But long before then, its power will fade, its instruments will fail, and we shall never hear from Voyager again.

Fig. 1.1   Artist's impression of the Voyager spacecraft. Credit: NASA/JPL.

1.2   The dream of travelling to the stars

Sad though the fate of Voyager may seem, it was not only a highly successful interplanetary mission. It has also paved the way for interstellar travel. For hundreds of years, people have dreamed of travelling not just to the Moon or the planets but also to the stars. Humans have walked on the Moon, and for robotic spacecraft interplanetary travel has moved from the realm of science fiction to science fact.

Interstellar travel is a whole new dimension, though. Proxima Cen lies just over 4×10^{13} km from the Sun (4.25 ly, 269 000 au). It is difficult to appreciate distances of this size. If we scale down the universe so that once around the real Earth becomes the distance to Proxima Cen, a scale factor of almost exactly 10^9, then the Earth–Sun distance becomes 150 m, the Sun has a diameter of 1.4 m, and the Earth has a diameter of 1.3 cm. The speed of light is $30\,\mathrm{cm\,s^{-1}}$. As of mid 2025, Voyager 1 in this scaled-down universe is 25 km from the Sun and moving away from it at 1.5 m per day. It's a long way to the nearest star.

Even with chemical rockets a mind-boggling hundred times larger than those we have now, we could not accelerate spacecraft to more than a few tens of kilometres per second. Getting to Proxima Cen would still take over ten thousand years. Even if humans were gifted with near-infinite patience, any instruments we sent at this speed would be hopelessly outdated by the time they arrived, if they still worked at all, and if civilization was still around to remember it had sent them. If we want to travel to the stars, we have to move faster.

Fortunately, there are technologies on the horizon that could get spacecraft to the nearest stars within a human lifetime. These include fusion propulsion and laser sails. If we could accelerate a spacecraft to 10% the speed of light, we could get to Proxima Cen in just over 40 years. There is no physical barrier to this. Yet the technological challenges are significant, not just in terms of propulsion but also in terms of navigation, onboard power, and communication, to name just some.

Serious research into interstellar travel has been going on for over half a century. A milestone study was undertaken by the British Interplanetary Society in the 1970s, the result of which was a 50 000-tonne fusion-powered rocket called Daedalus. It aimed to carry a 450-tonne spacecraft to Barnard's Star, the second nearest star to the Sun, in 50 years at an average of 12% the speed of light. More recent studies have focused on lower mass payloads and smaller rockets, and have explored other forms of propulsion. We probably won't leap to the stars in one step, though, and other work has examined precursor missions to explore nearby interstellar space.

1.3 The scope of this book

'It is hard to make predictions, especially about the future.' This maxim surely applies to space travel. Every generation has a tendency to predict future progress as an extension of what it currently has, rather than as something radically different. Futurists in the eighteenth century thought of massive balloons, those in the mid twentieth century of fast atomic spacecraft. Maybe one day we will have warp drives like those in Star Trek, hyperspace as in Star Wars, or wormholes like those in the film Interstellar. But for now, these just remain dreams or theoretical ideas without any hope of becoming reality.

The goal of this book is to examine the physics of getting a real spacecraft to a nearby star with a travel time below a human lifetime. I choose this travel time as being the longest that society would be willing to wait (although I will return to this in the final chapter). Given the range of challenges interstellar travel involves – propulsion, particle shielding, communication, etc. – we will touch on many aspects of physics. This includes classical mechanics, thermodynamics, special relativity, nuclear and particle physics, electromagnetism, optics, and astronomy. We will even look at some fairly extreme applications of physics, such as gigawatt lasers and antimatter propulsion. But we will only consider ideas based on existing proven physics. Things like faster-than-light travel and teleportation are not considered. On the other hand, we will look at solutions that involve some extreme technology that does not yet exist, or that appears absurdly expensive or difficult, but which is nonetheless plausible. It is worth remembering that the computers we have today were physically possible a few generations ago, but were technologically and economically unthinkable at the time.

The other bound on the scope of this book is that we shall only look into sending robotic, uncrewed spacecraft. Human interstellar spaceflight is a fascinating topic, but it involves much beyond physics, such as the biology and psychology of crews. Those interested in sending humans to the stars should nonetheless find most of this book relevant, not only because we will surely send robotic spacecraft first, but also because all of the physical issues still apply to sending humans.

1.4 Types of interstellar mission

When we speak of 'travelling to the stars' we need to be more specific about what we mean. It is useful to split the types of interstellar mission into four categories of increasing difficulty. These are: flyby; orbiter; lander; return. A flyby does exactly that: hardly slowing down, if at all, at the destination. An orbiter, in contrast, must almost completely decelerate to get into an orbit around the target star or one of its planets. As we shall see in later chapters, this requires considerably more rocket propellant or – in the case of sailcraft – an elaborate system for braking against the interstellar medium. A lander mission not only requires the spacecraft, or a probe deployed from it, to rendezvous. It must also land on the surface of a body in the star system. This requires an even higher level of autonomy than the first two scenarios. The final mission type is designed to return to Earth, carrying not only data but perhaps also physical samples, maybe even ones taken from a lander. A return mission requires the spacecraft to decelerate at the star, accelerate to return, and then decelerate back at the Earth. This requires considerably more energy or propellant than even an orbiter or lander, and makes the whole mission significantly more difficult.

As a spacecraft travelling at 10% the speed of light would travel 1 au in a little over an hour, a single flyby mission would gather far fewer data than a spacecraft that goes into orbit and could operate for many years. Even though a fast flyby could spend a long period of time sending its data back to Earth, an orbiter is highly preferred. But given how difficult any interstellar mission will be, a flyby seems the more likely choice for the first mission.

1.5   Units used in this book

This book covers a range of topics, so we will need a range of physical units.

In general I use SI units and multiples thereof, such as GW and μm. When discussing electromagnetism, I use SI units and not the still widespread cgs units, which can involve differences of factors of 4π, ϵ_0, and μ_0. I use the symbol 'g' for grammes and 't' for tonnes.

The main exception to the use of SI units is conventional units in astronomy, in particular for the large distances involved. Here I use the astronomical unit (au $= 1.496 \times 10^{11}$ m), the light year (ly $= 6324$ au), and occasionally the parsec (pc $= 3.262$ ly). The light year is rarely used by astronomers, but it is more convenient than the parsec in the context of travel at relativistic speeds. I also use units scaled to the Sun (subscript $\odot$), Jupiter (subscript Jup), and Earth (subscript $\oplus$) for quantities such as luminosity (e.g. $L_\odot$), mass (e.g. M_{Jup}), or radius (e.g. $R_\oplus$). It may be useful to know that $M_\odot = 1048\,M_{Jup}$, $M_{Jup} = 318\,M_\oplus$, $R_\odot = 9.7\,R_{Jup}$, and $R_{Jup} = 11.2\,R_\oplus$ (using equatorial radii).

For angles I use both radians and degrees, depending on context, as well arcseconds ($''$), milliarcseconds (mas), and microarcseconds (μas) as is conventional in astronomy. The conversions are $1° = 3600''$ and $1'' = 4.846\,\mu$rad.

There are various definitions of the year, such as tropical year or sidereal year. Normally the difference will not be relevant, but in the context of velocities I use the Julian year, which is 365.25 SI days. This is the case for the velocity unit au yr^{-1}, for example, which is 4.740 km s^{-1}.

I often use the term 'speed' to refer to the magnitude of the vectorial velocity. Vectors are written using bold font. Thus the velocity is $\mathbf{v}$ and the speed is $v = |\mathbf{v}|$. The scalar (dot) product of two vectors is written $\mathbf{x} \cdot \mathbf{v}$. I frequency use c as the unit of the speed of light.

As is standard in the context of nuclear and particle physics, I often use the electron volt (eV) as a measure of energy, where 1 eV $= 1.602 \times 10^{-19}$ J. The atomic mass unit is denoted m_u, which I take to be $1.66053904 \times 10^{-27}$ kg. In conventional energy units this is $m_u c^2 / e = 931.49$ MeV.

For dimensionless fractional measures, ppm denotes parts per million. A dimensionless but not unitless quantity is the bit of information, with symbol b, and its multiples kb, Mb, etc., as well as the byte, symbol B, which equals 8 b.

There is a myriad of terms that describe electromagnetic radiation according to whether we are considering power per unit area, per unit wavelength, and/or per unit solid angle, and no rigid convention. *Luminosity* is a power, units W, emitted by a source or received by a receiver. I use the term *intensity* to mean the power per unit area, units $W\,m^{-2}$. This is common in astronomy, although in other fields this is often called the 'irradiance' or 'flux density'. To describe the intensity per unit wavelength (or frequency), I use the term *spectral intensity*, units $W\,m^{-2}\,m^{-1}$ (or $W\,m^{-2}\,Hz^{-1}$). This is sometimes called the 'spectral irradiance' in other sources. The spectral intensity per unit solid angle, the power per unit area per unit wavelength (or frequency) per unit solid angle, is the *spectral radiance*, units $W\,m^{-2}\,str^{-1}\,m^{-1}$ (or $W\,m^{-2}\,str^{-1}\,Hz^{-1}$). The integral of this over all wavelengths (or frequencies), the power per unit area per unit solid angle, is the *radiance*, units $W\,m^{-2}\,str^{-1}$. Finally, in just one place we need the power per unit wavelength, the *spectral flux*, units $W\,m^{-1}$. I will generally define these less obvious terms again when I use them, but if confused, refer back here.

A brief word on notation. As this book covers a diverse set of topics, it is hard to achieve completely consistent notation across all chapters, so some symbols are used to mean different things in different contexts. For example, P indicates pressure in some places but power in others. Symbols are defined when they are introduced, so hopefully it will always be clear what is meant.

1.6 Chapter overview

The chapters of this book are best read in order because some earlier chapters provide background material for later ones, as shown in figure 1.2. Readers can nonetheless jump straight to chapters that interest them most, as earlier material is cross-referenced where required. I now give a brief overview of each chapter.

Chapter 2 looks at the motivation for interstellar travel, highlighting in particular how it can answer questions in astronomy, planetary science, and the search for life that are not answerable from the Earth (or nearby space) within the foreseeable future, or perhaps ever. I summarize nearby star targets and their known exoplanets.

Chapter 3 describes the basics of rocketry, introducing the all-important rocket equation. The essential concepts of thrust, effective exhaust velocity, and power are explained, and trade-offs between these in terms of travel time and engine requirements are examined.

Chapter 4 covers the fundamentals of Keplerian orbits and orbital manoeuvres. Understanding them is essential for rendezvous at the target planet as well as for solar sailing, for example. I explain the gravity assist and the Oberth effect, which are fundamental topics in astronautics.

The following chapter (5) is on thermal rockets, primarily chemical rockets, illustrating the basic thermodynamics and fluid concepts. We look at various real rockets that use liquid or solid fuels.

Ion engines, covered in chapter 6, are non-thermal rockets that provide a low but long-term continuous thrust. We examine both the principles and their use in the Dawn spacecraft that visited two asteroids.

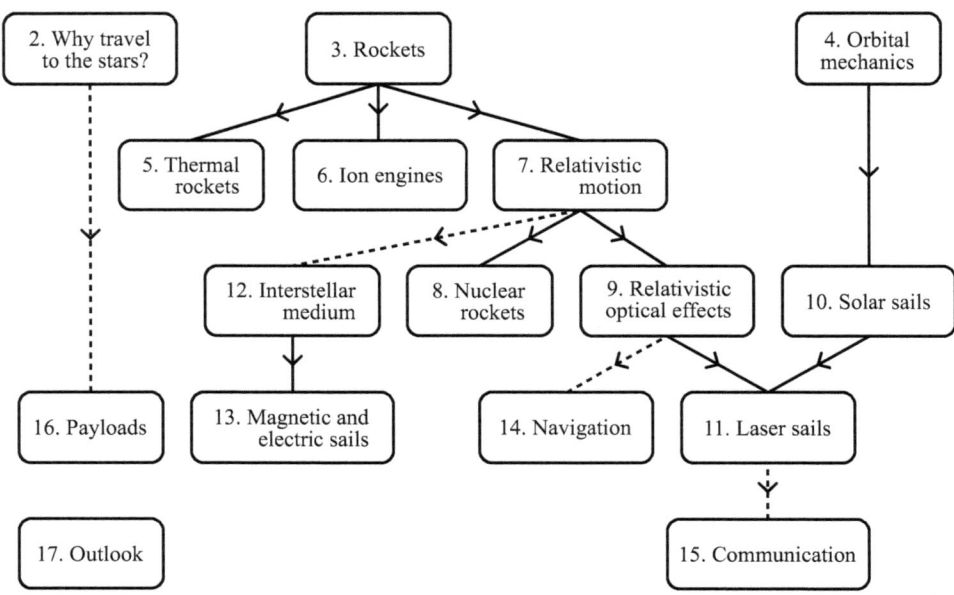

Fig. 1.2   Chapter dependencies. Dashed lines indicate a weaker dependence or one between specific sections.

We look at relativistic mechanics in chapter 7. After briefly revising the Lorentz transformations, we go over special relativity in accelerating systems. This chapter also covers the relativistic rocket equation, the photon rocket, and the mixed matter/photon rocket concept, which we will need in later chapters.

Armed with the appropriate background, we turn to nuclear rockets in chapter 8, covering fission, fusion, and antimatter as sources of energy and propellant. These come in the form of reactors or as miniature explosions coupled to the spacecraft via electromagnetic fields. We examine some prototypes that were built in the 1960s as well as the prototype mission design Daedalus and its successors. Fusion rockets are one of the two most promising ways of getting to the stars quickly.

Chapter 9 continues our exploration of relativity by looking at its effect on optics, in terms of the Doppler effect, aberration, and other consequences of the Lorentz transformations. We will need these in later chapters.

The big drawback of rockets is that they require a lot of their propellant to accelerate the remaining propellant, and so are intrinsically inefficient. Sailcraft overcome this by externalizing the propulsion, either as solar photons or as laser photons. Solar sails are an active area of research and development for interplanetary missions (they have already flown), which we look at in chapter 10 along with the basic principles of photon propulsion. In the following chapter (11) on laser sails, we look at the use of extremely powerful lasers to accelerate sails to relativistic velocities. Time dilation and finite photon travel times render the mechanics of laser-driven sails non-trivial. In addition to the challenging material properties required for the sailcraft themselves, we examine the lasers plus the optical arrays to focus them over large distances. Laser sailing is the other most promising means of travelling to the stars, and we look at some proposed missions.

In chapter 12 we examine the content and properties of the interstellar medium that the spacecraft must travel through. This presents both challenges, such as high energy particle impacts, and opportunities for scientific research. We also look at the Bussard ramjet, an ingenious if somewhat impractical idea for collecting and fusing interstellar protons as a source of both rocket propellant and energy.

In chapter 13 we return to sails, this time magnetic and electric sails. These exploit the charged particles in the interstellar medium or the heliosphere of the target star as a way of decelerating an incoming spacecraft.

To boldly go we need to know where we are and where we are heading, so chapter 14 on navigation explores how to determine one's orientation, location, and velocity in space. We start with ranging methods used for interplanetary spacecraft as well as inertial methods. For autonomous navigation in deep space, we examine two approaches. The first uses the three-dimensional positions of the stars; the second uses the timing and direction of pulsars. This latter method bears some relation to global navigation satellite systems, the principle of which I also briefly explain.

The whole point of travelling to a distant star is to return data. The problem of communicating over vast distances is taken up in chapter 15. Diffraction and noise are fundamental issues here. This takes us on a brief journey into information theory to cover concepts such as signal modulation, coding, and channel capacity. Photon fluxes may be so low that we need to consider quantum effects. We will consider the optimal wavelength to use for communication and the relative merits of Earth-based vs space-based receivers.

The penultimate chapter (16) considers what the spacecraft should take in terms of scientific instruments as well as how it can power and control itself. Here we are guided by instruments already used on interplanetary spacecraft. We also discuss the idea of sending relays or swarms of many smaller spacecraft to provide redundancy against failure or so they can operate collectively.

The final chapter (17) first takes a brief speculative look at some of the more outlandish ideas for achieving interstellar travel, such as warp drives and wormholes (those things I said I would not consider). I then conclude with an outlook of the most promising technologies for interstellar travel and how this may develop in the future.

So as not to disrupt the reading too much, I refer to literature sources mostly in the bibliographic essay in the appendix instead of in the chapters. There is a surprisingly large amount of technical literature on interstellar travel, so I don't try to be comprehensive. I instead mention the main sources I have used and those I think will be most useful for deeper study.

Without further ado, let us embark on our voyage by asking why we want to undertake it at all.

Why travel to the stars?

There are diverse motivations for travelling to the stars, both scientific and non-scientific. In this chapter we will examine primarily the scientific motivations, in particular what we can learn about stars and exoplanets as well as how to look for signs of extraterrestrial life. I will give a brief survey of the nearest stars and their planets, in particular the nearest star system, Alpha Centauri, which happens to be a triple star system with at least one planet. We will also look briefly at some of the science a spacecraft can do en route to a star as it moves through the interstellar medium.

2.1 Motivations

2.1.1 General scientific considerations

The main motivation for sending robotic spacecraft to another star system is arguably scientific discovery. As we shall explore below, getting up close to a star system will answer questions about not only the star itself but also its planets, their moons, and, perhaps most excitingly, extraterrestrial life. But as getting a spacecraft to another star will be difficult and expensive, we need to ask how much of this science we could already do from the Earth (including astronomical observatories in orbit or on the Moon). For this we have to look ahead to the time when a spacecraft would arrive at the target system, say in the early 2100s, and ask what Earth-based observational capabilities we may have by then. Perhaps some or all of the scientific objectives of an interstellar spacecraft could be achieved from the Earth more easily, better, or at lower cost?

The first obvious advantage of an interstellar spacecraft is spatial resolution. The best achievable angular resolution of an imaging system is set by the diffraction limit, which is of the order λ/d, where λ is the wavelength of light and d is the size of the collector. When used to image a source at distance r, the corresponding spatial scale is $\lambda r/d$. The diffraction limit of a 100 m diameter telescope on the Earth observing at 1 μm is 2 mas (10^{-8} rad), corresponding to a spatial scale at Proxima Cen of 0.0027 au or 63 $R_\oplus$. This is insufficient to spatially resolve even a giant planet (Jupiter's equatorial radius is 11.2 $R_\oplus$). The ratio of the spatial resolutions (inverse of spatial scale) achieved by two telescopes of different sizes and at different distances from the source is $(r_2/r_1) \times (d_1/d_2)$ at a common wavelength. So compared to a 100 m telescope observing from the Earth, a spacecraft with a 10 cm aperture observing from 0.1 au has a resolution $(0.1/100) \times (270\,000/0.1) = 2700$ times larger, and achieves a spatial scale of 0.023 $R_\oplus$, enough to map the surface. We could probably navigate nearer to the target during a flyby, and orders of magnitude nearer if the spacecraft went into orbit around a planet.

The second advantage of the close proximity of a spacecraft is light gathering power. Telescopes are built large not only to increase their resolution, but also to increase their light gathering power, which depends on the area of the aperture, which scales as d^2. The light from a distant source, on the other hand, spreads out on the surface of a sphere, so the observed intensity (power per unit area) of the source at distance r varies as $1/r^2$. Hence the ratio of the light gathering power of our two observing systems varies as $(r_2/r_1)^2 \times (d_1/d_2)^2$. Thus the 10 cm telescope observing from 0.1 au gathers $2700^2 = 7.3$ million times as much light as the 100 m telescope observing from 4.25 ly. This will enable the observation of surface features, small individual bodies, and low contrast ratios that will never be visible from the Earth.

A diameter of 100 m may be the limiting size for a single Earth-based telescope dish, but telescopes are not limited to being single large collectors. Interferometry is the means by which multiple telescopes are connected to achieve a resolution determined by the separation between them (the baseline), rather than the size of the individual dishes (see section 11.2.2). The four 8 m diameter optical telescopes of the European Southern Observatory's (ESO's) Very Large Telescope (VLT) can be combined with a baseline of around 100 m, for example.[1]

To achieve the resolution of the spacecraft configuration just mentioned, we need a much larger baseline at the Earth of $d_2 = d_1 r_2/r_1 = 270$ km. To keep the individual telescopes in phase (essential for interferometry), light has to be passed between the telescopes, and the wavefronts tracked over a fraction of a wavelength. This would be extremely difficult over such a large baseline with weak signals. Without going into the details, such long baseline interferometry has two other significant drawbacks. First, the field of view would be extremely small. Second, the individual telescopes would cover only a tiny fraction of the synthesized (270 km) aperture (sparse 'uv-plane coverage', in the jargon). This results in highly ambiguous images – one can barely even speak of this type of interferometry as imaging – so the data would be of much lower quality than that from the interstellar spacecraft. One can build up uv-plane coverage using multiple orientations, but this too has its limitations.[2]

Even if we lowered the requirements for Earth-based observations so as to only achieve 10 resolution elements at 1 μm wavelength across the diameter of an Earth-sized planet at Proxima Cen, the baseline would still have to be 32 km, which does not mitigate the difficulties. Interferometry itself also does not improve the light gathering power, so here the interstellar spacecraft remains vastly superior. Although this can be compensated to a small degree by using longer exposure times with Earth-based telescopes, there are limits to this, and it does not help for temporally resolving variable phenomena, such as planet rotation or cloud motions.

Beyond resolution and light gathering, there are at least two more major benefits of travelling to a star system. The third one is perspective. When we observe a stellar system from the Earth, we only ever see it from one direction. If we see a star or planet near pole-on, we will never see its equatorial features. If we view an object perpendicular to its spin axis, then

---

[1] This has already been used to image – but not resolve – giant planets around nearby young stars (ages below about 100 Myr). Such young giant planets are still hot from their formation and so are significantly brighter than older or lower-mass planets. This makes them much easier to pick out in the glare of their brighter host star.

[2] There are plans to build a space-based interferometer with much smaller baselines (up to a few hundred metres) that uses so-called nulling interferometry to block the starlight in order to obtain spectra of exoplanets. This would still not resolve a planet's surface, however.

in principle we see different longitudes as it rotates, but we never see the poles. A spacecraft that enters a planetary system, especially if it goes into orbit rather than flying by, has the opportunity to observe an object from many viewpoints and with different illuminations.

The fourth benefit of an interstellar spacecraft, and one that can never be fulfilled from afar, is in situ measurements of atoms, ions, dust particles, and magnetic fields. Just as interplanetary spacecraft already do in the solar system, an interstellar spacecraft could measure these directly.

Finally, if a spacecraft not only goes into orbit around an exoplanet, but actually dropped probes to fly through its atmosphere or land on its surface, then we enter a whole new realm of exploration that is not possible remotely.

From a scientific point of view, whether a specific interstellar mission is worth doing, even if technically feasible, depends on whether we think the extra science it can do is worth the price. This needs to be compared to what Earth-based astronomical capabilities we will have by the time the spacecraft would return its data. Recall that the spacecraft is likely to take at least 50 years to get to its destination, and that data would only start to arrive back five or more years after that. But even if we have kilometre-sized space telescopes by then, an interstellar spacecraft can do much more. One day there might be breakthroughs in photonics that allow the phase at optical wavelengths to be measured, and thus allow even larger interferometers to be built. Or there might be developments of metamaterials that overcome the diffraction limit. But such capabilities remain unknown, and even if realized, an in situ mission still has significant unique capabilities.

2.1.2 Non-scientific motivations

Beyond the scientific motivations for interstellar travel, there are also technological, economical, and cultural motivations.

Blue sky research in many areas of science, not least physics and astronomy, have historically led to numerous technological breakthroughs. Astronomical observations and theoretical developments in the sixteenth and seventeenth centuries powered the Copernican revolution and led to the theory of gravity. Research in nuclear physics led to medical imaging, quantum mechanics gave rise to computer chips, and without general relativity there could be no workable satellite navigation systems. Miniature cameras like those in mobile phones were simulated by the need for low-mass cameras on spacecraft, and other inventions such as contactless infrared thermometers, wireless headsets, and even advanced running shoes have their origins in space research.

There are many other technologies that have been improved or refined by their need for use in space, which have then led to Earth-based applications. As will become clear in later chapters, the development of interstellar spacecraft will inevitably have comparable technological spinoffs, for example in the fields of lightweight but strong materials, fusion power, high-power lasers, low photon count communications, and further miniaturization of instruments. Such technological spinoffs create economic activity. The investment in developing the capabilities for interstellar travel also has direct economic benefits in terms of job creation and skill development.

Achieving interstellar travel will be expensive. As there are still many uncertainties concerning which technologies are most suitable and how they could be developed, it is very hard to put a price tag on it. Moreover, some of the developments would be indirect or come from – or go into – other applications, such as nuclear fusion or laser development, making it unclear against which project a particular development should be budgeted. As an order of magnitude guess, the cost is probably hundreds of billions of Euros, so would be a megaproject on the scale of the Apollo programme, the international space station, some military projects, or the largest hydroelectric damns. Such a large science project is likely to be an international effort, involving people from many different countries and also benefiting those countries, and so has further value in fostering cooperation and the exchange of ideas.

Some people see the future of humanity in the stars. This can have a range of motivations, not all of which may be agreeable or compatible. One reason could parallel that of earlier colonists and settlers, such as the desire for more resources, or the wish to escape – or set up – autocratic governments and repression. Other reasons might be the inability to live sustainably on the Earth, or an attempt to ensure that human life survives possible destruction (natural or self-imposed) on the Earth. Sending people to the stars is a very far-fetched scenario, and technologically far beyond sending robotic spacecraft. Whether or not this always remains a pipe dream, an essential precursor is to send uncrewed spacecraft to demonstrate critical technologies and to reconnoitre possible habitats.

Finally, exploration of our universe is, for some people, a cultural necessity. Across history and societies, many people want to learn more about the world they live in. This can and does play a strong role in education, where a fascination with astronomy and space science can inspire curiosity-driven research in other areas. We don't need to point to any specific technological or economic spinoff to see this as a benefit. It raises the intellectual level. We can strive for health, security, freedom, wealth, and many other things, but what ultimately are these for? Knowledge, and the process of gaining knowledge, is arguably one of the most rewarding human activities. Exploring the stars is just another step in this process.

2.2   Stellar systems

We now look at the specific science cases for an interstellar spacecraft, starting in this section with stars, then moving to exoplanets, life, and the interstellar medium in subsequent sections.

Observed from the Earth, most stars other than the Sun are unresolved points of light. Astronomers have nonetheless managed to learn a lot about them, in particular from their intrinsic luminosities (coming from their apparent luminosities and distances), their spectra, and the time variations of these, as well as their position and motion in the Galaxy. A few stars are large, bright, and near enough to have been spatially resolved by interferometry showing, for example, that some are distorted into oblate shapes by their rotation (e.g. Regulus and Altair).

We know a lot more about the Sun because of our proximity to it. Most of our ideas about stellar chromospheres, coronae, and stellar winds come from the Sun. Although observations of various stars with different masses and ages have been essential to learn about the formation and evolution of stars, it is our knowledge of the Sun – for example accurate determinations of its mass, radius, luminosity, age, and rotation – that form the only well-determined points relative to which observations of other stars are compared and calibrated. Something as fundamental as the mass of a star is actually hard to measure. If it is not in a binary system, we can only infer its mass via a stellar model using observations of its brightness and spectrum. These models ultimately have to be calibrated using the Sun. A spacecraft visiting another star permits a direct measurement of its mass via Kepler's laws (section 4.1), provided we have good navigation (chapter 14). It could likewise determine other parameters of the star accurately and independently of the Sun. Detailed knowledge of another star obtained in this way would allow us to test and perhaps revise our understanding of stellar astrophysics.

Very young stars (a few million years old) are typically accompanied by protoplanetary disks of gas and dust, the remnants of star formation from which planets can form. A classic example of this is the star TW Hydrae, which is about 195 ly away. Some other stars are observed to have debris disks, which are dominated by dust rather than gas, and are believed to be the result of collisions between planetesimals, similar in concept to the solar system's Kuiper belt. Epsilon Eridani, a K2 dwarf just 10.5 ly away with an age of several hundred million years, has a debris disk, and is also known to harbour at least one planet.

Most stars are not alone, but exist in gravitationally bound binary systems (sometimes also triples or even higher order systems). Very close binaries can interact with each other and exchange mass, which significantly affects their evolution. For example, the more massive star in a pair will evolve more quickly from the main sequence onto the subgiant branch, expanding in the process. It if gets large enough (overfilling its so-called Roche lobe), some of its surface will be pulled by gravity onto the other star. This can lead to this other star then becoming the more massive one, even though it is still on the main sequence. An example of this is Algol (90 ly away), in which the two interacting stars are separated by only 0.06 au and orbit in under three days (a third star orbits this pair with a semi-major axis of 2.5 au). When the star receiving the mass is a white dwarf, the infalling matter (hydrogen) can form an accretion disk around the white dwarf. If the temperature becomes high enough, this hydrogen can fuse to helium, releasing a large amount of energy that we observe as a nova.

An interstellar spacecraft would allow these phenomena to be studied close up: spatially resolved and observed from different perspectives. For the star itself, it could observe flares and prominences on its surface, and characterize its magnetic fields and particle winds. For a stellar disk the spacecraft could measure its chemical composition in situ via mass spectroscopy. It would be able to measure the size distribution of dust and pebbles and determine how these vary with distance from the star, potentially also observing their accretion onto planets in the process of formation.

2.3 Exoplanets

Given our fascination with planets and our expectation that they could harbour life, it is difficult to imagine that a first interstellar mission would be sent to a stellar system that does not have any known planets. Many stars have planets, including the nearest star to the Earth, Proxima Cen.

2.3.1 Current status

The first exoplanets were discovered in 1992 from variations in the timing of pulsars. This gave clear evidence for two planetary-mass companions to a neutron star. The first exoplanet discovered around a main sequence star was announced in 1995, the Sun-like star 51 Pegasi.

The planet, 51 Peg b, has a mass of at least $0.5\,M_{\text{Jup}}$ and was detected by the radial velocity method. Both planet and star orbit their mutual barycentre, and so the small variations in the radial velocity of the star over its orbit (in this case of the order of $100\,\text{m s}^{-1}$) reveal the presence of a companion. As the orbital inclination i of the system cannot be determined from this measurement, it yields only $m \sin i$ ($i = 0$ is face-on, $i = 90°$ is edge-on), and thus a lower limit on the mass m.

51 Peg b has a very short period: just 4.2 days, corresponding to a semi-major axis of 0.05 au. Even though the radial velocity method is more sensitive to shorter periods and thus smaller orbits, 51 Peg b was a big surprise to astronomers. It proved not to be an exception, with a large population of what are now called hot Jupiters being discovered in subsequent years, and leading to a dramatic revision to theories of planet formation and evolution.

The early discoveries of exoplanets led to a boom of interest in a field that had been regarded with scepticism by many astronomers in the years before. As of mid 2025, we know of 5930 exoplanets. Most of these have been discovered by two methods: the radial velocity method already mentioned (1130 discoveries), and the transit method (4390 discoveries). If the inclination of the orbit is close to $i = 90°$, then the planet will pass in front of its host star once per orbit. Although this cannot be spatially resolved from the Earth, the transit will lead to a slight dimming of the starlight received, and this variation can be measured. Although these observations can be done from the Earth's surface for some targets, the presence of the Earth's atmosphere limits the accuracy to a few parts in a thousand. For this reason, several space observatories have been built specifically for measuring tiny stellar photometric variations in order to detect exoplanets. These include CoRoT, Kepler, TESS, and other upcoming missions like PLATO. General-purpose space observatories, in particular Spitzer, the Hubble Space Telescope (HST), and the James Webb Space Telescope (JWST) have also studied planets in this way, the latter achieving precision of a few tens of ppm.

Figure 2.1 shows the distribution of the orbital periods and masses (or $m \sin i$) of known exoplanets at the time of writing. Of the 5930 planets, 5610 have estimates of both the period and mass. The 'other' category of detections includes eclipse timing variation, transit timing variations, pulsar timing, astrometry, and a few others. There are strong observational biases in this figure. The radial velocity method is more sensitive to massive short-period planets; the transit method is more likely to identify large short-period planets; the imaging method is more sensitive to massive long-period planets.

Transit detections offer two big advantages over radial velocity detection: (1) the inclination of the orbit is more or less known, and so the mass of the planet can be determined; (2) the radius of the planet can be determined from the amount of dimming, if the radius of the star is known. From these we can then compute the average density of the planet, and thus get constraints on its bulk composition.

Transits can also be measured in dispersed light, which is known as transmission spectroscopy. Comparing the in-transit spectrum to the out-of-transit spectrum yields the absorption spectrum of the planet's atmosphere, from which we can learn about its chemical composition and physical condition. In particular with the advent of JWST, this is currently one of the main ways of learning about exoplanet atmospheres. Although a powerful tool, transmission spectroscopy is only possible for those planets that pass in front of their host star as seen from the Earth. Most planets are therefore not observable in this way.

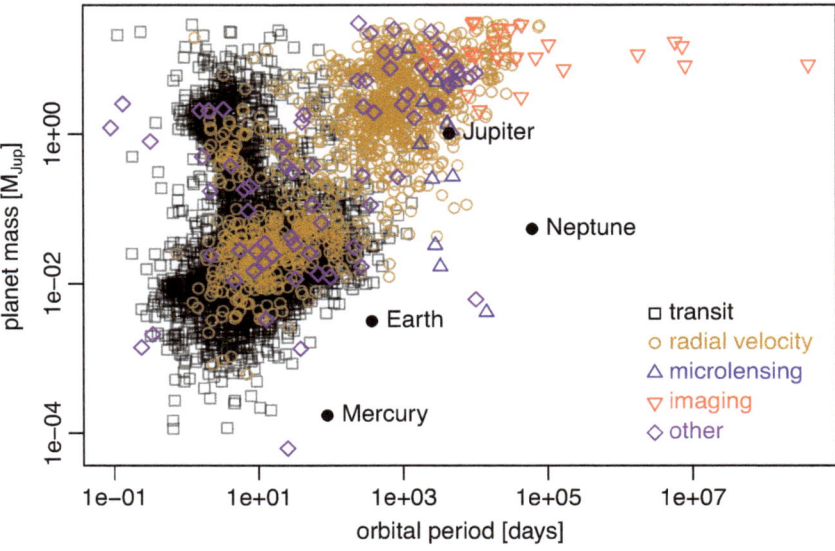

Fig. 2.1 The distribution of the 5610 exoplanets with estimated periods and masses as of mid 2025 taken from the NASA exoplanet archive. The label indicates the detection method. Error bars are not shown. Most of the radial velocity detections give $m \sin i$, where i is the unknown inclination of the orbit, and so give a lower mass limit. Four of the solar system planets are overplotted for comparison.

2.3.2 The circumstellar habitable zone

From the statistics of exoplanet detections, we now know that planets are common around stars. This includes both gas giant planets with masses – but not necessarily orbits – similar to or larger than Jupiter's, rocky planets with masses similar to or less than the Earth's, and much in between. As the detection signal is larger for larger masses, we know of comparatively few Earth-mass planets, but this is likely to change with the deployment of future telescopes and instruments with greater sensitivity.

Learning about the diversity of exoplanets, and how they could have formed and evolved, is of considerable interest in itself. But perhaps one of the most exciting questions is the presence of life. We will discuss this in section 2.4, but before we do, let us first look at the concept of the circumstellar habitable zone. This is the region of space around a star in which liquid water could exist on the surface of a planet. Liquid water is taken to be important because it is essential to almost all forms of life that we know of.

For water to be in its liquid form, both the pressure and the temperature have to be within a certain range. Planets that are too close to their host star will be too hot for water to remain liquid if the atmospheric pressure is not high enough. If the temperature is above the critical temperature of water (647 K), water will vaporize no matter how high the pressure. This high temperature limit marks the inner edge of the habitable zone. If the planet is too far from the host star then water will freeze, marking the outer edge of the habitable zone. There are actually many different definitions of the exact limits of the habitable zone. One example is shown in figure 2.2.

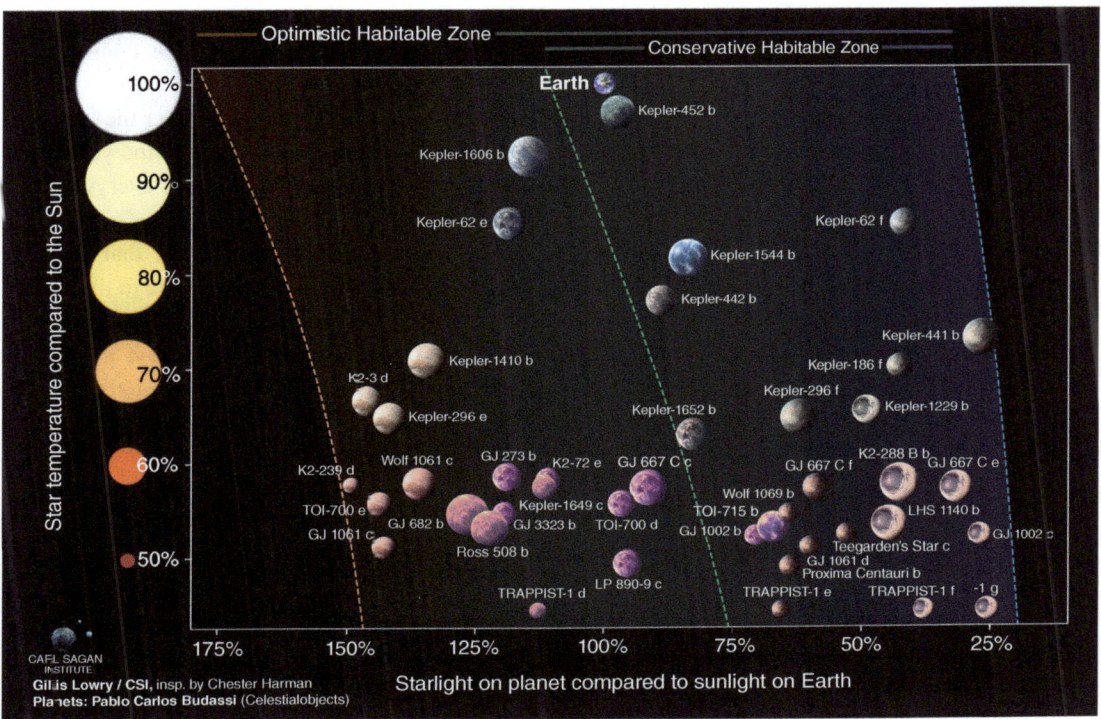

The circumstellar habitable zone showing the location of various known exoplanets as well as the Earth. Credit: adapted from Carl Sagan Institute and Gillis Lowry, licensed under CC-BY-4.0, Wikimedia Commons.

The location of the habitable zone depends not only on the luminosity of the star but also on the pressure and composition of the planet's atmosphere, for example the presence of greenhouse gases. A widely used definition for the habitable zone is the presence of an 'Earth-like' atmosphere consisting of a non-condensable background gas (nitrogen, N_2), a non-condensable greenhouse gas (carbon dioxide, CO_2), and a condensable greenhouse gas (water, H_2O). In this model the inner edge of the habitable zone is defined by the moist greenhouse effect, whereby large amounts of water evaporate into the stratosphere where it photodissociates, and the hydrogen is lost to space. Even closer to the star the runaway green house effect occurs, whereby the liquid water boils off entirely. The outer edge of the habitable zone is defined by the maximum greenhouse effect that can be provided by carbon dioxide to keep water liquid. Under this model the habitable zone for rocky planets around the Sun extends from 0.99 to 1.70 au, which places the Earth near to its inner edge, although this model neglects the effect of clouds.

There are other factors apart from atmospheric properties and distance to the star that determine the planet's surface temperature. These include internal heat left over from the planet's formation, radioactive decay, and tidal heating from a possible satellite. A planet very close to its host star may become tidally locked into a 1:1 resonance, whereby the orbital period equals the rotational period, meaning the planet has the same side facing the star all the time (e.g. the Earth–Moon system, bar the libration). In that case the permanent day side can become too hot, and the permanent night side too cold, even if the planet is inside the habitable zone. However, some such planets appear to have powerful winds that redistribute heat, and this may keep parts of the surface continuously habitable.

Most planets will have variations in temperature across the surface and over the year, and a planet on an elliptical orbit may be in the habitable zone for only some of the year. Thus being in the habitable zone does not mean the planet can support liquid water everywhere or all of the time.

Even when there is liquid water, this does not mean life could form, or that it would survive. The habitable zone of an M dwarf star, for example, is relatively close to the star on account of its low luminosity. But M dwarfs are very active, and the ionizing radiation from frequent flares could sterilize the planet even if water remains liquid. In contrast, subsurface liquid water may exist well beyond the circumstellar habitable zone, as appears to be the case for some satellites of Jupiter and Saturn. For all its many limitations, though, the habitable zone can be a useful concept for providing a first rough assessment of the suitability of a planet to potentially support life.

One can of course argue that life may exist without liquid water. This is valid, although water does have some unique physical and chemical properties that make it well-suited to acting as a solvent for organic reactions. Its constituents, hydrogen and oxygen, are also very abundant in the universe. The argument in favour of the habitable zone is that it directs our attention toward environments that we know are capable of supporting life. Other environments without water may also be supportive, but if we don't specify what those environments are, and where they lie, this won't help us to focus our search. Lacking observational data, one can try theoretically to define water-free life-supporting chemistries and environments. But given that the parameter space to search for life is already huge, and that our resources to search are limited, we should probably start looking in environments where we know life could exist, rather than in those where we have no supporting data.

2.3.3 What we can learn from an interstellar mission

How much we can learn about an exoplanet from a close-up visit depends on the type of mission (section 1.4). An orbiter would provide more information than a flyby or even a swarm of flybys (section 16.3). What we can learn depends on what type of planet it is, for example whether it is a gas-dominated planet like Jupiter or Neptune, a temperate rocky planet like the Earth, or a hot lava world, to mention just some of the possibilities. In section 16.1 we will look the instruments an interstellar spacecraft could carry. Here we take a brief look at the things we could learn.

The main benefit of in situ observations is the ability to observe different parts of the planet at high contrast and high spatial resolution, from different perspectives and illuminations, and to study how regions change over time. Observations can identify whether the planet has an atmosphere, and if it does, determine its chemical composition, and measure how its pressure, density, and temperature vary with altitude. We can look for clouds and study their composition and dynamics, and use these to study weather systems. We can measure the rotation speed of the planet as well as the orientation of its spin axis, both of which are central to the redistribution of heat on a daily and annual basis.

If the atmosphere is not completely opaque or permanently cloudy, we will be able to image the surface and look for mountains, deserts, oceans, ice caps, and other geological features. Reflectance and emission spectra of the surface can tell us about the composition of different regions, including possible signs of life, such as vegetation, which may vary with the seasons. We can look for surface water in the form of large oceans or small lakes or rivers. Both water and ice have characteristic colours and a reflectivity that depends on the angle of illumination and viewing. Variations in the ice cover at the poles may be observed over the course of a year. Even if surface water is not currently present, the spacecraft can look for geological signs of earlier water, such as channels or salt flats, or evidence of subsurface water, as has been done on Mars. Active radar or lidar ranging can map surface features, something that is impossible to do from the Earth. Direct imaging could even reveal large artificial structures that would be a sign of intelligent life.

Volcanoes and evidence of past volcanic activity tell us about subsurface processes and the internal structure of the planet. Measurements of chemical abundances in volcanic outgassing can reveal the internal composition. Plate tectonics is responsible for recycling material between the surface and subsurface, and so help us interpret surface features and rock types in light of this. Plate tectonics may also be relevant to the evolution of life through the cycling of carbon between the interior and the atmosphere. The presence or absence of impact craters tell us about erosion processes and timescales on the surface, as well as the existence of minor bodies in the target star system.

So far we have considered what we can learn from imaging and spectroscopy, but one thing we can never do from the Earth is take in situ measurements of gas, dust, magnetic fields, and gravity. Low orbits of the satellite allow it to sample the upper atmosphere directly using mass spectrometers and dust counters. By mapping the magnetic field of the planet, we can learn about the internal processes needed to create such a field, or the absence thereof if there is no magnetic field. The spacecraft can map the distribution of plasma around the planet to see how the stellar wind interacts with the planet and its magnetic field, and particle detectors can measure the composition of the stellar wind directly. If spectral lines of aurorae are observed from the interaction of the stellar wind with the planet's atmosphere, this too tells us about the composition of the atmosphere.

By observing the orbit of the planet around the star, as well as its own orbit about the planet, the spacecraft can accurately determine the masses of both bodies. Different inclination orbits can be used to measure the shape of the planet, its deviation from a perfect sphere due to rotation giving clues to the internal structure. If the spacecraft can go into a low orbit around the planet, small deviations in the spacecraft's orbit due to variations in gravity map variations in the planet's density, which help infer the inner structure and composition near the planet's surface.

The spacecraft can look for moons and rings around the planet. It can also observe further afield to look for additional planets, as well as comets, asteroids, and other minor bodies in orbit around the star that would not be detectable by Earth-based observers

If our spacecraft can release a probe to land on the planet's surface (when it has one), then many more investigations are possible, akin to those done with Mars landers and rovers. These include in situ measurements of the composition of the atmosphere and ground, as well as variations in temperature and pressure during the descent, and during the diurnal and annual cycles. Experiments could be carried out to identify microbial life in the atmosphere as well as in the soil, rock, or ocean.

2.4 Life

2.4.1 What is life?

To the best of our knowledge, life on Earth evolved only once, and all life today has evolved over the past 3.5 billion years or so from the last universal common ancestor (LUCA). Where, how, and from what LUCA itself evolved is unclear. It may have come from outside the solar system, perhaps in the form of DNA-carrying bacteria buried under the surface of a rock, and so protected from ionizing ultraviolet radiation when travelling between the stars (the panspermia hypothesis). Whether true or not, there may be other life in the universe, and this may or may not have an independent origin from life on Earth.

Searches for life in the solar system have so far identified no evidence for current or past life beyond the Earth. Mars shows geological evidence of once having had liquid water on its surface (which also means it must have had a thicker atmosphere in the past), but as yet no evidence for life. Most of the water is now probably frozen underground near the poles. It seems likely that subsurface oceans exist on several of the moons of Jupiter and Saturn, the best candidate being Jupiter's satellite Europa, which will be visited in the coming years by both NASA's Europa Clipper and ESA's Jupiter Icy Moons Explorer (JUICE).

To search for life we need to know what signatures to look for, and for this we need a concept of what life is. Life is notoriously difficult to define rigorously, not least because some properties are not as fundamental or as universal as we may think. For example, we might say that life needs energy, and while this is true, if an organism is not growing or reproducing, then it must put out as much energy as it takes in. A combustion engine also needs energy, but it is not alive.

Some general properties derived from common aspects of life we know are as follows:

1. Localized. All life occupies a well-defined limited spatial location, typically as one or many cells.
2. Out of equilibrium with the environment. If an organism came into equilibrium with its environment – the same temperature and chemical composition, for example – then it could not maintain itself as a distinct entity.
3. Uses free energy from the environment. This is necessary so that an organism can do work, for example to move, process information, or reproduce.
4. Metabolism. This includes the conversion of energy, the conversion of chemicals into building blocks, and the expulsion of waste products.
5. Self-organized based on a code. Organisms arrange their internal parts themselves based on a code, such as their genome. Examples are the organization of the components in a cell or of different cells in the body.
6. Ability to reproduce. Organisms can pass on their code/genome to offspring.
7. Evolution through natural selection. This implies that the organism will adapt to its changing environment.

Not all of these properties are independent. To maintain a low entropy and thus remain distinct from its environment (localized, out of equilibrium), an organism must use free energy from the environment. Some life on Earth does not fulfil all of the above properties: mules cannot reproduce, for example. Conversely, some things we would not consider alive possess some of the above properties: Fire can reproduce, and chemical systems can self-organize (although not based on a code). There are also borderline cases. Viruses are based on a code, use energy, can reproduce, and evolve via natural selection. But they can only do these things within a host, so they are not really independent life forms. Another ambiguous case might be artificial intelligence, by which I don't mean the label as currently used for sophisticated machine learning algorithms, but a potential future development with human-level intelligence and adaptability. Such computer life would use energy, be out of equilibrium, and could copy its software (reproduce), but it would not have to be localized, nor would it necessarily be subject to natural evolution.

One of the more widespread definitions, although still imperfect, is the one adopted by NASA: 'a self-sustaining chemical system capable of Darwinian evolution'. The above discussion notwithstanding, the lack of a robust definition need not hinder our search for life as we know it.

A lot of papers have been published trying to establish the probability of life emerging independently elsewhere in the universe. Many of these rely on statistical arguments, such as how long it took life to emerge after the Earth had formed. One interpretation is that because life emerged relatively quickly after the formation of the Earth (compared to the Earth's expected lifetime), life must form easily. The common drawback of most, if not all, of these kind of arguments is that they depend on just one data point – life on Earth – and so their conclusions are unavoidably dominated by assumptions. Different authors make different assumptions, and so come to different conclusions. We know that life emerged at least once in the universe, but this doesn't tell us whether it is common or a freak one-off event; in either case we observe the existence of our life. The rapid formation of life can therefore be interpreted in different ways, leading to opposite conclusions: life can form easily everywhere, and so is common; life can form easily only under the very specific (and still vague) circumstances we had on the Earth, and so is rare; life has a very low probability of forming in any specific environment, but there were many attempts on different planets, and it emerged quickly on at least one of them, in which case life could be common or rare.

The outcomes of most of these probability calculations also depend heavily on assumed, but often completely unknown, quantities. A classic example of this is the Drake equation, which expresses the expected number of communicating extraterrestrial civilizations in terms of the product of a number of factors. Some of these factors we can estimate with reasonable confidence, such as the star formation rate in the Galaxy, or the fraction of stars that have planets, because they are narrowly-defined and we have reasonably good data. But other factors, such as the average number of planets that could support life or the fraction of these that actually go on to develop life, we have essentially no estimate for. It is not sufficient to be apparently conservative and assign 'small' numbers to these, and then, having arrived at a number of communicating civilizations larger than one, state that because of this conservatism the number is probably much larger than one. We simply have no idea whether the fraction of life-conducive planets that go on to produce life is 0.01 or 10^{-100}, even if we adopt a narrow definition of life.

To know if there is other life in the universe, we have to go out and look.

2.4.2 Signatures of life

To detect life we don't have to directly image organisms. We can instead detect the chemical or physical signatures that accompany life. The general properties of life given in the previous section are too broad to help much here because they do not predict specific signatures.

To make progress on interpreting observable quantities, we need to adopt a narrower definition, and for this we orientate ourselves using life on Earth.[3]

We start by asking what signatures of life on Earth could be seen by an observer close to a planet. We consider first the biological signatures of life before considering the signatures of intelligent life.

In looking for biosignatures, it is important to distinguish between signatures that can be produced only by life – biotic processes – and those that can also be produced by abiotic processes. Take photosynthesis. This releases oxygen through the production of carbohydrates from carbon dioxide and water via the overall reaction[4]

$$CO_2 + 2H_2O + \gamma \ \rightarrow \ CH_2O + H_2O + O_2. \tag{2.1}$$

Oxygen is quite reactive, and so would be removed from the atmosphere on a short timescale if life died out (probably less than a few 10^5 years). The observation of atmospheric oxygen might therefore be interpreted as a indication of current life. Conversely, the lack of oxygen does not necessarily indicate a lack of (oxygen-producing) life. Oxygen probably did not build up to significant levels in the Earth's atmosphere until long after photosynthesis had started. Furthermore, although oxygen has no significant geological source, it is produced by abiotic processes, in particular the photodissociation (photolysis) of oxygen-bearing molecules such as water in the reaction

$$H_2O + 2\gamma \ \rightarrow \ 2H^+ + 2e^- + O. \tag{2.2}$$

This is a slow reaction in the Earth's atmosphere, but it could be faster in other atmospheres. There are other sources of oxygen. Jupiter's satellite Europa, for example, has a very thin oxygen atmosphere, produced not by photodissociation but by charged particles – accelerated by Jupiter's strong magnetic fields – splitting water ice molecules.

Thus the observation of oxygen by itself is not a secure indication of life. But observing it in the presence or absence of other certain molecules makes it more likely the oxygen was produced biotically. On an Earth-like planet, for example, a detection of oxygen along with water, or with water and methane (CH_4) but not significant amounts of carbon dioxide or carbon monoxide (CO), is a stronger indication of biotic oxygen.

The various molecules mentioned above can be searched for spectroscopically. Molecular oxygen has characteristic bands in the optical and near infrared. A much strong absorber is ozone (O_3), which is produced photochemically in the Earth's stratosphere by the photodissociation of molecular oxygen and the subsequent combination of atomic and molecular oxygen. Other possible biomarkers include nitrous oxide (N_2O), sulphur gases such as dimethyl sulfide (DMS), and maybe phosphine (PH_3).

Beyond searching for molecular biosignatures in the atmosphere as a whole, an interstellar spacecraft could look for signatures of plant life in spectral imaging. Terrestrial and aquatic

---

[3] Why limit our search to Earth-like life? With some imagination we could come up with a type of life that we don't have on Earth, make predictions about its signatures, and then search for these. But this could result in a very broad set of different signatures, the detection of which would still have an ambiguous interpretation. Searching comes with a cost, so it seems sensible to look first for the signatures of life we do know can exist, rather than those we don't.

[4] Water is not required for photosynthesis. A more general reaction using a different electron donor (reductant) A is
$$CO_2 + 2H_2A + \gamma \ \rightarrow \ CH_2O + H_2O + 2A.$$
Some bacteria use hydrogen sulphide (H_2S) instead of water in their photosynthesis.

plants have characteristic visual/near-infrared reflectance spectra. What is common to all of these is an increase in their reflectance at wavelengths longer than about 700 nm. This so-called vegetation red-edge is due to chlorophyll, which is responsible for the light absorption in photosynthesis. Photons with wavelengths longer than 700 nm have too low energy to be used in photosynthesis, and so plants have presumably evolved to reject this useless energy that could otherwise just burn the plant.[5] Observation of a red-edge in reflected spectra from parts of a planet might therefore be an indication of plant life.

Intelligent life, or at least civilization as we know it, may also be detectable through its technosignatures. Examples include surface structures such as agriculture, cleared land, or urban areas, as well emission from street lights or industrial activity, familiar from images of the Earth at night. For about a century, radio and television have used powerful transmitters, and some of this power has leaked into space. These emissions could be picked up by a sufficiently large radio telescope at the distance of a nearby star, or by a proportionally smaller receiver on an interstellar spacecraft nearer to the source. At least with human-made signals, the pattern is sufficiently complex and structured in time, frequency, or phase (depending on the modulation) as to be easily distinguishable from natural sources.

2.5 The nearest stars

Where shall our interstellar spacecraft go? Table 2.1 lists all known stars and brown dwarfs lying within 12.1 ly of the Sun. There are 35 objects in total, but as some are gravitationally bound, there is a total of 22 systems. Of those systems, 12 have one object, 7 have two objects (binaries), and 3 have three objects (triples). Astronomers have the unfortunate habit of assigning – and using – different names for the same star. The table lists the commonly used names, showing the historical name (e.g. Sirius) and Bayer name (e.g. Tau Cet = τ Ceti) where it exists, as well as the GJ (Gliese–Jahreiß) number for all the stars. Many stars in multiple systems have small angular separations so were originally identified as single stars. For this reason, most stars in multiple systems do not have unique GJ numbers. When they were later resolved into multiple components, these were normally then given letters, such as in the case of Alpha Cen A and B.

The three triple systems in the table are hierarchical binaries, in which two components form a tightly bound binary with the third object in a larger orbit around that pair. The closest of these, Alpha Cen, is also the nearest system to the Sun, so we will look at it in more detail below. In the triple system EZ Aquarii, the tight binary is made up of A and C. They are so close – a physical separation of 0.03 au and angular separation of $0.01''$ – that they have not been separately imaged; the magnitude and spectral type in the table is for the system as a whole. Nonetheless, the masses could be individually measured from spectroscopy, and are all around $0.1 \, M_\odot$. In the final triple system, Epsilon Indi, components B and C (sometimes also called Ba and Bb) are quite far from component A, which is why the table uses an upper limit of 12.1 ly for inclusion.

---

[5] At wavelengths below 700 nm, chlorophyll absorbs both blue and red light much better than green light, which is why most leaves look green. This is odd, because green is near the peak of the solar spectrum and so would be most efficient for photosynthesis.

Table 2.1 Nearest stars and brown dwarfs to the Sun out to 12.1 ly ordered by increasing distance. GJ is the Gliese–Jahreiß identification number. G is the apparent magnitude in the Gaia (broad optical) band, *r* is the distance from the Sun SpT the spectral type. The final column indicates planets (as of 7 July 2025). Those in parentheses are unconfirmed candidates; those crossed out were claimed but then refuted. In a few cases a planet was claimed, named, then refuted, and then new planets claimed with the same name.[1]

System name	Alternate or component name	GJ	*r* [ly]	G [mag]	SpT	Note[a]	Planets
Alpha Cen	C (Proxima Cen)	551	4.25	8.98	M5.5		Proxima Cen b (c,d)
	A	559	4.32	−0.1	G2		
	B	559	4.32	1.3	K1		~~Alpha Cen B b~~
Barnard's Star		699	5.96	8.19	M3.5		Barnard's Star b,c,d,e
Luhman 16	A	11551	6.52	16.95	L7.5	BD	
	B	11551	6.52	16.96	T0.5	BD	
WISEA J0855-0714		11286	7.43	21.9	Y4	BD	
Wolf 359	CN Leo	406	7.86	11.04	M6		(Wolf 359 b,e)
HD 95735	Lalande 21185	411	8.30	6.55	M1.5e		HD 95735 b,c,(d)
Alpha CMa (Sirius)	A	244	8.60	−1.49	A1		
	B	244	8.71	8.52	DA1.9	WD	
GJ 65	B (UV Cet)	65	8.72	10.82	M6		(GJ 65 A or B a)[b]
	A (BL Cet)	65	8.82	10.51	M5		
Ross 154	V1216 Sgr	729	9.70	9.13	M3.5e		
Ross 248	HH And	905	10.30	10.34	M5		
Epsilon Eri	Ran	144	10.50	3.47	K2	dd	Epsilon Eri b
HD 217987	Lacaille 9352	887	10.72	6.52	M2		GJ 887 b,c,(d)
Ross 128	FI Vir	447	11.01	9.60	M4		Ross 128 b
EZ Aqr	A	866	11.11	10.84[c]	M5[d]		
	B	866	11.11				
	C	866	11.11				
61 Cyg	B	820	11.40	5.45	K7		
	A	820	11.40	4.77	K5		
Alpha CMi (Procyon)	A	280	11.46	1.12	F5		
	B	280	11.46	10.7	DQZ	WD	
GJ 725	A	725	11.49	7.85	M3		(GJ 725 A b)
	B	725	11.49	8.52	M3.5		(GJ 725 B b,c,d)
GJ 15	A (GX And)	15	11.62	7.22	M1		GJ 15 A b,c
	B (GQ And)	15	11.62	9.69	M3.5e		
DX Cnc		1111	11.68	12.17	M6.5		
Epsilon Ind	A	845	11.87	4.32	K5		Epsilon Ind A b
Tau Cet		71	11.91	3.30	G8.5	dd	(Tau Cet ~~b,c,d,~~e,f,g,h)
GJ 1061		1061	11.98	11.00	M5.5		(GJ 1061 b,c,d)
Epsilon Ind	Ba (B)	845	12.05	18.06	T1	BD	
	Bb (C)	845	12.05	19.33	T6	BD	

[1]This table is based on the CNS5 catalogue of nearby stars of Golovin et al. (2023) with additional information from Reylé et al. (2021). The distances (derived from the parallaxes) and apparent G-band magnitudes are mostly from the third Gaia data release (Gaia DR3; Gaia Collaboration 2023). In a few cases the stars were too bright for Gaia DR3, in which case their magnitudes have been computed from other measurements. The uncertainty in the distance varies significantly with brightness, but in most cases is better than the two decimal places reported.

[a]BD = brown dwarf; WD = white dwarf; dd = has a debris disk.

[b]It is not known which of the two components of GJ 65 the planet orbits.

[c]Unresolved, so this is the system G magnitude.

[d]Unresolved, so this is the system spectral type.

Of the 35 objects, five are brown dwarfs (BD in the table) and two are white dwarfs (WD). Most of the other stars are late type stars of spectral type K or M, corresponding to masses in the range 0.1–0.9 $M_\odot$. There are only three stars with a temperature similar to or hotter than the Sun, namely Sirius A (the brightest star in the sky), Procyon A, and Alpha Cen A (which is very similar to the Sun).

Luhman 16 (WISE J104915 57-531906.1) and WISE J0855-0714 (full name WISEA J085510.74-071442.5) are interesting systems because although they are the third and fourth closest systems to the Sun, they were only discovered in the twenty-first century by a deep all-sky infrared survey satellite. They are much fainter in the optical than other nearby stars because they are very cool late-type brown dwarfs. Luhman 16 is still much brighter than more distant stars that are seen in other surveys, but it lies in a crowded field in the Galactic plane, so was only detected because of its large angular velocity (proper motion). It was later found to be a binary (each component mass around 0.03 $M_\odot$ or 30 M_{Jup}). WISE J0855-0714 is a much colder and fainter brown dwarf, with an effective temperature of just 285 K. It is in fact the coolest known brown dwarf, and spectral models indicate it has a mass below 13 M_{Jup}, making it a planetary mass object. The fact that the coolest brown dwarf is very nearby is not that surprising, because the cooler the brown dwarf, the fainter it is, so it has to be nearby to be seen at all. There are probably other similar objects nearby waiting to be discovered.

Before going on to discuss planets, the binary star 61 Cygni deserves a special mention because it was the first star for which significant stellar parallaxes were published, in the 1830s (see section 14.4.1). Parallaxes for Alpha Cen AB (not then known to be a double star) and Vega (at 25 ly) were published very soon after.

2.5.1 Planets around the nearest stars

The final column of Table 2.1 lists the presence of confirmed and candidate planets, as well as some claimed planets that have since been refuted. The terms 'confirmed', 'candidate', and 'refuted' are not rigorously defined or consistently applied. Ultimately it is the degree of evidence that matters, but there is no universally accepted standard. All of the confirmed (and most of the candidate and refuted) planets in the table except two have been discovered by the radial velocity method (section 2.3.1). None of them have yet been found to be transiting, so their orbital inclinations are unknown, and most of the masses are lower limits. The nearest stars are potentially ideal for finding planets through direct imaging because the angular separation from the host star is larger for a given physical separation. But the nearest stars are all relatively old, and so their planets intrinsically very faint, beyond the current levels of detection. As of mid 2025, the nearest star with a directly imaged exoplanet, and the only one in Table 2.1, is Epsilon Ind A b.

We now look briefly at the known planets in the table. The last two decades have revealed how common planets are, so it should not be too surprising that our nearest star, Proxima Cen, has a planet. It even lies in the habitable zone. This will be discussed in more detail in section 2.5.2 below.

The second closest system to the Sun, Barnard's Star (discovered in 1915), is of particular historical interest in the field of interstellar travel. A claimed detection of a Jupiter-mass planet around it in the 1960s (via the astrometrical method) led to the British Interplanetary Society choosing this star as the target for its Daedalus mission (which we will discuss in chapter 8). The existence of the planet was never widely accepted by the community, however. In 2018 a team claimed the detection of a super-Earth-mass planet on a 233-day period from radial velocity spectroscopy, but this too was later refuted. However, another 2024 radial velocity study claimed detection of a planet in a three-day orbit with a minimum mass of $0.4\,M_\oplus$, as well as three other possible planets also with sub-Earth masses and periods of a few days. A 2025 study confirmed all of these. Their orbits are all too small for them to be in the habitable zone.

The two claimed planets around Wolf 359 were listed in an unrefereed preprint reporting results of a radial velocity survey. The claimed inner planet, c, was later refuted, but the status of the outer b candidate is unclear.

The two confirmed planets around HD 95735 were discovered in 2019 and 2021. The closer-in one, b, has a minimum mass of about $3\,M_\oplus$ and an orbital period of 13 days. The other one, c, has a minimum mass of about $15\,M_\oplus$ and is in a much larger orbit with a period of eight years. The radial velocity data suggest there may be another perturbing object (d) with a 215-day period, but this has not been confirmed.

GJ 65 AB is an M dwarf binary star system with a period of 26 years in an elliptical orbit with a semi-major axis of 5.5 au. Astrometric measurements over seven years revealed perturbations of its orbit, which in 2024 was attributed to a $35\,M_\oplus$ planet candidate in a 156-day orbit around one of the stars; the data do not distinguish which. With a semi-major axis of around 0.28 au, the planet's orbit around its host star is tight enough to be stable to disruption from the other star.

Epsilon Eri b was initially discovered in 2000. More recent observations have confirmed the detection and refined its parameters for a planet with a minimum mass of $0.7\,M_{Jup}$ in a seven-year orbit (semi-major axis of 3.5 au). Before the planet had been discovered, the star was interesting because of the detection at millimetre and infrared wavelengths of a debris disk. It may also have an inner asteroid belt similar to ours.

GJ 887 b and c were discovered in 2020. They have minimum masses of 4 and $8\,M_\oplus$, and periods of 9 and 22 days, respectively. A further signal with a period of 50 days may indicate a third planet.

Ross 128 b was discovered in 2018 with a minimum mass of $1.4\,M_\oplus$ and a period of 10 days (semi-major axis of 0.05 au). It receives somewhat less flux from its star than the Earth does from the Sun, and consequently has an equilibrium temperature below 270 K (depending on its unknown albedo), placing it close to the inner edge of the habitable zone. The planet does not appear to transit its star, so it is not known if it has an atmosphere. There was brief excitement in 2017 when it was thought non-natural radio signals had been detected from this star system by the Arecibo telescope. But as has so often been the case with such signals, they could not be confirmed with independent observations and were soon attributed to a local source, probably an Earth-orbiting satellite.

A candidate planet around GJ 725 A was announced in late 2024. It has an 11-day orbit and a minimum mass of $2.8\,M_\oplus$.

Two separate claims have been made for a total of three planets around GJ 725 B. The first, from 2016, tentatively claims a single planet in a three-day orbit. The second claims two planets in longer orbits; this is the same unrefereed paper that claimed a planet around Wolf 359 mentioned above. The detections are perhaps too weak, as they appear not be have been followed up.

Two planets have been detected around GJ 15 A. The first, b, was found in 2014 with a period of 11 days with a minimum mass of 3–5 $M_\oplus$. The second planet, c, discovered in 2018, has a period of 21 years. It appears to be a super-Neptune with a minimum mass of 36 $M_\oplus$, although the uncertainty on this value is nearly 50%.

Epsilon Ind is a triple system. A pair of brown dwarfs (Ba and Bb) orbit each other with a period of 11 years (semi-major axis of 2.4 au). These in turn orbit component A (a star) at a distance of 11 600 au (0.18 ly), which makes it a very weakly bound system (their association was identified through their large common proper motion). A planet detection was initially claimed around Epsilon Ind A in 2019 using the radial velocity method. Curiously, direct infrared imaging with JWST in 2023 also identified a planet, but in a different orbit. It is nonetheless believed these two measurements correspond to the same planet. The combined data show the planet to have a mass of about 6 $M_{\rm Jup}$ with a semi-major axis of 30 au. This is the closest exoplanet to the Sun to have been directly imaged.

No fewer than seven planets have been claimed around Tau Ceti at some point. The first detections from 2013 – labelled b, c, and d – are probably not planets due to problems with assessing the noise (and so are listed as refuted in the table). In 2017, four more planet candidates with minimum masses of 2–4 $M_\oplus$ and periods of 20–640 days were claimed. The radial velocity amplitudes are very small, as low as 0.3 m s^{-1}, so they need to be confirmed.

The final star in the list, GJ 1061, has three candidate planets detected in 2020, all with minimum masses of around 1.5 $M_\oplus$ and periods of 3–13 days, making this a closely packed system (and all are expected to be tidally locked to the star). Planet d, and maybe also c, lies in the habitable zone.

In total, Table 2.1 lists 14 confirmed planets around the nearest stars, plus another 17 candidates. Most of the other stars listed in the table have been searched for planets, but nothing yet found. A non-detection limits the parameter space for possible planets, for example no planet above some mass within some orbit size range, but it does not rule out all possible planets. That most of the planets mentioned above have been discovered within the past few years of writing is a consequence of the growing timebase of surveys, plus their improved accuracy, both of which lead to increased sensitivity. More parameter space will be accessed by future surveys, and it is very likely that more planets will be found around the nearest stars in the coming years. Finding and characterizing planets is essential for selecting the best target for the first interstellar mission. New telescopes such as ESO's 39 m Extremely Large Telescope (ELT), due to start operations in 2030, is expected to find and characterize new rocky exoplanets around the nearest stars via direct imaging in the infrared. The proposed NASA Habitable Worlds Observatory would be a 6–8 m diameter visible/infrared space telescope, which might be capable of detecting and characterizing a few tens of nearby rocky exoplanets (a few in their habitable zones). If it goes ahead soon, it could launch in the 2040s.

2.5.2 The Alpha Centauri system

The nearest star to the Sun is a triple star system consisting of two stars in a binary, Alpha Cen A and B, orbited by a much more distant companion Alpha Cen C, better known as Proxima Cen.

Alpha Cen A is a main sequence star, quite similar to the Sun with a spectral type G2 and an effective temperature of 5800 K, but with a slightly larger mass (1.08 $M_\odot$) and luminosity

$(1.51 \, L_\odot)$. Alpha Cen B is a cooler K1 main sequence star with an effective temperature of 5300 K, a mass of $0.91 \, M_\odot$, and a luminosity of $0.50 \, L_\odot$. They orbit each other in a rather eccentric elliptical orbit ($e = 0.52$) with a periapsis of 11.2 au and an apoapsis of 35.6 au (see section 4.1 for a summary of orbital elements). The orbital period is 79.8 years. At a distance of 4.34 ly, the projection of the semi-major axis on the sky is $17.5''$. Alpha Cen AB is the third brightest star in the night sky after Sirius and Canopus.

Proxima Cen orbits this pair in a large eccentric orbit ($e = 0.50$) with a period of 510 kyr, a value known only to within 10%. The periapsis and apoapsis are about 4100 au and 12 000 au, respectively, the former constrained by observations only to about 20%. The star is currently close to apoapsis. The long period prevents an accurate determination of the orbit because this requires observations across most of the orbit. For this reason, it was not known for a long time whether Proxima Cen was gravitationally bound to Alpha Cen AB. Proxima Cen itself is a much later type star than A and B, with spectral type M5.5, an effective temperature of 3000 K, a mass of $0.12 \, M_\odot$, and a luminosity of just $1.5 \times 10^{-3} \, L_\odot$. At a distance of 4.25 ly, it is currently the closest star to the Sun.

There are no known planets around Alpha Cen A or B. There was a 2012 claim to have detected an Earth-mass planet around Alpha Cen B via the radial velocity method, but the signal was weak, and later reanalyses of the data showed it was probably a false detection. Various searches have been able to exclude the presence of certain types of planet, such as planets above some mass at a given distance from the star(s).

We do, however, know of a planet around Proxima Cen, called Proxima Cen b. This was discovered in 2016 by the radial velocity method. It has a period of just 11.2 days, which combined with the estimated mass of the star gives a semi-major axis of 0.049 au. Its minimum mass is 1.0–$1.3 \, M_\oplus$. We do not know its radius, but if it is similar to the Earth's, then it is probably a rocky planet. Its eccentricity is not yet well constrained, but it is so close to its host star that its orbit may well have been circularized by tidal forces. In that case it may also be tidally locked, either into a 1:1 resonance, or into another other spin–orbit resonance, such as the 3:2 resonance of Mercury with the Sun.

Even though the orbit of Proxima Cen b is very small, the star is very faint, putting the planet in the habitable zone (section 2.3.2). Unfortunately the planet is not transiting, so we cannot study its atmosphere using current technology. Indeed, we don't even know that it has an atmosphere. Even if it does, and this permits the existence of liquid water, the strong chromospheric activity of the star could produce strong magnetic fields, intense ultraviolet radiation, and strong solar winds that may not be conducive to the emergence of life, and could even erode the planet's atmosphere. Having said this, if the planet is tidally locked, the night side of the planet will be protected from direct irradiation by the star. Conditions near the day–night line could permit liquid water and/or support life, depending on the atmosphere and how it redistributes energy. This can be addressed with models, but lacking further data on the system, we cannot know which model applies. Some progress may be possible with future observing facilities. The semi-major axis corresponds to an angular separation of just 38 mas, which is six times larger than the diffraction limit of the 39 m ELT at 1 μm, but the large contrast ratio between the star and planet may still make detection a challenge. The fact that the closest star to the Sun has such an intriguing planet makes it a prime candidate target for an interstellar spacecraft.

'b' may not be the only planet around Proxima Cen. In 2020 a second planet was detected via the radial velocity method with $m \sin i = 6 \, M_\oplus$ in a five-year orbit, which was named

Proxima Cen c. This awaits confirmation by independent observations. A third planet, 'd', was then detected in 2022, also by the radial velocity method. This has a period of five days and a minimum mass of just $0.25\,M_\oplus$. If confirmed as a genuine planet, and if the inclination is high (so this mass is close to the true mass), it would be one of the lowest mass planets detected by the radial velocity method so far.

2.6 Science on the way

Although a star system and its planet(s) would be the main target of an interstellar mission, significant science can also be done by the spacecraft on its journey there. This includes the edge of the solar system as well as interstellar space, both of which we will look at in more detail in chapter 12.

Apart from the eight planets of the solar system, there are many other rocky or icy bodies of a range of masses and sizes that orbit the Sun. These include the asteroid belt between Mars and Jupiter, the Kuiper belt (which includes Pluto) beyond the orbit of Neptune, and the highly extended Oort cloud, no members of which have been directly imaged. The space density of all these populations is too low – and most members far too small and faint – for it to be likely that an interstellar spacecraft passes close enough to one by chance to usefully observe it. For this reason also it is extremely unlikely that the spacecraft would collide with one (see the calculation in section 12.4.1). But it might be possible to navigate an interstellar spacecraft to pass near a known object.

What the spacecraft could observe, however, are the various boundaries between the solar system and interstellar space. One is the heliopause, which is where the solar wind pressure drops to equal the pressure of the interstellar medium. At its closest point this lies about 120 au from the Sun. The spacecraft could measure the change in pressure and other properties, as Voyager has done, but now with more sophisticated instruments and in another direction. The interstellar spacecraft could also measure the equivalent of the heliopause at the target star.

Beyond the heliopause the spacecraft is in interstellar space. Using particle counters and mass spectrometers, it can measure the relative numbers and energies of different atoms and ions in space, and the sizes, masses, and compositions of dust particles. It can also observe how these, as well as the interstellar magnetic field, vary with distance from the Sun and target star.

For mass and cost reasons, the telescopes on an interstellar spacecraft are unlikely to rival anything on the Earth in terms of size and sophistication, so the spacecraft is unlikely to be competitive for general astronomy of distant objects. There are, however, a few possible niche cases.

The first is astrometry. Distances to celestial objects can be measured via parallaxes, a method that relies on observing how the apparent position of an object changes as the observer moves. To date, this observer motion has been achieved by the annual orbital motion of the Earth around the Sun (or an orbit very similar to this in the case of the Gaia spacecraft). Two opposite points on the orbit provide an observational baseline of 2 au. If a star appears to move by $2''$ (compared to very distant quasars, for example) between these two points, then by definition it lies at a distance of 1 pc (from trigonometry, $1'' = 1\,\text{au}/1\,\text{pc}$).

Instead of one observer observing the star from two different points, two different observers could observe the star from each of these points. All other things being equal, the accuracy of a parallax measurement depends on the separation of the observers. So if one of the two observers was on the Earth, and the other was an interstellar spacecraft receding from the Sun, the baseline would get ever larger and the parallaxes ever more accurate.

Key to parallax measurements is an accurate knowledge of the size of the baseline. This, in turn, requires knowing exactly when the spacecraft made its measurements and how far it was from the Earth at the time. This information could be obtained with the help of an accurate onboard clock that measures how long it takes signals to travel between the spacecraft and the Earth (see section 14.3 for further discussion). In practice, the distance out to which such parallax measurements could be made is rather limited, and because the spacecraft is moving fast, only a small set of parallaxes could be obtained in this way. But it might be useful for very distant objects, for example nearby galaxies, for which parallaxes have not yet been obtained.

A second example of astronomical science is use of the solar gravitational lens. Gravity bends the path of light, so a massive object (the lens) between an observer and a source distorts the image of the source. A gravitational lens is not the same as an optical lens, though; in particular it does not have a single focal point, but rather a focal line.

The Sun can be used as a gravitational lens to observe sources on the other side of the Sun from the observer. The Sun forms a image of such a source along a long line passing through the source, Sun, and observer starting at a distance of 548 au from the Sun. To image different parts of an extended source, the spacecraft needs to move laterally to this focal line. The image is hugely magnified in spatial resolution and intensity, far beyond what is achievable even with the next generations of telescopes. The technical challenges to exploit this lens are considerable, though. As we are observing at directions close to the Sun's surface, the Sun's light must be blocked. The spacecraft would also have to be carefully controlled to keep it close to the focal line. To observe a specific source in this way, the spacecraft has to fly out in a specific direction and then have a specific motion to keep everything aligned. This is unlikely to be compatible with a fast trajectory to a nearby star. The solar gravitational lens is therefore best exploited by a dedicated precursor mission instead.

A third possible astronomical objective of an interstellar spacecraft is observing transiting exoplanets. As noted earlier, a transit is one of the main tools for learning about the physical properties of exoplanets. But we can only view a planet in transit if we happen to observe along its orbital plane. Consequently most exoplanets will never be observed in transit from the Earth. But an interstellar spacecraft will have a different perspective, and one that changes over time. Over its journey, it could monitor selected nearby stars to look for transits either of known (non-transiting) exoplanets, or of new planets.

As a fourth and final example, our interstellar spacecraft can directly probe the properties of gravity on scales of hundreds to hundreds of thousands of astronomical units. Measurements of gravity so far confirm the predictions of general relativity, but these measurements have not been performed on all length scales. As the spacecraft recedes from the Sun in cruise mode (i.e. without propulsion), its motion will be dictated mostly by gravity. The variation of its position and velocity with time can be compared to predictions of general relativity. This is not straightforward, however, as other small forces will also act on the spacecraft, such as radiation pressure, solar wind, and magnetic fields. These all have to be accounted for. Discrepancies between the expected and measured positions and velocities of Pioneer 10 and

11 were initially unexplained (the so-called Pioneer anomaly), and one suggestion was that it might be a modification of the law of gravity on large scales. Needless to say, this observation generated a lot of ideas and theories. It was eventually found that the anomaly could be explained by the anisotropic radiation of heat (photons, which have momentum) from the spacecraft's radioisotope thermoelectric generators (RTGs). This does not rule out that there may be detectable variations in gravity, however, and any future interstellar spacecraft should certainly be tracked for as long as possible in order to test this.

2.7 Summary

There are diverse motivations for sending a robotic spacecraft to another star, including intellectual and economic stimulation, technological spinoffs, and making the next big step in exploration. But the primary motivation is the science. A spacecraft travelling to a distant star system can do astronomy, geology, chemistry, and exobiology that are either impossible or extremely hard to do with Earth-based telescopes. In terms of both spatial resolution and light gathering power, an interstellar spacecraft will probably be superior to Earth- or space-based telescopes (including interferometers) by orders of magnitude, even in a century from now. A spacecraft also opens up domains that will never be attainable from the Earth, such as a variable observing perspective and in situ measurements of particles.

The primary science objectives are learning about another star and its exoplanets, as well as rings, moons, and other minor bodies in the star system such as comets. Of the 35 stars within 12.1 ly of the Sun, there are 14 confirmed planets, 17 more candidates, and probably additional planets that have not yet been detected (as of mid 2025).

Secondary science objectives for an interstellar mission include studying the edge of the solar system and interstellar space, and possibly making astronomical observations such as parallax measurements, tests of gravity, or exploiting the solar gravitational lens.

Rockets

Space exploration to date relies heavily on rockets. Although most rockets have been driven by chemical propellants, there are many other types of rocket, such as ion engines and nuclear rockets. In this chapter we look at the general principles of a rocket, regardless of the mechanism used to generate thrust. We derive the important rocket equation and examine its consequences for space travel.

3.1 The rocket equation

The principle of a rocket is to eject something in one direction in order to move the spacecraft in the opposite direction. This occurs on the basis of the conservation of momentum, and a mathematical description of this leads to a fundamental equation in astronautics.

Consider a rocket of mass M moving with velocity V relative to some inertial reference frame (figure 3.1). The rocket ejects a packet of propellant of mass dm at velocity v_e relative to the rocket, with the result that the rocket's speed increases to $V + dV$ relative to the reference frame. Assuming there are no external forces, in particular no gravity, then from the law of conservation of momentum, the initial momentum equals the final momentum. From figure 3.1 this is

$$MV = dm(V - v_e) + (M - dm)(V + dV) \tag{3.1a}$$

$$0 = MdV - v_e dm \tag{3.1b}$$

where $v_e > 0$ is directed to the left, and we've neglected the $dmdV$ term in the second line on the grounds that we are dealing with differentials. Assuming mass is conserved, then the change in the mass of the rocket is

$$dM = -dm. \tag{3.2}$$

Putting together these two equations to eliminate dm we get

$$\frac{dV}{v_e} = -\frac{dM}{M}. \tag{3.3}$$

This tells us how the instantaneous velocity of the rocket changes as a function of the propellant velocity and change in rocket mass. We integrate this from an initial condition when the rocket is moving at velocity V_i with total mass M_i (spacecraft plus propellant) to a final condition when it is moving at velocity V_f with mass M_f, after having ejected propellant of mass $M_p = M_i - M_f$. This gives

$$\Delta v = v_e \ln\left(\frac{M_i}{M_f}\right), \tag{3.4}$$

where $\Delta v = V_f - V_i$ is the change in velocity of the rocket in the reference frame.

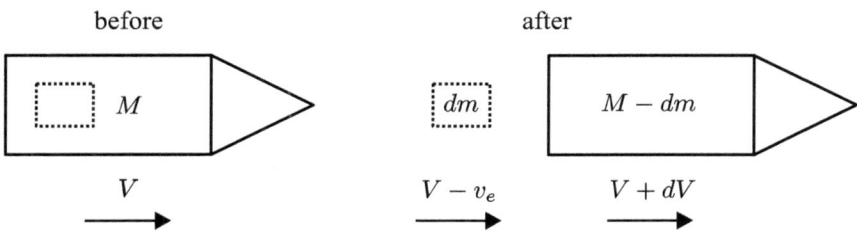

before   after

M dm $M - dm$

V $V - v_e$ $V + dV$

Derivation of the rocket equation. The rocket and propellant together initially have mass M and move at velocity V to the right ('before'). The rocket then ejects propellant of mass dm at velocity v_e relative to the rocket ('after'). This accelerates the rocket to velocity $V + dV$, leaving the propellant moving at velocity $V - v_e$. All quantities are positive and all velocities are relative to the observer's inertial reference frame.

Fig. 3.1

Equation 3.4 is the rocket equation. It shows that the change in velocity of the rocket depends only on its initial-to-final mass ratio and on the exhaust velocity of the propellant. The equation is plotted in figure 3.2. We see that Δv increases only very slowly with mass ratio for a given v_e. To achieve a Δv equal to v_e we need a mass ratio of 2.7, i.e. a propellant mass 1.7 times the rest of the spacecraft. To change the velocity of the rocket by $4v_e$, a propellant mass around 55 times the mass of the empty rocket is required.

We shall see in chapter 5 that the maximum value of v_e attainable with chemical propellants is about $5 \, \mathrm{km \, s^{-1}}$, so even with a mass ratio of 55 a rocket would only achieve a Δv of $20 \, \mathrm{km \, s^{-1}}$. Assuming our rocket started in the Earth's orbit around the Sun, with a velocity of $30 \, \mathrm{km \, s^{-1}}$, the rocket's final velocity (after a rapid engine burn) would be $50 \, \mathrm{km \, s^{-1}}$. This is only just enough to escape from the solar system (see section 4.1.4), leaving the rocket to travel to the next star at just a few $\mathrm{km \, s^{-1}}$. Even a mass ratio of 1000 would yield a Δv of only $35 \, \mathrm{km \, s^{-1}}$, and gravity assists, discussed in section 4.3, do not substantially improve this. To get to the nearest star in a human lifetime, a spacecraft needs to achieve a velocity of about $\Delta v = 4.25 \, \mathrm{ly}/50 \, \mathrm{yr} = 25\,000 \, \mathrm{km \, s^{-1}}$. From equation 3.4, we see that the mass ratio required to achieve this with $v_e = 5 \, \mathrm{km \, s^{-1}}$ is $M_i/M_f = \exp(5000)$, or 10^{2170}. Even with a payload mass of only $1 \, \mathrm{g}$, this is far more mass than in the known universe (a mere $10^{53} \, \mathrm{kg}$).

The rocket equation tells us that we need exponentially more propellant to achieve linear increments in velocity. The physical reason for this is that the rocket is using some of its propellant to accelerate the not-yet-ejected propellant. The pursuit of high velocity rockets must therefore focus on achieving a large v_e. This is not attainable with chemical rockets (see section 5.4).

It is important to appreciate the assumptions made when deriving the rocket equation. First, we assumed no external forces are acting, so this rocket equation does not apply for launch from the surface of a planet. We shall relax this assumption in section 3.10. Second, we assumed that the propellant is ejected continuously, at both a constant rate and at a constant velocity relative to the rocket. Third, we derived the rocket equation using classical mechanics, not special relativity, and so are implicitly assuming that Δv is much less than the speed of light. This will be addressed in section 7.5.

We have not, however, made any assumptions about what the propellant is or what is causing it to be ejected. The rocket equation therefore applies equally well to a nuclear or ion drive as it does to a chemical thermal rocket, and even applies to just spraying an inert gas through a hole. We will, however, have to pay attention to the definition of v_e when there is

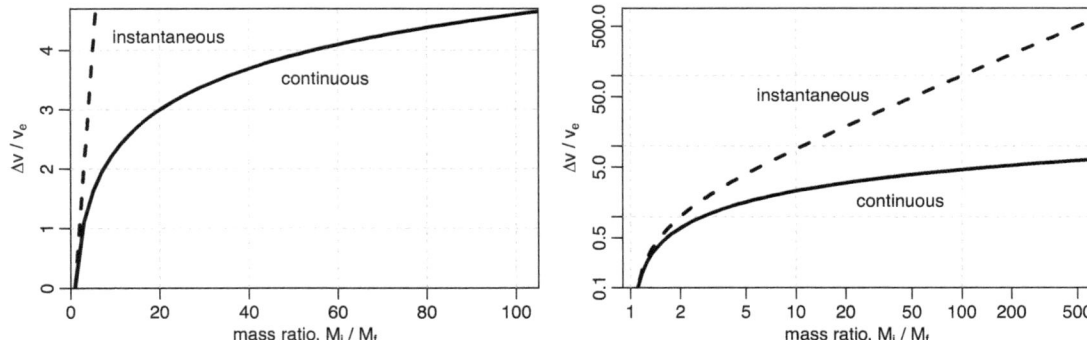

Fig. 3.2 The solid line shows the classical rocket equation, $\Delta v = v_e \ln(M_i/M_f)$, where Δv is the change in velocity of the rocket when ejecting propellant continuously at velocity v_e such that the rocket's mass decreases from M_i to M_f. The dashed line shows the change in velocity when the propellant is ejected in a single instantaneous event (equation 3.5). The left panel uses linear–linear axes whereas the right one uses log–log axes over a wider range.

a significant conversion of mass into energy, or when there are mass/energy losses. We will look at these in section 7.8.

To relax the second assumption mentioned above we could re-derive the rocket equation with time-dependent values of v_e and $\dot{m}$. It turns out that if v_e can be increased as the rocket accelerates, then this minimizes the kinetic energy given to the exhaust. This increases the kinetic energy of the rocket and thus the energy efficiency (defined in section 3.6). This so-called 'perfect' rocket is not achievable in practice with a chemical thermal rocket. It could be more straightforward with other types of rocket that we will discuss in later chapters.

We can also ask what would change if we ejected the propellant not continuously, but in a single discrete chunk. Because we then do not have to use propellant to accelerate any remaining propellant, it is clear that this 'instantaneous' rocket will achieve a larger Δv for a given v_e and mass of propellant. This follows from momentum conservation (equation 3.1a) with $M = M_i$, $dV = \Delta v$, and $dm = M_i - M_f$, which gives

$$\Delta v = v_e \left(\frac{M_i}{M_f} - 1 \right) \quad \text{(instantaneous rocket).} \tag{3.5}$$

The velocity is now linear in the mass ratio and so increases much faster than the logarithm in the usual rocket equation. This is shown in figure 3.2 by the dashed line. With $v_e = 5\,\text{km}\,\text{s}^{-1}$, a mass ratio of 55 would now yield a Δv of $270\,\text{km}\,\text{s}^{-1}$. Ejecting all the mass in one go produces considerably larger velocity increments and – as we will see in section 3.9 – a higher efficiency. Unfortunately, such an approach is impractical and could not be achieved with a conventional engine.

3.2 Braking at the target star

The rocket equation tells us how much propellant is needed to accelerate by Δv. But how much propellant is needed to also decelerate back to the original velocity, in order to rendezvous at a target star, for example? For this we need two separate boosts. First we accelerate by Δv, then we turn the rocket around and decelerate by Δv. Let M_m be the mass of the rocket after the first boost. From the rocket equation we have for the first boost

$$\Delta v = v_e \ln \left(\frac{M_i}{M_m} \right). \tag{3.6}$$

For the second boost we have

$$\Delta v = v_e \ln \left(\frac{M_m}{M_f} \right). \tag{3.7}$$

Combining these we get

$$\frac{M_i}{M_f} = \left(\frac{M_m}{M_f} \right)^2. \tag{3.8}$$

The mass ratio required to decelerate back to the initial velocity is the *square* of that required to accelerate in the first place. Decelerating at the target system is therefore very costly in terms of mass budget. Note that the mass after the first boost is the geometric mean of the initial and final masses, $M_m = \sqrt{M_i M_f}$.

In terms of propellant required, it of course does not matter in which direction the engine is pointing. The propellant required for the journey is the same as that required to accelerate to $2\Delta v$, which of course requires the same mass ratio,

$$2\Delta v = 2v_e \ln \left(\frac{M_i}{M_m} \right) = v_e \ln \left(\frac{M_i}{M_m} \right)^2. \tag{3.9}$$

In general, if we require a mass ratio R to change the velocity by Δv, then to change the velocity by $n\Delta v$, we need a mass ratio of R^n.

3.3 The thrust equation

Newton's second law tells us that the force exerted by the rocket on the propellant – and therefore also by the propellant on the rocket (Newton's third law) – is equal to the rate of change of momentum. When propellant is ejected at mass flow rate $\dot{m}$ at velocity v_e relative to the rocket, the magnitude of this force, or thrust, is

$$F = \dot{m} v_e, \tag{3.10}$$

where in general both $\dot{m}$ and v_e are time dependent. Thrust is important for two reasons. First, when manoeuvring in a gravitational field, a minimum thrust is required to overcome the force of gravity. This is most apparent when launching from the surface of a planet, where the thrust must exceed the weight of the rocket for it to get off the launch pad (see section 3.10). Second, even in the absence of gravity, a larger thrust means a larger instantaneous acceleration (for a given mass), and so the faster the rocket will be after a certain period of time. How can we reconcile this with the rocket equation, which tells us how fast the rocket will be moving after a certain amount of propellant has been expended? At any instant, the acceleration of a rocket of mass M is

$$\frac{dV}{dt} = \frac{F}{M} = \frac{\dot{m} v_e}{M} \tag{3.11a}$$

$$= -\frac{dM}{dt} \frac{v_e}{M}, \tag{3.11b}$$

which reduces to equation 3.3, the differential form of the rocket equation. Hence the rocket equation is the integral of the thrust equation when assuming constant v_e (but not necessarily constant $\dot{m}$). This should not be surprising given that both arise from Newton's second law. Nonetheless, the two equations are still separately useful because they describe different aspects of the rocket. The rocket equation tells us what Δv is ultimately achievable, whereas the thrust equation tells us how fast it can be attained. Both aspects are relevant to getting to a destination quickly, as we will now see.

3.4 Travel time and distance for a constant mass rate

The rocket equation gives the velocity of the rocket, $\Delta v = V(t)$, as a function of the current mass of the rocket, $M_f = M(t)$, which comes from integrating the variable acceleration (equation 3.11b) over time. Integrating again over time gives the distance travelled by the rocket,

$$d(t) = \int_0^t V(t')dt'. \tag{3.12}$$

Adopting a constant rate of mass ejection, the mass decreases linearly with time. Specifically, if $M_i - M_f$ is the total mass of propellant ejected over some time interval, then the mass of the rocket at any time t within this interval is

$$M(t) = \begin{cases} M_i - \dot{m}t & \text{if} \quad \dot{m}t < M_i - M_f \\ M_f & \text{otherwise.} \end{cases} \tag{3.13}$$

Putting this into the rocket equation gives

$$V(t) = \begin{cases} v_e \ln\left(1 - \dfrac{\dot{m}t}{M_i}\right)^{-1} & \text{if} \quad t < (M_i - M_f)/\dot{m} \\ v_e \ln\left(\dfrac{M_i}{M_f}\right) & \text{otherwise.} \end{cases} \tag{3.14}$$

This is plotted in figure 3.3. The horizontal axis is the fraction of the initial mass ejected as propellant, $\dot{m}t/M_i$. This is shown for two cases: one in which the propellant is 99% of the total mass ($M_i/M_f = (99+1)/1 = 100$), and the other in which the propellant is 90% of the total mass ($M_i/M_f = (9+1)/1 = 10$). We see that most of the change in velocity comes later in the mass ejection, when the rocket is lightest.

Inserting equation 3.14 into equation 3.12 and integrating, we get the following expression for how far the rocket travels during its acceleration, once its mass has reduced to M_f (see exercises):

$$d_{\text{acc}} = v_e \frac{M_i}{\dot{m}}\left[1 - \frac{M_f}{M_i}\left(1 + \ln\frac{M_i}{M_f}\right)\right] \quad \text{or} \tag{3.15a}$$

$$= v_e \frac{M_f}{\dot{m}}\left[\frac{M_i}{M_f} - 1 - \ln\left(\frac{M_i}{M_f}\right)\right]. \tag{3.15b}$$

Although equation 3.15b is a function of four variables, we can consider just three: the exhaust velocity v_e, the mass ratio $R = M_i/M_f$, and $M_i/\dot{m}$, which is the time it would take to eject a mass equal to the initial mass of the rocket.

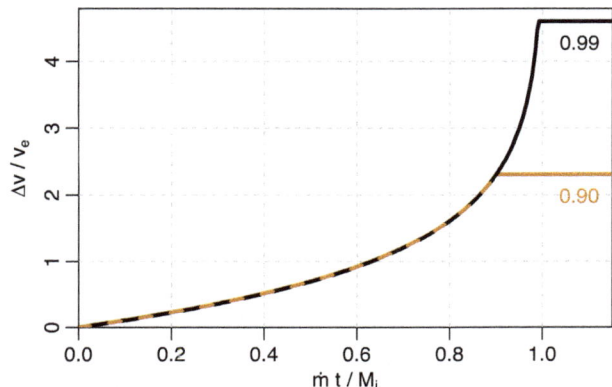

The variation in velocity of a rocket (in units of v_e) of initial mass M_i with time t as it ejects mass at constant rate $\dot{m}$, shown as a function of the fraction of mass ejected so far (equation 3.14). The upper line is for a propellant mass fraction of 0.99, so the rocket accelerates up until time t given by $\dot{m}t/M_i = 0.99$. The lower line is for 0.90.

Fig. 3.3

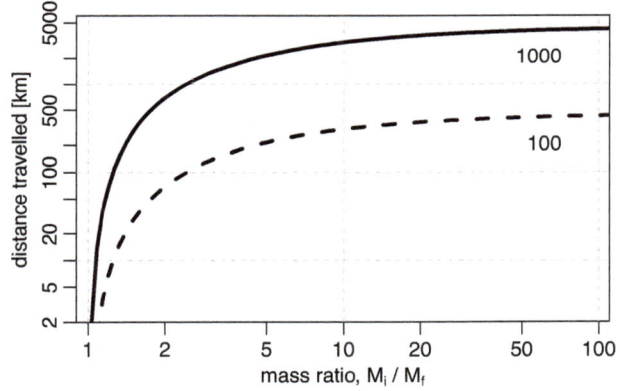

The distance travelled by a rocket during its acceleration when ejecting propellant at a constant velocity $v_e = 4.5 \text{ km s}^{-1}$ and constant rate $\dot{m}$ (equation 3.15b). The solid line is for $M_i/\dot{m} = 1000$ s and the dashed line is for $M_i/\dot{m} = 100$ s, where M_i is the initial mass of the rocket.

Fig. 3.4

The distance travelled is plotted in figure 3.4 as a function of the mass ratio for fixed v_e and two values of $M_i/\dot{m}$. Consider the behaviour along one of the curves. The larger the mass ratio R, the more propellant is ejected, so the larger the change in velocity. The time taken to accelerate is

$$t_{\text{acc}} = \frac{M_i - M_f}{\dot{m}} = \frac{M_i}{\dot{m}}\left(1 - \frac{1}{R}\right) = \frac{M_f}{\dot{m}}(R - 1). \tag{3.16}$$

Increasing the mass ratio at constant $M_i/\dot{m}$ extends the acceleration time, so a larger distance is covered.

We see from equation 3.15b that the distance travelled during the acceleration is *inversely* proportional to the mass ejection rate, $\dot{m}$. This might seem counter-intuitive, because for given v_e, a smaller $\dot{m}$ means a smaller thrust and so a smaller acceleration. But if M_i and M_f are fixed, then the final Δv is the same regardless of $\dot{m}$, from the rocket equation. If the rocket uses up the propellant more slowly, it spends more time accelerating and so travels

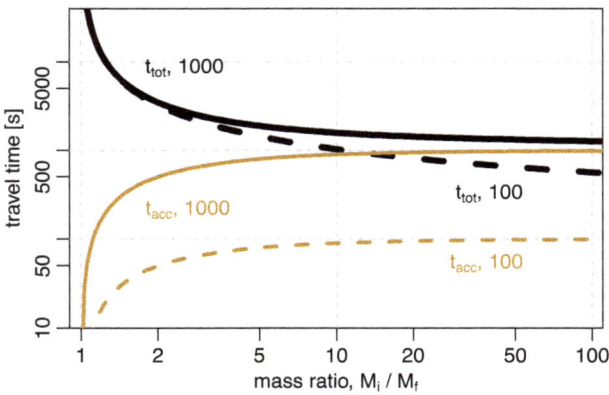

Fig. 3.5 The time taken by a rocket to travel a distance of 10 000 km (equation 3.17a). The rocket ejects propellant at constant velocity $v_e = 4.5$ km s^{-1} and constant rate $\dot{m}$. It accelerates until it has exhausted its propellant, then continues at constant velocity. The thick upper line shows the total time (t_{tot}) taken to reach the distance. The thin lower line shows how much of this time the rocket spends accelerating (t_{acc}) to achieve its cruise velocity of $v_e \ln(M_i/M_f)$. The solid line is for $M_i/\dot{m} = 1000$ s and the dashed line is for $M_i/\dot{m} = 100$ s, where M_i is the initial mass of the rocket.

further during this acceleration. In the limit of an infinitesimally small mass ejection rate, the rocket will spend an arbitrarily long time at a non-zero velocity and so travel arbitrarily far. Of course, it would take such a rocket much longer to travel that distance, and for travelling to a given star we are particularly interested in minimizing the time.

When we want to travel further than the acceleration distance, the remainder of the journey will be spent cruising at a constant velocity Δv relative to the initial rest frame. If the total distance is d_{tot}, then the time spent cruising is $(d_{\text{tot}} - d_{\text{acc}})/\Delta v$. The total travel time is therefore

$$t_{\text{tot}} = \frac{M_i}{\dot{m}}\left(1 - \frac{1}{R}\right) + \frac{d_{\text{tot}} - d_{\text{acc}}}{v_e \ln R} \quad \text{or} \tag{3.17a}$$

$$= \frac{M_f}{\dot{m}}(R - 1) + \frac{d_{\text{tot}} - d_{\text{acc}}}{v_e \ln R} \tag{3.17b}$$

for $d_{\text{tot}} \geq d_{\text{acc}}$, where again $R = M_i/M_f$. This is plotted as the thick line in figure 3.5 as a function of R for two values of $M_i/\dot{m}$. The larger the mass ratio, the quicker the rocket reaches the destination, because the cruise velocity is larger. The smaller $M_i/\dot{m}$, the faster the propellant is ejected, and so the quicker the rocket reaches Δv. The rocket spends more of the journey at a larger velocity, compared to when it has a lower mass ejection rate. The thin lines in figure 3.5 show the amount of time spent accelerating (equation 3.16). For a larger mass ratio (for given $M_i/\dot{m}$) the rocket spends not only more time accelerating, but also a larger fraction of the total time. Finally, even though v_e and $\dot{m}$ are constant, and so the thrust is constant (equation 3.10), the rocket's acceleration is not because its mass is varying. The time dependence of the acceleration is given by equations 3.11a and 3.13.

Figure 3.6 shows the total time taken for a journey to Proxima Cen ($d_{\text{tot}} = 4.25$ ly) as a function of v_e for different configurations. The thick line is for a rocket with a mass ratio $R = 10$ and time to exhaust all the propellant at constant $\dot{m}$ set to $t_{\text{acc}} = 10$ years. With $v_e = 0.043\,c$, for example, the journey takes 50 years, and the cruise velocity is $0.043 \ln(10) = 0.099\,c$. If the time spent accelerating is increased by a factor of four to 40 years (the dotted

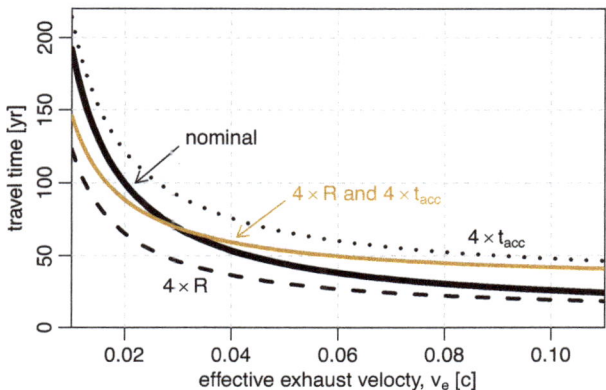

Fig. 3.6

The time taken by a rocket to travel to Proxima Cen ($d_{tot} = 4.25$ ly) as a function of v_e. The thick solid line labelled 'nominal' is for a mass ratio $R = 10$ and acceleration time $t_{acc} = 10$ yr. The dashed and dotted lines increase each of these by a factor of four, respectively, holding the other constant. The thin solid line increases both of these by a factor of four over the nominal.

line), the total travel time is increased, although not by much at low v_e (at $v_e = 0.043\,c$ the time is now 72 years). If instead the mass ratio is increased by a factor of four (the dashed line), the travel time is shorter, but not by much at high v_e (at $v_e = 0.43\,c$ the time is now 34 years). If we increase both acceleration time and mass ratio by a factor of four, then the travel time is given by the thin solid line in figure 3.6. The travel time is shorter than the nominal case at low v_e, but longer at high v_e.

3.5 Specific impulse and effective exhaust velocity

The term *specific impulse* is often encountered in astronautics, and although I will rarely use it, it is worth knowing that it has two mathematical definitions. The first is the change of momentum of the rocket per unit mass of propellant expelled. The change of momentum is the force integrated over time, so when keeping all terms in the thrust equation $F = \dot{m}v_e$ constant, this is

$$F\delta t = \delta m\, v_e. \tag{3.18}$$

The specific impulse is therefore just v_e, and has units of velocity. For a rocket that derives its impulse entirely from ejected mass, the specific impulse is just the velocity of its propellant. However, in some cases, such as when the propellant contains a mixture of particles moving at different speeds, the exhaust velocity must be replaced with a representative value known as the *effective exhaust velocity*. I will often use this term, rather than simply 'exhaust velocity', to remind us that they are not always the same.

The second definition of specific impulse is the change of momentum per unit *weight* of propellant. In this case, the specific impulse is v_e/g, where g is the acceleration due to gravity, and this version has units of time. Normally we take g as the value at the Earth's surface. As this is irrelevant to interstellar travel, I will not use this definition of specific impulse. It tends to be used in the context of jet engines operating in the Earth's atmosphere, where air as well as fuel make up the propellant. In that case the effective exhaust velocity is very different from the actual exhaust velocity.

3.6 Energy and energy efficiency

How much energy do we need to accelerate a rocket to a certain speed? Consider first the view from the instantaneous rest frame of the rocket during its acceleration. At every moment the rocket ejects a packet of propellant of mass δm with velocity v_e relative to the rocket. Each packet has an energy of $(1/2)\delta m v_e^2$. The total mass ejected is $M_i - M_f$, so the total energy that the rocket must supply to eject this (assuming no losses) is

$$E_{\text{required}} = \frac{1}{2}(M_i - M_f)v_e^2 = \frac{1}{2}M_f v_e^2(R - 1), \qquad (3.19)$$

where $R = M_i/M_f$. As the rocket is the only source of energy, E_{required} is the energy stored in the rocket fuel at the beginning, when the rocket is at rest in the initial rest frame. From the point of view of the continuously changing instantaneous rest frame of the rocket, all of this fuel energy goes into the propellant (the kinetic energy of the rocket in this frame is zero). But from the point of view of the initial rest frame, some of the fuel energy goes into the propellant and some into the rocket. Due to conservation of momentum in the initial rest frame, not all of the energy can go into the rocket's motion. Once the rocket has exhausted its propellant, it has reached a velocity Δv in the initial rest frame, so has a kinetic energy of

$$E_{\text{rocket}} = \frac{1}{2}M_f(\Delta v)^2 = \frac{1}{2}M_f v_e^2 \ln^2 R, \qquad (3.20)$$

having used the rocket equation (equation 3.4) in the second step. Seen from the initial rest frame, propellant is moving with a range of (signed) velocities from $-v_e$ to $\Delta v - v_e$ (and not uniformly distributed over this range because the rocket did not have constant acceleration). From conservation of energy, the total kinetic energy of the propellant in the initial rest frame is $E_{\text{required}} - E_{\text{rocket}}$.

We need to be careful when considering conservation of energy in different reference frames because (kinetic) energy is a relative quantity. What we have done above is equate the 'before' and 'after' energy in the initial rest frame, having determined the energy required in the rocket's rest frame.

A useful measure of the energy efficiency is the ratio of the rocket's kinetic energy to the energy it supplied. This is

$$\frac{E_{\text{rocket}}}{E_{\text{required}}} = \frac{\ln^2 R}{R - 1} \qquad (3.21)$$

and is plotted as the solid line in figure 3.7 (the dashed line will be derived in section 3.9). For low mass ratios, only a small fraction of the total energy ends up as kinetic energy of the rocket. At a mass ratio of $R = 4.92$, or equivalently $\Delta v/v_e = 1.59$, the efficiency reaches its maximum value of 0.65. It is not possible to put all of the energy into the rocket. In order to conserve both energy and momentum, at least a third of the energy supplied must end up in the propellant. At higher mass ratios than this optimum, the rocket will attain a larger velocity, but the efficiency is lower. Note that this is an ideal theoretical efficiency which does

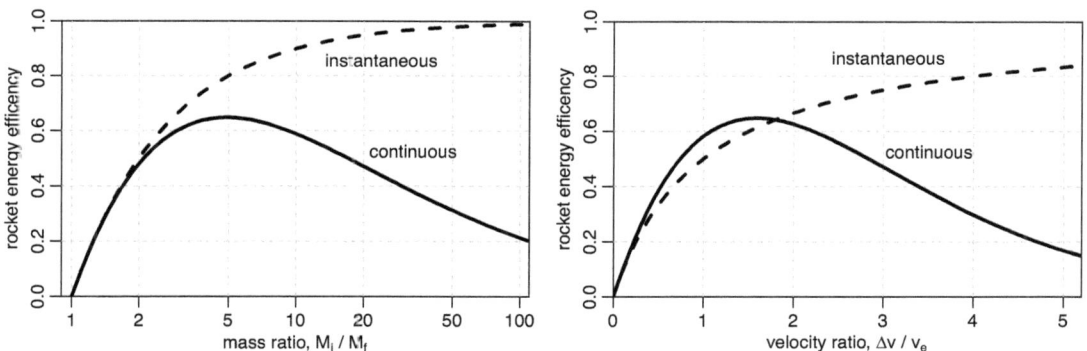

The ratio of the kinetic energy of the rocket (in the initial rest frame) to the energy expended by the rocket (to accelerate it and the propellant) plotted against the mass ratio (left) and the velocity ratio (right). The solid line is for the standard rocket equation, with efficiency given by equation 3.21. The dashed line is for the instantaneous rocket as given by equation 3.32. Note that the two types of rocket do not have the same relationship between mass ratio and velocity ratio.

Fig. 3.7

not take into account any practical losses that a real engine would have, for example due to a finite thermodynamic efficiency in a heat engine, or neutron losses in a fission or fusion engine.

3.7 Power

The power of a rocket is the rate at which energy flows out via the propellant. From the expression for kinetic energy for a constant mass flow rate $\dot{m}$, it follows that the power is

$$P = \frac{1}{2}\dot{m}v_e^2 .$$

(3.22)

Like energy and velocity, power is not absolute but is defined relative to a reference frame, here the instantaneous rest frame of the rocket. Using the thrust equation 3.10 we can relate the power to the thrust as

$$F = \frac{2P}{v_e} .$$

(3.23)

This tells us that lowering the exhaust velocity for a given power increases the thrust. This might seem counter-intuitive until we realize that by pushing the propellant out more slowly at constant power we must increase the mass flow rate, from equation 3.22. When working with these various equations – for velocity, thrust, and power – we must always be aware which quantities are being held constant. Equation 3.23 tells us that in order to maxmize the thrust for a given power (for example at launch), we should expel the propellant relatively slowly (small v_e). Equation 3.22 then tells us that this requires a high mass flow rate, which for a low exhaust velocity implies a large engine or a heavy propellant.

Note that P above is the power of the propellant in the instantaneous rocket rest frame, and is therefore the power that the rocket must provide. The power of the engines themselves must inevitably be higher due to efficiency limitations.

It can also be informative to consider the power per unit mass flow rate, $P/\dot{m}$. From equation 3.22 this is

$$\frac{P}{\dot{m}} = \frac{1}{2}v_e^2, \tag{3.24}$$

which is the kinetic energy per unit mass (specific energy) of the propellant. For a chemical rocket the specific energy is determined by the fuel and the engine. A liquid hydrogen-oxygen propellant mix that provides $v_e = 4.5\,\text{km s}^{-1}$ has a specific energy of $10.1\,\text{MJ kg}^{-1}$.

3.8 Staging

The tanks that hold the rocket propellant have a mass that can be a significant fraction of the complete fuelled rocket. If the propellant is stored in multiple tanks, each tank can be jettisoned as soon as it is empty to avoid wasting propellant on accelerating it further. Dropping surplus mass is the concept of staging.

For reasons that will be discussed in chapter 5, each stage of a thermal rocket often has its own engines too (figure 3.8). Once the propellant in that stage is exhausted, the empty tanks together with the engines are jettisoned, and then the engines in the next stage fire. We are most familiar with this in launches from the Earth's surface.

Figure 3.9 demonstrates the principle with two stages. From the rocket equation we can compare two alternative scenarios: (a) using the entire rocket as a single stage (i.e. without

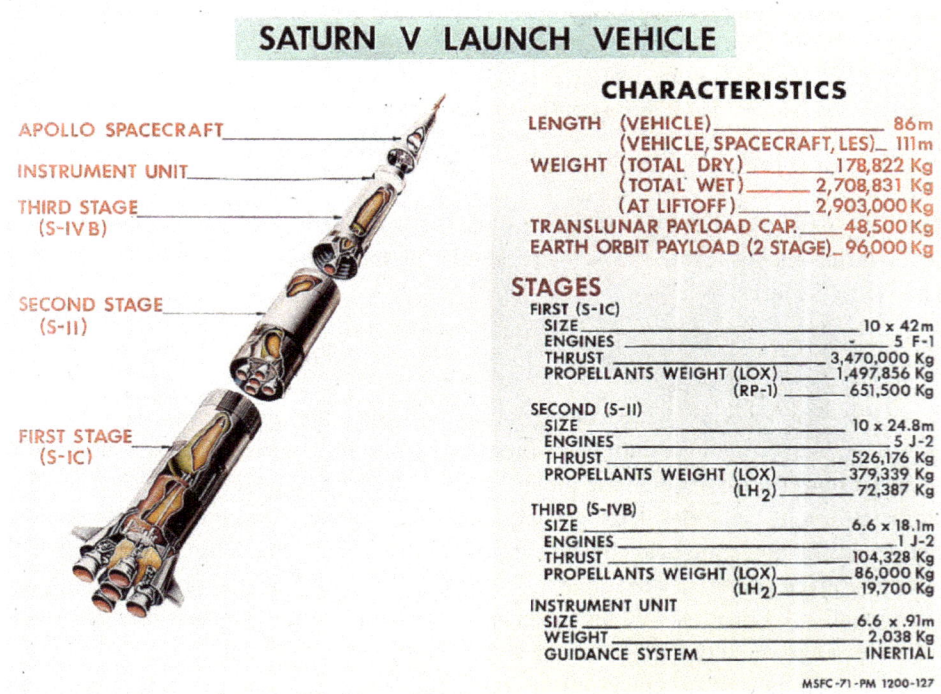

Fig. 3.8   The different stages of the Saturn V rocket that carried the Apollo spacecraft to the Moon. Credit: NASA.

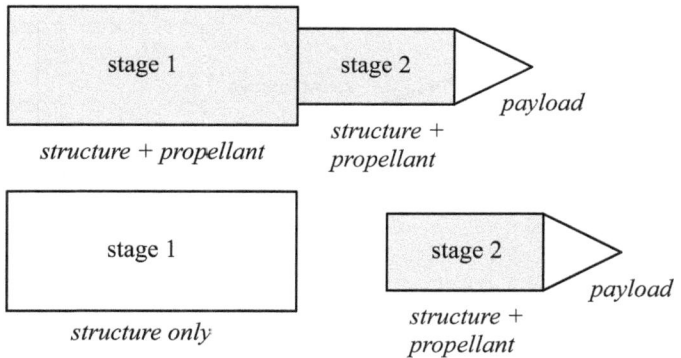

Principle of rocket stages. Top: The rocket is initially full of propellant. Bottom: Once a stage has exhausted all of its propellant, it is ejected to remove the mass of the empty tanks.

Fig. 3.9

jettisoning stage 1), and (b) using up the propellant first in stage 1, jettisoning it, then using the propellant in stage 2. Let us assume the engines in each stage have the same exhaust velocity v_e. We then only need to consider the initial-to-final mass ratio in order to compare the Δv in these two scenarios. We define f as the ratio of the mass of the structure (empty tanks and engines) to the propellant in each stage; for simplicity we'll make f the same in both stages too. The mass of the propellant in the two stages is m_1 and m_2, respectively, and m_{pl} is the mass of the payload.

(a) The initial-to-final mass ratio of the rocket when using it as a single stage is

$$R_s = \frac{m_{pl} + m_2(f+1) + m_1(f+1)}{m_{pl} + m_2 f + m_1 f}. \tag{3.25}$$

From the rocket equation the velocity change Δv_s is

$$\frac{\Delta v_s}{v_e} = \ln R_s. \tag{3.26}$$

(b) When using the stages separately, the mass ratio of stage 1 is

$$R_1 = \frac{m_{pl} + m_2(f+1) + m_1(f+1)}{m_{pl} + m_2(f+1) + m_1 f} \tag{3.27}$$

and the mass ratio of stage 2 is

$$R_2 = \frac{m_{pl} + m_2(f+1)}{m_{pl} + m_2 f}. \tag{3.28}$$

From the rocket equation we compute the Δv given by each of these stages separately and then sum them to get the total Δv, which is

$$\frac{\Delta v_1 + \Delta v_2}{v_e} = \ln R_1 + \ln R_2. \tag{3.29}$$

To compare scenarios (a) and (b), let us adopt specific values of $m_1 = 30 m_{pl}$ and $m_2 = 5 m_{pl}$ and examine how the velocities vary as a function of f. This is plotted in figure 3.10. As expected, the two-stage scenario (thick solid line) gives a larger Δv than the single-stage scenario (thin solid line) for all $f > 0$, even though both use the same amount of propellant. This difference of course drops to zero in the limit of a zero mass structure, as then nothing is gained by jettisoning it. The advantage of staging can be quite significant: for $f = 0.1$,

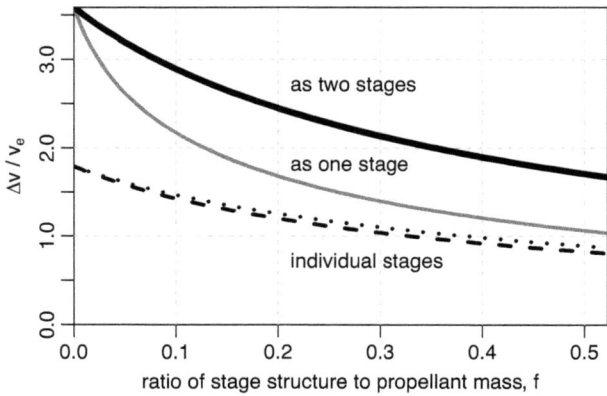

Fig. 3.10 Two-stage rocket performance for the case $m_1 = 30m_{pl}$ and $m_2 = 5m_{pl}$. The thick solid line shows the Δv delivered using two stages according to equation 3.29 as a function of the ratio of the mass of the structure (empty tanks and engines) to the propellant in the stages. The dashed and dotted lines show the Δv delivered by stages 1 and 2, respectively. The thin solid line shows, for comparison, the Δv delivered when using the same rocket as a single stage (i.e. without jettisoning the first stage when empty), as given by equation 3.26.

$\Delta v/v_e$ of the two stages is 2.89 compared to 2.17 when using them as one stage, a gain of a factor of 1.33. As the structure gets more massive (increasing f), Δv of course decreases for both scenarios, but the advantage of two stages over a single stage is more pronounced: at $f = 0.5$, $\Delta v/v_e$ is 1.71 vs 1.06, a gain of a factor of 1.61.

This illustrates the advantage of staging. When designing a real mission, we would be given m_{pl} and know the $\Delta v_1 + \Delta v_2$ required, and would then find the value of m_1/m_2 that minimizes $m_1 + m_2$. In practice f, and possibly v_e, would not be the same for each stage.

We have examined only two stages in order to illustrate the concept. A larger Δv for a given mass is achieved using more stages, and in principle the more stages, the better. There are practical and economic issues with using an ever larger number of stages, however, not least because each requires its own engines, plus the overall risk of failure increases. For launches from the Earth's surface that escape Earth's gravity, three stages has proven to be optimum, also for reasons of thrust and nozzle optimization in chemical rockets, as will be discussed in chapter 5.

3.9 Instantaneous rockets

At the end of section 3.1, we saw that by ejecting all of the propellant as a single chunk we achieve a higher Δv for a given mass ratio and effective exhaust velocity (figure 3.2 and equation 3.5). This physical situation is shown in figure 3.11. Suppose the rocket provides an energy E in order to simultaneously accelerate the rocket of mass M_f to a velocity Δv and the propellant of mass M_p to a velocity v_e, both velocities being relative to the initial rest frame of the rocket (and both positive). The equations of conservation of momentum and energy are

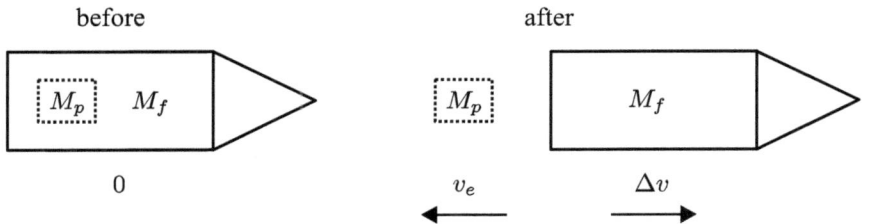

before after

M_p M_f M_p M_f

0 v_e Δv

The instantaneous rocket. The rocket of mass M_f and its propellant of mass M_p are initially at rest. The rocket then ejects the propellant at velocity v_e relative to the rocket, which accelerates the rocket to velocity Δv. All quantities are positive.

Fig. 3.11

$$M_p v_e = M_f \Delta v, \tag{3.30a}$$

$$E = \frac{1}{2} M_p v_e^2 + \frac{1}{2} M_f (\Delta v)^2 . \tag{3.30b}$$

Combining these to eliminate M_p gives the rocket's velocity for a given energy:

$$(\Delta v)^2 = \frac{2E}{M_f} \left(\frac{1}{1 + M_f/M_p} \right) = \frac{2E}{M_f} \left(1 - \frac{1}{R} \right) . \tag{3.31}$$

The energy of the rocket (in the initial rest frame) is $(1/2)M_f(\Delta v)^2$ and the energy required is E. Thus the efficiency of the instantaneous rocket is

$$\frac{E_{\text{rocket}}}{E_{\text{required}}} = 1 - \frac{1}{R} . \tag{3.32}$$

The variation of this efficiency with the mass ratio R is shown in the left panel of figure 3.7 as the dashed line. Whereas the efficiency of the continuous rocket (solid line) has a maximum and then decreases at large mass ratios and velocity ratios ($\Delta v/v_e$), the efficiency of the instantaneous rocket continues to increase, asymptoting to unity. Note that in the left panel, the efficiency of the instantaneous rocket is always higher than that of the continuous rocket, which is not the case in the right panel. This is because a given mass ratio does not correspond to the same velocity ratio for both types of rockets. They have different rocket equations, so a given mass ratio in the left panel corresponds to two different velocity ratios in the right panel.

3.10 Launching from the surface of a planet

So far we have considered a rocket in space, free from external forces such as gravity. The derivation of the rocket equation in section 3.1 needs to be modified to determine the motion of a rocket launched from a planet's surface. If g is the local acceleration due to gravity, then for a rocket of mass M moving vertically (parallel to the gravitational force), Newton's second law gives

$$\frac{dV}{dt} = \frac{F}{M} - g, \tag{3.33}$$

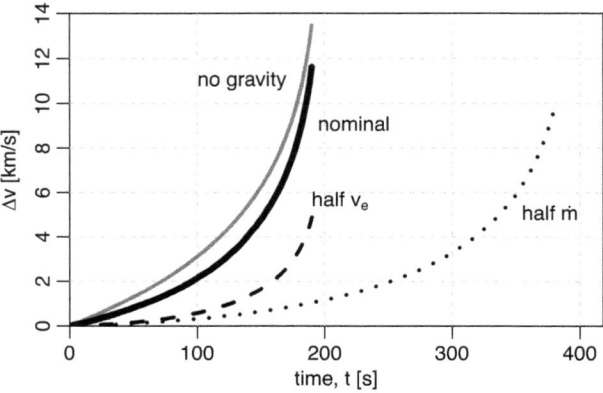

Fig. 3.12 The variation of velocity with time for a rocket with $M_f = 10$ t, $M_i = 200$ t launched vertically in a constant gravitational field of $g = 9.8$ m s^{-2} up until it has exhausted all of its propellant. The thick solid line labelled 'nominal' is for $v_e = 4.5$ km s^{-1}, $\dot{m} = 1000$ kg s^{-1}. The dashed line shows the case for half v_e, and the dotted line shows the case for half $\dot{m}$. The thin solid line shows the gravity-free case ($g = 0$) for comparison.

where F is the rocket thrust. From the thrust equation (equation 3.10) and recalling that $\dot{m} = -dM/dt$, this becomes

$$dV = -\frac{dM}{M}v_e - gdt. \qquad (3.34)$$

Let us assume g remains constant as the rocket rises – which is not such a bad approximation for a launch into a very low orbit (but see below) – and take v_e as constant. The integral of the above equation from the initial mass M_i at time $t = 0$ to the mass $M(t)$ at time t is

$$V = v_e \ln\left(\frac{M_i}{M(t)}\right) - gt. \qquad (3.35)$$

If we further assume $\dot{m}$ is constant, then using equation 3.13 we get the following expression for the variation of velocity with time

$$V = v_e \ln\left(\frac{M_i}{M_i - \dot{m}t}\right) - gt. \qquad (3.36)$$

The thick solid line in figure 3.12 shows how the velocity varies during launch when $g = 9.8$ m s^{-2}, $v_e = 4.5$ km s^{-1}, $\dot{m} = 1000$ kg s^{-1}, $M_f = 10$ t, and the mass ratio is $R = 20$. The rocket starts at rest at $t = 0$. The end point of the line is at the time when all the propellant has been exhausted, which is $t = (M_i - M_f)/\dot{m} = 190$ s. The final velocity is 11.62 km s^{-1}. Comparing this to the no gravity case (thin solid line), we see the effect of gravity reducing the velocity throughout the launch. The dashed line in the figure shows the launch profile when halving the effective exhaust velocity: the final velocity is only 4.88 km s^{-1}, less than half the nominal case. The dotted line shows the launch profile for half the mass flow rate. As the propellant mass is unchanged, the travel time is now twice as long. However, the final velocity is now 9.76 km s^{-1}, less than in the nominal case. This is because gravity has acted for longer and so causes a larger overall deceleration.

To get into a low Earth orbit a spacecraft needs a Δv of around 10 km s^{-1}. The above example shows that rockets with too low an effective exhaust velocity, for a given mass ratio, will not reach orbit at all. Indeed, for a rocket to even get off the ground, the thrust must exceed its weight; that is, it must have $\dot{m}v_e > M_i g$. Some values of the parameters in

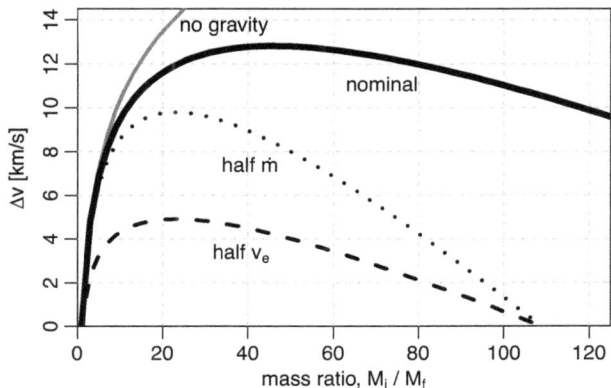

The rocket equation (equation 3.37b) for a rocket launched vertically in a constant gravitational field with constant v_e, $\dot{m}$, and M_f. The thick solid line labelled 'nominal' is for $g = 9.8 \text{ m s}^{-2}$, $v_e = 4.5 \text{ km s}^{-1}$, $\dot{m} = 1000 \text{ kg s}^{-1}$, and $M_f = 10 \text{ t}$. The dashed line shows the case for half v_e, and the dotted line shows the case for half $\dot{m}$ (or equivalently for twice g). The thin solid line shows the force-free rocket equation for comparison.

Fig. 3.13

equation 3.36 yield negative velocities. These correspond to less thrust than weight, and so the rocket would fall if the ground were not there to hold it.

We can rewrite equation 3.36 in terms of the final mass to get an expression for the Δv attained once all the propellant has been exhausted:

$$V = v_e \ln\left(\frac{M_i}{M_f}\right) - g\frac{M_i}{\dot{m}}\left(1 - \frac{M_f}{M_i}\right) \quad \text{or} \tag{3.37a}$$

$$V = v_e \ln\left(\frac{M_i}{M_f}\right) - g\frac{M_f}{\dot{m}}\left(\frac{M_i}{M_f} - 1\right). \tag{3.37b}$$

This is the constant gravity rocket equation. The first term on the right is the familiar one from the force-free rocket equation. The second term is the deceleration caused by gravity.

The variation of V with the mass ratio $R = M_i/M_f$ is shown in figure 3.13. The thick solid line is for $g = 9.8 \text{ m s}^{-2}$, $v_e = 4.5 \text{ km s}^{-1}$, $\dot{m} = 1000 \text{ kg s}^{-1}$, and $M_f = 10 \text{ t}$, which gives $gM_f/\dot{m} = 98 \text{ m s}^{-1}$. As R is varying and M_f is fixed, M_i is varying. We see there is a maximum Δv that can be achieved. By differentiation of equation 3.37b, we find that this occurs at a mass ratio of $R = v_e\dot{m}/gM_f$, which in this case is 45.9. This variation is different from the gravity-free rocket equation, in which the velocity increases monotonically with the mass ratio (the thin solid line). A consequence of this maximum is that there are now two possible values of the mass ratio that will deliver a specified Δv (as long as that is not too large), as we can see from the figure. We can find these by solving equation 3.37b for M_i/M_f. This has to be done numerically. For example, to achieve $\Delta v = 10 \text{ km s}^{-1}$ for the case shown, we require either $M_i/M_f = 11.6$ or 118. These correspond to initial masses of $M_i = 116 \text{ t}$ and 1180 t, respectively, and to travel times of 106 s and 1170 s, respectively. The lower mass solution is obviously more desirable. The higher mass one is actually too heavy to get off the ground initially. Mathematically the velocity from equation 3.35 is initially negative, before reaching a minimum and then increasing to positive values. In practice the ground prevents the rocket from falling, so this corresponds to the rocket sitting on the launch pad burning off fuel for a while before taking off.

The thin solid line in figure 3.13 shows the gravity-free rocket equation for comparison, which is independent of $gM_f/\dot{m}$. As expected, the mass ratios required when considering gravity are higher than when neglecting it. Whereas the gravity rocket needs a mass ratio of 11.6 to reach $\Delta v = 10\,\text{km s}^{-1}$, the gravity-free rocket equation requires only $\exp(10/4.5) = 9.2$.

It must be stressed that these values are not applicable to real launches, because the derivation of the above gravity rocket equation has ignored several practical and important aspects of launching. The first is the assumption of constant gravity. In the nominal example in figure 3.12, the rocket will move of the order of 1000 km in its 190 s journey, a distance over which the Earth's gravity drops from $9.8\,\text{m s}^{-2}$ at the surface to $7.3\,\text{m s}^{-2}$. Second, air resistance is quite significant. Third, rockets are not launched vertically, but instead are given a significant horizontal component in order to get them into a Keplerian orbit. Such factors mean that the required mass ratios are larger (and flight times longer) than estimated here.

3.11 Summary

The rocket equation gives the change in velocity Δv of a rocket when it expels propellant at constant velocity relative to the rocket (equation 3.4). The Δv is independent of both how quickly the propellant is ejected and the means used to accelerate it.

The logarithmic dependence of Δv on the ratio of the initial mass (spacecraft including propellant) to final mass (spacecraft without propellant) means that in practice Δv is limited to a few times the effective exhaust velocity v_e of the propellant. To reach $25\,000\,\text{km s}^{-1}$ $(0.08\,c)$ with a mass ratio of 100, for example, we need $v_e = 5400\,\text{km s}^{-1}$. This logarithmic dependence also means that the mass ratio required to accelerate to velocity Δv and then decelerate back to zero is the square of that required to accelerate to Δv. Rockets intended for a rendezvous mission are therefore much more massive than those intended for a flyby.

Staging is a common way to increase Δv, by jettisoning expended mass such as empty fuel tanks.

To get to the stars quickly we not only need large v_e, but we need to achieve the Δv quickly. This means expelling the propellant rapidly (large $\dot{m}$), which corresponds to a large thrust (equation 3.10). In later chapters we will see what technologies can be used to achieve a large v_e and/or large $\dot{m}$.

3.12 Exercises

1. Show by integration that equation 3.11b leads to equation 3.4 when assuming that mass is exhausted at a constant rate.
2. Derive equation 3.15b, the expression for the distance a rocket travels while exhausting propellant.
3. Compute the time taken to travel from Earth to Proxima Cen for $v_e = 0.05\,c$ and mass ratio $R = 5$, with the constant mass flow rate $\dot{m}$ adjusted so that the spacecraft accelerates from rest to its maximum velocity (having exhausted all its fuel) at the destination.

4. In section 3.2, a rocket with constant v_e and mass ratio $R = M_i/M_f$ first accelerates to Δv then decelerates, arriving at its destination at rest. Show that if the rocket turnaround point is placed midway between origin and destination, this requires that $\dot{m}$ (the mass ejection rate) in the acceleration phase be R times larger than $\dot{m}$ in the deceleration phase.

Orbital mechanics

Even though an interstellar spacecraft would spend most of its time away from the gravity of stars and planets, knowledge of orbital mechanics is important if a spacecraft is to make a rendezvous. In this chapter we examine the essentials of orbits as well as orbital manoeuvres, in particular the Hohmann transfer orbit. We will see that the efficiency of a rocket impulse depends on where in an orbit it is performed, something known as the Oberth effect. Finally, we examine how spacecraft can accelerate or decelerate in a star system using gravity assists around planets.

4.1 Two-body motion under gravity: Keplerian orbits

Two point masses interacting via only their mutual gravity undergo Keplerian orbits about their barycentre (centre of mass). Consider two bodies of different masses moving in closed (elliptical) orbits, as shown in figure 4.1. Viewed from their barycentre, which is also the focus F of their orbits, the more massive body (1) has a smaller orbit, but both orbits have the same shape. It can be shown that object 1 also has a Keplerian orbit *relative* to object 2, and vice versa. These relative orbits are not in inertial reference frames – the bodies accelerate one another – but the same equations of motion apply in the rest frame of each body. The two relative orbits have all the same properties as each other – size, shape, period, etc. – except for the phase, because the two bodies are on opposite sides of the barycentre. As an example of this, the orbit of the Sun viewed from the Earth is the same as the orbit of the Earth viewed from the Sun.[1]

Both the total energy and angular momentum of a two-body system interacting under gravity is conserved. The latter means that both the orbits and barycentre are all in a common plane.

Below I summarize various properties of Keplerian orbits that will be useful in our considerations of space travel. They follow from the inverse square law of gravity assuming point masses and Newtonian mechanics. The derivations can be found in all good textbooks on orbital mechanics. Extended objects and special and general relativity introduce deviations from the inverse square law and some interesting effects (e.g. orbital precession), but those will not be considered. Furthermore, as we are interested in spacecraft orbiting a much more massive planet or star, I will consider the former to have negligible mass, so the barycentre becomes coincident with the central body. The standard gravitational parameter is then

---

[1] The statement 'the Sun orbits the Earth' is therefore not entirely wrong. This equivalence of the relative motion partly explains why it took us so long to realize that the description of the Earth orbiting the Sun, rather than vice versa, is a better *dynamical* description of the heavens.

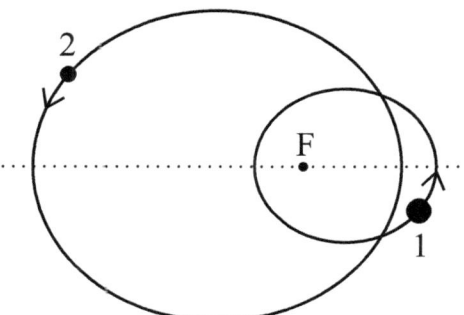

The elliptical orbits of two bodies, 1 and 2, about their barycentre F. In this example body 1 has twice the mass of body 2, so its orbit is half the size.

Fig. 4.1

$\mu = GM$, rather than $G(M + m)$, where G is the gravitational constant and M the mass of the central body. The quantities r and v are, respectively, the distance and velocity of the orbiting body relative to the central body.

4.1.1 All orbits

For any Keplerian orbit, the total energy per unit mass (specific energy) of the orbiting body is the sum of its kinetic and gravitational potential energies,

$$E = K + U \tag{4.1a}$$

$$= \frac{v^2}{2} - \frac{\mu}{r} \quad \text{(all orbits)}, \tag{4.1b}$$

taking the usual convention of zero potential energy at $r = \infty$. The nature of a Keplerian orbit depends on the value of E. If the energy is low enough, the objects will remain bound and the orbit will be an ellipse (the circle being a special case). With our potential energy convention, an elliptical orbit is one with negative energy E. If the energy is positive, the object is not bound and the orbit is a hyperbola. If the system has zero energy, then the orbit is a parabola. The central mass is at the focus of the ellipse, parabola, or hyperbola. For all orbits we can write the energy per unit mass in terms of the semi-major axis a (defined later),

$$E = -\frac{\mu}{2a} \quad \text{(all orbits)}. \tag{4.2}$$

From now on I will refer to the 'energy per unit mass' as simply 'energy' (you can always check the units to be sure). Combining the above two equations 4.1 and 4.2 we get the *vis-viva equation*

$$v^2 = \mu \left(\frac{2}{r} - \frac{1}{a} \right) \quad \text{(all orbits)}, \tag{4.3}$$

which is useful for computing the velocity at any point in the orbit.

All Keplerian orbits can be described geometrically by their semi-major axis a (size) and eccentricity e (shape). Solving the equations of motion, we arrive at the following equation for the motion in polar coordinates,

$$r = \frac{a(1 - e^2)}{1 + e \cos \theta} \quad \text{(all orbits)}, \tag{4.4}$$

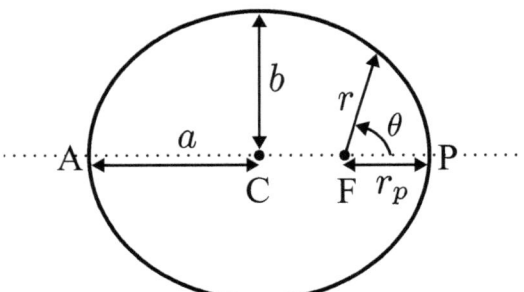

Fig. 4.2 An elliptical orbit. A = apoapsis, C = centre, F = focus, P = periapsis. AC = CP = a is the semi-major axis, b is the semi-minor axis, AF = r_a = $a(1 + e)$ is the apoapsis distance, and FP = r_p = $a(1 - e)$ is the periapsis distance. This ellipse has an eccentricity of $e = \sqrt{1 - b^2/a^2}$ (= 0.5 in this example). The true anomaly θ is measured from the periapsis.

where the angular coordinate θ is known as the true anomaly. Interestingly, although we can also solve the equations of motion for the time as a function of r, there is in general no closed-form solution for r as a function of time. The resulting equation – Kepler's equation – has to be solved numerically to get $r(t)$. The exceptions to this are circular and parabolic orbits.

Let us now look at elliptical and hyperbolic orbits separately.

4.1.2 Elliptical orbits

Elliptical orbits are those with $0 \leq e < 1$ in equation 4.4 and are closed orbits. See figure 4.2. The semi-major axis is half the length of the long axis. The semi-minor axis – half the length of the short axis – is given by $b = a\sqrt{1 - e^2}$, so $0 < b \leq a$. The eccentricity measures how elongated the ellipse is. When $e > 0$, the distance of the orbiting object from the focus (central mass) varies over the orbit. It follows from equation 4.4 that the closest approach (periapsis, distance r_p) occurs at $\theta = 0°$, and the most distant point (apoapsis, distance r_a) occurs at $\theta = 180°$, so

$$
\begin{aligned}
r_p &= a(1 - e) \\
r_a &= a(1 + e)
\end{aligned}
\quad \text{(elliptical orbits).}
\tag{4.5}
$$

For the Earth (actually the barycentre of the Earth–Moon system), $a = 1.496 \times 10^8$ km (approximately 1 au) and $e = 0.0167$, so the periapsis is 1.471×10^8 km and the apoapsis is 1.521×10^8 km, a difference of five million kilometres. Figure 4.3 shows examples of orbits with different eccentricities. Note that an orbit with $e = 0.1$ has almost the same shape as a circular orbit ($e = 0$), but the focus is notably shifted.

The specific angular momentum h of the orbiting body about the central mass is directly related to the semi-major axis and eccentricity via

$$
h^2 = \mu a(1 - e^2) \quad \text{(all orbits).}
\tag{4.6}
$$

Thus for a given energy, a larger eccentricity has a smaller angular momentum. Also useful is Kepler's third law, which relates the semi-major axis and total mass to the orbital period T via

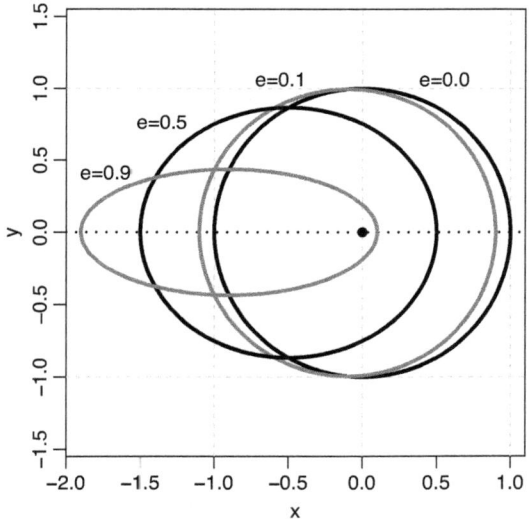

Elliptical orbits with different eccentricities. All have $a = 1$ and a common focus at $(x, y) = (0, 0)$.
$x = r \cos \theta, y = r \sin \theta$ are in units of the semi-major axis.

Fig. 4.3

$$a^3 = \frac{\mu}{4\pi^2} T^2 \quad \text{(elliptical orbits).} \tag{4.7}$$

4.1.3 Circular orbits

Circular orbits are special cases of elliptical orbits with $e = 0$. From equation 4.3, $r = a$ on all points of the orbit and so

$$v^2 = \frac{\mu}{a} \quad \text{(circular orbits).} \tag{4.8}$$

This can also be found by equating the magnitudes of the gravitational and centripetal forces (both time independent). Thus objects in larger orbits have smaller velocities. The outer planets in the solar system move more slowly than the inner ones, for example. They also have smaller angular velocities, v/a. The Earth at 1 au orbits at about 29.8 km s^{-1}, although this varies slightly due to the non-zero eccentricity. As the left side of equation 4.8 is twice the kinetic energy K, and the right side is the negative of the potential energy U, we can write

$$2K + U = 0 \quad \text{and} \tag{4.9a}$$

$$E = -K = \frac{U}{2} \quad \text{(circular orbits).} \tag{4.9b}$$

This is an example of the *virial theorem*. Recall that $E < 0$. The virial theorem is instructive because it tells is that if we increase (towards zero) the total energy of an object in a circular orbit, we *decrease* the kinetic energy by this amount and *increase* the potential energy by twice this amount. Adding energy puts the object into a higher orbit but also slows it down, which might seen counter-intuitive. Viewed from the perspective of thermodynamics, such a system has a negative heat capacity, because adding energy reduces the temperature (a proxy for kinetic energy). This is true for all systems of gravitationally bound bodies. The virial theorem relates time-averaged energies, but always holds for circular orbits since r and v are constant.

4.1.4 Escape velocity

A useful concept is the escape velocity. For a spacecraft a distance r from a star or planet, what initial velocity does it need to just escape the gravitational field, that is, to reach infinite distance at zero velocity? We can compute this by equating the initial energy, given by equation 4.1b, with the final energy, which is zero. This gives

$$v_{\text{escape}} = \sqrt{\frac{2\mu}{r}} = \sqrt{2}\, v_{\text{circ}}(r), \tag{4.10}$$

where $v_{\text{circ}}(r)$ is the circular orbital velocity at distance r (equation 4.8). The escape velocity from the Earth's surface (ignoring the Sun and all other bodies) is $11.2\,\text{km}\,\text{s}^{-1}$. The escape velocity from the Sun starting at the distance of the Earth is $42.1\,\text{km}\,\text{s}^{-1}$, but when starting at the surface of the Sun it is $617\,\text{km}\,\text{s}^{-1}$ (ignoring the Earth and other bodies in both cases). The above equation shows us that to escape to infinity, a spacecraft initially on a circular orbit must increase its velocity in the tangential direction by $\Delta v = (\sqrt{2} - 1)v_{\text{circ}}(r)$. So a spacecraft in the Earth's orbit about the Sun ($v_{\text{circ}} = 29.8\,\text{km}\,\text{s}^{-1}$) but far from the Earth will need a prograde boost of $\Delta v = 0.414 \times 29.8\,\text{km}\,\text{s}^{-1} = 12.3\,\text{km}\,\text{s}^{-1}$ to escape the solar system. As gravity is a conservative force, once the spacecraft has acquired its escape velocity it will reach infinity no matter what direction it was initially moving in. The orbit of a spacecraft with escape velocity is a parabolic orbit, as this is the orbit with zero energy.

At $11.2\,\text{km}\,\text{s}^{-1}$, the escape velocity of the Earth is more than the maximum specific exhaust velocity available with chemical rockets, which is why large mass ratios are required for a rocket to escape from the Earth (see chapter 3). If we rewrite equation 4.10 in terms of the mean density ρ and radius r of the planet we get

$$v_{\text{escape}} = r\sqrt{\frac{8\pi G\rho}{3}}. \tag{4.11}$$

For a given density, the larger the planet, the larger the escape velocity from its surface. Eventually this will exceed the Δv that can be provided by a chemical rocket with any plausible mass ratio. In the best case, a rocket exhausts all of its propellant near the surface of the planet, in which case Δv is given by the force-free rocket equation (equation 3.4). Setting this Δv equal to the escape velocity, we get an expression for the maximum radius a planet can have such that a rocket with effective exhaust velocity v_e and mass ratio R can still escape from it:

$$r_{\text{max}} = v_e \ln R \sqrt{\frac{3}{8\pi G\rho}}. \tag{4.12}$$

With $v_e = 5\,\text{km}\,\text{s}^{-1}$ and ρ the density of the Earth, then if we adopt a maximum plausible mass ratio of $R = 1000$, the maximum radius is $3.1\text{R}_\oplus$. Rocky exoplanets larger than the Earth have a range of densities, so the maximum radius would deviate somewhat from this. But this calculation shows that potential aliens on a very large rocky planet could not escape from its surface using chemical rockets that we know of.

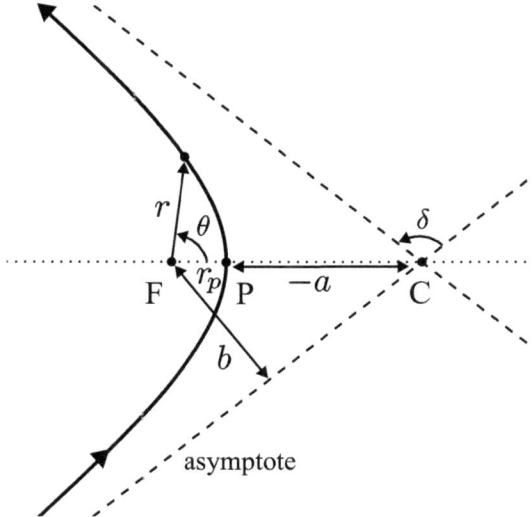

A hyperbolic orbit. F = focus, P = periapsis, C = centre. FP = r_p, the periapsis distance, PC = $-a$, the semi-major axis ($a < 0$). If the central body at F had zero mass, then the orbiting body would move on a straight line along the asymptote with impact parameter b. The mass instead turns the orbiting body through an angle δ, such that at infinity it is travelling along the outgoing asymptote. In this example $a = -2.50$ and $r_p = 0.70$ (arbitrary units), $\delta = 103°$, and $e = 1.28$.

Fig. 4.4

4.1.5 Hyperbolic orbits

When an orbiting object has sufficient energy, it will not be bound to the central mass, but its path will be deflected by it. This is a hyperbolic orbit: see figure 4.4. Hyperbolic orbits have $E > 0$ and $e > 1$. It follows from equation 4.2 that they therefore have negative semi-major axes[2], $a < 0$.

If an object has an asymptotic velocity v_∞, and so a total energy $E = v_\infty^2/2$, then from the vis-viva equation (equation 4.3) we see that

$$a = -\frac{\mu}{v_\infty^2} \quad \text{(hyperbolic orbits).} \tag{4.13}$$

The eccentricity is defined in terms of this and the semi-minor axis b,

$$e^2 = 1 + \frac{b^2}{a^2} \quad \text{(hyperbolic orbits),} \tag{4.14}$$

which differs from the definition for elliptical orbits. The sign of b is not defined by this equation; we take it to be positive. The quantity b is also the impact parameter, which is the closest approach the object would have to the central body in the absence of gravity. This can also be seen by substituting equations 4.14 and 4.13 into equation 4.6, from which we find that the angular momentum is $h = bv_\infty$. As with elliptical orbits, the periapsis of the hyperbolic orbit $\theta = 0$ is $r_p = a(1 - e)$, but its apoapsis is infinite.

A spacecraft making a flyby of a planet or star is actually in a hyperbolic orbit about that body (ignoring all other masses). At a large distance from the body, the energy and speed of

---

[2] Some sources define hyperbolic orbits to have $a > 0$ and then modify the equations.

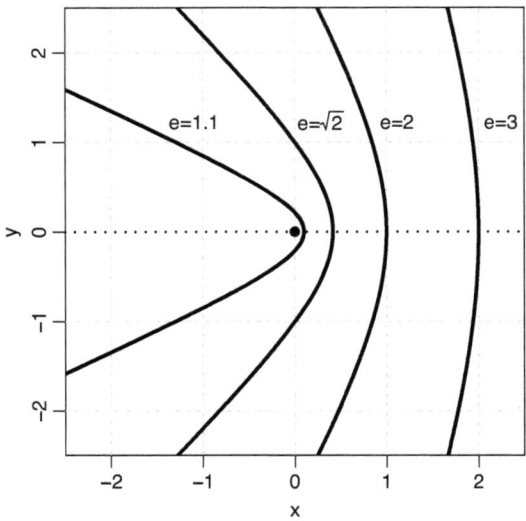

Hyperbolic orbits with different eccentricities, but all with $a = -1$ and a common focus at $(x, y) = (0, 0)$. The Cartesian coordinates $x = r \cos \theta$, $y = r \sin \theta$ are in units of the semi-major axis.

the spacecraft is the same before and after the flyby, but its direction of travel has changed (figure 4.4). The deflection angle of the orbit is a property of the hyperbola and is given by

$$\delta = 2 \sin^{-1} \frac{1}{e}. \tag{4.15}$$

Because hyperbolic orbits are not closed, θ in equation 4.4 varies between only $-(\delta + \pi)/2$ and $+(\delta + \pi)/2$.

Knowing the impact parameter b and the asymptotic velocity v_∞ relative to a central body, we can derive the geometric properties (a, e) of the orbit. The example in figure 4.4 is for an orbit around an Earth-mass object with $b = 2R_\oplus$ and $v_\infty = 5\,\mathrm{km\,s^{-1}}$, which corresponds to $e = 1.28$ (a and r_p in the caption are then also in units of $R_\oplus$). Examples of orbits with other eccentricities are show in figure 4.5. The larger the eccentricity, the closer the orbit is to a straight line.

A limiting case of a hyperbolic orbit has $e = 1$ and is a parabolic orbit. This has zero energy and is on the boundary of bound/unbound, elliptical/hyperbolic. Its velocity at periapsis is therefore equal to the escape velocity. The deflection angle is then $\delta = 180°$. After passing the central mass, the spacecraft eventually returns in the direction it came from, albeit only after infinite time. If two bodies are headed straight at each other, this can also be considered as a limiting case of a hyperbolic orbit with $b = 0$, and therefore $e = 1$. This is not a parabolic orbit but a radial orbit, for which the above equations break down due to the singularity at $r = 0$.

The velocity at infinity – also called the hyperbolic excess velocity – is used by mission planers for specifying the energy of a spacecraft relative to a body. It is sometimes specified via the characteristic energy, defined as $C_3 = v_\infty^2 = 2E$, which is twice the specific energy the spacecraft would have if it escaped to infinity. The New Horizons spacecraft to the outer solar system, for example, had $C_3 = 157.8\,\mathrm{km^2\,s^{-2}}$ with respect to the Sun at launch, the largest launch specific energy of a spacecraft so far. Voyager 1 started with a lower characteristic energy, but after its gravity assists it attained $v_\infty = 16.6\,\mathrm{km\,s^{-1}}$, making it the fastest spacecraft to leave the solar system so far. It is not the fastest spacecraft with respect to the Sun, however. That record goes to the Parker Solar Probe, which reached $192\,\mathrm{km\,s^{-1}}$ during a perihelion passage at a distance of $9.8\,\mathrm{R_\odot}$ ($8.8\,\mathrm{R_\odot}$ from the surface) in December 2024. As it is bound to the Sun, its characteristic energy is negative.

4.2 Changing orbits

Although Newtonian gravity appears simple (it's just $1/r^2$ after all), Keplerian orbits are non-trivial, and changing orbits can be surprisingly complicated. We won't dive deeply into the fascinating details here but will look briefly at some simple ways of moving from one orbit to another using impulsive manoeuvres. These are manoeuvres that take place instantaneously, for example via a short burst of a rocket engine. The spacecraft therefore moves instantaneously from one Keplerian orbit to another with a different energy and/or angular momentum. The change in the speed at this instant is the scalar quantity Δv. From the rocket equation (chapter 3), the total amount of propellant in a rocket determines the total Δv a rocket can deliver, and so the Δv budget of a manoeuvre is a critical aspect of mission design.

Before we get to this, let us briefly ignore gravity for a useful insight. Consider a spacecraft moving with a vectorial velocity $\mathbf{v}$ in some reference frame. The velocity now changes by $\Delta\mathbf{v}$. If the initial and final (specific) kinetic energies are K_i and K_f respectively, and the change in kinetic energy is ΔK, then we see that

$$K_i = \frac{1}{2}|\mathbf{v}|^2, \tag{4.16a}$$

$$K_f = \frac{1}{2}|\mathbf{v} + \Delta\mathbf{v}|^2, \tag{4.16b}$$

$$\Delta K = \frac{1}{2}|\Delta\mathbf{v}|^2 + \mathbf{v}\cdot\Delta\mathbf{v}. \tag{4.16c}$$

The largest change in the kinetic energy is therefore achieved when $\Delta\mathbf{v}$ is made parallel to $\mathbf{v}$, to give $\Delta E = (1/2)(\Delta v)^2 + v\Delta v$. If we instead applied the burst perpendicular to the direction of travel, we would only get $\Delta E = (1/2)(\Delta v)^2$. Thus to maximize the change in energy of the spacecraft, we should apply a given Δv in the direction of travel. As energy is a measure of the semi-major axis of an orbit, this is also the way to make the largest change in the size of an orbit. This consideration is involved in the Oberth effect, which we look at in section 4.2.3.

4.2.1 Hohmann transfer

Suppose that a spacecraft is initially in a circular orbit of radius r_1. We want to lift it onto a larger orbit of radius r_2 using an impulsive rocket. This will require at least two bursts of the rocket: one to move it from the inner orbit onto a transfer orbit, and a second to move it from the transfer orbit onto the outer orbit. Following the discussion in the previous section, the most energy efficient way to do this will be with two tangential bursts, as shown in figure 4.6.

To move out from orbit 1, we need to increase the energy of the spacecraft, so the first burst must be in the prograde direction. This will put the spacecraft on an elliptical transfer

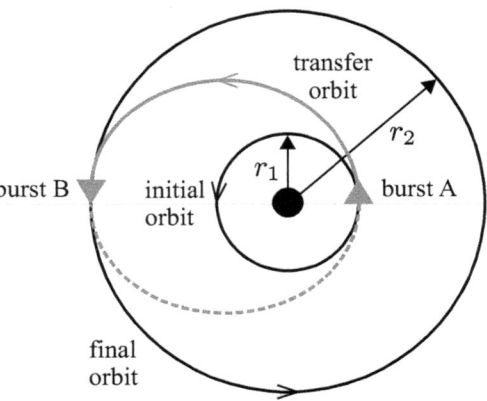

Fig. 4.6 A Hohmann transfer orbit between an inner circular orbit of radius r_1 and an outer circular orbit of r_2. The Hohmann transfer orbit is half of an elliptical orbit with periapsis r_1 and apoapsis r_2.

orbit with semi-major axis a and eccentricity e (values to be determined). The point of the burst (A) will be the periapsis of the transfer orbit, which is equal to r_1. If we arrange that the apoapsis of this transfer orbit is r_2, then apply a second Δv there (point B), we can get the spacecraft onto the outer circular orbit. What values of Δv do we need at A and B to achieve this?

This can be solved by considering the (specific) angular momenta of the three orbits involved. The angular momenta of orbits 1 and 2 are, from equation 4.6, $h_1 = \sqrt{\mu r_1}$ and $h_2 = \sqrt{\mu r_2}$, respectively. The semi-major axis of an elliptical orbit is half the sum of the periapsis and apoapsis distances, which for the transfer orbit is $a = (r_1 + r_2)/2$. Using this together with the expressions for periapsis and apoapsis from equation 4.5, equation 4.6 gives the angular momentum of the transfer orbit to be

$$h_t = \sqrt{2\mu \frac{r_1 r_2}{r_1 + r_2}}. \tag{4.17}$$

As $r_2 > r_1$ it follows that $h_2 > h_t > h_1$. As the spacecraft moves tangentially at periapsis and apoapsis, the angular momenta at these points can also be written $h = rv$, where v is the velocity. The velocity on orbit 1 before the burst is $\sqrt{\mu/r_1}$, and the velocity immediately after the burst (on the transfer orbit) is h_t/r_1. Thus the required change in velocity by the first burst is

$$\Delta v_A = \sqrt{\frac{\mu}{r_1}} \left(\sqrt{\frac{2r_2}{r_1 + r_2}} - 1 \right). \tag{4.18}$$

This is positive and so a prograde burst, as anticipated. When the spacecraft gets to the apoapsis of its transfer orbit, its velocity is h_t/r_2. To move onto the circular orbit, it must change its velocity to $\sqrt{\mu/r_2}$, so the required change is

$$\Delta v_B = \sqrt{\frac{\mu}{r_2}} \left(1 - \sqrt{\frac{2r_1}{r_1 + r_2}} \right). \tag{4.19}$$

This is also positive, so the second burst is also prograde. This makes sense because when it gets to apoapsis, the spacecraft needs to increase both its angular momentum and its semi-major axis (energy).

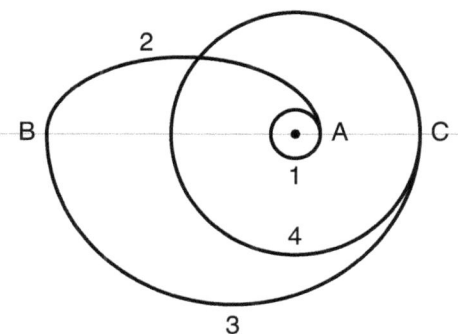

Fig. 4.7

Bielliptic transfer orbits between two circular orbits. To get from the inner orbit (1) to the outer orbit (4), a prograde boost is made at A to put the spacecraft on the elliptical orbit 2. At its apoapsis (B), another prograde boost is made to put the spacecraft on orbit 3. At its periapsis (C), a retrograde boost is made to put the spacecraft on the outer orbit. To go from the outer orbit to the inner one, the procedure is reversed.

Let's put in some numbers. Suppose we want to raise a satellite from a low Earth orbit (LEO) at $r_1 = R_\oplus + 400$ km to a geosynchronous orbit (GEO) at $r_2 = R_\oplus + 35\,800$ km. The velocity on the lower circular orbit is 7.67 km s^{-1} and the first burst is $\Delta v_A = 2.40$ km s^{-1}. This puts the spacecraft on an elliptical orbit of semi-major axis 24 500 km and eccentricity 0.72. The second burst is $\Delta v_B = 1.46$ km s^{-1}, putting the satellite on a GEO with a velocity of 3.07 km s^{-1}. The transfer duration – half the period of the transfer orbit – is 5.3 hr. The angular momenta of the initial, transfer, and final orbits are 5.2×10^4, 6.8×10^4, and 1.3×10^5 km^{2}s^{-1}, respectively.

4.2.2 Other transfer orbits

It turns out that the Hohmann transfer of the previous section is the most energy efficient two-impulse manoeuvre between any two coplanar elliptical orbits. There is, however, a more energy efficient way of achieving the same change in orbit using three impulses. This again uses tangential boosts at an apoapsis or periapsis, but not all are prograde.

Suppose again that we want to move from an inner circular orbit of radius r_1 to an outer circular orbit of radius r_4, as shown in figure 4.7. The idea of the bielliptic transfer orbit is to again do a prograde boost (at point A), but now a larger one than in the Hohmann case in order to take the spacecraft to a larger apoapsis distance than that of the target orbit. At the apoapsis (point B), the rocket does another prograde boost to put it onto a higher energy (larger semi-major axis) and lower eccentricity (larger angular momentum) orbit 3. Point B is also the apoapsis of orbit 3. The size of this boost is chosen such that the periapsis of orbit 3 (point C) is the radius of the target circular orbit (4). As this target orbit has a smaller semi-major axis than orbit 3, a *retrograde* boost is then required at point C to put the spacecraft onto the circular orbit.

It turns out that this manoeuvre can use less energy than the Hohmann transfer. The idea is to place point B so far away from the central mass that the spacecraft is moving very slowly at this point; the boost required to put it on orbit 3 then is very small. The sum of the three Δv can then be less than the two in the Hohmann case. It turns out that if $r_4 < 11.94 r_1$, then the Hohmann transfer is more energy efficient for any r_B, whereas if $r_4 > 15.58 r_1$, then the bielliptic transfer is more energy efficient for any r_B. In between these values, the bielliptic transfer is more energetically favourable for larger values of r_B. Energy efficiency may not be the only consideration, of course. As it involves moving out to a larger intermediate outer orbit, the bielliptic transfer lasts much longer.

A simpler variation of this transfer orbit is when we want to drop into the Sun from an initial circular orbit of radius r_1. To achieve this with a single boost, we need to apply a retrograde boost equal to the circular velocity, which is $\Delta v_0 = \sqrt{\mu/r_1}$. Alternatively, we could do a prograde boost into an elliptical orbit with an apoapsis distance r_2. The velocity change required for this we already calculated for the Hohmann transfer as Δv_A from equation 4.18. At the apoapsis of this elliptical orbit (point B in figure 4.6), the Δv required to remove the angular momentum and so drop into the Sun is equal to the velocity at that point. This is the term we subtracted in equation 4.19 (or we can use the vis-viva equation), and is $\Delta v'_B = \sqrt{\mu}\sqrt{2(r_1/r_2)/(r_1 + r_2)}$. Thus the ratio of the total Δv from this two-burst manoeuvre to the Δv needed in the one-burst 'direct drop' manoeuvre is

$$Q = \frac{\Delta v_A + \Delta v'_B}{\Delta v_0} = \sqrt{2\left(\frac{r_1}{r_2} + 1\right)} - 1. \tag{4.20}$$

By inspection we see that for $r_2 > r_1$ we get $Q < 1$, that is, the two-burst manoeuvre requires a smaller Δv. Thus by boosting to a larger orbit first, we require less energy to drop into the Sun. The smallest value of Q is achieved as r_2/r_1 tends to infinity, in which case $Q = \sqrt{2} - 1$: the two-burst manoeuvre only needs 41% the Δv as a direct drop into the Sun. It is somewhat counter-intuitive that the more energy efficient way to get to the Sun is to first boost to an arbitrarily large orbit. This will, however, require an arbitrarily long time.

The manoeuvres looked at so far have only involved a Δv perpendicular to the radial vector (and so only at apoapsis and periapsis) and in the orbital plane. Boosts in other directions are of course possible and have a range of applications, for example to intercept an object on another orbit, or to change the inclination of the orbit. Even more complicated manoeuvres occur when we consider not instantaneous bursts but long-duration burns of an engine (e.g. an ion engine; chapter 6), or the application of other external forces (e.g. a solar sail; chapter 10). The problems can become quite complicated and will often require numerical methods to solve.

4.2.3 The Oberth effect

We saw from equation 4.16c that the largest change in the kinetic energy of a spacecraft is achieved when applying a boost (of magnitude Δv) in the direction of travel. This equation also shows that a given Δv corresponds to a larger change in kinetic energy the larger the initial velocity. The Δv is the same for all inertial reference frames (neglecting special relativity), but the corresponding change in kinetic energy is not. This follows from the quadratic dependence of kinetic energy on velocity. Consequently, the faster a spacecraft is moving relative to a star when the Δv is applied, the larger its change in kinetic energy. This is the Oberth effect. It is particularly relevant when changing orbits. For example, when a spacecraft is in an elliptical orbit and wants to transfer to the highest possible orbit using a given Δv, it should apply the boost when it is moving fastest, which is at periapsis.

An illustration of the Oberth effect relevant to interstellar travel is when a spacecraft is in a parabolic orbit about a star and wants to use a single boost to escape and achieve the maximum velocity at infinity. Suppose the periapsis distance is r_i, where the velocity is v_i. The total energy of a parabolic orbit is zero, so it follows from the vis-viva equation that $v_i^2 = 2\mu/r_i$. This is the escape velocity (section 4.1.4), meaning the spacecraft would reach infinity with zero velocity. If this spacecraft flew to infinity and applied its boost, its final velocity would be Δv. If it instead applies the same boost at periapsis, then the total energy immediately after the boost is the sum of the potential energy $(-\mu/r_i = -v_i^2/2)$ and the now-increased kinetic energy, which is

$$E_i = -\frac{v_i^2}{2} + \frac{1}{2}(v_i + \Delta v)^2 = v_i \Delta v + \frac{(\Delta v)^2}{2}. \tag{4.21}$$

Upon reaching infinity in this case, the final energy is just the kinetic energy $(1/2)v_f^2$, where v_f is the final velocity. By conservation of energy, this final energy equals E_i, so

$$v_f = \Delta v \sqrt{1 + \frac{2v_i}{\Delta v}}. \tag{4.22}$$

This tells us that doing the boost at periapsis rather than at infinity increases the final velocity by a factor of $\sqrt{1 + 2v_i/\Delta v}$.

Suppose we arrange for a rocket to make a parabolic flyby of the Sun at 10 solar radii (9 solar radii from the surface), which corresponds to $v_i = 195.4\,\mathrm{km\,s^{-1}}$ (the escape velocity). If the rocket can provide $\Delta v = 5\,\mathrm{km\,s^{-1}}$, then applying this boost at perihelion results in a velocity at infinity of $44.5\,\mathrm{km\,s^{-1}}$, as opposed to $5\,\mathrm{km\,s^{-1}}$ if we do the boost at infinity. With $\Delta v = 10\,\mathrm{km\,s^{-1}}$ applied at perihelion, the final velocity is $63.3\,\mathrm{km\,s^{-1}}$. Thus the Oberth effect can give large boosts and so allow fuel to be used more efficiently.

We can apply the same idea to escaping from an elliptical orbit. Both the orbital velocity and the escape velocity are largest at periapsis and smallest at apoapsis. If we compute the difference between the escape velocity and orbital velocity at each apsis, we find that the Δv required to escape is always smaller at periapsis than at apoapsis (see exercises). Furthermore, the difference increases the larger the eccentricity. Thus to escape from an elliptical orbit using minimum Δv, we should apply the boost at periapsis.

The above effect is of course reversible. If a spacecraft is initially unbound to a star and wants to go into orbit about it, it should apply its braking Δv at periapsis in order to minimize the Δv required.

As a parabolic orbit can be thought of as the limiting case of an elliptical orbit with an infinite semi-major axis and high eccentricity, the parabolic example above is the best we could achieve in trying to escape from the Sun with the Oberth effect. In practice we would start in an elliptical orbit, in which case both the perihelion velocity and final velocity would be lower. But not by much (see exercises). For example, if we start at the Earth's orbit and do a large retrograde boost to drop to 10 solar radii, the perihelion velocity is $193.2\,\mathrm{km\,s^{-1}}$. Following a $5\,\mathrm{km\,s^{-1}}$ boost, the velocity at infinity is $44.2\,\mathrm{km\,s^{-1}}$, hardly any smaller than in the parabolic case above.

4.3 Gravity assists

An unbound spacecraft flying close to a planet will have its direction changed by their gravitational interaction. Ignoring all other masses, this two-body interaction is a hyperbolic orbit of the spacecraft about the planet (section 4.1.5). Energy is conserved, so the magnitude

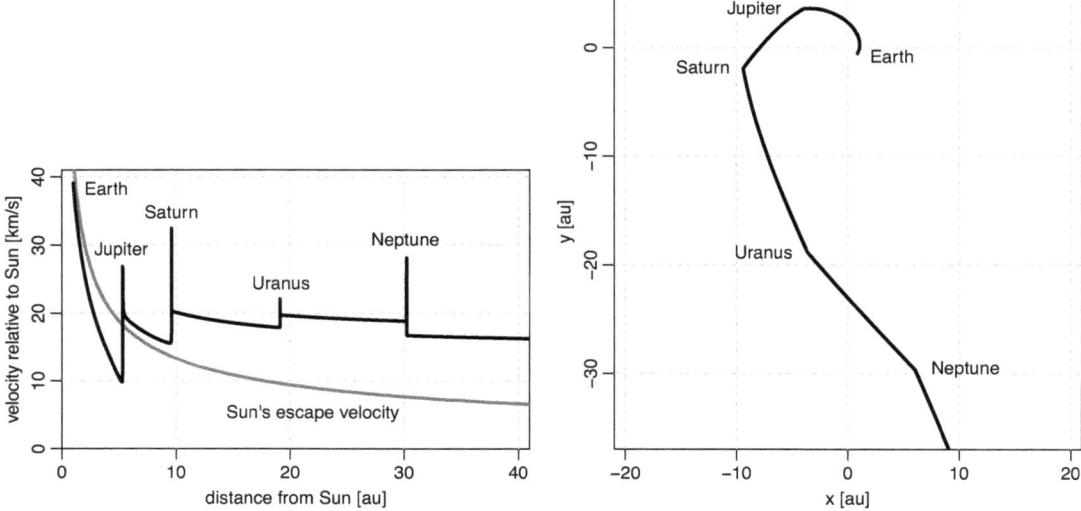

Fig. 4.8 Gravity assists of Voyager 2. The left panel shows how the magnitude of the velocity of the spacecraft relative to the Sun is modified by close encounters of the four outer planets. The lower smooth line shows the Sun's escape velocity (equation 4.10). Already after its gravity assist at Jupiter, Voyager 2 had enough velocity to escape from the solar system. The right panel shows the trajectory in the plane of the solar system, with the Sun at (0, 0) and the positive x-axis the line of zero ecliptic longitude. Data obtained from JPL Horizons.

of the asymptotic velocity of the spacecraft relative to the planet will be unchanged by the encounter, but the direction of travel will change. When both planet and spacecraft are moving relative to a star, then the magnitude of the velocity of the spacecraft *relative to the star* can also change. This is the principle of the gravity assist, and it has been used many times in interplanetary flight. Famously, the gravity assist was used to give the Pioneer and Voyager spacecraft enough velocity to escape from the solar system through flybys of the outer planets, as shown in figure 4.8 for Voyager 2. This figure shows how the spacecraft's heliocentric velocity varies with its distance from the Sun. At each planet, the spacecraft received a significant velocity boost (left) and change in direction (right). Between these close encounters only the Sun's gravity is acting, so the velocity drops according to the vis-viva equation (equation 4.3). Note that the velocity after the Neptune encounter is actually lower than before the encounter; gravity assists can decelerate spacecraft too. This was done to allow Voyager to make a close pass of Neptune's moon Triton.

4.3.1 Concept

Before examining in detail how the gravity assist works, it is useful to look at a simpler system that demonstrates the same principle. Consider standing at rest on a station platform as a train approaches at velocity u_t. A ball is thrown at velocity u_b along the track towards the approaching train. The ball hits the front of the train and bounces off elastically. What is the final velocity of the ball?

The situation is illustrated in figure 4.9. The initial situation in the rest frame of the station is shown in the top left. In the rest frame of the train (top right), the ball approaches the train

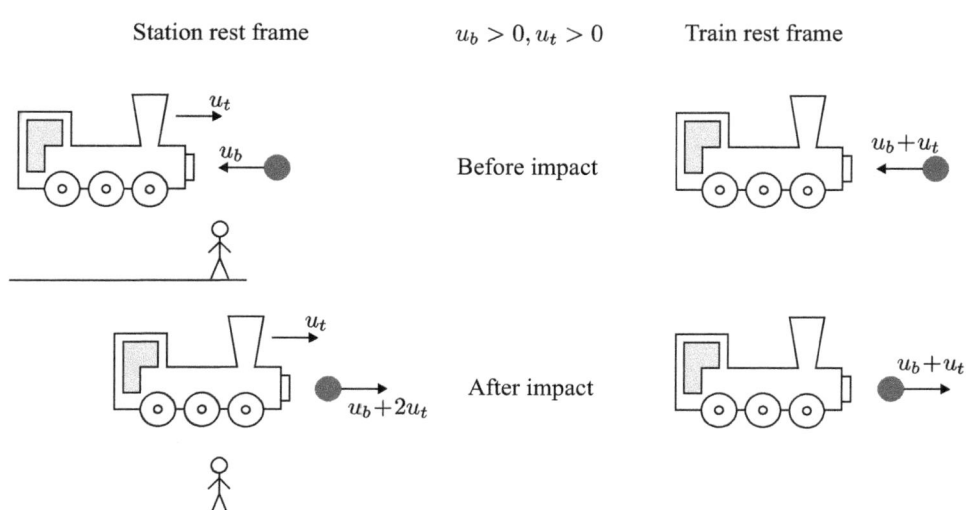

Station rest frame $\quad u_b > 0, u_t > 0 \quad$ Train rest frame

Before impact

After impact

The concept of gravity assist in one dimension is the same as how a ball gains speed when bounced elastically off a (massive) moving train. In the rest frame of the station (left), the train is initially moving to the right at speed u_t and the ball to the left at speed u_b. After the bounce, the speed of the train is essentially unchanged, but the ball picks up twice the speed of the train. In the rest frame of the train (right) the bounce is symmetric.

Fig. 4.9

at a velocity $u_b + u_t$. After bouncing elastically, the ball recedes at the same velocity but in the opposite direction (bottom right). Back in the station's rest frame (bottom left), the train is still moving at velocity u_t, so the ball is now moving at velocity $u_b + 2u_t$. Thus we see that in the station's rest frame, the ball changes direction and picks up an additional velocity of twice that of the train.[3]

The principle of the gravity assist is very similar to this, with the spacecraft playing the role of the ball and the planet the role of the train. The interaction is viewed from the rest frame of the star (station). Instead of the deflection occurring by direct impact, that is, electrostatic forces in the material of the ball and train, it occurs by gravity.

In what follows I neglect the gravity of the star, so the planet is taken to be moving on a straight, unaccelerated path. This is a reasonable approximation when the timescale of the flyby is short. The spacecraft is not bound to the planet, so its flyby is a hyperbolic orbit about the planet as seen from the planet's rest frame. This encounter changes the direction of travel of the spacecraft, but not its asymptotic velocity, just as we saw with the ball in the train's rest frame. But in the star's rest frame, because the planet is moving, the velocity of the spacecraft is changed.

In the star's rest frame, let the planet's two-dimensional velocity be $\mathbf{u}_p = (u_{p,x}, u_{p,y})$, and the initial velocity of the spacecraft well before the encounter be $\mathbf{u}_s = (u_{s,x}, u_{s,y})$. In the planet's rest frame, the initial velocity of the spacecraft is $\mathbf{u}_{sp} = \mathbf{u}_s - \mathbf{u}_p$, and the encounter serves just to turn the spacecraft's velocity by an angle δ (see figure 4.4). This rotation can be represented by the rotation matrix

---

[3] I have assumed that the train is much more massive than the ball, and so the change in the train's momentum is negligible.

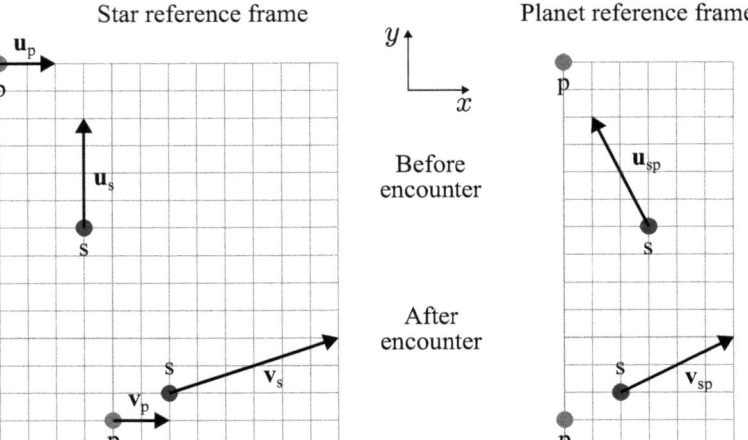

Fig. 4.10 The gravity assist of a spacecraft (s) around a planet (p) as viewed in the star's reference frame (left) and the planet's reference frame (right). The arrows show the magnitude and direction of the velocity. The upper diagrams show the situation 1.5 time units before the encounter, and the lower diagrams 0.5 time units after the encounter. In the planet's reference frame, the encounter changes the direction of the spacecraft by 90°, but it does not change the magnitude of its velocity. In the star's reference frame both change.

$$\mathbf{R} = \begin{bmatrix} \cos\delta & -\sin\delta \\ \sin\delta & \cos\delta \end{bmatrix}. \qquad (4.23)$$

Thus the asymptotic velocity of the spacecraft in the planet's rest frame after the encounter is given by the matrix product $\mathbf{v}_{sp} = \mathbf{R}\mathbf{u}_{sp}$. As the velocity of the planet is unchanged (we assume it to be massive), the post-encounter asymptotic of the spacecraft in the star's rest frame is $\mathbf{v}_s = \mathbf{R}\mathbf{u}_{sp} + \mathbf{u}_p$.

Consider the case $\mathbf{u}_s = (0, u_{s,y})$, $\mathbf{u}_p = (u_{p,x}, 0)$ shown in figure 4.10. The flyby is arranged to give a deflection angle of $\delta = -90°$ (a clockwise rotation), so

$$\mathbf{R} = \begin{bmatrix} 0 & +1 \\ -1 & 0 \end{bmatrix}. \qquad (4.24)$$

In the planet's rest frame, the spacecraft's velocity is $\mathbf{u}_{sp} = (-u_{p,x}, u_{s,y})$ before the encounter and $\mathbf{v}_{sp} = \mathbf{R}\mathbf{u}_{sp} = (u_{s,y}, u_{p,x})$ after the encounter. In the star's rest frame, the velocity of the planet after the encounter is $\mathbf{v}_s = (u_{s,y} + u_{p,x}, u_{p,x})$. With $u_{s,y} = 2$ and $u_{p,x} = 1$, we get $\mathbf{u}_{sp} = (-1, 2)$ and $\mathbf{v}_{sp} = (2, 1)$, drawn to scale in the figure. The final velocity of the spacecraft in the star's rest frame is $\mathbf{v}_s = (3, 1)$ giving a speed of $\sqrt{10} = 3.2$, compared to an initial speed of 2. The gravity assist has increased the speed of the star.

The direction of travel of the spacecraft in the star's rest frame after the encounter is $\theta' = \operatorname{atan}(v_{s,y}/v_{s,x})$, where $v_{s,x}$ and $v_{s,y}$ are the x and y components of the spacecraft's velocity, respectively. Angles are measured anticlockwise from the x-axis. Thus the deflection angle in the star's rest frame is $\theta' - \theta$, where $\theta = \operatorname{atan}(u_{s,y}/u_{s,x})$ is the initial direction of travel. For the numerical example just given, the initial direction of travel is $\theta = +90°$ and the final direction of travel is $\theta' = \operatorname{atan} 1/3 = +18°$. Hence the deflection angle in the star's rest frame is $-72°$.

Note that the gravity of the star has played no part in this calculation. The star is just the frame of reference from which we observe the encounter. Planning a real gravity assist is of course more complicated because it takes place over a finite time during which the planet will move on its orbit. But this is not the cause of the 'gravity' assist. Indeed, we have not even explicitly considered the gravity in the spacecraft–planet interaction. The curved hyperbolic orbit will appear in the closest parts of the encounter, but it does not affect the asymptotic velocities considered here.

The interaction has changed the spacecraft's energy and momentum in the star's rest frame. These are taken from the energy and momentum of the planet in this same frame. In reality that corresponds to the orbital energy and angular momentum of the planet about the star. A gravity assist around a planet that accelerates the spacecraft therefore leads to the planet losing energy (relative to the star) and going into a smaller orbit. In practice this change is tiny, due to the enormous difference in mass of the planet and the spacecraft.

In the above example, the clockwise turn of the spacecraft would be achieved by the spacecraft encountering the planet from behind its direction of travel, just as it flies past (top left of figure 4.10). Gravity pulls the spacecraft to give it a velocity component in the direction the planet is moving and so the spacecraft picks up speed relative to the star. If the spacecraft instead passed in front of the planet, it would turn in the opposite direction (to the left in figure 4.10) and so would lose speed. To compute this, with $\delta = +90°$ (an anticlockwise rotation), the final velocity of the spacecraft in the planet's rest frame is $\mathbf{v}_{sp} = (-2, -1)$. In the star's rest frame it is $\mathbf{v}_s = (-1, -1)$, giving a speed of $\sqrt{2} = 1.4$, a direction of $\theta' = \mathrm{atan}\, 1/1 = -45°$ and thus a deflection angle of $-135°$. This is a general feature of gravity assists: to increase its speed (in the star's rest frame), the spacecraft needs to pass behind the planet in its orbit about the star; to decrease its speed the spacecraft needs to pass in front of the planet.

4.3.2 Gravity assists on real bodies

The size of the spacecraft's deflection in the planet's rest frame is given by equation 4.15, which depends only on the eccentricity e. The smaller the eccentricity – closer to 1 – the larger the deflection angle (apparent in figure 4.5). From equations 4.13 and 4.14 we see that

$$e^2 - 1 = \frac{b^2 v_\infty^4}{\mu^2}. \tag{4.25}$$

To achieve a given eccentricity (and thus deflection angle), then the larger the velocity, the smaller the impact parameter must be. Similarly, increasing the mass of the planet, decreasing the impact parameter, or decreasing the velocity of the spacecraft all produce a smaller eccentricity and thus a larger deflection angle.

Suppose we want to use a gravity assist to turn a spacecraft by $\delta = 90°$ (in the planet's rest frame). This corresponds to an orbit with $e = \sqrt{2}$ and so, from equation 4.25, $b = \mu/v_\infty^2 = -a$. With $v_\infty = 5\,\mathrm{km\,s^{-1}}$, and using Jupiter for the gravity assist (mass of $1.9 \times 10^{27}\,\mathrm{kg}$), we require an impact parameter of $b = 5.1 \times 10^6\,\mathrm{km}$, which is 71 times Jupiter's equatorial radius. Because of Jupiter's gravity, the spacecraft will approach closer to Jupiter than the impact parameter ('gravitational focusing'). The closest approach is the periapsis distance,[4] $r_p = a(1 - e) = \mu(e - 1)/v_\infty^2$, which in this case is $(\sqrt{2} - 1)b = 2.1 \times 10^6$ km, or 29 times Jupiter's equatorial radius.

If we recompute what we need to get this same deflection at an asteroid of radius 100 km and density 2500 kg m^{-3} (mass of just over 10^{19} kg) we get $b = 28$ m and $r_p = 12$ m. The spacecraft would hit the asteroid, so such a turn is not possible with such a low mass object. In fact, at this density, and taking into account the gravitational focusing, such an asteroid would have to have a radius of at least 9300 km (see exercises). Gravity assists are possible at large asteroids (the largest is Ceres with a radius of around 470 km and with a mean density of 2200 kg m^{-3}), but only for smaller deflection angles. The Earth is smaller than the above limit, but it has a higher density. The impact parameter to achieve a 90° turn around the Earth at $v_\infty = 5$ km s^{-1} is 15 900 km, and results in a closest approach to the Earth of just 6606 km. This is not much more than the Earth's mean radius of 6371 km. The Cassini–Huygens spacecraft did a flyby of the Earth in August 1999 at an altitude of just 1170 km, giving it a Δv of 5.5 km s^{-1} to help it get it to Saturn.

The final velocity of the spacecraft relative to the star depends on the velocity of the planet through the vectorial addition of the velocities. The maximum change in velocity occurs when the spacecraft and planet are moving (anti-)parallel and the deflection angle is 180°. This corresponds to a parabolic orbit with $e = 1$. When planet and spacecraft are moving anti-parallel (near head-on collision) with (positive) speeds u_p and u_s, respectively, the spacecraft's speed increases by twice the planet's speed, to $2u_p + u_s$. This is exactly analogous to the ball bouncing off the train in figure 4.9. When they are instead moving parallel – so that one catches up with the other – the final velocity of the spacecraft relative to the star is $2u_p - u_s$, which could be negative.

Gravity assists can provide very large changes in velocity, of several km s^{-1}. A spacecraft orbiting the Sun at the Earth's distance needs $\Delta v = 12.3$ km s^{-1} to escape from the Sun (section 4.1.4). To achieve this with a chemical rocket with $v_e = 4.5$ km s^{-1} would require a mass ratio of 15.5, from the rocket equation. If we can instead achieve some of this using one or more gravity assists about planets, the rocket can be smaller and cheaper, or the scientific payload larger. For this reason, many interplanetary missions have used gravity assists. As shown earlier in figure 4.8, Voyager 2 used them at each of the planets – Jupiter, Saturn, Uranus, and Neptune – that it visited in its grand tour.[5] At 720 kg launch mass, Voyager 2 was a relatively light spacecraft and went directly to Jupiter without a gravity assist. Cassini–Huygens, in contrast, had a launch mass of 5700 kg, so to get to Jupiter (on its way to Saturn) it needed to do two gravity assists at Venus (both Δv around 7 km s^{-1}) and one at the Earth (5.5 km s^{-1}). Its trajectory and change in velocity are shown in figure 4.11. As the gravity assist may not put the spacecraft on the exact trajectory required, rocket boosts are often used in between to make fine corrections to the path. This was the case with Cassini, which used

---

[4] For hyperbolic orbits we can also write this as

$$\frac{r_p}{b} = \sqrt{\frac{e - 1}{e + 1}}.$$

[5] Voyager 2 relied on a rare alignment of the outer planets to allow it to visit these planets in succession without needing a large rocket. Such an alignment – by which we do not mean all planets lying in a straight line – only occurs every 175 years.

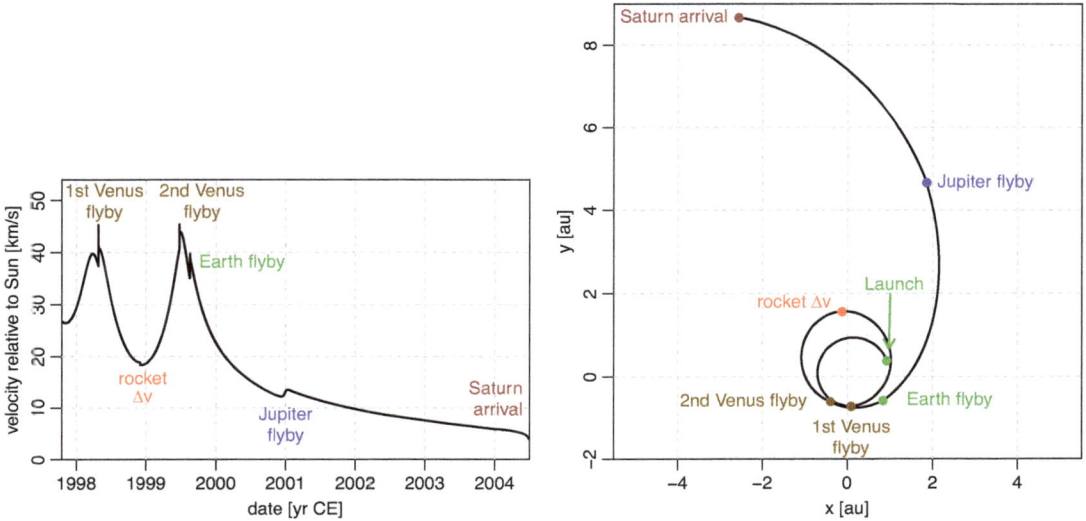

Gravity assists of Cassini–Huygens at Venus (twice), the Earth, and Jupiter. The spacecraft rockets also gave it a Δv
between the two Venus flybys. The right panel shows the trajectory in the plane of the solar system, with the Sun at
(0, 0) and the positive x-axis the line of zero ecliptic longitude. The motion of the spacecraft is anticlockwise. Data
obtained from JPL Horizons.

Fig. 4.11

a relatively large boost after its first Venus encounter to put it on course for a second Venus
encounter.

As a large Δv is also required to get to the inner solar system, gravity assists are also
used for missions to Mercury and the Sun. The BepiColombo mission to go into orbit about
Mercury, for example, has done nine gravity assists in total at the Earth, Venus, and Mercury
between its launch in October 2018 and its orbital insertion planned for November 2026.

Are gravity assists useful for interstellar travel? Gravity assists about planets in the solar
system could theoretically give a spacecraft a Δv of up to twice the orbital velocity of the
planet in the case of a head-on encounter and 180° turn. The largest velocity change would
be with Mercury, which could be up to twice its orbital speed, $2 \times 48\,\mathrm{km\,s^{-1}}$. To avoid hitting Mercury this could only be achieved with a large impact parameter with a small initial
velocity, and the latter is the opposite of what we want for interstellar travel. Furthermore,
the spacecraft would then lose a lot of velocity escaping from the deep gravitational potential close to the Sun. In practice, gravity assists off solar system planets will only increase
the velocity at infinity of a spacecraft by a few tens of kilometres per second, so are of no
benefit for getting to the nearest stars quickly. But they could be important for manoeuvring
a spacecraft in a target star system after rendezvous.

4.4 Summary

We have looked at some of the properties of orbits and how to move between them using
impulsive tangential boosts. The main orbits of interest are bound elliptical orbits and
unbound hyperbolic orbits.

The Oberth effect shows that the maximum change in the kinetic energy of a rocket in some reference frame is achieved by firing its rockets when it is moving fastest (equation 4.22). For a spacecraft moving on an elliptical orbit, this means applying the Δv at periapsis. The Oberth effect is simply a consequence of the quadratic dependence of kinetic energy on velocity, and is useful for maximizing the efficient use of propellant.

The flyby of a spacecraft past a planet is a hyperbolic orbit. When this is viewed from the perspective of a third body, specifically the central star, we get the concept of the gravity assist: what is a symmetric inbound–outbound motion in the planet's rest frame is asymmetric in the star's frame because the planet is moving (figure 4.10). The net effect is a change in the velocity of the spacecraft relative to the star. In order for this flyby to accelerate the spacecraft, it must pass behind the planet; to decelerate it, the spacecraft must pass in front of the planet. The spacecraft exchanges energy and momentum with the planet, but the latter is so massive that the change in its orbit is unmeasurable. Gravity assists are frequently used in interplanetary missions to manoeuvre spacecraft and to boost spacecraft above the solar system's escape velocity. In practice, gravity assists are limited to velocity changes of a few tens of kilometres per second in the best cases, so are of little use for reaching high interstellar velocities.

4.5 Exercises

1. In the biellptical transfer orbit of figure 4.7, the inner orbit has a radius of 1 au and the outer orbit a radius of 5 au. If we place point B at 10 au from the central mass, what are the semi-major axes and eccentricities of orbits 2 and 3?

2. In section 4.2.2, we consider the orbit transfer required to drop into the Sun efficiently. Derive equation 4.20.

3. In section 4.2.3 on the Oberth effect, it is stated that it is best to escape from an elliptical orbit by applying Δv at periapsis rather than apoapsis. Derive an expression for the velocities at infinity when applying Δv at (i) periapsis and (ii) at apoapsis. Plot how these vary with eccentricity and prove that the velocity in case (i) is never smaller than the velocity in case (ii).

4. We discussed in section 4.2.3 how to use the Oberth effect to maximize the velocity at infinity for a given boost by first diving close to the Sun to increase the pre-boost velocity (v_i in equation 4.22). For a given perihelion distance r_p, this can be achieved with a highly elliptical orbit about the Sun, that is, one with a large aphelion distance r_a. Show, however, that the perihelion velocity is more or less independent of r_a when $r_a \gg r_p$.

5. Show that for a gravity assist to achieve a turn angle δ around a planet of mean density ρ when the asymptotic spacecraft velocity is v_∞, the planet must have a minimum radius of

$$r_{\min} = v_\infty \sqrt{\frac{3}{4\pi G\rho(e-1)}}$$

to avoid colliding with the planet, where $e = 1/\sin(\delta/2)$. What is the maximum value of v_∞ for achieving $\delta = 90°$ at Jupiter?

The general principle of a rocket is to expel propellant in one direction, thereby forcing the rocket to move in the opposite via conservation of momentum. Most rockets built so far are propelled by burning a fuel to produce a gas jet. In this chapter we will see how a convergent–divergent nozzle is used to accelerate the gas to a large velocity by maximizing the conversion of its thermal energy (random motion) into kinetic energy (linear motion). We will examine the characteristics of liquid and solid fuels, and look at a few examples of real rockets that use them.

5.1 Rocket thermodynamics

The basic idea of a thermal rocket is to heat a gas to a high temperature and pressure in a combustion chamber that has a hole at one end (figure 5.1). Beyond the hole is either the near-vacuum of space or the atmosphere of a planet. Provided the pressure of the gas in the combustion chamber is higher than the ambient pressure, the gas escapes through the hole. This gas has momentum relative to the rocket, so by conservation of momentum the rocket acquires equal and opposite momentum. The microscopic mechanism for this momentum transfer is the collisions of the gas particles with the walls of the combustion chamber. For an axisymmetric combustion chamber, the collisions perpendicular to the axis impart no net momentum on the rocket. Because the hole is only at one end of the rocket, there is a net momentum exchange along the axis.

Although this approach imparts momentum on the rocket, we will see that passing the gas through a specially shaped nozzle can accelerate the gas to a larger velocity. It does this by extracting more thermal energy from the gas.

Let us first consider how the properties of a gas change as it moves along a nozzle or pipe without yet specifying its cross-sectional shape. Inside the combustion chamber, the hot gas is at high pressure, whereas at the end of the nozzle the gas expands into a lower pressure environment. If this change takes place rapidly, we can approximate it as an adiabatic process, one in which there is no heat exchange with the environment (in particular the walls of the nozzle). We also assume there is no energy dissipation, so it is also a reversible process, and thus isentropic.

As the gas moves along the nozzle, then depending on the nozzle's shape, the gas will be compressed or expanded, resulting in a change in its pressure p, temperature T, and density ρ (a compressible flow). Assuming the gas moves steadily along the nozzle without turbulence, we can also consider its bulk velocity u at any point along the nozzle. To determine how these quantities relate, we consider the total specific energy E (i.e. per unit mass) of the gas. This is the sum of all the different specific energies involved. These are the internal energy U, the

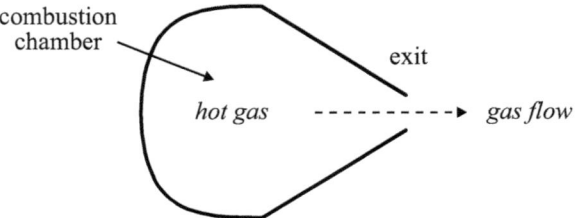

A simple thermal rocket consists of a combustion chamber in which a gas is heated and can escape from a hole at one end.

kinetic energy $K = \frac{1}{2}u^2$, and the ability to do work (arising from the finite entropy of the gas) $\int p\,dV$, which we write as pV, where V is the specific volume (inverse density). Thus

$$E = U + pV + K. \tag{5.1}$$

As we are considering an adiabatic process and assuming the gas is not doing work on any external bodies (the sides of the nozzle are rigid), then E is conserved along the flow. At this point it is useful to introduce the specific enthalpy, which is defined as

$$H = U + pV. \tag{5.2}$$

The change in energy between any two points along the nozzle is then

$$\Delta E = \Delta H + \Delta K = 0. \tag{5.3}$$

The reason for introducing the enthalpy is that it is directly related to a quantity we can measure. If the gas changes its temperature by ΔT then

$$\Delta H = c_p \Delta T, \tag{5.4}$$

where c_p is the specific heat capacity of the gas at constant pressure.

The gas in the combustion chamber is well mixed with a single temperature T_c and is on average stationary relative to the rocket, that is, $u_c = 0$. As the pressure in the combustion chamber is higher than the ambient pressure at the exit, the gas will flow along the nozzle. Let the temperature and velocity of the gas at some arbitrary point be T and u, respectively. It follows from equations 5.3 and 5.4 that

$$c_p(T_c - T) + \frac{1}{2}u_c^2 - \frac{1}{2}u^2 = 0 \tag{5.5}$$

so

$$u^2 = 2c_p(T_c - T). \tag{5.6}$$

This tells us that if the gas temperature drops along the nozzle, the gas will accelerate. The gas gains kinetic energy (K) at the expense of its internal energy (U) through conservation of energy. The hotter the gas in the combustion chamber, and the more we can get the gas to cool by the time it leaves the nozzle, the faster the gas, and thus the larger the change in momentum of the rocket. Indeed, the velocity at the exit of the nozzle is, in an ideal world, the effective exhaust velocity of the rocket that appears in the rocket equation (equation 3.4). In reality, the exhaust velocity will be lower due to losses in the nozzle arising from heat transfer, viscosity, and imperfect collimation of the exiting gas flow.

It is neither immediately obvious what the temperature T is, nor whether it has to drop along the nozzle, so it is more insightful to write equation 5.6 in terms of other properties of the gas. To do this we assume an ideal gas, the equation of state for which is $p = \rho RT/\mathcal{M}$, where $\mathcal{M}$ is the molar mass (kg mol^{-1}) of the gas and R is the universal gas constant ($\text{J mol}^{-1}\,\text{K}^{-1}$). For an ideal gas undergoing reversible, adiabatic changes we also have the relation

$$Tp^{(1-\gamma)/\gamma} = \text{constant}, \qquad (5.7)$$

where γ is the ratio of the specific heat capacity at constant pressure to that at constant volume, which is always larger than one. This relation tells us that if the gas pressure decreases from combustion chamber to the outside, so does the gas temperature. To avoid having both γ and c_p in our equations, we can write the latter as

$$c_p = \frac{\gamma}{\gamma-1}\frac{R}{\mathcal{M}}. \qquad (5.8)$$

Substituting equations 5.7 and 5.8 into equation 5.6 we can now write the latter as

$$u^2 = \frac{2\gamma}{\gamma-1}\frac{RT_c}{\mathcal{M}}\left[1 - \left(\frac{p}{p_c}\right)^{(\gamma-1)/\gamma}\right]. \qquad (5.9)$$

This gives the velocity of an ideal gas under the assumption of a steady isentropic flow along a nozzle when the pressure has decreased from p_c to p. $\mathcal{M}$ and γ are constant along the flow: the gas is assumed not to undergo any chemical changes after the combustion (and we assume γ does not vary with temperature). Figure 5.2 plots how this gas velocity varies as a function of the pressure ratio for a given gas. With all other quantities held constant, this shows that the exit velocity is maximized by reducing the pressure at the exit to the lowest value possible. This can be achieved by operating the rocket in a vacuum.

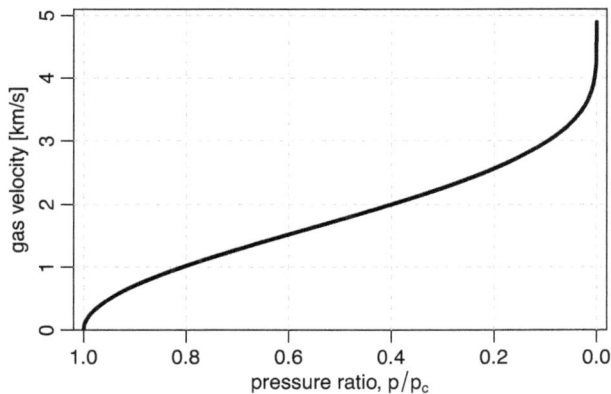

The variation of gas velocity in a nozzle as a function of the ratio of the pressure in the gas to the pressure in the combustion chamber (equation 5.9) for $\gamma = 1.25$, $\mathcal{M} = 0.0113\ \text{kg mol}^{-1}$, $T_c = 3250\ \text{K}$ (applicable for the combustion of hydrogen in oxygen). Note that the horizontal axis is flipped. The pressure p could represent either the pressure at the exit of the nozzle or the pressure at some point along the nozzle.

Fig. 5.2

When the ambient pressure is essentially zero and the gas pressure drops to this value at the exit of the nozzle, then the exit velocity is

$$u_{max} = \sqrt{\frac{2\gamma}{\gamma - 1} \frac{RT_c}{\mathcal{M}}}. \tag{5.10}$$

This expression shows that there are various ways to maximize the exit velocity:

• raise the temperature in the combustion chamber (T_c);
• use a lighter propellant ($\mathcal{M}$);
• use a propellant with a smaller value of γ, because $\gamma/(\gamma - 1)$ increases with decreasing γ and $\gamma > 1$. The quantity γ is determined by the number of degrees of freedom f of the gas via $\gamma = 1 + 2/f$. For an ideal monatomic gas $f = 3, \gamma = 5/3 (= 1.67)$ and for a diatomic gas $f = 5, \gamma = 7/5 (= 1.4)$. More complicated molecules have higher values of f (lower values of γ). To maximize u we therefore want to have a gas with lots of degrees of freedom. Rocket fuel exhausts are often polyatomic with an average value of $f = 8, \gamma = 5/4 (= 1.25)$, although for a real gas γ varies with temperature.

Note that γ and $\mathcal{M}$ refer to properties of the exhaust gases, the products of combustion, rather than the fuel that is burned to make them.

An ideal gas has two independent properties, for example p and T, which then determine ρ. Under the additional assumption that the gas behaves adiabatically, these two properties are no longer independent. We see from equation 5.7 how they vary, although this is more explicit if we differentiate:

$$\frac{dT}{T} = \left(\frac{\gamma - 1}{\gamma}\right) \frac{dp}{p}. \tag{5.11}$$

As $\gamma > 1$, this means that changes in p and T for an adiabatic process are positively correlated. We can also write the adiabatic relationship in terms of pressure and density as

$$\rho p^{-1/\gamma} = \text{constant} \tag{5.12}$$

and differentiating this gives

$$\frac{dp}{p} = \gamma \frac{d\rho}{\rho}, \tag{5.13}$$

showing that pressure and density are also positively correlated. Thus p, T, and ρ all vary in a common direction.

5.2 The convergent–divergent nozzle

Equation 5.9 shows that the lower the gas pressure at the nozzle exit, the larger the gas velocity. How can we achieve this low gas pressure? Just because the ambient pressure is low (or even zero), this does not mean the gas pressure at the exit equals this. The exiting gas could have a larger pressure, in which case its velocity would not be as large.

To arrange for a low exit gas pressure, we need to consider the geometry of the nozzle. To do this we examine the mass flow rate (kg s^{-1}) of the gas at any point along the nozzle,

$$\dot{m} = \rho A u, \tag{5.14}$$

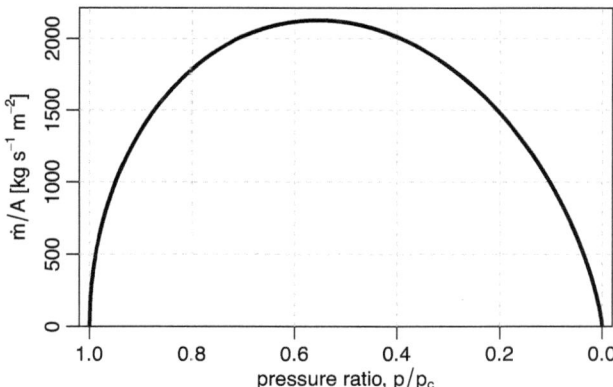

The variation of the mass flow rate per unit cross-sectional area in a nozzle as a function of the gas pressure expressed as a fraction of the combustion chamber pressure (equation 5.17) for $\gamma = 1.25$, $\mathcal{M} = 0.0113$ kg mol^{-1}, $T_c = 3250$ K, and $p_c = 5$ MPa. Note that the horizontal axis is flipped.

Fig. 5.3

where A is the cross-sectional area of the nozzle. We can write the density in terms of the pressure using the adiabatic gas relation

$$\frac{\rho}{\rho_c} = \left(\frac{p}{p_c}\right)^{1/\gamma}. \tag{5.15}$$

Using the ideal gas law, the density in the combustion chamber is

$$\rho_c = \frac{\mathcal{M}}{R}\frac{p_c}{T_c}. \tag{5.16}$$

Putting these and the expression for u from equation 5.9 into equation 5.14 allows us to write the mass flow rate per unit cross-sectional area as

$$\frac{\dot{m}}{A} = p_c \sqrt{\frac{2\gamma}{\gamma-1}\frac{\mathcal{M}}{RT_c}\left(\frac{p}{p_c}\right)^{2/\gamma}\left[1-\left(\frac{p}{p_c}\right)^{(\gamma-1)/\gamma}\right]}. \tag{5.17}$$

This is plotted as a function of the pressure ratio in figure 5.3. We see that as the pressure ratio drops, $\dot{m}/A$ initially increases, hits a maximum, and then decreases. The pressure ratio on the horizontal axis could be interpreted as the exit to chamber ratio or as the ratio at any point along the nozzle. In either case the mass of the propellant is conserved, so $\dot{m}$ must be constant along the nozzle. Thus equation 5.17 tells us how the cross-sectional area of the nozzle must vary in order to achieve a low gas pressure – hence large velocity – at the exit.

Moving away from the combustion chamber, the cross-sectional area of the nozzle must initially decrease (a converging nozzle). As it does so, the gas pressure decreases and the gas velocity increases (figure 5.2). The smallest area corresponds to the maximum in figure 5.3. If this was the exit of this nozzle (a purely converging nozzle), then the corresponding pressure ratio determines the velocity of the gas at the exit. Even if the ambient pressure were lower than this gas pressure, the gas velocity would be no higher. (The gas would then expand further outside the rocket nozzle, but this would not transfer any momentum to the spacecraft.)

To achieve a lower pressure ratio, figure 5.3 shows that the area of the nozzle must now increase. Continuing along this now diverging nozzle, shown in figure 5.4, the pressure of

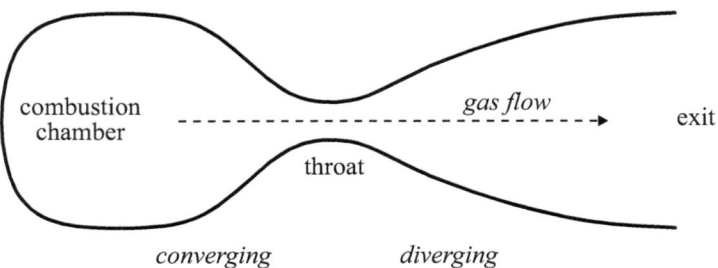

combustion
chamber

gas flow

exit

throat

converging   diverging

Fig. 5.4 A convergent–divergent nozzle, also known as a de Laval nozzle.

the gas continues to decrease, and the velocity increases further (figure 5.2 again). Ideally the exit of the nozzle would be at the point where its cross-sectional area corresponds to the gas pressure equalling the ambient pressure.

Thus to achieve the largest drop in pressure (and temperature) of the gas, and therefore achieve the largest exit velocity of the gas, we should use a convergent–divergent nozzle together with a sufficiently large combustion chamber to exit pressure ratio. If we use only a convergent nozzle, then the gas will not expand by a large amount within the nozzle – only up to the maximum in figure 5.3, even if the ambient pressure is zero – and so the exit velocity will be limited.

I have not said anything about how the nozzle area should vary with physical distance along the axis. Determining that requires a more detailed analysis, and a design which ensures that the flow remains steady and isentropic without turbulence. The key thing for us is that it converges then diverges.

Let us explore in more detail how the nozzle shape is producing a high velocity gas. The narrowest part of the nozzle, where the mass flow rate per unit area is largest, is called the *throat*. Differentiating equation 5.17 with respect to the pressure ratio we find that the pressure at the throat p_t is

$$\frac{p_t}{p_c} = \left(\frac{2}{\gamma+1}\right)^{\gamma/(\gamma-1)}. \tag{5.18}$$

This ratio only depends on γ. The ratio does not vary much: for $\gamma = 1.25$ (polyatomic gas) it is 0.555 and for $\gamma = 1.4$ (diatomic gas) it is 0.528. The corresponding mass flow rate per unit area at the throat is

$$\frac{\dot{m}}{A_t} = p_c \sqrt{\gamma \left(\frac{2}{\gamma+1}\right)^{(\gamma+1)/(\gamma-1)} \frac{\mathcal{M}}{RT_c}}. \tag{5.19}$$

It is instructive to compute also the velocity of the gas at the throat. Putting the expression for c_p (equation 5.8) into equation 5.6, the squared gas velocity at any point along the nozzle is

$$u^2 = \frac{2\gamma}{\gamma-1} \frac{R}{\mathcal{M}} T \left(\frac{T_c}{T} - 1\right). \tag{5.20}$$

Using equation 5.7 we can convert the pressure ratio at the throat in equation 5.18 into a temperature ratio:

$$\frac{T_c}{T_t} = \frac{\gamma+1}{2}. \tag{5.21}$$

Putting this into equation 5.20 with $T = T_t$, we find that the gas velocity in the throat is

$$u_t = \sqrt{\frac{\gamma R T_t}{M}}. \tag{5.22}$$

This is equal to the local speed of sound in the gas, given in general by

$$a = \sqrt{\frac{\gamma R T}{M}}. \tag{5.23}$$

This means that if the pressure ratio in the gas drops low enough such that $\dot{m}/A$ reaches the maximum in figure 5.3, then the gas velocity in the throat will reach the speed of sound. This in turn can only occur if the throat is narrow enough. If the maximum of the curve is not reached, then the gas will remain subsonic at the throat, and equations 5.18, 5.19, 5.21, and 5.22 do not apply.

Whether or not the gas continues to accelerate after the throat depends on whether it reaches the speed of sound at the throat. To see this, we need to relate changes in the velocity of the gas to changes in the area of the nozzle. As the mass flow rate $\dot{m}$ is constant, the derivative of equation 5.14 is

$$\frac{d\rho}{\rho} + \frac{dA}{A} + \frac{du}{u} = 0. \tag{5.24}$$

To eliminate the density from this we first combine equations 5.11 and 5.13 to get

$$\frac{dT}{T} = (\gamma - 1)\frac{d\rho}{\rho}. \tag{5.25}$$

Differentiating equation 5.6 gives $u\,du = -c_p dT$, which we then use together with equation 5.8 to eliminate dT in the above equation to get

$$\frac{d\rho}{\rho} = -\frac{u^2}{(\gamma - 1)c_p T}\frac{du}{u} = -\frac{u^2}{a^2}\frac{du}{u} = -M^2\frac{du}{u}, \tag{5.26}$$

where $M = u/a$ is the *Mach number*, the ratio of speed of the gas to the speed of sound in the gas. Substituting this into equation 5.24 finally gives an expression relating velocity to area anywhere along the nozzle in terms of just the Mach number,

$$\frac{du}{u} = \frac{1}{M^2 - 1}\frac{dA}{A}. \tag{5.27}$$

Equation 5.27 shows that when $M < 1$ (subsonic flow) then $M^2 - 1$ is negative, so an increase in the area causes a decrease in the velocity, and a decrease in area causes an increase in the velocity. This is characteristic also of incompressible fluids (as can be seen from equation 5.24 when $d\rho = 0$). In contrast, when $M > 1$ (supersonic flow) then $M^2 - 1$ is positive, so an increase in the area causes an increase in the velocity, and a decrease in area causes a decrease in the velocity. Recall that the velocity always has the opposite trend from the pressure (figure 5.2).

In the converging part of the nozzle, corresponding to the left of the maximum in figure 5.3, the velocity is always subsonic and so the velocity increases as the gas moves down the nozzle towards the throat. There are then two alternative outcomes arising from equation 5.27:

1. If the pressure at the throat is low enough that the gas attains the maximum in figure 5.3, then the gas reaches $M = 1$. When it infinitesimally exceeds this speed, the flow is supersonic, and so now an increase in the area of the nozzle (the diverging part) will result in a further increase in the velocity. This corresponds to the part of figure 5.3 to the right of the maximum.

2. If instead the pressure at the throat is too high, such that the gas remains subsonic, then in the diverging part of the nozzle the gas velocity decreases again (and the pressure increases). This corresponds to the part of figure 5.3 to the left of the maximum again.

Which of these two conditions occurs for a given nozzle depends on the pressure ratio at the throat.

Thus a convergent–divergent nozzle will accelerate the gas to its highest velocity only if the gas reaches the speed of sound at the throat. This in turn occurs only if the nozzle is narrow enough and if the ambient pressure is low enough that the maximum in figure 5.3 can be attained. As we saw from equation 5.18, this requires the pressure at the throat to drop to about half the pressure in the combustion chamber.

Provided the gas hits the speed of sound at the throat, it continues to accelerate in the diverging part of the nozzle. The temperature of the gas continues to drop – this is the source of the increasing velocity of the gas – so the speed of sound also drops, and thus the gas becomes increasing supersonic.

When the system is configured such that the gas reaches $M = 1$ at the throat, the gas will not exceed $M = 1$ at the throat, even if the combustion chamber pressure is increased or the exit pressure decreased. If the chamber pressure is increased, then the pressure at the throat will increase and so the actual velocity at the throat will increase, but this is still the local speed of sound (equations 5.21 and 5.22). In this condition the flow is said to be *choked*. The velocity of the gas at the exit can still be increased by lowering the exit pressure (from equation 5.9). But once the flow is choked, the mass flow rate cannot be increased by lowering the exit pressure, as we see from equation 5.19. Because the flow is choked, changes in the nozzle downstream of the throat do not affect anything upstream of it. The exit velocity and mass flow rate can still be raised by increasing the pressure in the combustion chamber.

To help illustrate the above concepts, figure 5.5 summarizes various scenarios for gas flow along the nozzle. The figure shows how the pressure ratio p/p_c varies along the nozzle for several values of the ambient pressure p_a, which is lowered monotonically as we proceed from scenario (a) to scenario (g).

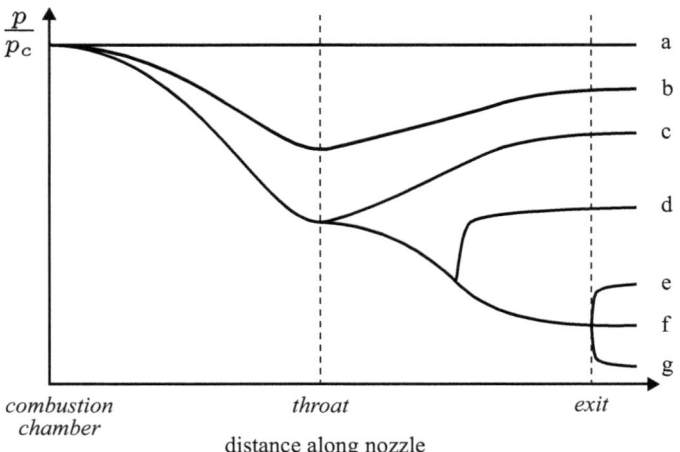

Fig. 5.5 A sketch of the variation of the gas pressure p along a nozzle relative to the pressure in the combustion chamber p_c for various scenarios (a) to (g) described in the text. In all scenarios the flow upstream of the throat is subsonic.

(a) $p_a = p_c$. The pressure is constant everywhere. There is no flow.

(b) $p_a < p_c$. The gas now flows. The velocity increases in the converging part of the nozzle, but it remains subsonic throughout, so the pressure increases (and velocity decreases) again in the diverging part of the nozzle in accordance with equation 5.27.

(c) p_a is lowered further to the critical value that attains $M = 1$ in the throat, the maximum in figure 5.3. The flow is choked. Downstream of the throat, the flow is still subsonic and the pressure rises. In all following scenarios, the flow in the converging part of the nozzle remains the same as for this choked flow.

(d) Lowering p_a further puts the downstream flow into the supersonic condition. The pressure continues to decrease initially in the diverging part of the nozzle. But the ambient pressure is too large for the isentropic conditions to hold to the end of the nozzle, so at some point a shock wave forms, after which there is a sharp increase in the gas pressure and a decrease to subsonic velocities.

(e) p_a is now low enough that isentropic conditions hold throughout the nozzle, so the gas remains supersonic up until the exit (this is also the case for the next two scenarios). But the ambient gas pressure is still larger than the gas exit pressure – the flow is said to be 'over-expanded' – so undergoes a sudden increase in pressure at the exit, causing a sharp contraction of the gas.

(f) The gas exit pressure matches the ambient pressure and so flows smoothly out with the width of the nozzle at a supersonic speed.

(g) The ambient pressure is lower than the emerging gas pressure, so the gas will continue to expand outside the nozzle. The flow is said to be 'under-expanded'.

A rocket nozzle is designed to operate in condition (f). When the ambient pressure is negligible, the exhaust velocity is given by equation 5.10 and the mass flow rate by equation 5.19. These are the two key parameters for the rocket and thrust equations discussed in chapter 3.

Rockets launched from the Earth's surface experience a large ambient pressure at first. A nozzle used here should have a relatively large exit pressure and thus be relatively short in order to achieve condition (f). Higher in the atmosphere where the pressure is lower, the exit pressure should be lower and thus a longer nozzle is optimal. As a nozzle cannot be continuously adjusted during flight, it may be too wide and long at launch, in which case the exit gas pressure is too low. In this case we see the exhaust emerging in a column that is narrower than the nozzle exit, due to the over-expanded condition (e). In contrast, when the rocket is at high altitudes where the atmospheric pressure is lower, the nozzle may be too narrow and short, so the exit gas pressure is too high. We now see the exhaust expanding into a broad bell-shaped plume that is wider than the nozzle exit due to the under-expanded condition (g). This is another reason why rocket staging increases the efficiency of the rocket: the nozzle for each stage can be optimized for the average ambient pressure in which it operates.

5.3 The rocket as a heat engine

As a rocket converts thermal energy into mechanical energy it is a type of heat engine. Figure 5.6 shows the idealized pressure–volume diagram for a thermal rocket. Point 1 indicates where the fuel enters the combustion chamber. The fuel ignites, releasing heat and

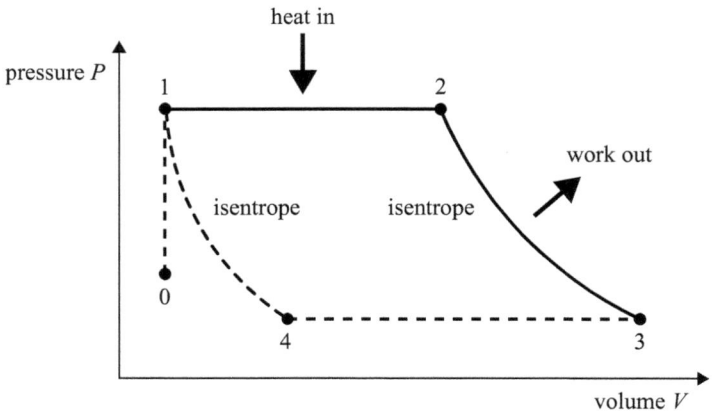

Fig. 5.6 The pressure–volume diagram of a thermal rocket engine. The engine performs the open sequence 1–2–3. A gas turbine (jet engine) performs the open sequence 4–1–2–3. If this is closed by returning isentropically to 4, then we have the Brayton cycle.

expanding at the constant chamber pressure to point 2. From points 2 to 3 the gas is expelled via the nozzle at constant entropy, doing work to propel the rocket and cooling in the process.

The rocket engine has an open cycle, in contrast to the closed cycles we may be more familiar with for internal combustion engines, such as the Otto or Carnot cycles. If, at point 3, the gas were compressed at constant pressure to point 4 (removing heat from the engine in the process), and then compressed at constant entropy back to point 1, we would have the closed Brayton cycle. An important difference between the open cycle of a rocket and the closed cycle of a four-stroke internal combustion engine is that in the rocket the PV diagram sequences occur simultaneously, whereas in the internal combustion engine they occur sequentially.

The open sequence 4–1–2–3 in figure 5.6 is that of an ideal gas turbine jet engine, where 4–1 corresponds to the intake and compression of ambient air. In a rocket, the step 0–1 can be considered as the compression of the fuel in the pumps before it reaches the combustion chamber, which is done at almost constant volume.

5.4 Chemical rockets

In discussing the principle of a thermal rocket above, we started with a hot gas in the combustion chamber. How this gas is heated is not relevant to the behaviour of the nozzle. All significant rockets launched from the Earth to date have used chemical engines, in which chemicals are combusted to produce a hot propellant. This requires that an oxidant be carried in addition to the fuel.

A common fuel is hydrogen, which when mixed with oxygen and combusted produces water. This water vapour is the exhaust that is then accelerated by the nozzle. In chemical rockets it is usual that the fuel (and oxidizer) is the source of both the energy and the mass that propels the rocket. This does not have to be the case, as we shall see with ion engines in chapter 6. In some rockets, additional hydrogen is mixed into the propellant after the combustion. This hydrogen is not burned, but rather acts to lower the mean molar mass of the propellant. Because of equation 5.9, this increases the exhaust velocity (assuming T_c is kept constant).

A brief note on terminology. With chemical thermal rockets it is common to refer to the fuel and oxidizer together (prior to combustion) as the propellant, and to refer to the product of the reaction that is exhausted through the nozzle as the exhaust. In more general contexts it is common to refer to the exhausted product as the propellant (for example in the context of nuclear rockets). I mostly use the latter definition, but it should be clear from the context what is meant.

There are various types of chemical rocket fuel, best divided into liquid and solid fuels. One of the most important criteria of a rocket fuel is its energy per unit mass, in order to achieve the largest exhaust velocity for the lowest possible mass. Also important is the mean molar mass. Although a larger mass lowers the exhaust velocity, it increases the mass flow rate, and so can increase the thrust overall (see equation 3.10). This is particularly important for launch when the gravity of the Earth must be overcome (see section 3.10). Other criteria are important too. These include: a high density so that the fuel does not take up too much volume, because large containers require more mass; complexity and cost of production; ease and safety of handling and storage.

5.4.1 Liquid fuels

In liquid engines, the fuel and oxidizer are stored in separate tanks. They are pumped into the combustion chamber where they are mixed and ignited. The rate of pumping can be used to control the mass flow rate (throttling) and thus the thrust of the rocket.

A common fuel–oxidizer combination is liquid hydrogen and liquid oxygen (LOX). The temperature achieved by their combustion depends on their mixture ratio. With an oxygen-to-hydrogen mass mixture ratio of 4.8, the reaction achieves $T_c = 3250$ K.[1] If the exhausted propellant were entirely water vapour, with molar mass $\mathcal{M} = 0.018$ kg mol^{-1} and ratio of specific heats $\gamma = 1.31$, then equation 5.9 tells us that the exhaust velocity into a vacuum ($p/p_c = 0$) would be 3.56 km s^{-1}. However, at this mixture ratio, some of the hydrogen remains uncombusted.[2] This lowers $\mathcal{M}$ and raises γ. For real gases, values of $\mathcal{M} = 0.0113$ kg mol^{-1} and $\gamma = 1.25$ are more representative for imperfect combustion (there is also OH and atomic O and H in the exhaust). With these values and $T_c = 3250$ K, equation 5.9 then predicts an exhaust velocity into a vacuum of 4.89 km s^{-1}, but in practice it is closer to 4.55 km s^{-1} (also because the uncombusted hydrogen has to be heated, which lowers the temperature). A mixture ratio of eight would increase the combustion temperature – to about 3750 K – but the change in γ and $\mathcal{M}$ results in a lower exhaust velocity. The value of 4.55 km s^{-1} is more or less the maximum exhaust velocity that can be achieved with a hydrogen–oxygen mixture.

---

[1] The combustion temperature is actually higher than the value given, but some of the water produced dissociates, lowering the temperature.

[2] The oxygen atom is 16 times more massive than the hydrogen atom, so an O_2/H_2 mass ratio of eight is needed for complete combustion via $2H_2 + O_2 \rightarrow 2H_2O$.

Hydrogen is a favoured fuel because it has a very large energy per unit mass of $120\,MJ\,kg^{-1}$.[3] However, hydrogen and oxygen are gases at room temperature and pressure, with low densities, so would require large and massive fuel tanks. They are therefore liquefied by cooling, increasing their density, although this makes them more complicated to store and handle.

Another commonly used liquid fuel is the hydrocarbon kerosene, or rather its highly refined variant RP-1. This too is combusted with liquid oxygen. Although this achieves a higher stable combustion temperature than hydrogen, it provides a lower exhaust velocity on account of the larger molar mass of its exhaust, a significant component of which is carbon dioxide. Typical values are 2.6–$3.4\,km\,s^{-1}$. This larger mass is useful for obtaining a larger thrust. Another advantage of RP-1 over hydrogen is that although also highly flammable, it does not need cooling, so is overall easier to handle. Although RP-1 has a lower energy per unit mass than hydrogen ($43\,MJ\,kg^{-1}$) it has a much larger mass density, giving it a higher energy per unit volume than hydrogen ($35\,MJ$ per litre vs $8.5\,MJ$ per litre.) This means that smaller tanks can be used, which both have less mass and cause less atmospheric drag.

A third widely used liquid propellant is hydrazine, N_2H_4. This dissociates exothermically in the presence of a catalyst, such as iridium, so does not need an oxidizer. Such a monopropellant makes the engine design simpler. Typical exhaust velocities are 2.2–$2.4\,km\,s^{-1}$. Like RP-1, hydrazine is liquid at room temperature, so is easier to handle and store than liquid hydrogen and oxygen. It still needs to be handled carefully, however, as it is very toxic. Hydrazine can also be used as a bipropellant, typically with nitrogen tetroxide (N_2O_4) as the oxidizer (also toxic). This gives a higher exhaust velocity, up to about $2.9\,km\,s^{-1}$. The reaction is hypergolic, meaning that combustion takes places spontaneously on contact between the fuel and oxidizer.

One disadvantage of hydrazine is that it freezes at around $2\,^{\circ}C$, and partly for this reason derivatives of hydrazine with lower freezing points are more commonly used. These include monomethylhydrazine (MMH) and unsymmetric dimethylhydrazine (UDMH), which when oxidized with N_2O_4 can produce exhaust velocities of around $3.4\,km\,s^{-1}$.

While hydrogen–oxygen produces the largest exhaust velocity of rocket engines in general use, higher exhaust velocities can be achieved. Replacing oxygen with fluorine produces a higher combustion temperature (around $4200\,K$) and an exhaust velocity of around $4.8\,km\,s^{-1}$. Tripropellants offer even more possibilities. A mixture of lithium, fluorine, and hydrogen produces an exhaust velocity of $5.3\,km\,s^{-1}$, for example. However, fluorine is expensive and toxic, and the exhaust contains highly toxic hydrogen fluoride, making it very difficult to use for launch from the ground.

Other chemical reactions can release even more energy. Energy is required to dissociate molecular hydrogen, so the recombination of atomic hydrogen into its molecular form releases this energy. In principle this can give an exhaust velocity of nearly $21\,km\,s^{-1}$. However, the hydrogen needs to be stored at extremely low temperatures (a few kelvins) and in a strong magnetic field (several teslas), to prevent premature recombination, and this renders it impractical as a fuel. A more speculative idea is the use of metallic hydrogen. Under extremely large pressures, probably several hundred gigapascals, hydrogen is expected to

---

[3] As we need eight times as much mass of oxygen to completely burn hydrogen, the energy per unit mass for such a hydrogen–oxygen mixture is nine times lower.

become metallic. When this pressure is released, the work energy put into the compression is also released, and is theoretically enough to achieve an exhaust velocity of $17\,\mathrm{km\,s^{-1}}$. Another idea is to store helium in an excited state. The energy released from the decay back to the ground state would theoretically achieve an exhaust velocity of $31\,\mathrm{km\,s^{-1}}$. However, even when stored at very low temperatures, the lifetime of the excited state is only a few hours, making this effectively unusable.

5.4.2 Solid fuels

Liquid fuels need to be pumped into the combustion chamber at the right rate to achieve the required combustion conditions. This requires not only separate fuel tanks but also a propellant delivery system. Some of the fuels are also quite complicated to handle, such as cryogenic hydrogen and oxygen.

Solid fuels overcome these problems. The fuel and oxidizer are mixed in solid form in a long tube and then ignited. The fuel burns from the inside out along the length of the fuel block, so at any one time the inside surface around the axis is undergoing combustion (see figure 5.7). Thermodynamically, solid fuels behave just like liquid ones, in the sense that they produce a hot gas that is expelled through a hole into the convergent–divergent nozzle at the bottom of the fuel block.

A typical solid propellant is a mixture of ammonium perchlorate, aluminium powder, and a polymeric binder, which together contains fuel and oxidant in one material. This is often referred to as ammonium perchlorate composite propellant (APCP). Aluminium has a slightly lower energy per unit mass than RP-1, but a higher mass density and so a larger energy per unit volume (84 MJ per litre vs 35 MJ per litre). APCP can achieve an exhaust velocity of around $2.7\,\mathrm{km\,s^{-1}}$. Other materials can also be used, such as zinc–sulphur, although this gives a lower exhaust velocity.

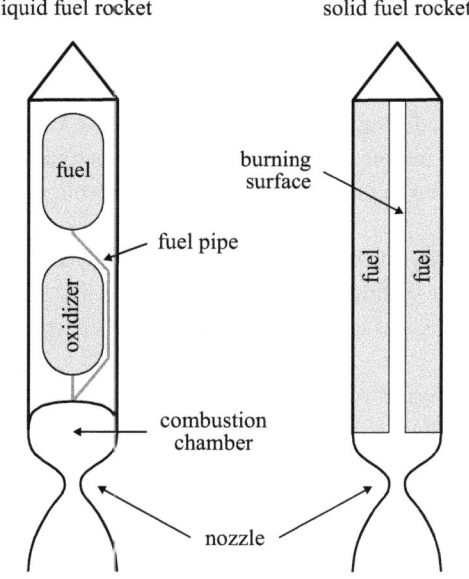

A comparison of a liquid rocket engine with a solid rocket engine.

Fig. 5.7

Solid-fuelled rockets predate liquid ones by many centuries. The earliest known solid-fuelled rockets were used in thirteenth-century China, propelled by gunpowder, which is a mixture of sulphur, charcoal, and potassium nitrate (the latter contains the oxidizer). The first liquid-fuelled rocket (using petrol and liquid oxygen), in contrast, was not launched until 1926.

As it does not need any pumps and pipelines, a solid-fuelled rocket engine is simpler than a liquid-fuelled one, making it cheaper and more reliable. One disadvantage of solid-fuelled rocket engines is that they normally cannot be throttled or switched off. One ignited, they burn until exhausted, meaning they also cannot be individually tested. The lack of pumps also means it is harder to achieve a stable thrust. For this reason the geometry of the fuel pellets and combustion surface must be carefully designed.

Prior to ignition the solid fuel is stable, so it can be stored for long periods. This makes solid fuels suitable for applications where the rocket may not be used for a long period, or where it needs to be launched quickly without having to fuel it. One such application is strategic nuclear weapons; the USA replaced its hydrazine-fuelled Titan II missiles with the solid-fuelled Minuteman partly for this reason.

5.5 Real rockets

Let us now look briefly at two examples of real rockets used in spaceflight.

One of the most powerful rockets ever to launch was the Saturn V, which sent the Apollo missions to the Moon (figure 3.8). It had a height of 111 m (with payload), a maximum diameter of 10 m, a dry mass of 179 t plus 50 t payload, and a launch mass (including propellant) of around 2900 t. It comprised three stages, all powered by liquid fuels. The first stage used 650 t of RP-1 together with 1500 t of LOX. Five F-1 engines achieved a combined thrust of 34.5 MN. Each engine provided an effective exhaust velocity of 2.6–3.0 km s^{-1} (depending on altitude) and burned propellant – fuel and oxidizer – at a rate of 2.6 t s^{-1}, operating at a combustion chamber pressure of 7 MPa. The burn time was around 165 s and took the rocket to an altitude of about 60 km. The second stage used five J-2 engines powered by liquid hydrogen. They provided a vacuum exhaust velocity of 4.1 km s^{-1}, a combined thrust of 5.1 MN, and burned for six minutes. The third stage used a single J-2 liquid hydrogen engine. This used two burns: the first to put the rocket into Earth orbit, and the second to put it on a translunar injection orbit.

This phenomenal capacity of the Saturn V rocket was necessary to send 50 t of payload to the Moon. To get a sense of how powerful this was, the first stage had a power of around 45 GW (equation 3.23). This was about 0.6% of the rate of energy use across the entire world at the time.

The Ariane 6 rocket (figure 5.8) also has three stages, one of which is a set of two or four solid rocket boosters (SRBs). At launch, the rocket is lifted by both the core stage and the SRBs operating in parallel. The core stage is powered by a Vulcain 2.1 engine running on liquid hydrogen and LOX (140 t propellant). This operates at a combustion chamber pressure of 12 MPa, provides a maximum thrust of 1.4 MN, and delivers a vacuum exhaust velocity of 4.3 km s^{-1}. The burn time is around 470 s. In parallel to this, two or four SRBs (depending on payload mass) provide significantly more thrust for

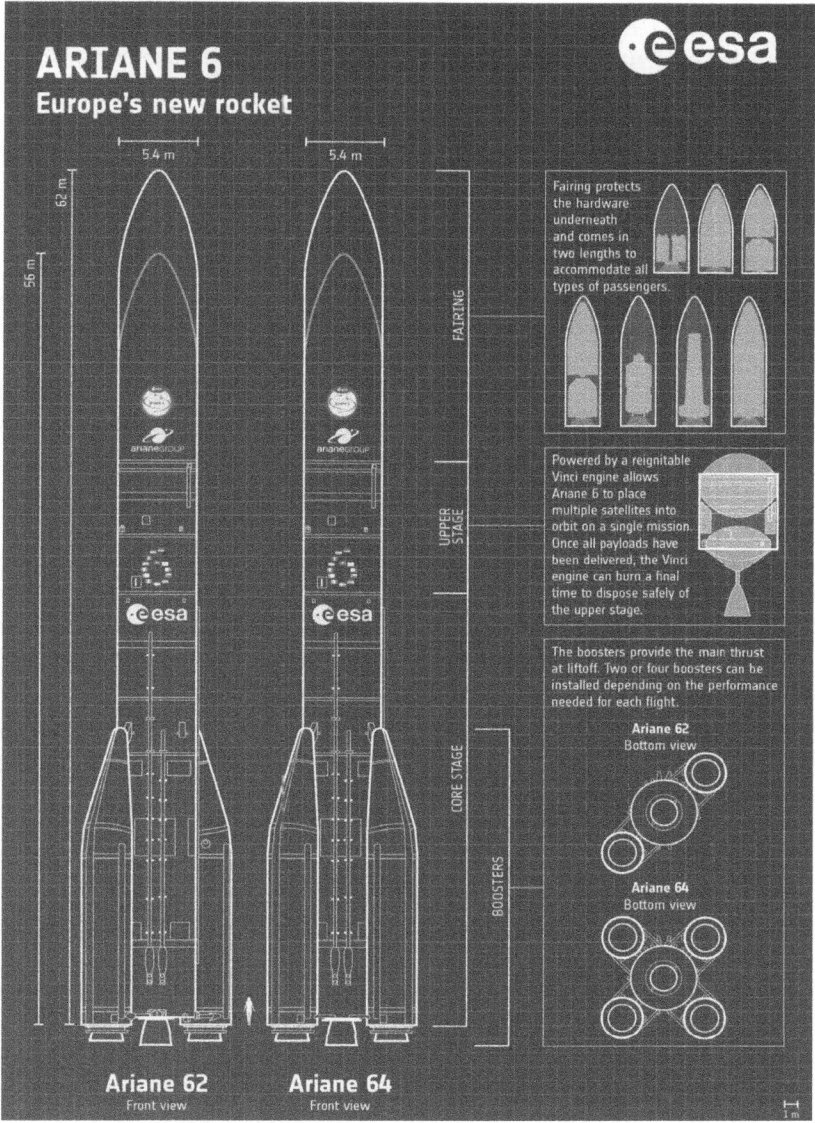

Ariane 6. Credit: European Space Agency (ESA), ESA Standard Licence.

Fig. 5.8

the initial part of the launch. Each SRB provides a thrust of 3.5 MN and has 142 t of propellant. The exhaust velocity is 2.7 km s^{-1}. They deplete their propellant within 135 s, after which they are ejected, leaving the core stage to continue to propel the rocket. When this has burned out, it too is ejected and the upper stage kicks in. This is a single Vinci liquid hydrogen engine that provides an exhaust velocity of 4.5 km s^{-1} and a thrust of 0.18 MN. This stage can be turned off and restarted up to three times in order to perform orbital injections.

5.6 Summary

A thermal rocket works by heating a gas to a high temperature and expelling it through a convergent–divergent nozzle. The nozzle maximizes the conversion of the internal energy of the hot gas (random motion) into kinetic energy (directed motion). When the throat of the nozzle has the right size, and the ratio of the pressure in the combustion chamber to the ambient pressure is large enough, the gas hits the speed of sound at the throat. The gas then continues to accelerate through the diverging part of the nozzle to supersonic speeds by converting more thermal energy into kinetic energy. If we only used a convergent nozzle, then the maximum exhaust velocity would be the local speed of sound of the gas.

The two critical terms in the description of rockets are the exhaust velocity and the mass flow rate. They are given by equations 5.9 and 5.19, respectively, for a thermal rocket. The exhaust velocity can be increased by raising the temperature in the combustion chamber or by using propellants (combustion products) that have a small molar mass $\mathcal{M}$ or a small ratio of specific heat capacities γ.

Chemical thermal rockets may use liquid or solid fuels. Liquid fuels deliver higher exhaust velocities: a liquid hydrogen–oxygen mix can provide up to 4.55 km s^{-1}, about as high as can be practically achieved with chemical reactions. Solid fuels provide lower exhaust velocities (around 2.7 km s^{-1}), but as solid-fuelled rocket engines do not need separate fuel tanks, pipes, or pumps, they are simpler and cheaper.

Because their effective exhaust velocities are low, chemical rockets cannot be used for fast interstellar travel. Even with an exhaust velocity of 5 km s^{-1} and an unfeasibly large mass ratio of $10\,000$, they could still only achieve $\Delta v = 46 \text{ km s}^{-1}$.

5.7 Exercises

1. Once the choked condition has been achieved in a convergent–divergent nozzle, what happens if the throat area is (a) increased, or (b) decreased?
2. What mass of liquid hydrogen–oxygen propellant is required to get a 100 t spacecraft (dry mass) into orbit? Assume $\Delta v = 11 \text{ km s}^{-1}$ is required, and that the entire chemical energy of the propellant is converted into the kinetic energy of the exhaust.
3. As the gas expands in the nozzle, some of its thermal energy is converted into kinetic energy. When this expansion is adiabatic, it does not transfer any energy to the rocket by definition. So how does the gas propel the rocket?

Ion engines

In this chapter we look at a type of non-thermal rocket, the ion engine. This uses electric or magnetic fields to ionize and accelerate a propellant. Although there are many types, we focus here on the so-called gridded ion engine to illustrate the general principles, and to identify how the effective exhaust velocity, thrust, and power depend on characteristics of the engine. We will examine a real interplanetary mission, Dawn, that uses this engine. We finish by looking briefly at some of the other types of ion engine.

6.1 Principle

In a chemical thermal rocket, the fuel energy and the propellant mass come from the same source. For example, hydrogen is burned in oxygen to release energy that then propels the product of this reaction, water. This is convenient, but it is not the only way of producing rocket propulsion.

In an ion engine the fuel and propellant are separate. The principle is as follows. A gas is ionized into its constituent ions and electrons. An electric field is then used to accelerate the much more massive ions into a collimated beam out of the back of the rocket (figure 6.1). By conservation of momentum, this accelerates the rocket in the opposite direction, just like in a chemical thermal rocket. The electrical energy required to ionize and accelerate the propellant could come from any source, such as solar panels, a nuclear reactor, or even chemical reactions.

If the potential difference between the anode and the cathode of the accelerating field is U, then the kinetic energy of each ion of mass m and charge q, which is $(1/2)mv^2$, is equal to the work done by the electric field, qU. Thus the velocity of the ions is

$$v = \sqrt{\frac{2qU}{m}} . \tag{6.1}$$

When we consider the ions at the cathode leaving the spacecraft, v is the (effective) exhaust velocity of the ion engine. Intuitively, a larger potential, a larger charge, or a smaller ion mass yield a larger velocity.

Several ion engines have been built and used for interplanetary missions. The NSTAR engine (discussed below) uses a potential of around $U = 1250\,\text{V}$. Although equation 6.1 implies we want to use ions with small masses, effective exhaust velocity is not the only parameter of interest. We also want sufficient thrust and, for efficiency reasons, a high thrust-to-power ratio. The latter is given by $2/v_e$ (equation 3.23), which suggests we want massive ions. In practice there is a trade-off.

Noble gases are preferred as the source of ions because they have low first ionization potentials, and because they are relatively massive yet gaseous. With full outer electron shells

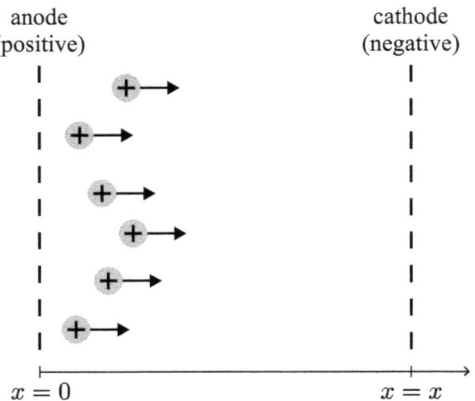

anode
(positive)

cathode
(negative)

$x = 0$ $x = x$

Fig. 6.1 The basic principle of an ion engine.

in their neutral state, noble gases cannot build molecules, so energy is not wasted during the ionization process exciting internal rotational or vibrational states.

A common choice of gas is xenon (atomic number 54), which has seven stable isotopes with a mean nucleon number in nature of 131.3. When this is singly ionized ($q = e$), equation 6.1 gives $v = 38.3\ \mathrm{km\,s^{-1}}$. This compares to about $4.5\ \mathrm{km\,s^{-1}}$ for liquid hydrogen–oxygen chemical engines. In practice there will be some losses, so the effective exhaust velocity of the engine will be lower than the simple calculation above, but still considerably higher than a chemical thermal rocket.

Let us now compute the mass flow rate so we can calculate the engine's thrust. If there were no ions in the space between the anode and cathode in figure 6.1, then the electric potential would vary linearly from U at the anode to zero at the cathode. But the positive ions being accelerated partially shield the negative cathode from the anode, so the anode experiences a lower potential difference in the downstream direction. Thus the more ions there are between the two grids (anode and cathode), the smaller the potential gradient dU/dx. Once the ion density is large enough, this gradient drops to zero. As the magnitude of the force on the ions is $F = qE = -q\,dU/dx$, $E = |\mathbf{E}|$ being the magnitude of the electric field, this means no more ions can be accelerated. Hence there is a maximum charge density that can be attained between the grids, and this places an upper limit on the mass flow rate.

To solve for this maximum flow rate we use Poisson's equation. This relates the gradient of the electric field to the charge density nq, where n is the ion number density,

$$-\nabla \cdot \mathbf{E} = \frac{d^2 U}{dx^2} = -\frac{nq}{\epsilon_0}.$$ (6.2)

If v is the velocity of the ion at some position between the grids, then the current density (amperes per unit cross-sectional area of the flow) is $j = nqv$. Using equation 6.1 we can rewrite the above equation as

$$\frac{d^2 U}{dx^2} = -\frac{j}{\epsilon_0}\sqrt{\frac{m}{2q}}\,U^{-1/2} = -CU^{-1/2},$$ (6.3)

which defines C. Multiplying both sides by dU/dx and integrating, we can rewrite this as[1]

$$\int_0^{(dU/dx)^2} d\left(\frac{dU}{dx}\right)^2 = -4C \int_{U^{1/2}}^0 d(U^{1/2}).$$ (6.4)

The lower integration limit refers to the maximum charge density at the anode, where the potential is U and the potential gradient is zero. The upper integration limit is at some arbitrary downstream point x where we take the potential to be zero, for example at the cathode. The result of the integration is

$$\left(\frac{dU}{dx}\right)^2 = 4CU^{1/2}.$$ (6.5)

Taking the square root of this – and retaining the negative solution because the potential gradient is negative – we get

$$\frac{dU}{dx} = -2C^{1/2}U^{1/4},$$ (6.6)

which we integrate again between $U = U$ at $x = 0$ and $U = 0$ at $x = x$ to get

$$\frac{4}{3}U^{3/4} = 2C^{1/2}x.$$ (6.7)

Substituting for C (defined in equation 6.3) we find that the maximum achievable current density is

$$j_{max} = \frac{4\epsilon_0}{9}\left(\frac{2q}{m}\right)^{1/2}\frac{U^{3/2}}{x^2}.$$ (6.8)

U is the potential difference between anode and cathode of separation x. The above equation has $U \propto x^{4/3}$. In a vacuum the dependence is $U \propto x$, because the electric field is constant. In the ion engine the presence of the ions increases the gradient of the potential.

The mass flow rate in the ion engine is $\dot{m} = jAm/q$, where A is the cross-sectional area of the grids that make up the engine. The thrust is

$$F = \dot{m}v_e = jA\sqrt{\frac{2mU}{q}}$$ (6.9)

using equation 6.1. From equation 6.8, the maximum mass flow rate is

$$\dot{m}_{max} = \frac{4\epsilon_0}{9}A\left(\frac{2m}{q}\right)^{1/2}\frac{U^{3/2}}{x^2}$$ (6.10)

and so the maximum thrust, achieved at the charge density limit, is

$$F_{max} = \frac{8}{9}\epsilon_0 A\left(\frac{U}{x}\right)^2.$$ (6.11)

Although the effective exhaust velocity depends on the mass and charge of the ions, the maximum thrust is independent of both. This thrust depends only on the gradient of the

---

[1] I have used

$$\frac{1}{2}\frac{d}{dx}\left(\frac{dU}{dx}\right)^2 = \frac{dU}{dx}\frac{d^2U}{dx^2} \quad \text{and} \quad 2\frac{d}{dx}(U^{1/2}) = U^{-1/2}\frac{dU}{dx}.$$

Also, I somewhat sloppily do not distinguish between the integration variable and the limits in equation 6.4.

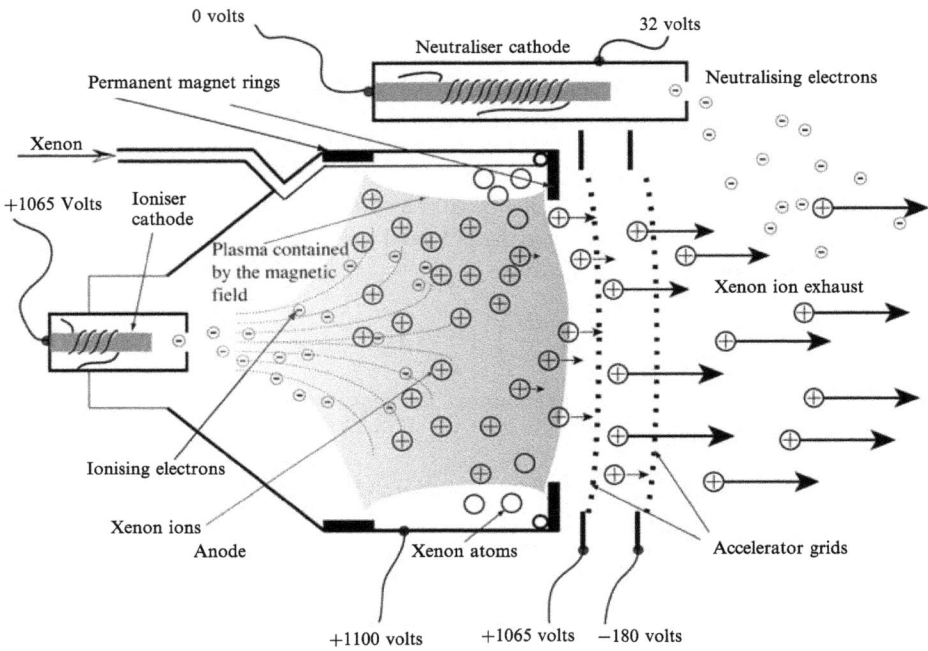

Fig. 6.2   A schematic of the NSTAR gridded ion engine on Deep Space 1. Credit: Used with permission of Springer Nature BV, from Turner (2009); permission conveyed through Copyright Clearance Center, Inc.

electric potential and the size of the engine. This might seem counter-intuitive, and will be discussed further in section 6.2.

An example of a real ion thruster is shown in figure 6.2. This is the NSTAR ion engine, which was used on the NASA Deep Space 1 and Dawn missions (see section 6.3). The process in the engine is as follows. In the left part of the engine, a hot cathode releases electrons by the process of thermionic emission, the same process that is used in low pressure mercury lamps. These electrons are accelerated to a few tens of electronvolts by an electric field into a region where the xenon atoms are pumped in. These xenon atoms are ionized by collisions with the electrons. To increase the probability that any one atom is ionized, an axial magnetic field is used to make the electrons move on long spiral paths. The ions drift to the right end of the ionization chamber. Here they are accelerated by a large positive potential $(+1065 - -180\,\text{V} = 1245\,\text{V})$ held between the two grids through which ions pass (these are the anode and cathode of the earlier description). The ions escape through the cathode with a large velocity relative to the spacecraft, thus providing the thrust.

This positive charge current leaving the spacecraft would cause the spacecraft to become negatively charged over time, which among other things would result in an attractive force between the spacecraft and the departing ions, which would in turn decelerate the spacecraft. To avoid this, electrons are injected into the ion beam after it has passed through the cathode in order to neutralize the spacecraft.

Unlike a chemical rocket, there is no nozzle, because we do not need to extract internal thermal energy from the ions. In fact we are not even able to, because the ion flow is already ordered. The ion engine is, in some sense, an electromagnetic nozzle.

With an accelerating potential of $U = 1000\,\text{V}$, a grid separation $x = 1\,\text{mm}$, and an area $A = \pi 0.15^2 = 0.07\,\text{m}^2$, the above expressions give $\dot{m}_{\text{max}} = 1.5 \times 10^{-5}\,\text{kg s}^{-1}$ and $F_{\text{max}} = 0.56\,\text{N}$. Mass flow rates and thus thrusts are typically low from ion engines, despite their high effective exhaust velocities. The density of the ion gas is also very low (see exercises), resulting in a very low exit pressure. The engines therefore need to be operated in a very low pressure environment, otherwise the surrounding gas would flow in and prevent it from operating. Ion engines are therefore not suitable for launching a spacecraft from the surface of the Earth. But on account of their large effective exhaust velocities, they can be used over long periods to acquire a significant Δv on a low-mass spacecraft.

6.2 Power, potential, and propellant

It may seem odd that the maximum thrust of the gridded ion engine is independent of the ion mass (equation 6.10). After all, if we double the mass of the ion, v_e will decrease by a factor of $\sqrt{2}$ (equation 6.1), but wouldn't $\dot{m}$ double and so the thrust ($\dot{m} v_e$) increase by a factor of $\sqrt{2}$ also? With such apparent contradictions we need to careful about what quantities are being kept constant, and what is forced to change when we vary some parameter. In this case, decreasing v_e by a factor of $\sqrt{2}$ lowers the ion flow rate (number per second) by the same factor, and so even if we double the individual ion masses, $\dot{m}$ – mass times rate – increases by a factor of $\sqrt{2}$. This agrees with equation 6.10 (the other terms are fixed). Thus the thrust is unchanged. Note in particular that we kept U constant.

We can get a bit more insight if we consider the power. The energy of an ion is qU, and the ion flow rate is $\dot{m}/m$, so the power of the engine output is their product:

$$P = \frac{q}{m} U \dot{m} . \tag{6.12}$$

Using the expression for the maximum mass flow rate (equation 6.10) we can write this power at the maximum thrust in terms of the quantities we directly control in the engine:

$$P_{\text{max}} = \frac{4\epsilon_0}{9} A \left(\frac{2q}{m} \right)^{1/2} \frac{U^{5/2}}{x^2} . \tag{6.13}$$

Thus if we double the mass of the ion at constant potential – keeping also A, q, and x constant – the power decreases by a factor of $\sqrt{2}$. If we instead want to keep the power constant while doubling the ion mass, then we have to increase the potential by a factor of $2^{1/5}$. It follows that v_e then decreases by a factor of $2^{2/5}$ (from equation 6.1), $\dot{m}$ increases by a factor of $2^{4/5}$ (from equation 6.12), and so the thrust increases by a factor of $2^{2/5}$.

Causally, the power is not something we fix, but is rather what we get out having fixed the potential and the mass flow rate. This is just like with an electric current (not least because this *is* an electric current): we fix the potential U and current jA, and these determine the power, $P = UjA$. The power we need to provide to the engine is of course closely related to P, but is higher, due to losses.

Another important parameter for ion engines is the thrust-to-power ratio, F/P. For any type of rocket this is $2/v_e$ (equation 3.23), and for an ion engine this equals $\sqrt{2m/qU}$ (equation 6.1). A more powerful engine will be more massive, and the thrust must also accelerate that mass. Chemical rockets tend to have initial propellant-to-spacecraft mass ratios of a few tens, so most of the thrust is used to propel the remaining propellant at first. But with ion engines, the mass flow rates are so small that there is relatively little propellant; the initial propellant-to-spacecraft mass ratio is often much less than one. The total mass is then dominated by the mass of the spacecraft including its engine. This makes it more important to achieve a high thrust-to-power ratio, even though this means a low v_e. In general there will be an optimal value of v_e that minimizes the engine mass.

During a mission, m, q, A, and x cannot be changed. The only control parameter we have is U, and varying this will vary both the effective exhaust velocity and the maximum thrust, because $v_e \propto U^{1/2}$ and $F_{\mathrm{max}} \propto U^2$. There is a limit to how large U can be, because too large a potential gradient (U/x) can lead to the acceleration grid being distorted, as the propulsion force acts on the grid. We can of course operate the engines below maximum thrust by lowering the current. We could also install multiple ion engines in parallel, then adjust the thrust by varying the number of engines in operation, which is effectively equivalent to varying A.

With chemical rockets we sometimes choose a propellant with a low molar mass to raise v_e (following equation 5.10), in order to achieve a larger Δv for a given mass ratio, in accordance with the rocket equation. As mass flow rates tend to be large in chemical rockets, the propellant can be exhausted quickly and the required Δv achieved quickly (of the order of minutes). In ion engines the mass flow rate is typically so low that the Δv takes a long time to achieve (of the order of months). If we lower the ion mass at constant power to increase v_e (following equation 6.1), this lowers the mass flow rate even further, and so it takes even longer to achieve the desired velocity. Depending on the distance to the destination, this may increase the total journey time. Thus depending on the mission profile, maximizing v_e may not be optimal. Indeed, ion engines are not necessarily employed to get somewhere quickly, but rather to provide a continuous, long-term thrust that can provide more flexible trajectories (more on this below).

From the above discussion we see that the choice of propellant, in particular the ion mass, will affect the effective exhaust velocity and thrust, depending on whether we allow the grid potential and/or the power to vary. As we have just seen, we do not necessarily want the highest v_e possible, and so in practice heavier ions such as xenon are often used (mercury and caesium were used in early engines). There are other criteria in the choice of propellant. For example, xenon has a relatively low ionization potential, and so is easier to ionize than the lighter noble gas helium. Xenon is also non-toxic, making it easier to handle (unlike mercury), can be liquefied and stored for long periods (harder with helium), and is less reactive than caesium or mercury, which tended to corrode the grid more rapidly.

6.3   An example application: Dawn

The NSTAR engine shown in figure 6.3 has been used on two NASA interplanetary spacecraft, Deep Space 1 launched in 1998 and Dawn launched in 2007, both of which visited minor bodies in the solar system.

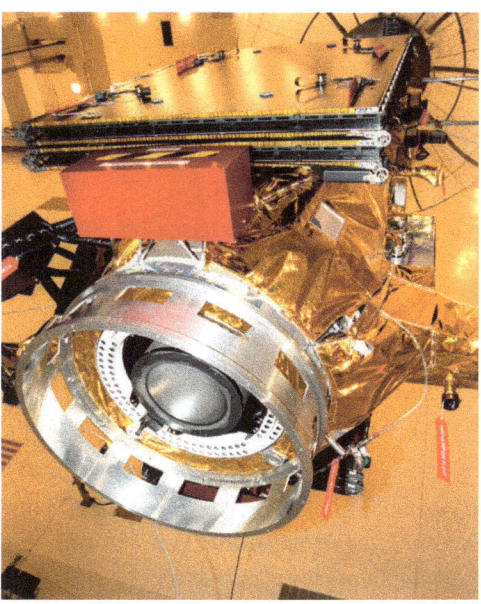

An NSTAR ion engine on the interplanetary mission Deep Space 1. Credit: NASA, cropped by the author.

Fig. 6.3

When used on the Dawn spacecraft, NSTAR delivered an effective exhaust velocity of $30\,\mathrm{km\,s^{-1}}$ and a thrust of up to 90 mN, implying $\dot{m} = 3 \times 10^{-6}\,\mathrm{kg\,s^{-1}}$. Dawn had a dry mass of 747 kg plus 425 kg of xenon propellant that enabled it to achieve a total[2] Δv of $11.5\,\mathrm{km\,s^{-1}}$. This was used to put it into orbit, first around the asteroid Vesta (in 2011) and then around the asteroid Ceres (in 2015), making it the first spacecraft to orbit two different extraterrestrial bodies. The power for the ion engines and the spacecraft as a whole was provided by $36.4\,\mathrm{m^2}$ of photovoltaics (figure 6.4) providing in excess of 10 kW of power at 1 au, but nine times less at 2.99 au (the aphelion of Ceres). The power of the ion beam when operating at the above thrust and effective exhaust velocity is $P = (1/2)\dot{m}v_e^2 = 1350\,\mathrm{W}$. The input power must be larger than this for reasons we will discuss below.

Using the above data, equation 6.1 suggests that the accelerating potential was 610 V. In practice it must have been larger than this, due to imperfect energy transfer to the ions.

With such a low thrust and an initial mass of 1218 kg, the acceleration from the ion engine is extremely small, initially just $0.09/1218 = 7.4 \times 10^{-5}\,\mathrm{m\,s^{-2}}$. Using equation 3.14 and the data above, we can estimate that in 10 days of continuous thrust, the spacecraft changes its velocity by just $64\,\mathrm{m\,s^{-1}}$ during which it expels 2.6 kg of propellant. (As the mass has hardly changed, we can also estimate the change in velocity as $v = at$ in this case.) The engine clearly has to be used for prolonged periods to achieve a large change in velocity. If run continuously at the above rate, the engine could have run for 4.5 years. It actually operated for a total of 5.9 years, indicating that it was sometimes used at lower values of $\dot{m}$. These long durations mean ion engines cannot be approximated as giving impulsive thrusts, as we assumed when computing orbit changes with the short-burst, high-thrust chemical engines in chapter 4. Thus, assuming that an ion engine is not always thrusting radially away from the

---

[2] The rocket equation implies $\Delta v = 30 \times \ln[(747+425)/747] = 13.5\,\mathrm{km\,s^{-1}}$. The discrepancy comes in part from the fact that Dawn also had hydrazine thrusters and fuel.

An artist's impression of the Dawn spacecraft. Credit: NASA/JPL.

Sun, it provides a continuous non-radial force, so the spacecraft will not move on a Keplerian orbit. This opens up interesting opportunities for unusual trajectories, something we shall return to when we discuss another means of achieving this with solar sails in chapter 10.

Dawn had three NSTAR thrusters, but only used one at a time. An advantage of ion engines is that not only can the mass flow rate be varied, but also the effective exhaust velocity, by varying the potential between the anode and cathode. In the case of Dawn, the effective exhaust velocity could be varied between 7 and 30 km s^{-1} and the thrust between 19 and 90 mN.

The power we calculated above is the output power of the ion beam. The power that must be provided to achieve this is higher because we first need to ionize the atoms. If the energy required to ionize each atom is E_i, then the power required for ionization at the output mass flow rate is $P_i = E_i \dot{m}/m$. The ionization energy of xenon is 12.1 eV, but significantly more energy is required on average. One reason is that electron collisions may not ionize the atom, but instead excite the electron, which then radiates the energy as photons as it decays back to its ground state. In practice of the order of $E_i = 500$ eV is required per ion. Using the above data for Dawn's nominal mass flow rate, and with $m = 131\,m_u$, this implies $P_i = 1100$ W. This is comparable to the power of the ion beam itself calculated earlier. It would be inefficient if we spent most of the available power to ionize the atoms but then comparatively little to accelerate them. The ratio of the ion beam power (equation 6.12) to P_i is qU/E_i. So for high energy efficiency, we want the acceleration potential to be much higher than the potential needed for the ionization.

Additional energy losses include electrons that fail to ionize any atoms and simply pass straight to the anode in the ionization chamber, as well the energy that must be expended to accelerate the neutralizing electrons at the back of the engine.

If we define an energy efficiency as the ratio of the rocket's kinetic energy to the energy in the ion beam, then we arrive at the same expression for efficiency as that for a generic rocket given in equation 3.21 (see exercises). This is arguably not a very useful expression in the present application, however, because it ignores the energy needed to ionize the atoms. Note, though, that thermal engines are limited in their efficiency to convert heat into work by the second law of thermodynamics. Ion engines, in contrast, are not heat engines, and so by making qU much greater than E_i we can in principle achieve high efficiencies. Whether that is important depends on the main purpose of the ion engine.

6.4 Other types of ion engine

There are many variants on gridded ion engine. They vary in how they use electric and/or magnetic fields to ionize and accelerate the plasma, and have a range of performances in terms of effective exhaust velocity and thrust. Here we look very briefly at a few alternatives; many more variants exist.

Field emission electric propulsion is a variant that uses an electric field to extract ions from the surface of a metal. The metal is usually in liquid form; indium or gallium is often used. Once extracted from the metal, the ions are accelerated by an electric field to velocities of order $100 \, \text{km s}^{-1}$. The thrusts are very low, however, typically in the region of micronewtons or millinewtons, so they are mostly used for attitude control and orbit adjustment.

One of the disadvantages of the gridded ion engine is that the cathode tends to erode over time. A Hall effect thruster overcomes this by using a non-physical cathode. Electrons fired out of the back of the engine are attracted into the chamber towards the anode by an axial electric field. A radial magnetic field puts these electrons on circular paths around the axis. Atoms released near the anode are ionized by these electrons through collisions. As the ions are much more massive than the electrons, they are deflected little by the magnetic field. The axial electric field then accelerates the positive ions to produce thrust. The electrons act both as a virtual cathode and serve to ionize the atoms, with ion generation and acceleration occurring in the same region of the chamber. Some electrons fired from the back of the engine neutralize the exiting ion beam and thus keep the engine neutral overall. Effective exhaust velocities of around $15 \, \text{km s}^{-1}$ have been achieved with Hall effect thrusters, and like gridded ion engines they produce very small thrusts (typically less than a newton). The Psyche spacecraft, for example, has four Hall effect thrusters operating with xenon, each with $v_e = 18 \, \text{km s}^{-1}$ and $F = 0.28 \, \text{N}$, which corresponds to an ion beam power of $2.5 \, \text{kW}$, although they can operate at up to $4.5 \, \text{kW}$. Hall effect thrusters are more compact than gridded thrusters, and tend to be the choice of ion engine in newer spacecraft.

Some ion engines use a magnetic field to accelerate the ions via the Lorentz force, rather than an electric field to accelerate them via the Coulomb force. One particular implementation of this concept is the magnetoplasmadynamic (MPD) thruster. This has a cylindrical anode surrounding a central cathode. A large current between them ionizes the gas. The current interacts with a strong magnetic field to accelerate the plasma out of the cylinder via the Lorentz force. This may be able to achieve an effective exhaust velocity as high as $100 \, \text{km s}^{-1}$ when using hydrogen propellant.

Another proposed variant is the Variable Specific Impulse Magnetoplasma rocket (VASIMR). This uses radio waves to ionize and heat a plasma, which is then accelerated by a magnetic field. A solenoid directs the plasma, behaving as the magnetic equivalent of the convergent–divergent nozzle (see section 8.4 for more details). Effective exhaust velocities of $50 \, \text{km s}^{-1}$ and powers of $100 \, \text{kW}$ have been achieved in laboratory tests which, if achieved together, imply a relatively large thrust of $4 \, \text{N}$. One advantage over gridded ion engines is that VASIMR has no electrode grids, so does not suffer from problems of their corrosion. Indeed, the plasma is confined electromagnetically, so could be heated to extremely high temperatures of millions of kelvins.

In none of these descriptions have we been concerned with the source of the energy (electricity) to accelerate the ions. For interplanetary spacecraft such as Dawn and Psyche, the source has often been solar power collected by photovoltaic cells. To achieve a higher intensity radiation than the Sun, and thus a larger output power for a given collecting area, one could consider using laser energy beamed from the Earth. This could involve similar technology and infrastructure on the Earth as that required for propelling a laser sail, which we look at in chapter 11. The main difference is that whereas for the laser sail we use the energy and momentum of the beamed photons, for the ion engine we would only use the energy.

Another source of energy is a nuclear reactor. A heat engine can be used to convert heat into mechanical motion and then into electricity, as done in a conventional power station. This could potentially provide a large current that could give larger thrusts than have been considered so far for most ion engines. This approach to accelerating ions is often referred to as nuclear electric propulsion.

6.5 Summary

A gridded ion engine works by ionizing a propellant and then accelerating it using an electric field. Variants on this use a magnetic field or a combination of electric and magnetic fields to ionize and accelerate the propellant. The propellant depends on the type of engine, but is often a heavy noble gas such as xenon. In contrast to chemical thermal rockets in which the fuel (and oxidizer) is the source of both the energy and the propellant, in an ion engine a separate energy source is used to generate electromagnetic fields that then accelerate an inert propellant. The energy source is often solar electric power, but could be anything in principle.

Ion engines typically provide high exhaust velocities (10–$100\,\mathrm{km\,s^{-1}}$) but low thrusts (often below $1\,\mathrm{N}$) on account of their low mass flow rates. The acceleration is therefore low but can be maintained for a long time. The mass ratios (spacecraft plus propellant to spacecraft) used to date are typically low, so the total Δv achieved is often less than the effective exhaust velocity.

Ion engines are not, at present, an option for achieving fast interstellar travel. Even if a very large mass ratio (for an ion engine) of 100 could be achieved with $v_e = 100\,\mathrm{km\,s^{-1}}$, this would still only provide $\Delta v = 460\,\mathrm{km\,s^{-1}}$. But because they can be operated continuously for long periods of time (years), ion engines are suited for long-duration deep space missions where multiple manoeuvres or non-Keplerian orbits are required.

6.6 Exercises

1. Derive an expression for the number density of the accelerated ions at the exit of an ion engine, and compute this for an engine with $U = 1000\,\mathrm{V}$ and $x = 1\,\mathrm{mm}$ using single ionized molecules. Compare this to the number density of air molecules at the Earth's surface. What does this imply?
2. Derive an expression for the ratio of an ion engine's final kinetic energy to the total energy put into the ion beam. Is this a useful measure of efficiency?

Relativistic motion

The special theory of relativity rests on just two axioms: the identity of the laws of physics in all inertial reference frames, and the invariance of the speed of light. After briefly reviewing the basics of special relativity, including the Lorentz transformations and time dilation, we look at the effects of relativity during an accelerated journey. Time dilation has implications even for a spacecraft moving at low velocities. We examine the relativistic rocket equation and then rockets that are propelled by photons or a combination of matter and photons. General relativistic effects are not considered. In a later chapter we look at the impact of relativistic motion on optical phenomena.

7.1 Basic relations

Throughout this and later chapters I adopt the standard notion of $\beta = v/c$ to indicate the velocity of an object moving at velocity v as a fraction of the speed of light c, and

$$\gamma = \frac{1}{\sqrt{1 - \beta^2}} \quad \text{where} \quad 0 \leq |\beta| < 1, \tag{7.1}$$

which maps β to γ as shown in figure 7.1. At low velocities, then using a Taylor expansion and neglecting terms of the order of β^3 and higher, we see that

$$\gamma \simeq 1 + \frac{\beta^2}{2} \quad \text{when} \quad |\beta| \ll 1. \tag{7.2}$$

From equation 7.1 we can derive the following useful relations:

$$\beta = \sqrt{1 - \frac{1}{\gamma^2}}, \tag{7.3}$$

$$\gamma\beta = \sqrt{\gamma^2 - 1}, \tag{7.4}$$

$$\gamma(1 - \beta) = \sqrt{\frac{1 - \beta}{1 + \beta}}, \tag{7.5}$$

$$\gamma(1 + \beta) = \sqrt{\frac{1 + \beta}{1 - \beta}}. \tag{7.6}$$

A body of mass m moving at a velocity v has a total energy relative to the observer of

$$E = \gamma mc^2 = \gamma E_0, \tag{7.7}$$

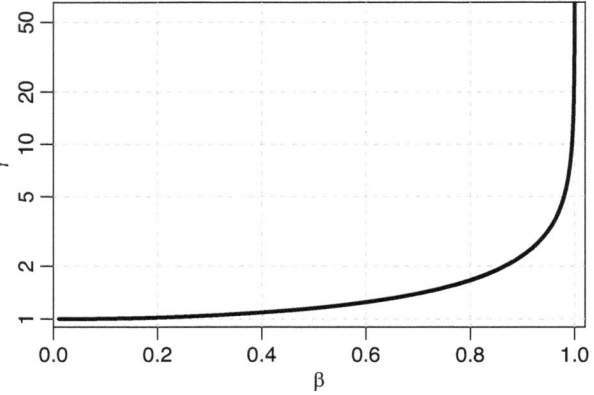

Fig. 7.1 The relationship between $\gamma = (1 - \beta^2)^{-1/2}$ and $\beta = v/c$.

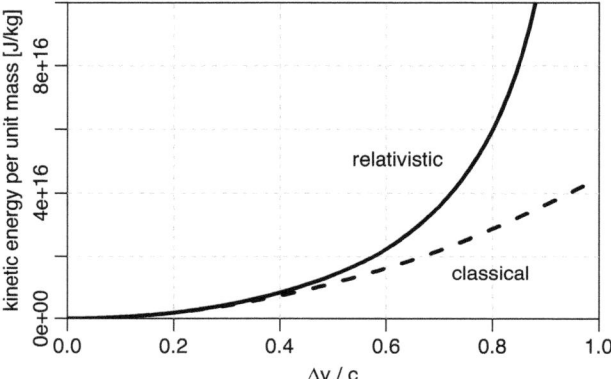

Fig. 7.2 The kinetic energy per unit mass of a relativistic rocket (solid line), $(\gamma - 1)c^2$, and the classical Newtonian case (dashed line), $(1/2)v^2$.

where $E_0 = mc^2$ is the rest energy. The mass m is invariant. The momentum of the body is

$$p = \gamma m v. \tag{7.8}$$

Combining the previous two equations and using equation 7.4 we can write

$$E^2 = p^2 c^2 + m^2 c^4. \tag{7.9}$$

The kinetic energy per unit mass of the body is $(\gamma - 1)c^2$. This is plotted in figure 7.2 in SI units to show the large amounts of energy involved at relativistic speed. For comparison, a 1 GW power station running continuously for one year produces 3×10^{16} J of energy, which is the energy of a 1 kg mass moving at $0.66\,c$, or a 1000 kg mass moving at $0.026\,c$.

7.2 Lorentz transformations

We will need to transform time, position, velocity, and acceleration between inertial reference frames. Consider motion in two reference frames, S and S', whereby S' is moving with a

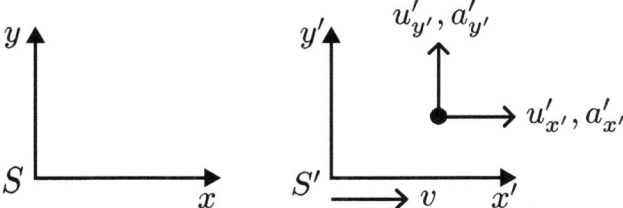

Reference frames for Lorentz transformations. The reference frame S' moves with a velocity v along the positive x-axis relative to the reference frame S.

Fig. 7.3

velocity v along the x-axis relative to S, as shown in figure 7.3. The x'-axis is parallel to the x-axis. The time in S is t and the time in S' is t'.

The frames coincide at $t = t' = 0$. Using the axioms of special relativity one can derive the following Lorentz transformations between the coordinate systems:

$$x = \gamma(x' + vt'),$$ (7.10a)
$$y = y',$$ (7.10b)
$$t = \gamma\left(t' + \frac{vx'}{c^2}\right).$$ (7.10c)

The transformations for the z-axis, being perpendicular to the x-axis, behave in the same way as y. The same is true for its velocity and acceleration transformations (below).

We can rearrange equations 7.10 to get the primed quantities in terms of the unprimed ones. As S is moving with a velocity $-v$ relative to S', those rearranged equations can be attained more simply by taking equations 7.10 and swapping prime and non-prime symbols and changing the sign of v.

Consider two events taking place at the same position $x'_1 = x'_2$ but at different times t'_1 and t'_2 in S'. What is the time between these events in S? Let their corresponding times be t_1 and t_2 in S. We can relate these to the times in S' using equation 7.10c. Defining $\Delta t = t_2 - t_1$ and $\Delta t' = t'_2 - t'_1$, this gives

$$\Delta t = \gamma \Delta t'.$$ (7.11)

Thus any time interval in S' will appear longer in S. If we have two identical clocks at rest in the two frames, then as seen from S, the clock in S' takes longer for its hands to move. In other words, the moving clock appears to tick more slowly. This is known as *time dilation*. Time measured within the rest frame is called the *proper time*.

Note that $\Delta x = v\Delta t$ from equation 7.10a. The two events are not recorded at the same position in S. Time dilation is not caused by light travel time effects.

One consequence of time dilation is that if a spacecraft covers some distance Δx in the Earth's rest frame S at an arbitrarily large velocity (below c, of course), then it can cover this in an arbitrarily short time as measured in its frame S'. But this time will not be short in the Earth's frame: that will always be at least $\Delta x/c$.

Consider now a rod of length $\Delta x'$ as measured in S' lying parallel to the x'-axis (and therefore also the x-axis). What is the length of the rod in S? To determine this we measure the positions of the two ends of the rod at the same time $t_1 = t_2$ in S. The inversion of equation 7.10a is $x' = \gamma(x - vt)$. With $\Delta x' = x'_2 - x'_1$ and $\Delta x = x_2 - x_1$, then taking the difference of this equation for the two events gives

$$\Delta x = \frac{1}{\gamma} \Delta x'. \tag{7.12}$$

Thus an object in S' will be measured to be shorter in S in its direction of travel. This is known as *length contraction*. Distance measured within the rest frame is called the *proper distance*. Note that lengths perpendicular to the direction of travel are unaffected because $y = y'$.

We see there is an asymmetry in the factor γ in the time dilation and length contraction equations. This arises because for the former we consider two events at the same position in S', whereas for the latter we consider two events at the same time in S. In both cases the effects are independent of the sign of the velocity: it doesn't matter whether the source is moving towards or away from the observer.

We could equally well consider what an observer in frame S' would measure for a clock and ruler at rest in frame S. The result is identical in the sense that they too would see a clock and ruler moving at a speed v, and so they too would measure a time dilation and length contraction. Naturally, when we consider the situation from S' instead of S, we have to change also that the events are then at the same position in S (time dilation) and same time in S' (length contraction).

Consider now an object moving with a velocity $(u'_{x'}, u'_{y'})$ in S' (figure 7.3). To find the velocities that an observer at rest in S observes, we differentiate the Lorentz transformations to get

$$u_x = \frac{dx}{dt} = \frac{u'_{x'} + v}{1 + \frac{vu'_{x'}}{c^2}}, \tag{7.13a}$$

$$u_y = \frac{dy}{dt} = \frac{u'_{y'}}{\gamma \left(1 + \frac{vu'_{x'}}{c^2} \right)}, \tag{7.13b}$$

where $u'_{x'} = dx'/dt'$ and $u'_{y'} = dy'/dt'$.

Differentiating again we get the accelerations

$$a_x = \frac{du_x}{dt} = \frac{a'_{x'}}{\gamma^3 \left(1 + \frac{vu'_{x'}}{c^2} \right)^3}, \tag{7.14a}$$

$$a_y = \frac{du_y}{dt} = \frac{1}{\gamma^2 \left(1 + \frac{vu'_{x'}}{c^2} \right)^3} \left[\left(1 + \frac{vu'_{x'}}{c^2} \right) a'_{y'} - \frac{vu'_{y'}}{c^2} a'_{x'} \right], \tag{7.14b}$$

where $a'_{x'} = du'_{x'}/dt'$ and $a'_{y'} = du'_{y'}/dt'$. We see that acceleration is not absolute, but depends on the observer. Furthermore, a constant acceleration in S' does not imply a constant acceleration in S, as $u'_{x'}$ may be varying as a result of the acceleration. Note also that the acceleration a_x parallel to the relative direction of travel of the two observers (in S' and S) depends only on the velocities and acceleration in that parallel direction. In contrast, the perpendicular acceleration depends on the velocities and accelerations in both the parallel and perpendicular directions.

Below we will use one particular case of this acceleration equation, namely when we consider the acceleration of the frame S' itself along the x-axis. We get this when $u'_{x'} = 0$, which gives

$$a_x = \frac{1}{\gamma^3} a'_{x'}. \tag{7.15}$$

When an acceleration $a'_{x'}$ is measured within frame S', the acceleration measured by an observer moving at a velocity v relative to it and in the same direction as the acceleration is smaller by a factor of γ^3.

7.3 Constantly accelerating spacecraft

Consider a spacecraft initially at $x = x_0$ at time $t = t_0$ moving at speed $v = v_0$ in reference frame S. The spacecraft then starts to accelerate in its direction of travel. The acceleration has a constant value a' in its own instantaneous rest frame. Even though the spacecraft is accelerating, and so no longer provides an inertial reference frame over a non-zero time duration, at any instant we can define a frame that is momentarily at rest relative to the spacecraft. We want to find the motion of the spacecraft – acceleration, velocity, position – in S as a function of time in S. It is important to remember that in this section, because the velocity of the spacecraft in S is not constant, γ is not constant either.

The acceleration in S is given by equation 7.15. Because γ increases with time, this acceleration will decrease with time.

The velocity in S is obtained by integrating equation 7.15 over t to give

$$v = \frac{a't + \gamma_0 v_0}{\sqrt{1 + (a't + \gamma_0 v_0)^2/c^2}} \quad \text{where} \quad \gamma_0 = \left(1 - \frac{v_0^2}{c^2}\right)^{-1/2}, \tag{7.16}$$

where I have dropped the x subscript because the velocity and acceleration are parallel. Integrating again over t we get the position of the spacecraft in S:

$$x = \frac{c^2}{a'} \left[\sqrt{1 + \left(\frac{a't + \gamma_0 v_0}{c}\right)^2} - \gamma_0^2 \right] + x_0. \tag{7.17}$$

It will also be useful to express v and x in terms of the spacecraft time t'. The relation between the two times is given by the equation for time dilation, which in differential form is

$$dt = \gamma dt'. \tag{7.18}$$

If we substitute for v (via γ) from equation 7.16, and then integrate, we get the relationship between the two times:

$$t = \frac{c}{a'} \left[\sinh\left(\frac{a't'}{c} + \text{arcsinh}\left[\frac{\gamma_0 v_0}{c}\right]\right) - \frac{\gamma_0 v_0}{c} \right]. \tag{7.19}$$

When the spacecraft starts at $x_0 = 0$ and $v_0 = 0$, these expressions are simplified; they are summarized in Table 7.1 in terms of both t and t'. Hyperbolic tangent functions occur frequently in special relativity so they are plotted in figure 7.4.

The expressions are plotted in figure 7.5 for two accelerations: $1\,g$ (solid black line) and $0.5\,g$ (dashed black line), where g is the standard acceleration due to gravity at the Earth's surface. The top two plots show the relation between the times in the two reference frames. This difference is not a simple time dilation, because one frame has accelerated with respect to the other, and so the speed has varied. After 10 years in the initial rest frame, the clock on

Table 7.1 Expressions for the time t, position x, velocity v, and acceleration a of a spacecraft as observed from inertial frame S, in terms of the time t in S in the left column, and in terms of the time t' in the spacecraft's instantaneous rest frame in the right column. The spacecraft starts at $t = t' = 0$ at $x = 0$ and $v = 0$ and undergoes constant acceleration a' in its instantaneous rest frame.

	t		t'
t	$= t$	$=$	$\dfrac{c}{a'} \sinh\left(\dfrac{a't'}{c}\right)$
x	$= \dfrac{c^2}{a'}\left[\sqrt{1+\left(\dfrac{a't}{c}\right)^2} - 1\right]$	$=$	$\dfrac{c^2}{a'}\left[\cosh\left(\dfrac{a't'}{c}\right) - 1\right]$
v	$= a't\left[1+\left(\dfrac{a't}{c}\right)^2\right]^{-1/2}$	$=$	$c \tanh\left(\dfrac{a't'}{c}\right)$
a	$= a'\left[1+\left(\dfrac{a't}{c}\right)^2\right]^{-3/2}$	$=$	$a' \operatorname{sech}^3\left(\dfrac{a't'}{c}\right)$

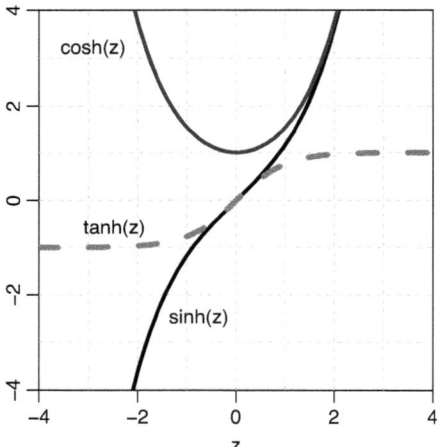

Fig. 7.4 The three main hyperbolic functions ($\operatorname{sech}(z) = 1/\cosh(z)$).

the spacecraft accelerating at 1 g has advanced by only 2.9 years, whereas after 10 years of spacecraft time, nearly 15 000 years have passed in the initial rest frame (note the log scale in the right panel).

The second row shows the distance travelled by the spacecraft. After 10 years in the initial rest frame, the spacecraft accelerating at 1 g has travelled 9.08 ly. At half the acceleration it has travelled nearly as far, 8.25 ly. The Newtonian relation is shown by the orange lines in figure 7.5, and yields distances of 51.6 ly and 25.8 ly, respectively, which are in the same ratio as the two accelerations (because $x = (1/2)at^2$), a factor of two. In special relativity that is not the case because, as shown in the third row, the spacecraft is travelling close to the speed of light after just a few years as measured in either reference frame. The relationship

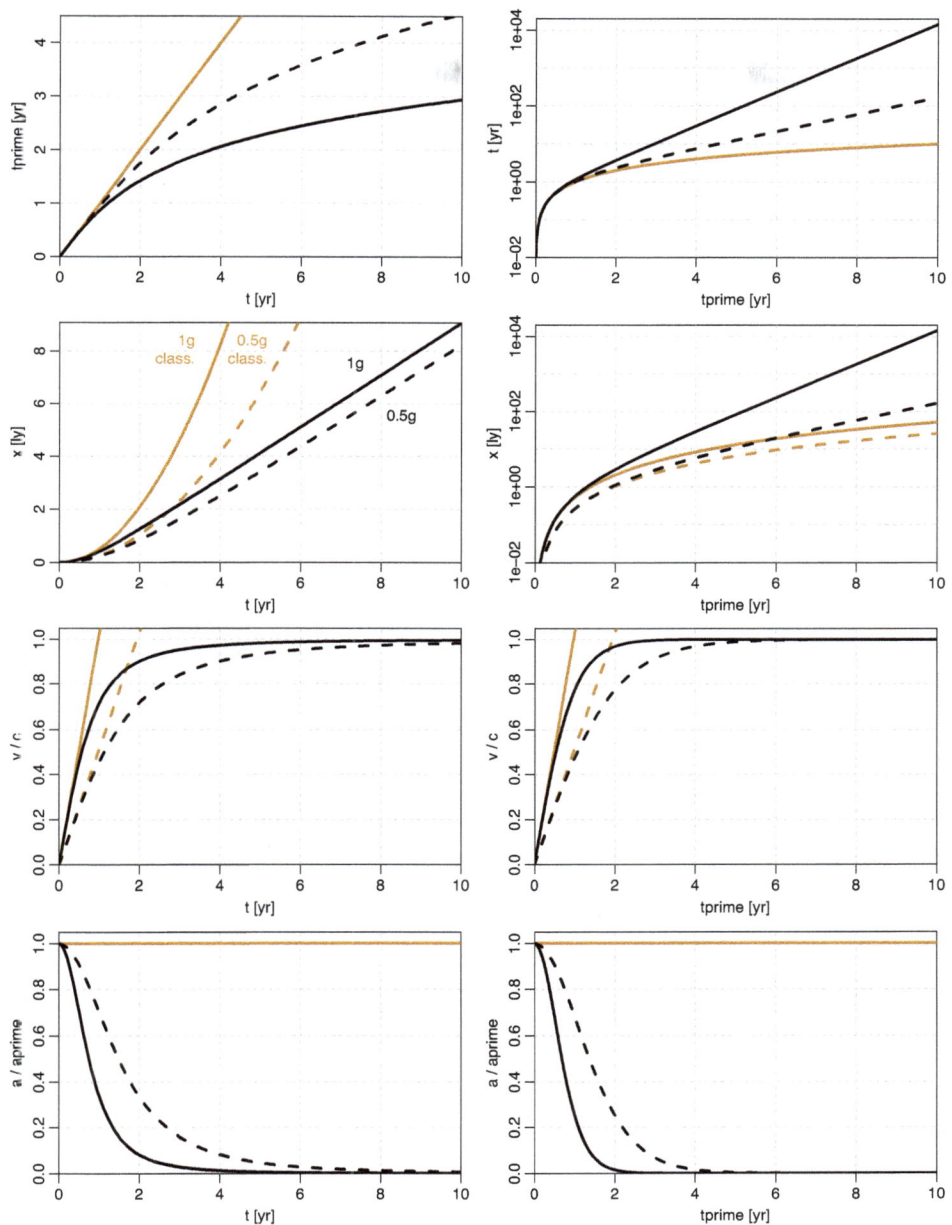

The time t, position x, velocity v, and acceleration a of a spacecraft observed from inertial frame S in terms of the time t in that frame t (left column) and the time t' in the spacecraft's instantaneous rest frame (right column). The spacecraft starts at $t = t' = 0$ at $x = 0$ and $v = 0$ and undergoes constant acceleration a' in its instantaneous rest frame. The solid black line is for an acceleration of 1 g, the dashed black line is for an acceleration of 0.5 g. The orange lines are the corresponding results for classical Newtonian mechanics. Other than the top-left panel, which is just an inversion of the top-right panel, the panels correspond to the equations in Table 7.1.

Fig. 7.5

between distance and acceleration then becomes very nonlinear (bottom row). In particular, the closer the spacecraft gets to the speed of light, the slower its velocity increases.

For 1 g acceleration, the Newtonian model predicts velocities larger than the speed of light after about a year (it so happens that $c \simeq g \times \mathrm{yr}$). It therefore predicts far larger distances travelled at a given initial rest frame time than the relativistic calculation. But in terms of spacecraft time, the Newtonian model predicts shorter distances because it assumes that time is universal.

Returning to the relativistic calculations, we saw that because of the extreme difference between the timescales in the two reference frames after several years, the spacecraft travels far further in 10 years of spacecraft time than in 10 years of initial rest frame time: the right plot in the second row shows that at 1 g, it travels nearly 15 000 ly. Yet at 0.5 g it travels only 167 ly within the same time. In both cases the spacecraft has been travelling close to the speed of light (in the initial rest frame) for most of its travel time. But for the higher acceleration spacecraft, those 10 years of spacecraft time correspond to a far longer period of time in the initial rest frame (15 000 years vs 169 years), and so the former travels much further.

Once the spacecraft has reached nearly the speed of light, it moves at nearly 1 ly per year, where both time and distance are measured in the initial rest frame. This is the *coordinate velocity*, which we have been denoting with v. But for each year of spacecraft time, the spacecraft will cover far more than 1 ly in the initial rest frame. This quantity, $w = \Delta x/\Delta t'$, where Δx is measured in S and $\Delta t'$ is measured in S', is called the *proper velocity*. More formally,

$$w = \frac{dx}{dt'} = \gamma v, \tag{7.20}$$

the relation to the coordinate velocity following from the equation for time dilation (equation 7.11).[1] The proper velocity is numerically larger than the speed of light once $\beta > 1/\sqrt{2}$. Nothing physically moves at this velocity in any inertial frame, however, so it is not in conflict with special relativity. It just tells us how far the spacecraft moves per unit of spacecraft time.

The dimensionless version of the proper velocity is $\omega = w/c = \gamma\beta$. Other useful conversions are

$$\beta = \frac{\omega}{\sqrt{1 + \omega^2}} \quad \text{and} \tag{7.21a}$$

$$\gamma^2 = 1 + \omega^2 . \tag{7.21b}$$

When $|\beta| \ll 1$, $\omega \simeq \beta$. When $\beta \simeq 1$, $\omega \simeq \gamma$.

7.4 Force and Newton's second law

In special relativity, the momentum of a body of mass m moving at velocity v in frame S is $p = \gamma m v$. If force F, as measured in S, acts on the body, its rate of change of momentum is given by Newton's second law:

$$F = \frac{dp}{dt} = \frac{d(\gamma m v)}{dt} = \frac{d(mw)}{dt} = \frac{1}{\gamma}\frac{d(mw)}{dt'}, \tag{7.22}$$

---

[1] Unfortunately, the attribution of the descriptor 'proper' in special relativity is not really consistent.

where in the penultimate expression I have used the proper velocity from equation 7.20, and in the last one the proper time, $dt = \gamma\, dt'$. If the mass is constant, the above expression becomes

$$F = m\frac{d(\gamma v)}{dt} = m\frac{1}{\gamma}\frac{dw}{dt'}\,. \tag{7.23}$$

With $a = dv/dt$, we find by differentiation that

$$\frac{dw}{dt} = \frac{d(\gamma v)}{dt} = \gamma^3 a = a'\,, \tag{7.24}$$

with the last step given by the Lorentz transformation from equation 7.15. Thus we can write

$$F = \gamma^3 ma = ma'\,. \tag{7.25}$$

The acceleration a' is called the *proper acceleration*. It equals the derivate of the proper velocity with respect to the coordinate time. Newton's second law for constant mass looks like its classical counterpart when we write it in terms of the proper acceleration. In terms of the proper time, it follows from equation 7.24 that

$$a' = \frac{1}{\gamma}\frac{dw}{dt'}\,. \tag{7.26}$$

We saw at the end of section 7.2 that the proper acceleration is the acceleration experienced within the instantaneous rest frame. This is the physical acceleration that an astronaut feels in a rocket when it is firing its engines, for example. This is conceptually different from the *coordinate acceleration a*, which is the rate of change of velocity of an object as measured by an observer in their rest frame on their clock.

7.5 The relativistic rocket equation

Once a rocket starts to move at relativistic speeds, the derivation of the classical rocket equation using Newtonian mechanics in section 3.1 is no longer valid. An equivalent derivation in special relativity can be performed (I'll outline one in section 7.8). This again gives the change in velocity of a rocket of initial mass M_i after it has ejected $M_i - M_f$ of propellant at velocity v_e relative to the rocket, but the expression is now

$$\Delta v = c\tanh\left[\frac{v_e}{c}\ln\left(\frac{M_i}{M_f}\right)\right]\,. \tag{7.27}$$

As the tanh function is limited to the range ± 1 (figure 7.4), and we are only dealing with positive quantities, Δv lies in the range 0 to c. This function is shown in figure 7.6. The inverse of this is

$$\frac{M_i}{M_f} = \exp\left[\frac{c}{v_e}\operatorname{atanh}\left(\frac{\Delta v}{c}\right)\right]\,. \tag{7.28}$$

With $v_e = 0.01\,c$, achieving a velocity of $0.2\,c$ requires a mass ratio of 640 million. When v_e is increased to $0.05\,c$, the required mass ratio drops to 58, and when $v_e = 0.1\,c$, it is just 7.6. To achieve relativistic speeds clearly requires a large v_e.

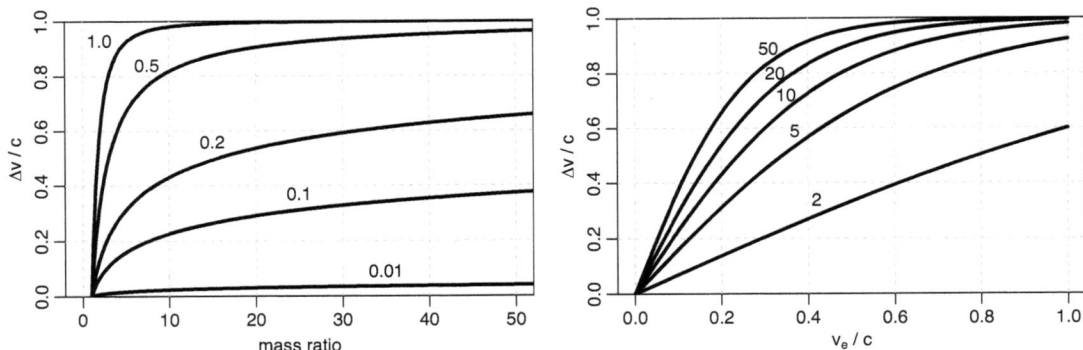

Fig. 7.6 The relativistic rocket equation (equation 7.27) as function of mass ratio for several values of v_e/c (left), and as function of v_e/c for several mass ratios (right).

The tanh function can be written as a Taylor expansion, $\tanh z = z - z^3/3 + \mathcal{O}(z^5)$. When z is small, $\tanh z \simeq z$. Thus for $v_e \ll c$ and/or small mass ratios, the relativistic rocket equation becomes the classical rocket equation. The ratio of Δv from the relativistic to classical rocket equations is $\tanh(z)/z$ with $z = (v_e/c) \ln(M_i/M_f)$. Only once $z > 0.4$ do the two velocities differ by more than 5%.

With the classical rocket equation, we found that in order to accelerate from zero to Δv and then decelerate back to zero again required the square of the mass ratio needed for the acceleration part (see section 3.2). From equation 7.27 we see that this is still the case with the relativistic rocket equation, because

$$\frac{\Delta v}{c} = \tanh\left[\frac{v_e}{c} \ln\left(\frac{M_i}{M_m}\right)\right] = \tanh\left[\frac{v_e}{c} \ln\left(\frac{M_m}{M_f}\right)\right], \tag{7.29}$$

where M_m is the mass after the first boost, from which it again follows that $M_i/M_f = (M_m/M_f)^2$. However, unlike with the classical rocket equation, squaring the mass ratio does not generally double Δv. We will see an example of this in the next section.

Given that tanh is defined in terms of exponentials, the relativistic rocket equation can also be written as

$$\Delta v = c \frac{Q^2 - 1}{Q^2 + 1}, \quad \text{where} \quad Q = \left(\frac{M_i}{M_f}\right)^{(v_e/c)}, \quad \text{or} \tag{7.30a}$$

$$\frac{M_i}{M_f} = \left(\frac{1 + \Delta v/c}{1 - \Delta v/c}\right)^{(c/2v_e)}. \tag{7.30b}$$

We tend to think of the rocket equation as describing a situation in which a mass $M_i - M_f$ is ejected at velocity v_e. However, once nuclear reactions are involved that convert a non-negligible fraction of the initial mass into energy, not all of that mass is ejected at the speed of the propellant, so we need to adopt a different *effective* exhaust velocity to accommodate this. We will address this in section 7.8.

7.6 A trip to Alpha Centauri

Let us consider a trip from the Earth to Alpha Centauri, a distance we take to be 4.4 ly in this section. For simplicity, we will assume that the origin and target are stationary relative to one another, and refer to this as the Earth's rest frame.

Consider first a flyby mission. Starting at rest, a spacecraft accelerates continuously at 1 g in its instantaneous rest frame until it reaches Alpha Cen. Using the expressions in Table 7.1, we can compute that the spacecraft time required for this mission is 2.32 years, the Earth time is 5.28 years, and the velocity at encounter is 0.984 c.

We can compare this flyby to a rendezvous mission, in which the spacecraft stops accelerating at the halfway point (2.2 ly) and then decelerates at 1 g in its instantaneous rest frame, arriving with zero velocity in the Earth's rest frame. The spacetime diagram for this mission is shown as the solid line in figure 7.7 (ignore the return journey for now). We can compute that the Earth time to the halfway point is 3.02 years (spacecraft time 1.80 years), when the velocity is 0.952 c. The second half of the outward journey is identical to the first half, just played out backwards, so the total journey time to Alpha Cen is double these values. The top three rows of Table 7.2 summarize these results.

For a rocket with fixed effective exhaust velocity, we can use the relativistic rocket equation (section 7.5) to determine the mass ratios required for the two types of journey. Suppose $v_e = 0.2 c$. The rocket's mass decreases with time as it exhausts propellant, so in order to maintain constant acceleration in its instantaneous rest frame, the mass flow rate will have to

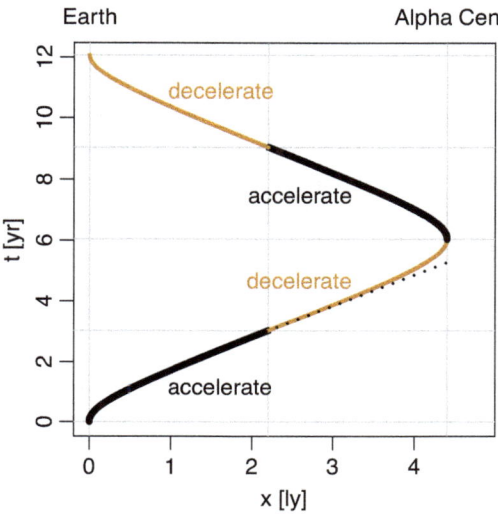

The relativistic distance–time diagram for a round trip mission covering the 4.4 ly from the Earth to Alpha Centauri and back. The solid line shows the profile for a spacecraft that accelerates continuously at 1g in its instantaneous rest frame, changing the direction of the acceleration (i.e. to decelerate) at the halfway point: the thick parts of the line are acceleration and the thin deceleration. The return journey is the mirror image of the outward trip. The dotted line shows how the spacecraft would continue its outward journey if it continued to accelerate rather than decelerate at the halfway point, that is, a flyby rather than a rendezvous at Alpha Centauri.

Fig. 7.7

Table 7.2 Parameters of various relativistic trips for a spacecraft (S/C) with constant acceleration/deceleration. The mass ratios in the final column are for $v_e = 0.2\,c$.

Trip profile	Distance travelled [ly]	Duration Earth t [yr]	Duration S/C t' [yr]	Total $\Delta v/c$	Mass ratio M_i/M_f
1 g accel (flyby)	4.4	5.28	2.32	0.984	160 000
1 g accel (to halfway point)	2.2	3.02	1.80	0.952	10 600
1 g accel + decel (rendezvous)	4.4	6.03	3.59	0.952 + 0.952	110 × 10⁶
1 g accel	18.8	19.73	3.59	0.9988	110 × 10⁶
0.1 g accel (flyby)	4.4	10.23	8.92	0.726	100
0.1 g accel (to halfway point)	2.2	6.89	6.41	0.579	27
0.1 g accel + decel (rendezvous)	4.4	13.78	12.83	0.579 + 0.579	750
0.1 g accel	9.8	16.91	12.83	0.868	750

decrease with time too (equation 3.10). Note, though, that the (relativistic) rocket equation makes no assumption about the mass flow rate.

For the flyby mission we can simply plug the Δv value into equation 7.28 to find a mass ratio of about 160 000. This is a huge value, unachievable in practice. This shows how infeasible it is to get a rocket close to the speed of light, even if v_e is high (higher than we could achieve even with fusion engines, as we shall see in chapter 8).

For the rendezvous mission, the mass ratio required to accelerate at $1\,g$ to the halfway point is, using the required $\Delta v = 0.952\,c$, 10 600. As shown in section 7.5, the mass ratio required for the entire mission is the square of this, about 110 million. This is 700 times larger than the mass ratio for the flyby mission.

Why does the rendezvous mission require so much more propellant than the flyby mission? It is not because of decelerating rather than accelerating: the spacecraft does not 'know' in which direction it is firing its rocket; nor can this matter. It is also not because of the difference in maximum velocity: the flyby mission, which uses less propellant, attains a higher velocity. Instead, it is because for the rendezvous mission more spacecraft time elapses, and so the rockets have to be fired for longer. It is only about 1.55 times longer (3.59 years vs 2.32 years), but this actually implies a lot more than 1.55 times as much propellant.

It must be the case that if we continuously accelerate the rocket for 3.59 years of spacecraft time, it will require the same mass ratio as the rendezvous mission. From the $v(t')$ equation in Table 7.1, we see that a continuous acceleration at $1\,g$ for this time gives $\Delta v = 0.9988\,c$. Putting this into the relativistic rocket equation shows that we require a mass ratio of exactly the same value as for the rendezvous mission (110 million). This can be seen directly by substituting $\Delta v = v(t')$ from the equation in Table 7.1 into equation 7.28, to give $M_i/M_f = \exp(a't'/v_e)$. The mass ratio required to maintain constant proper acceleration depends exponentially on the spacecraft time.

The results of the above calculations are summarized in Table 7.2, which also shows the results for a lower acceleration of 0.1 g. If we adopt more plausible values of $v_e = 0.08\,c$ and $a' = 0.005\,g$, then we find that a flyby mission to Alpha Cen takes 41.5 years of Earth time, requires a mass ratio of 14.3, and achieves an encounter velocity of 0.21 c. A rendezvous mission with the same v_e and a' would take 58.6 years and require a mass ratio of 43. Its velocity at the turn-around point is $0.15\,c$. As we shall see in chapter 8, $v_e = 0.08\,c$ is around the theoretical maximum for a fusion rocket.

7.6.1 The twin paradox

We saw in the previous section that the time elapsed onboard the spacecraft for the trip to Alpha Cen is less than the time elapsed on the Earth for the same trip. This can be see as a consequence of time dilation, or more generally the Lorentz transformations (we have not used anything else). To get back to the Earth, the spacecraft just repeats the same journey in reverse (figure 7.7), and so the durations in both the spacecraft and Earth frames are twice as long.

Suppose one of two twins – let's call her Sabina – travelled on the spacecraft, and the journey was done with constant 1 g acceleration as in Table 7.2. By the time Sabina gets back to Earth, $2 \times 3.59 = 7.18$ years of time have passed on the spacecraft; she has aged by this amount. Her twin sister Emily, who worked at ground control on Earth for the entire journey, has instead aged by $2 \times 6.03 = 12.06$ years. So Emily is older. From Emily's perspective (in the Earth's rest frame), Sabina flew to a point 4.4 ly away and then turned around and came back. But coordinate velocity and acceleration are relative. So from Sabina's perspective (in the spacecraft rest frame), Emily also flew to a point 4.4 ly away and then turned around and came back. So wouldn't Emily be the younger one? Their perspectives of the journey appear to be symmetric – same distances, velocities, and coordinate accelerations – so why is there a difference in their ages at all?

This is the notorious twin paradox, and like most (all?) paradoxes in physics, it isn't one. The resolution is that the situation is not symmetric for the twins: Emily has been in the same inertial frame all of the time, whereas Sabina has undergone various proper accelerations to move between a continuous set of inertial frames before finally returning to Emily's frame where the age comparison is done. Once Sabina arrived at Alpha Cen (and came to rest), she would have been $6.03 - 3.59 = 2.44$ years younger than Emily (both twins are in the same rest frame, just physically separated). If Emily then got in a spacecraft and did the same journey to Alpha Cen, then when she arrived, she would again be the same age as her sister.

A lot of time and ink has been spent trying to resolve the paradox, and the reader will have no problem finding many discussions on the internet. Some explanations are not really correct, for example the claim that the twin paradox is a consequence of general relativity (perhaps understandable because gravity also introduces ageing differences).[2] Some explanations use modifications of the above scenario to overcome plausible objections or to show that some claimed solutions do not work. Many identify acceleration as the key, although in some sense it is only indirectly responsible as a way of inducing changes in the inertial frame of one twin. Different people will be more or less satisfied by different explanations.

---

[2] Having said that, the twin paradox can also be understood as a consequence of the strong equivalence principle. The accelerating twin can be considered as experiencing the effect of a strong gravitational field which, through gravitational time dilation, causes that twin's clock to tick more slowly than the non-accelerating twin.

7.7 Photon rockets

The rocket equation suggests we should aim to use a propellant with a high exhaust velocity. The highest possible is the speed of light, achievable only with photons. Although they have no mass, photons do have momentum, so can propel a spacecraft. Let us work out the rocket equation for this case.

Consider the situation shown in figure 7.8. The rocket initially has a mass of $M_f + M_p$ and is at rest in some inertial reference frame S. The propellant mass M_p is converted into photons that propel the rocket to a velocity Δv in S. When we derive the classical or relativistic rocket equations, we normally consider differentials of mass and velocity because the velocity of the propellant in the initial rest frame varies. But now the propellant is photons, which always have a velocity c, so we can consider just the initial and final states. In S, the initial mass–energy is $(M_f + M_p)c^2$ and the final mass–energy of just the rocket is $\gamma M_f c^2$, where $\gamma = \gamma(\Delta v)$. Suppose that the total energy of the photons emitted (as measured in frame S) is E_p. By conservation of energy,

$$(M_f + M_p)c^2 = \gamma M_f c^2 + E_p. \tag{7.31}$$

In S, the final momentum of the rocket is $\gamma M_f \Delta v$ and the momentum of the photons is E_p/c. Thus by conservation of momentum,

$$0 = \gamma M_f \Delta v - \frac{E_p}{c}. \tag{7.32}$$

Eliminating E_p between these we get the photon rocket equation,

$$\frac{M_p}{M_f} = \gamma(1 + \beta) - 1, \tag{7.33}$$

where $\beta = \Delta v/c$. Writing $(M_p + M_f)/M_f = M_i/M_f$ and then rearranging equation 7.33, using also the relation in equation 7.6, we find that the photon rocket equation is just a special case of the relativistic rocket equation (equation 7.30b) when $v_e = c$.

Matter can in principle be converted entirely into photons via electron–positron annihilation (equation 8.11 in the next chapter). If anything other than photons is produced, then the above derivation has retained this 'waste' mass–energy on board the spacecraft as part of M_f. (We will modify this approach in the next section.)

Regardless of how the photons are produced, we have assumed that all of the mass M_p is converted into photons. Note, however, that $E_p < M_p c^2$, as can be seen from the energy conservation equation above. In the final rest frame of the rocket, the energy expended is indeed $M_p c^2$. But in frame S, some of the rest energy from M_p has gone into the kinetic

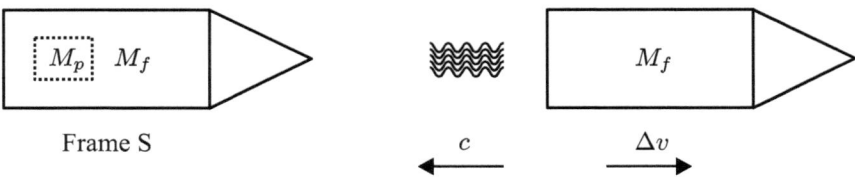

Frame S c Δv

Fig. 7.8 The photon rocket. Left: Initially the rocket of mass $M_p + M_f$ is at rest in frame S. Right: The mass M_p is converted into photons that leave the rocket at the speed of light c, accelerating the rocket to velocity Δv.

energy of the rocket. That is the whole point of a rocket, after all. If we write $E_p = \alpha M_p c^2$, then solving for α using the above equations gives

$$\alpha = \frac{\beta}{1 + \beta - 1/\gamma} .$$

(7.34)

This decreases monotonically from 1 at $\beta = 0$ to 0.5 at $\beta = 1$. This means that at very low velocities, almost all of the propellant rest energy goes into the photons, and so virtually none into the rocket. But as the rocket approaches the speed of light, half of the energy goes into the photons, and the other half into the rocket.

If at $\beta = 0$, all of the energy goes into photons and none into the rocket, how can the rocket ever get going? The answer is that the velocity in all these expressions is the velocity of the rocket after it has released at least one photon. Equation 7.34 therefore does not make sense when $\beta = 0$ exactly, because this would correspond to not having emitted a photon. So at the very beginning, only a tiny fraction of the energy is put into the rocket, but it is not zero. The same applies for a matter rocket too.

Another way to look at the energy partition between the rocket and the exhaust is to consider the energy efficiency, the ratio of the rocket's kinetic energy to the energy expended. The energy E_r that the photon rocket expends in its rest frame is $M_p c^2$. Substituting for M_p from equation 7.33 we get

$$E_r = [\gamma(1 + \beta) - 1]M_f c^2.$$

(7.35)

The kinetic energy of the rocket, $(\gamma - 1)M_f c^2$, divided by E_r is the efficiency. We will look at this in detail in the next section when we consider a more general rocket, but one may peek ahead to see the efficiency for this photon rocket in figure 7.11 (right panel, curve marked $\epsilon = 1$). The maximum efficiency is 0.5, occurring at $\beta = 1$, and the efficiency goes to zero at $\beta = 0$.

7.8 Mixed matter–photon rockets

A nuclear reaction used to power a rocket often produces massive particles and massless photons, both of which have momentum and could potentially be used to propel a spacecraft. As the reaction must conserve energy and momentum, only a fraction of the energy released by the reaction can go into the kinetic energy of the spacecraft. Here we use those conservation laws to compute the effective exhaust velocity and energy efficiency of a mixed matter–photon rocket.

Consider a small mass of fuel/propellant dm on board a rocket. This undergoes a reaction that converts a fraction $(1 - \epsilon)$ of its mass–energy into massive particles that can be used for propulsion ('propellant mass'); see figure 7.9. Of the remaining part ϵdm, a fraction η is converted into useable energy, and the other fraction $1 - \eta$ is lost. This lost fraction may include heat (via photons) to the spacecraft, which is later radiated isotropically, or neutral particles resulting from the reactions, such as neutrinos or neutrons that cannot form part of the propellant (discussed below in section 7.8.2). Of the useable energy portion η, a fraction ζ goes into the kinetic energy of the massive particles (the propellant mass), and the rest is emitted as collimated photons that also propel the spacecraft. We can therefore think of four

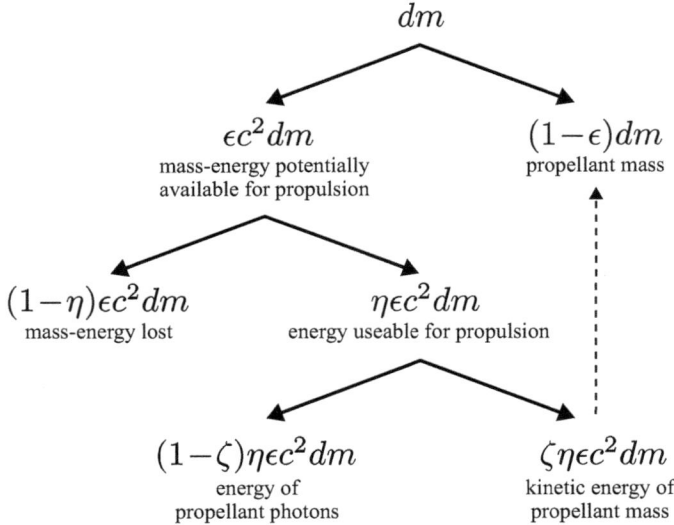

Partition of mass–energy derived from an initial fuel mass dm in a mixed matter–photon rocket. Adapted from Westmoreland (2010).

distinct mass–energy end products: the mass of the propellant particles; the kinetic energy of the propellant particles; the energy of the propelling photons; and mass–energy losses.

Consider the instantaneous reference frame of the spacecraft defined at the moment before the reaction. The spacecraft has mass M. After the reaction it has mass $M + dM$, where $dM = -dm$ and $dm > 0$, velocity dV, and therefore momentum[3] $(M + dM)dV$. This must balance the momentum of the particles and photons in the propellant, which we assume have been collimated so that they are ejected in the same direction as one another. Assuming that the particles are moving at common velocity v_m, they have momentum $\gamma_m v_m (1-\epsilon)dm$, where $\gamma_m = (1 - v_m^2/c^2)^{-1/2}$. The photons have momentum E/c, where $E = (1 - \zeta)\eta\epsilon c^2 dm$ is their energy (figure 7.9). Equating the momentum before and after the reaction we get

$$0 = (M + dM)dV - \gamma_m v_m(1 - \epsilon)\, dm - (1 - \zeta)\eta\epsilon c\, dm \qquad (7.36\text{a})$$

$$= (M + dM)dV + \gamma_m v_m(1 - \epsilon)\, dM + (1 - \zeta)\eta\epsilon c\, dM. \qquad (7.36\text{b})$$

Rearranging this and neglecting the higher-order term $dMdV$ we get

$$dV = -v_e \frac{dM}{M}, \quad \text{where} \qquad (7.37\text{a})$$

$$v_e = (1 - \epsilon)\gamma_m v_m + (1 - \zeta)\eta\epsilon c. \qquad (7.37\text{b})$$

Equation 7.37a has the same form as the classical rocket equation (equation 3.3), except now the velocity V is relative to the instantaneous reference frame of the rocket. This velocity is sometimes called the *proper speed*,[4] denoted by σ. The proper speed is the integral of the proper acceleration over the proper time, and is related to the coordinate velocity (in the

[3] Strictly this is the relativistic momentum $\gamma_{dV}(M + dM)dV$ where $\gamma_{dV} = (1 - (dV/c)^2)^{-1/2}$. But in the limit as $dV \to 0$, $\gamma_{dV} \to 1$.

[4] The proper speed divided by the speed of light, σ/c, is called the *rapidity*. The proper speed should not be confused with proper velocity defined in section 7.3.

original rest frame) by $\beta = \tanh(\sigma/c)$. Using this expression (with V in place of σ) in equation 7.37a, we get the relativistic rocket equation (see exercises).

The above derivation has given us an expression for the effective exhaust velocity v_e, although here it is better referred to as a specific impulse because it involves a mixture of particles moving at velocity v_m and photons at velocity c. Note that even when we have no propellant photons ($\zeta = 1$), the effective exhaust velocity v_e is not equal to the velocity of the particles v_m. Even in the classical limit ($\gamma_m = 1$) with no losses ($\eta = 0$), we have $v_e = (1 - \epsilon)v_m$. The factor $(1 - \epsilon)$ arises because some of the fuel mass has been converted into energy and so cannot be ejected as propellant mass.

7.8.1 Effective exhaust velocity

We get a deeper insight into v_e when we use energy conservation to equate the kinetic energy of the particles to the mass–energy converted and used to accelerate them:

$$(\gamma_m - 1)(1 - \epsilon)c^2\, dm = \zeta\eta\epsilon c^2 dm, \tag{7.38a}$$

$$\gamma_m = \frac{\zeta\eta\epsilon}{1 - \epsilon} + 1. \tag{7.38b}$$

As we would expect, the velocity of the particles depends on the three conversion factors.[5] Using equation 7.4 to rewrite $\gamma_m\beta_m$ in terms of γ_m in equation 7.37b, and then using equation 7.38b, we get an expression for the effective exhaust velocity in terms of just the three conversion factors:

$$\frac{v_e}{c} = \sqrt{\zeta\eta\epsilon(\zeta\eta\epsilon - 2\epsilon + 2)} + (1 - \zeta)\eta\epsilon. \tag{7.39}$$

Rockets with only matter as their propellant have $\zeta = 1$. If there are no mass-energy losses ($\eta = 1$),

$$\frac{v_e}{c} = \sqrt{\epsilon(2 - \epsilon)} \quad \text{(matter rocket with } \eta = 1), \tag{7.40}$$

where ϵ is now the fraction of the mass converted into energy that is then used to propel the remaining fraction $(1 - \epsilon)$ of the mass. In principle, if $\epsilon = 1$, the effective exhaust velocity is the speed of light, but then no mass is available for ejection. The other limiting case of $\epsilon = 0$ is also impractical because then no energy is available to accelerate the propellant. In the limit of small ϵ, equation 7.40 reduces to[6]

$$\frac{v_e}{c} \simeq \sqrt{2\epsilon} \quad (\zeta = 1, \eta = 1, \epsilon \ll 1). \tag{7.41}$$

For a photon rocket, $\zeta = 0$, so

$$\frac{v_\epsilon}{c} = \eta\epsilon \quad \text{(photon rocket)}. \tag{7.42}$$

Such a rocket still produces a 'propellant' mass of $(1 - \epsilon)dm$, but this mass receives no kinetic energy. This waste product is simply dumped at zero velocity relative to the rocket. When $\epsilon = 1$ there is no propellant mass at all. The propellant photons are generated by 100%

---

[5] As we shall see in section 8.5.2, a nuclear reaction generally produces more than one type of massive particle, which move with different velocities. In that case v_m is the velocity corresponding to their combined kinetic energy.

[6] This is also what we would derive using classical mechanics and equating the energy released, $\epsilon M_p c^2$, with the kinetic energy of the remaining mass, $(1/2)(1 - \epsilon)M_p v_e^2$, for $\epsilon \ll 1$.

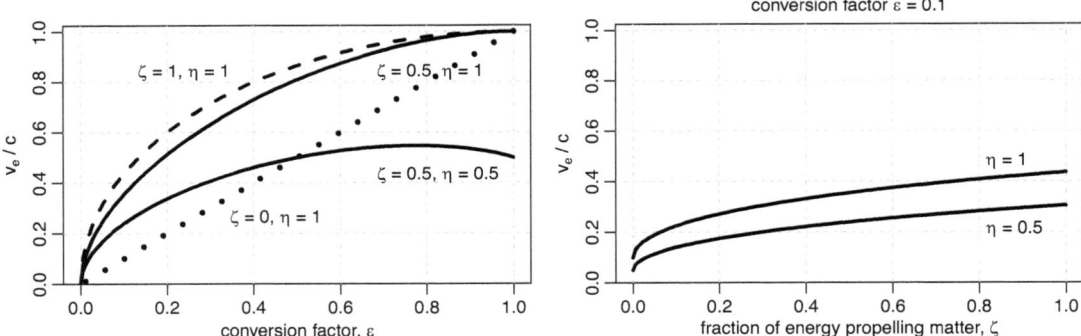

Fig. 7.10 Variation of the effective exhaust velocity for a mixed matter–photon rocket with ϵ (left) and with ζ (right), according to equation 7.39. In the left panel, the dashed line shows the variation for a matter rocket ($\zeta = 1$) with no energy losses ($\eta = 1$) (equation 7.40), and the dotted line the variation for a photon rocket ($\zeta = 0$) also with no energy losses (equation 7.42).

conversion of mass into energy, so $v_e/c = \eta$, where η is the fraction of the energy which is converted into photons that propel the spacecraft. I refer to this as the *pure* photon rocket. If we further have $\eta = 1$ (no energy losses), then $v_e = c$, which is the situation we derived in section 7.7.

Figure 7.10 shows how the effective exhaust velocity varies with some of the parameters for the general case of a mixed matter–photon rocket. The two solid lines in the left panel show how v_e varies with ϵ for two different values of η at fixed $\zeta = 0.5$. When there are no losses ($\eta = 1$), v_e increases monotonically with increasing ϵ, or equivalently with a decreasing amount of propellant mass. But when there are losses ($\eta = 0.5$), v_e reaches a maximum before decreasing. This panel also shows that the matter rocket (dashed line) gives a larger v_e than the photon rocket (dotted line) at all ϵ. The right panel shows the variation of v_e with ζ when 90% of the initial mass is converted to propellant mass. As the proportion of energy going into the particles, as opposed to the photons, increases, the larger the value of v_e. In terms of effective exhaust velocity, a matter rocket is preferred over a photon rocket.

7.8.2 Mass losses

In deriving equation 7.39 for the effective exhaust velocity, we assumed that the momentum gained by the spacecraft was equal and opposite to the momentum gained by the particles and the photons, excluding the losses. In reality, the products of numerous nuclear reactions used to propel a spacecraft move in random directions in all three dimensions. By conserving momentum in one dimension above, we assumed that all the useable products of the nuclear reaction were narrowly collimated in one direction. This could be achieved with strong magnetic fields for the charged particles, and with reflectors for the photons, but it may not be achieved for neutrons, which are neutral, or for neutrinos, which barely interact.[7] Yet both these particles can carry away significant amounts of energy and momentum. How has our derivation taken these losses into account?

---

[7] It might be possible to absorb some of the neutrons through collisions in the engine, but this can cause structural problems.

The important thing to realize is that the portion $(1 - \epsilon)dm$ is the mass of the *useable* particles that emerge from the nuclear reaction, that is, those that can be part of the massive propellant. This is usually just the charged particles. The other portion, the mass ϵdm, or equivalently the mass–energy $\epsilon c^2 dm$, comprises everything else, some of which is useable and some which is not (see figure 7.9). The useable part is the kinetic energy of the massive propellant particles plus the energy of the photons we can collimate. The unusable part is the mass and kinetic energy of particles that are not part of the propellant (typically neutrinos and neutrons), as well as photons we cannot collimate.

There is a potential source of confusion here, because the symbol ϵ is sometimes used in the context of nuclear reactions to indicate just the fraction of the initial mass that is converted into kinetic energy. We are instead using ϵ to include also the mass of any unusable particles, such as neutrinos or neutrons. The fraction of initial rest energy available for propulsion is $\eta\epsilon$, of which a fraction ζ goes into accelerating the massive particles.

We will use the theory developed above in sections 8.5.2 and 8.8 to compute ϵ, η, and ζ and therefore v_e for some real fusion and antimatter reactions.

7.8.3 Energy efficiency

Let us now consider how efficiently energy is used in a mixed matter–photon rocket. Consider a rocket of initial mass M_i initially at rest in frame S. Once it has attained a velocity v in S, its mass is M_f, and some of the energy released has gone into the rocket and some into the propellant. There are different ways in which we could define an efficiency. I will define it as

$$\xi = \frac{\text{kinetic energy of rocket}}{\text{energy put into propellant}}, \tag{7.43}$$

which parallels the definition used in section 3.6 for classical rockets. The numerator is $(\gamma_v - 1)M_f c^2$ and the denominator is $\eta\epsilon M_p c^2$ (see figure 7.9), where $M_p = M_i - M_f$. Thus

$$\xi = \frac{(\gamma_v - 1)M_f}{\eta\epsilon M_p} = \frac{(\gamma_v - 1)}{\eta\epsilon \left(\frac{M_i}{M_f} - 1\right)}. \tag{7.44}$$

The velocity and the mass ratio M_i/M_f are related via the relativistic rocket equation, equation 7.28. Using this we can write the efficiency as

$$\xi = \frac{(\gamma_v - 1)}{\eta\epsilon \left(\exp\left[\frac{c}{v_e} \operatorname{atanh} \beta_v\right] - 1\right)}, \tag{7.45}$$

where v_e/c is given by equation 7.39. The energy efficiency depends on the three conversion factors – ϵ, η, ζ – as well as the final speed of the rocket β_v in the initial rest frame (the one in which we are measuring the kinetic energy of the rocket).

The above energy efficiency is for the rocket as a whole. It should not be confused with η, which is an 'internal' efficiency of the conversion of mass into energy available for propelling the spacecraft. Note also that the 'energy put into propellant' in the denominator of equation 7.43 is the energy as measured in the rest frame of the rocket. This makes sense, because this is the energy that the rocket has to provide. The kinetic energy in the numerator, in contrast, is measured in the initial rest frame. This also makes sense, because we are interested in how fast the rocket is moving with respect to that initial rest frame (the kinetic energy in its own rest frame is zero).

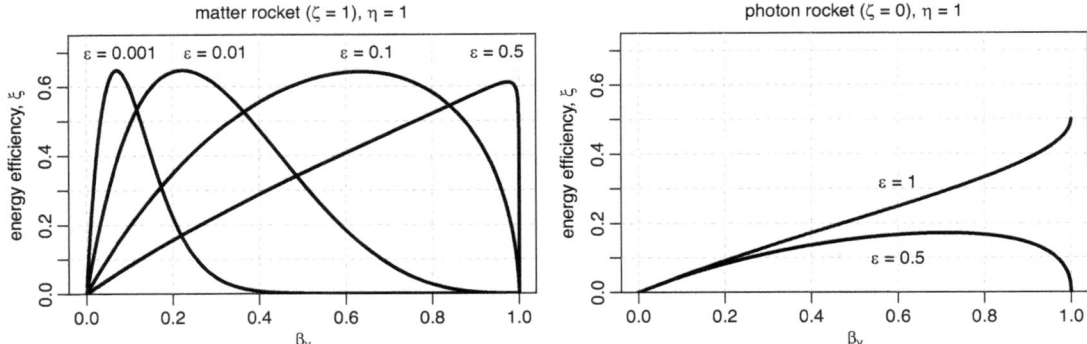

Fig. 7.11 The energy efficiency of a relativistic matter rocket (left), and a photon rocket (right), as defined by equation 7.43 and expressed by equation 7.45, as a function of the final velocity of rocket. In both panels we assume there are no energy losses when mass is converted into energy ($\eta = 1$). The matter rocket is shown for four different values of the fraction of the mass that is converted into energy to propel the remaining propellant. The photon rocket is shown for a pure photon rocket ($\epsilon = 1$), in which all of the mass is converted into photons, and one in which only half the mass is converted into useful photons ($\epsilon = 0.5$). In the latter case the remaining mass receives no kinetic energy, so contributes nothing to the propulsion.

Consider a matter rocket, $\zeta = 1$, with perfect conversion efficiency, $\eta = 1$. The left panel of figure 7.11 shows how the energy efficiency ξ varies with the final velocity of the spacecraft for four different values of ϵ. As $\eta = 1$, these values of ϵ correspond to the fraction of the initial mass that is converted into kinetic energy of the propellant. We see that for a wide range of ϵ, an energy efficiency of around 0.65 can in principle be achieved. That is, 65% of the energy available for propulsion goes into the rocket, with the rest going into the propellant. For all values of ϵ, the efficiency drops to zero as the velocity tends to 0 or c. Thus as a spacecraft accelerates, the energy efficiency will initially be very low. If the cruise velocity is low, say around 0.05 to 0.1 c, then the plot shows that to maximize efficiency we should use a small ϵ, of the order of 0.001: it is more efficient to convert only a small fraction of the propellant mass into energy and use that to accelerate the rest. But with this value of ϵ it is very inefficient to accelerate the spacecraft much beyond 0.3 c. At larger velocities, a larger ϵ is more efficient. Thus to accelerate efficiently from zero to large velocities, we would ideally use different reactions with different values of ϵ.

Figure 7.12 plots the efficiency against ϵ for three values of β. This shows more clearly that at low velocities (e.g. $\beta = 0.1$) there is a narrow range of small values of ϵ that maximize the efficiency for the matter rocket. To maximize the energy efficiency with a single reaction, we should try to match ϵ to the velocity we wish to achieve. There is only limited freedom to do this in practice, however, because ϵ is determined by the specific nuclear reactions available for energy conversion. For some of these, η is less than unity due to the production of non-interacting particles, as we shall see in chapter 8.

Let us turn now to the photon rocket, again for $\eta = 1$. The variation of its energy efficiency with velocity is shown in the right panel of figure 7.11. The line labelled $\epsilon = 1$ is a pure photon rocket, in which all of the mass–energy of the propellant is converted into photons. In contrast to the matter rocket, the efficiency of the pure photon rocket increases monotonically with the velocity of the rocket. The maximum efficiency, only achieved at the speed of light,

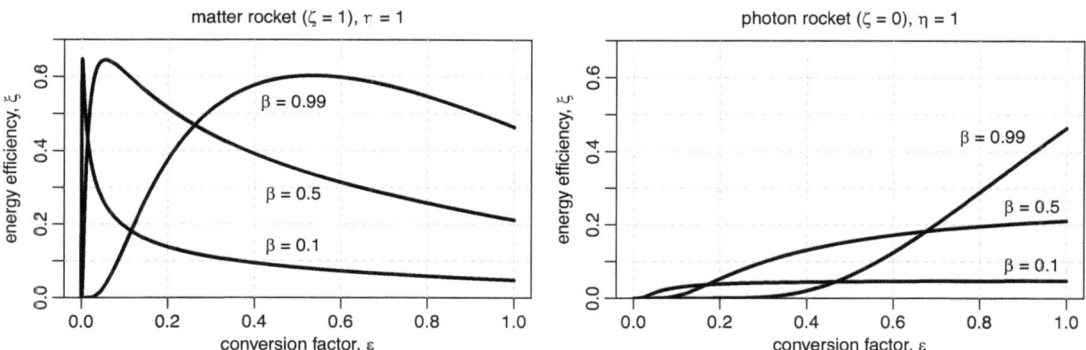

As figure 7.11, but now showing the efficiencies as a function of ϵ for three different values of the final rocket velocity β.   Fig. 7.12

is 0.5, which is less than the maximum efficiency of the matter rocket. The expression for the efficiency simplifies for a pure photon rocket as already found at the end of section 7.7.

The lower line in the right panel of figure 7.11 labelled $\epsilon = 0.5$ corresponds to only half of the initial mass being converted into photons. The other half is massive particles, but they receive no kinetic energy (by definition of being a photon rocket). As there is no point keeping this on board and wasting energy accelerating it, it is simply dumped from the spacecraft at zero relative velocity. In this case there is a maximum efficiency at a velocity below the speed of light and, as we would expect, this efficiency is lower than for the pure photon rocket. The right panel of figure 7.12 shows how the efficiency varies with ϵ. The efficiency increases monotonically with ϵ, in stark contrast to the matter rocket (shown in the left panel). The efficiency of the photon rocket only increases slowly with ϵ at low velocities, however.

In the classical Newtonian limit, $\gamma_v \simeq 1 + v^2/2c^2$. For a matter rocket with $\eta = 1$ and $\epsilon \ll 1$, which is also the classical limit, we get $\epsilon = v_e^2/2c^2$ (equation 7.41). This allows us to write equation 7.44 as

$$\xi = \frac{v^2/v_e^2}{\left(\frac{M_i}{M_f} - 1\right)} \quad \text{(classical limit)}, \tag{7.46}$$

which is the same expression for the energy efficiency of the classical rocket in equation 3.21 and figure 3.7.

7.9 Summary

When dealing with large relative velocities, the classical laws of motion have to be replaced by those derived from special relativity. The velocity and distance travelled by a constantly accelerating spacecraft as seen by an external observer are less in special relativity than in classical mechanics. Time dilation means that the time measured by a clock on a moving spacecraft is less than that measured on a clock in the initial rest frame (equation 7.11). The 'proper' acceleration experienced by the spacecraft is greater than the 'coordinate' acceleration measured in the initial rest frame (equation 7.24).

The rocket equation must be modified to accommodate special relativity (equation 7.29). The difference with respect to the classical rocket equation only becomes large when the rocket velocity is a significant fraction of the speed of light.

As photons have momentum, they can be used as a means of propulsion: the photon rocket. In later chapters we will look at nuclear and antimatter reactions for propulsion, which in general provide a mix of particles and photons. These reactions convert a variable fraction of their rest mass–energy into kinetic energy, some of which is lost, and the rest is divided among photons and particles (figure 7.9). We derived a general expression for the effective exhaust velocity (equation 7.39). Examining the energy efficiency (equation 7.45), we found that a single reaction does not have optimal efficiency over a wide range of spacecraft velocities.

7.10 Exercises

1. Show that integrating equation 7.15 over t yields equation 7.16.
2. Plot the ratio of Δv from the classical rocket equation (Δv_{cl}) to that from the relativistic rocket equation (Δv_{rel}) as a function of $\Delta v_{rel}/c$. For what values of Δv_{rel} is this ratio greater than 1.05? Can the classical rocket equation ever apply if v_e is near the speed of light? If so, what condition would have to apply? If not, why not?
3. In section 7.6 we looked at rather extreme relativistic journeys involving a large v_e ($0.2\,c$) and large accelerations ($1\,g$ and $0.1\,g$). Compute the quantities in Table 7.2 for a flyby and rendezvous to Alpha Cen assuming an effective exhaust velocity of $v_e = 0.05\,c$ and a constant acceleration of $0.005\,g$.
4. In section 7.8 just after equation 7.37, I introduced the proper speed. From its definition, and using equation 7.15, show that the proper speed equals $c\,\mathrm{atanh}\,\beta$. Then show that equation 7.37a leads to the relativistic rocket equation (equation 7.27).
5. Why does the energy efficiency of a rocket depend on an external observer?

Nuclear rockets

In a chemical rocket, the kinetic energy of the propellant is obtained from chemical bonds by burning the fuel. The available energy per unit mass is limited to about $10\,\text{MJ}\,\text{kg}^{-1}$, which limits the effective exhaust velocity to about $4.5\,\text{km}\,\text{s}^{-1}$ (equation 3.24). Nuclear reactions, on the other hand, can release the much larger binding energy between nucleons, of the order of $10^8\,\text{MJ}\,\text{kg}^{-1}$, implying effective exhaust velocities of around $0.05\,c$. In this chapter we look at the basics of nuclear fission and fusion, including reactors and the optimal reactions for propulsion. We then examine the various ways in which these can be used to propel a rocket, which includes both sustained reactions and miniature explosions. Although nuclear-propelled rockets have not been used in space (as far as we know), there has been quite some research and development. We will also look at the principles of antimatter propulsion and see what effective exhaust velocity this could achieve.

8.1 Nuclear binding energy

Nucleons – protons and neutrons – are held together in the nucleus of an atom by the strong nuclear force. The attractive nature of this force means that as individual nucleons are brought closer together, the force does work, and this energy – the binding energy – is released. As more nucleons are added, not only does the total binding energy increase, but so does the binding energy per nucleon, because each new nucleon interacts with all of the existing ones. However, in addition to this attractive nuclear force there exists also the electromagnetic force that repels the positively charged protons, and this works to lower the binding energy. For small nuclei the nuclear force is the dominant one. But as the nuclear force has only a very short range, the electromagnetic force lowers the binding energy per nucleon for larger nuclei.[1]

The exact physics is more complicated than this, but the overall result is that as the number of assembled nucleons increases, the binding energy per nucleon at first increases, and then decreases, as shown in figure 8.1. The combination of two small nuclei therefore leads to a net release of binding energy, because the resulting product will have more binding energy than the sum of the original components. This is the process of nuclear fusion. The maximum binding energy per nucleon occurs at nucleon number $A = 62$, corresponding to nickel-62, with $8.79\,\text{MeV}$ per nucleon. Fusion of nuclei larger than this no longer releases energy, because the resulting product would have less binding energy than the sum of the original components. Instead, breaking up a single larger nucleus into two or more products releases the excess binding energy. This is the process of nuclear fission.

---

[1] At very short distances the nuclear force is repulsive, which is what prevents the nucleus from collapsing.

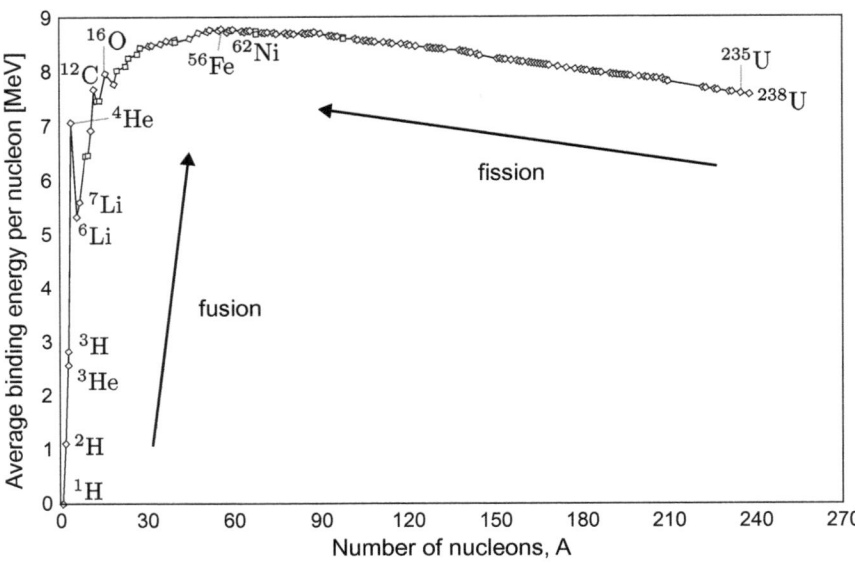

Fig. 8.1 The variation of the binding energy per nucleon in atomic nuclei with the number of nucleons (*A*, also known as the atomic mass number) for a selection of nuclei. Assembling small nuclei releases energy (fusion), as does breaking up large nuclei (fission). Credit: Public domain, from Wikimedia Commons, modified by the author.

Figure 8.1 shows that the typical binding energies per nucleon are of the order of several MeV per nucleon; 1 MeV per nucleon is equivalent to about $0.96 \times 10^8 \, \mathrm{MJ \, kg^{-1}}$. For comparison, the rest energy ($E_0 = mc^2$) of a proton is 938.27 MeV and that of a neutron is 939.57 MeV.

8.2 Nuclear fission

Nuclear fission is the process by which a heavy nucleus splits into two or more lighter nuclei. It can occur spontaneously, but the probability of it occurring is often larger if a neutron with a certain range of energies interacts with the nucleus. This is relevant because fission usually releases unbound neutrons in addition to nuclei. When, on average, more than one of the released neutrons goes on to cause another fission event, there will be an exponential increase in the number of fission events for as long as there is still fissionable material available.

The key to extracting energy from a fission reaction is therefore to control the number of neutrons. The *neutron multiplication factor k* is the ratio of the number of neutrons released in a generation of reactions to the number in the previous generation. When $k > 1$ we get an exponential increase; the reaction is said to be supercritical. As the timescale between fission events is very short, this leads to a very rapid release of energy. This is the goal of a fission bomb.

It would seem that in order to control the rate of energy release, we would need to keep *k* very close to one for an extended period of time. Fortunately for the peaceful extraction of energy, neutrons are released not only during the rapid fission event itself. The product

Table 8.1 Example uranium-235 fission products showing the amount of energy released that ends up in the kinetic energy of these products as well as neutrons and photons. Data from De Sanctis et al. (2016). Decays to three or more product nuclei are also possible, but are much less probable.

Product	Product	No. neutrons released	Energy released [MeV]
$^{144}_{54}$Xe	$^{90}_{38}$Sr	2	176
$^{141}_{56}$Ba	$^{92}_{36}$Kr	3	173
$^{141}_{55}$Cs	$^{93}_{37}$Rb	2	180
$^{137}_{53}$I	$^{96}_{39}$Y	3	179
$^{136}_{52}$Te	$^{97}_{40}$Zr	3	182
$^{127}_{50}$Sn	$^{105}_{42}$Mo	4	178

nuclei resulting from the fission are generally unstable and undergo radioactive beta decay. The resulting nucleus is sometimes excited and may then emit a neutron. Some of these neutrons have the right energy to cause fission, and this contributes to the neutron budget and thus the value of k. The fortunate aspect is that because beta decay is controlled by the weak interaction, its timescale is orders of magnitudes longer than the timescale on which the so-called 'prompt' neutrons are released directly from the fission. The 'delayed' neutrons following beta decay are emitted of the order of 0.01–100 s after the fission. This is long enough to implement a control mechanism to react to the number of neutrons available and thus keep the reaction near critical ($k = 1$). This is achieved in a fission reactor with control rods, which are made from strong neutron absorbers such as boron or cadmium. By moving these in and out of the volume in which the fission is taking place, the reaction rate can be controlled. It is the existence of the delayed neutrons that makes it possible to control nuclear fission.

The most widely used element for harnessing energy from fission is uranium-235, which contains 92 protons and 143 neutrons. When a neutron hits the nucleus, it can be absorbed to create uranium-236. This is unstable and splits into two (occasionally more) product nuclei. There are many possible paths for this decay, as shown in table 8.1, and these occur randomly in different proportions. The product nuclei tend to have unequal masses, the heavier one with a nucleon number A between 125 and 150, and lighter one with A between 80 and 105. Each fission typically releases two, three, or four neutrons, plus around 200 MeV of energy. This energy is the difference between the binding energy of the parent nucleus and that of the products. The parent has the order of 7.6 MeV per nucleon, whereas the products have the order of 8.5 MeV per nucleon, a difference of about 0.9 MeV per nucleon with 235 nucleons. This binding energy is released in the form of both high-energy photons (gamma rays), which receive about 7 MeV, and kinetic energy of the product nuclei (165 MeV) and neutrons (5 MeV). The rest of the energy comes from the beta and gamma decays of intermediate fission products in the form of photons (6 MeV), electrons (7 MeV), and antineutrinos (10 MeV). This released energy of 200 MeV compares to a rest energy of

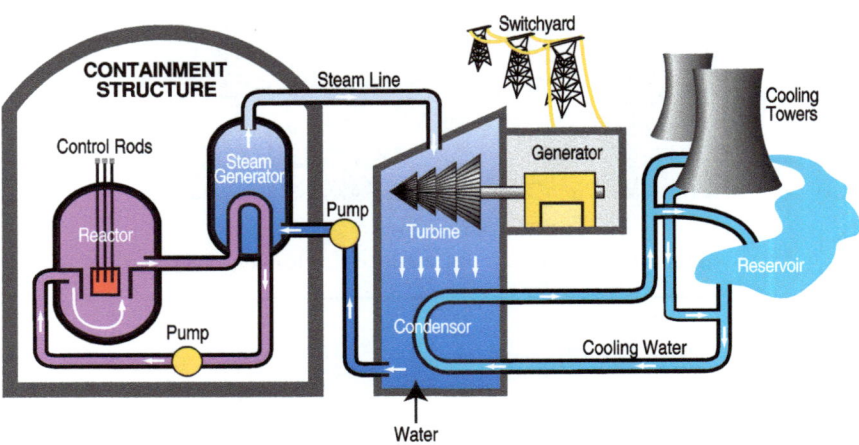

Fig. 8.2 A pressurized water nuclear fission reactor. Credit: Tennessee Valley Authority, public domain, from Wikimedia Commons.

the original uranium-235 nucleus of 2×10^5 MeV. Thus about 0.1% of the initial mass–energy is converted into kinetic energy of the fission products.

In a nuclear reactor, the energy of the gamma rays and the kinetic energy of the product nuclei are absorbed by a working fluid (often water). This drives a heat engine, usually a steam turbine, to produce electricity in a closed cycle. The typical layout of such a nuclear power station is shown in figure 8.2.

The neutrons released by a uranium-235 fission typically have kinetic energies of 2 MeV each. At this energy they have a high probability of causing fission in uranium-235, so most could contribute to the value of k. However, naturally occurring uranium on the Earth is a mixture of isotopes, with 99.3% uranium-238 and only 0.7% uranium-235 (plus 0.006% uranium-234). Not only does uranium-238 have a very low probability of undergoing induced fission, it also strongly absorbs neutrons with energies around 2 MeV. Thus most of the neutrons released by a uranium-235 fission would be absorbed by the uranium-238 isotopes, and very few would go on to cause further fission in uranium-235, resulting in a subcritical reaction. For this reason a sustainable nuclear fission reaction cannot spontaneously occur in natural uranium.

To achieve sustainable fission we have two options. First, we could increase the proportion of uranium-235 – by removing uranium-238 – so that enough neutrons remain unabsorbed and so attain $k > 1$. In principle – in an infinite mass of uranium such that neutron losses from the surface can be ignored – this can be achieved once the uranium is enriched to 5.5% uranium-235. In practice, though, an enrichment of 20% or more is used in what is then called a 'fast reactor', because it uses the high-energy neutrons directly.

A second option for achieving sustainable fission is to exploit the fact that the probability of a neutron causing fission, as opposed to being captured, depends on the neutron's energy. Neutrons with an energy significantly lower than 2 MeV are more likely to cause fission in uranium-235 than be captured by uranium-238. We can therefore use a *moderator* to slow down the neutrons, via elastic collisions with other atoms or molecules, to energies of around 0.1 eV (a temperature of 800 K) or less. Graphite (carbon) and heavy water (where deuterium, ^{2}H, takes the place of ^{1}H) are good moderators, and some nuclear reactors use these to achieve fission with natural uranium. Normal water ('light water') is a good neutron moderator, but unfortunately it is also a good neutron absorber, so we cannot achieve a stable fission reaction with natural uranium and normal water. However, if we enrich natural uranium to have more than 3% uranium-235, this is sufficient to produce enough neutrons to compensate for the absorption.[2]

Most commercial reactors today use low-enriched (3–5%) uranium as the fuel and light water as the moderator. Uranium enriched to about 20% is used in the reactors of nuclear submarines, because such reactors are more compact and allow the submarine to operate for longer without refuelling. Fission bombs use uranium enriched to 80% or more, to provide a much larger k value and thus a very rapid release of energy.

Uranium is not the only fissile material available. Plutonium (atomic number $Z = 94$), specifically the isotope plutonium-239, has been used for both reactors and bombs. It is barely naturally occurring (half-life of 24 kyr), but is produced in uranium reactors when uranium-238 captures a neutron. In fact, as the fuel ages in a normal uranium reactor, plutonium-239 becomes responsible for a significant fraction of the energy production.

Controlling the neutron flux is central to controlling a fission reaction, and here the shape of the fissile material plays an important role. If the material is arranged in a long thin rod, it has a large surface area for its volume, so a lot of neutrons escape from the surface before they can cause a fission event. If we instead arrange the fuel into a shape that minimizes the surface area for a given volume, which is a sphere, then a larger proportion of the neutrons are captured. The surface area to volume ratio of a sphere of radius r is $1/r$, and the rate of neutron loss is proportional to this. Thus for material that in principle could achieve criticality ($k > 1$), a sufficiently small sphere will actually be non-critical. The *critical mass* is the minimum mass that a sphere of fissile material (of given density) must have in order to achieve criticality. For pure uncompressed uranium-235, the critical mass is about 50 kg, corresponding to $r = 8.5$ cm. This means that two isolated spheres each slightly below this mass are both subcritical. But if we bring them close together, they will turn critical and explode uncontrollably. This is in fact the basic approach to detonating a fission bomb (see section 8.3.3).

Even without the goal of generating an explosion, the loss of neutrons from the surface of fissile material means that the most compact reactors should aim to minimize their surface area for a given volume. A spherical reactor is not very practical, however, so they tend to be cylindrical instead. A light water reactor produces energy at a rate of about 30 MW per tonne of fuel, with one tonne generating around 40 gigawatt days of energy. This corresponds to 3.5×10^6 MJ kg^{-1}. This is much less than the 80×10^6 MJ kg^{-1} (200 MeV/235 m_u) released by the fission reaction itself, because most of the fuel is uranium-238 that is not consumed. A typical reactor contains about 100 tonnes of enriched uranium fuel and produces around 3 GW of thermal power, of which around 1 GW can be converted into electricity by a steam turbine.

---

[2] Around 1.7 billion years ago, uranium-235 – which has a half-life of 704 Myr – made up about 3.1% of naturally occurring uranium, and so with a normal water moderator fission would have been possible. This actually occurred at that time in what was once a uranium-rich deposit in Oklo, Gabon, after groundwater seeped in to act as a moderator. This was identified after geological samples showed a significantly lower level of uranium-235 than the global average and other nearby mines, indicating fission had already occurred.

8.3 Fission rockets

We saw in the previous section that the fission of one uranium-235 nucleus puts about 165 MeV of kinetic energy (K) into the product nuclei. If all of these nuclei (of combined mass $m \simeq 233m_{\mathrm{u}}$) could be directly used as rocket propellant, they would provide an effective exhaust velocity of $\sqrt{2K/m} = 11\,700\,\mathrm{km\,s^{-1}}$, or $0.039\,c$. How this translates into the Δv of a rocket depends on how this energy is actually harnessed. Here we look at different ways to do this.

8.3.1 Thermal fission rockets

Perhaps the simplest use of the fission energy is to use it to heat an inert propellant, as shown in figure 8.3. The fission occurs continuously in a controlled manner in a reactor similar to a commercial reactor for generating electricity. But instead of the reactor heating a working fluid in a closed cycle to drive a turbine, the energy heats a propellant that is ejected. This delivers a much lower effective exhaust velocity than using the fission products directly, but a much larger thrust. The difference with respect to a chemical rocket is the source of the heat. After that, the rocket behaves just like any other thermal rocket discussed in chapter 5, in particular by using a nozzle to maximize the conversion of heat into kinetic energy.

A typical propellant is hydrogen on account of its low molar mass, thus giving a large effective exhaust velocity (equation 5.9). The hydrogen does not burn, and the exhaust is not radioactive provided all of the fission decay products remain within the fuel rods. In practice, however, corrosion of the fuel rods puts radioactive contaminants in the exhaust.

To keep the mass of the rocket as low as possible, the engine should use highly enriched uranium as fuel to avoid carrying non-fissile uranium-238. This has the further advantage that we do not need a moderator to sustain a reaction (a fast reactor), giving a further mass saving.

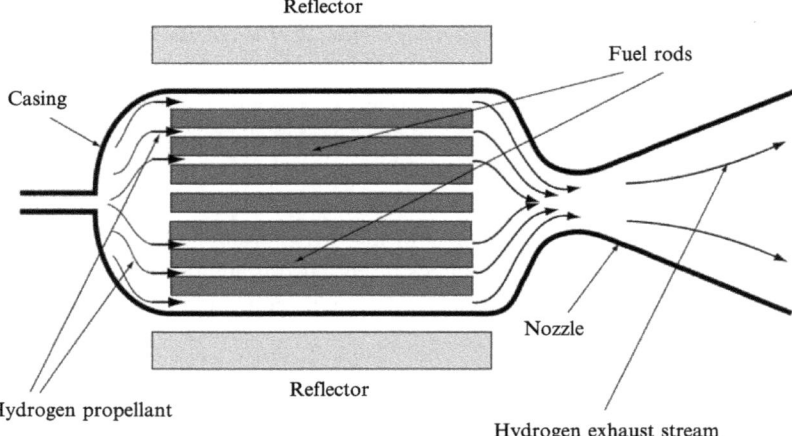

Fig. 8.3 A thermal fission rocket, in which a fission reactor heats an inert (non-combusted) propellant. Credit: Used with permission of Springer Nature BV, from Turner (2009), conveyed through Copyright Clearance Center, Inc.

A NERVA nuclear thermal rocket engine in testing. Credit: Photo courtesy of National Nuclear Security Administration / Nevada Site Office, public domain, via Wikimedia Commons.

Fig. 8.4

Although this makes the reactor smaller, it also increases the neutron loss as discussed above, lowering the reaction rate. For this reason, neutron reflectors are often employed in small reactors to compensate. A neutron reflector is a material that scatters neutrons elastically – but not specularly – so it needs to have a low absorption cross-section and not be fissionable. Various materials fulfil these requirements. Graphite and beryllium are frequently used in light water reactors, and have the advantage of low density. But they are also moderators that lower the energies of some of the neutrons, which is undesirable in a fast reactor. High-atomic-number materials are also good reflectors, but their high density makes them less desirable in a rocket.

To maximize the effective exhaust velocity we should try to achieve the largest reactor temperature possible (T_c in equation 5.9). This is often limited in practice by the melting temperature of the materials used to make the engines, and in the case of a fission reactor also the melting temperature of the fuel. Pure uranium melts at around 1400 K, which is rather low, so the fuel is usually made into a compound of uranium dioxide, which melts at 3075 K.

Thermal nuclear engines have been developed by both the USA and the Soviet Union/Russia. Between 1955 and 1972 the USA ran a programme called NERVA, Nuclear Engine for Rocket Vehicle Applications (figure 8.4). This developed working engines that were tested on the ground, although they never flew in space. Various versions were developed. One of them used a 1.5 GW reactor to heat a hydrogen propellant to $T_c = 2500\,\mathrm{K}$ and achieved $v_e = 8.7\,\mathrm{km\,s^{-1}}$. This velocity is only about twice that of a chemical engine, and far below the maximum achievable when directly using the fission products. The temperature is actually lower than that often achieved in a chemical thermal rocket engine. But because the propellant is much lighter – H_2 with a molar mass of two, instead of H_2O with a molar mass of 18 – the effective exhaust velocity is nonetheless higher; equation 5.10 gives $v_e = 8.5\,\mathrm{km\,s}^{-1}$.

From the above velocity and engine power we compute the mass flow rate (equation 3.22) and thrust (equation 3.10) to be $40\,\mathrm{kg\,s}^{-1}$ and 345 kN, respectively, assuming all of the power is used. The engine itself had a mass of about 8000 kg; another 6000 kg was required for shielding the radiation. Although it used highly enriched uranium (over 20%), the reactor nonetheless used a moderator made from graphite and zirconium hydride. The engine was designed to operate for a few hours.

This approach to nuclear propulsion has a number of challenges. One is that even if we stop the reaction to turn off the engine, the fission decay products still produce heat. To avoid the core melting, we would have to pass hydrogen over it to cool it, which would still generate thrust. This makes it difficult to operate as a restartable engine. The decay product xenon, which is a strong neutron absorber, can also prevent the reaction being restarted (neutron poisoning).

The temperature of the nuclear core, and therefore the velocity of the propellant, is limited in the above designs by the use of solid fuel. Using a melted, liquid fuel permits higher temperatures. One approach is to confine the fuel within rotating tubes that prevent the hottest parts of the fuel touching the surface of the tubes. In this way a hydrogen propellant temperature of 5500 K could be achievable, providing $v_e = 17\,\mathrm{km\,s}^{-1}$. Even higher temperatures are in principle achievable with a gas core reactor, in which the fuel is vaporized into a plasma that heats the propellant by radiation. Exhaust velocities of $30\text{–}70\,\mathrm{km\,s}^{-1}$ might then be achievable.

For a rocket to overcome Earth's gravity, it must have a thrust larger than its weight. Nuclear rockets tend to be heavy, with a lower thrust-to-weight ratio than chemical rockets. For this reason, plus the fact that their exhaust can contain radioactive isotopes, they are more suited for upper stages high in the Earth's atmosphere or for orbital transfers to other planets.

NASA's NERVA programme did a lot of development, tested several designs, and was not far from flight tests before the programme was cancelled for a combination of political and financial reasons, after having cost around 1.4 billion US dollars. There have since been further developments and plans to use such engines for missions to the Moon and Mars, but nothing concrete. As of mid 2025 no such thermal fission rocket has been publicly launched. However, it appears that the Russian military is developing a missile powered by such an engine, called the 9M730 Burevestnik (NATO name SSC-X-9 Skyfall). It was apparently tested in 2025, but details of its design and status remain unclear.

8.3.2 Fission fragment rockets

Fission reactions in the solid thermal nuclear engine only achieve propellant velocities of the order of $10\,\mathrm{km\,s}^{-1}$, just twice that of a chemical thermal rocket. The large energy density does allow a large amount of propellant to be heated, however, thus generating a large thrust. Liquid or gas cores could increase v_e, but it would still be at least two orders of magnitude below the theoretical limit of $11\,700\,\mathrm{km\,s}^{-1}$ achievable from uranium-235 fission.

The obvious way to attain this limit is to use the fission product nuclei directly as propellant. This is challenging because in the rest frame of the reaction, the product nuclei are moving in random directions. As the nuclei are charged, a magnetic field can

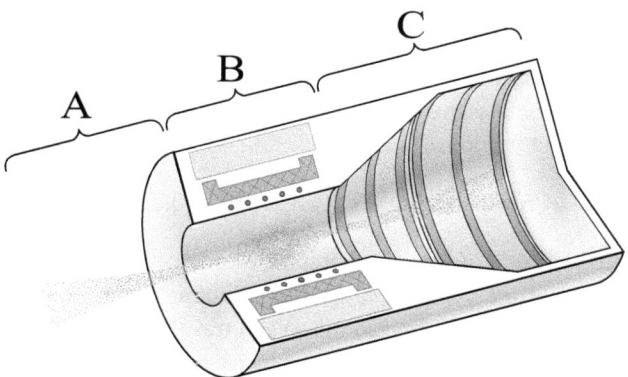

A fission fragment rocket. The fission reactions take place in region B. The high-energy fission products are collimated by magnetic fields and propelled out of the back of the spacecraft in region A. Some of the product nuclei could be used within the spaceship to provide energy for onboard power (region C). Credit: Duckysmokton, licensed under CC-BY-2.5, from Wikimedia Commons, modified by the author; following the design of Clark and Sheldon (2005).

Fig. 8.5

be used to direct them into a collimated propellant beam, something we will examine in section 8.4.

Unlike in a nuclear reactor, where the fissile material is usually arranged into multiple thin rods, the fuel in a fission fragment rocket is in the form of small dust-sized grains ($10\,\mu$m or less) to allow the fission fragments to escape from the fuel (figure 8.5). By ionizing the grains, electric fields can be used to contain them within the combustion chamber. As the fuel grains move slowly, they have a much lower kinetic energy to charge ratio than the fission products. The electric field can be tuned to contain the fuel without having much effect on the fission products, which then move off to be collimated by the magnetic nozzle and exhausted as high velocity propellant. At any one time there needs to be a critical mass of fuel dust in the reactor to sustain the reaction, so fuel must be continuously replenished to balance what is ejected as propellant. Control rods moderate the reaction rate as in a normal reactor. The positively charged fission fragments must be neutralized as they leave the spacecraft to avoid the spacecraft becoming negatively charged with time, as also done with the ion engine of chapter 6.

A specific design by Werka et al. (2012) uses a 1 GW reactor several metres in size with an engine mass of 110 t to produce an effective exhaust velocity of $v_e = 5170\,\text{km s}^{-1}$, or $0.017\,c$. About half of the engine mass is the moderator, and much of the rest is the magnetic/electric field generators. The engine thrust is low, just 43 N, on account of the low mass flow rate of $\dot{m} = 8.3 \times 10^{-6}\,\text{kg s}^{-1}$, or 0.72 kg per day. This corresponds to a power of $(1/2)\dot{m}v_e^2 = 110\,\text{MW}$, so about 11% of the energy of the reactor is going into the kinetic energy of the propellant. Even though the fuel is presented in dust-sized grains to maximize the escape probability of the fission fragments, some fragments will inevitably collide with unburned fuel, thereby dissipating some of their kinetic energy as heat. A cooling system is therefore essential. Some of this heat could be captured and used to drive a generator in a closed cycle heat engine, thus generating electrical energy for the spacecraft. But the heat flux is too large for all of it to be captured in this way, so most has to be radiated into space as infrared radiation. The study authors calculate that 700 MW has to be radiated in this way, 70% of the reactor power.

As the thrust of this fission fragment engine is very low and its mass large, the engine has to be operated for an extended period in order to achieve a large Δv. If we optimistically assume an initial spacecraft mass of 110 t (i.e. dominated by the engine; in practice the cooling system is more massive) and operate it for n days, then from the rocket equation (equation 3.14) the velocity achieved is

$$\Delta v = 5170 \ln(1.1 \times 10^5/[1.1 \times 10^5 - 0.72n]) \, \text{km s}^{-1} . \tag{8.1}$$

For $n = 1$ this is $0.034 \, \text{km s}^{-1}$; for $n = 365$ it is $12.3 \, \text{km s}^{-1}$. Even after 100 years the rocket is only moving at $1410 \, \text{km s}^{-1}$ ($0.0047 \, c$). After this time it has consumed 26 t of fissile material (more or less equal to the ejected mass) and travelled 0.22 ly (equation 3.15b). If it cruised at constant velocity from this point on, the spacecraft would need another 860 years to travel the remaining distance to Alpha Cen. Even though v_e is high, because the mass ratio is low, it takes a long time to achieve the final Δv, which is also not very high. In reality the initial mass would be much larger to accommodate the propellant and spacecraft structure, so the mass ratio would be even lower and the travel time longer. This engine was not designed for an interstellar mission, but this calculation nonetheless shows that even such a large powerful fission fragment engine will not get us to the nearest stars very quickly.

The engine could be scaled up in power to try to shorten the mission length. Suppose we increase the power of the engine by a factor of 10 at constant v_e using the same fission reactions and efficiency (so the propellant power also increases by a factor of 10, to 1100 MW). This increases the mass flow rate and hence the thrust by a factor of 10. The 26 t of fuel will now be used up 10 times faster, within 10 years. After this time the spacecraft is moving at the same velocity as before (v_e and the mass ratio are unchanged) and it has moved a tenth as far (0.022 ly). Although it has achieved its cruising velocity earlier, the spacecraft now has a larger distance left to Alpha Cen, and in fact still needs 899 years to get there. The total travel time is 909 yr, not much less than the 960 years in the previous case, despite a 10-fold increase in power. The limiting factor is the amount of propellant, not how quickly we can expel it.

Suppose we instead double v_e to $10\,340 \, \text{km s}^{-1}$ ($0.035 \, c$), which is only slightly less than the maximum possible with uranium-235. The spacecraft and propellant mass, as well as the mass flow rate, are unchanged, so the thrust doubles. It still takes 100 years to exhaust all the propellant, after which time the spacecraft is travelling at $2820 \, \text{km s}^{-1}$ ($0.0094 \, c$) and has moved 0.45 ly. Cruising at this speed, the spacecraft needs another 404 years to get to Alpha Cen, for a total journey time of 504 years. As most of the journey is spent cruising at twice the speed as the nominal case, the journey time is nearly halved.

If we varied the propellant mass or accelerated for a significant part of the journey, then the scalings are not as straightforward as above. This is explored in an exercise at the end of the chapter.

Returning now to original engine design above, Werka et al. (2012) used it to outline a crewed return mission to Jupiter. In this, the total mass is around 300 t, a large part of which is $6000 \, \text{m}^2$ of radiators to remove the excess heat from the engine. This mass is about a third that of a comparable mission they outline using a thermal fission rocket (section 8.3.1), but the round trip time is 16 years as opposed to 4.5 years, on account of the much lower thrust.

The exhaust of a fission fragment engine is highly radioactive, so we would not want to operate such a rocket in the Earth's atmosphere. It would anyway be useless for getting into space, not only because of the low thrust, but also because the fission fragments would be absorbed by the air before they could be magnetically coupled to the rocket. In practice such a rocket would be launched into space by conventional chemical rockets (perhaps in parts and then assembled there) and the fission engine only started once well away from the Earth.

Fissile materials other than uranium-235 could be used in rockets and may be preferred. Americium-242, for example, has a higher fission cross-section and produces more neutrons per fission, which lowers the critical mass. It does not occur naturally, however, so would have to be created by neutron capture in americium-241, which in turn is produced by beta decay of plutonium-239. Americium-242 has a half-life of 141 years, so for many missions it could be synthesized on the ground before launch.

8.3.3 Fission pulse rockets

A nuclear fission explosion occurs when a critical mass of fissionable material is brought together rapidly, such that the neutron capture rate becomes supercritical. If supercriticality can be retained for long enough, there will be a rapid release of large amounts of energy. The fission bomb dropped on Hiroshima by the USA in 1945 consisted of two pieces of uranium with a combined mass of 64 kg and a uranium-235 enrichment of 80%. When brought together they became supercritical, and this was retained long enough to release 15 kt TNT-equivalent of energy. A kiloton (kt) TNT-equivalent is a unit of energy equal to 4.184×10^{12} J often used to measure explosive strength. The bomb was very inefficient, as only about 1% of the uranium-235 underwent fission. The second bomb, dropped on Nagasaki, comprised two pieces of plutonium-239 with a combined mass of 6.2 kg. This is actually subcritical; the weapon used neutron reflectors to achieved supercriticality. This bomb used a shell arrangement (as opposed to a gun) leading to 20% of the plutonium undergoing fission during the explosion, releasing 21 kt TNT-equivalent of energy.

The concept of nuclear pulse propulsion is straightforward: explode a small nuclear bomb near a spacecraft and let the momentum of the fragments propel it. In its simplest incarnation the spacecraft is equipped with a 'pusher plate', which absorbs the momentum of an explosion that takes place outside the spacecraft. This plate is attached to the rest of the spacecraft by what is essentially a large spring, so that the impulses provided by a series of explosions can be smoothed out into a more continuous acceleration of the vehicle as a whole (figure 8.6). Provided the plate is massive enough and of the right materials, then a series of small explosions should individually not heat the plate enough to destroy it, and also allow enough time between explosions for it to cool. A thin layer at its surface will be vaporized by each explosion, but this could be built into the design. Effective exhaust velocities of 30–100 km s^{-1} could be achievable in this way. By using a magnetic field to couple the plate to the fission products of the explosion, the ablation at the surface could be reduced and more powerful explosions used, increasing v_e to perhaps 1000 km s^{-1}.

A variation on this idea is for the explosions to occur within a combustion chamber inside the rocket. The motivation is to transfer more of the momentum from the explosion by capturing also the fragments that were moving away from the spacecraft and not intercepting the pusher plate. This could be done by using the explosion to heat up a propellant, such as hydrogen, which is then ejected through a nozzle just as with any other thermal rocket. This would again be pulsed, with propellant injected into the combustion chamber for each explosion. This approach was deemed infeasible, however. Part of the problem is that some of the energy from the nuclear explosions is carried by neutrons. These mostly pass through

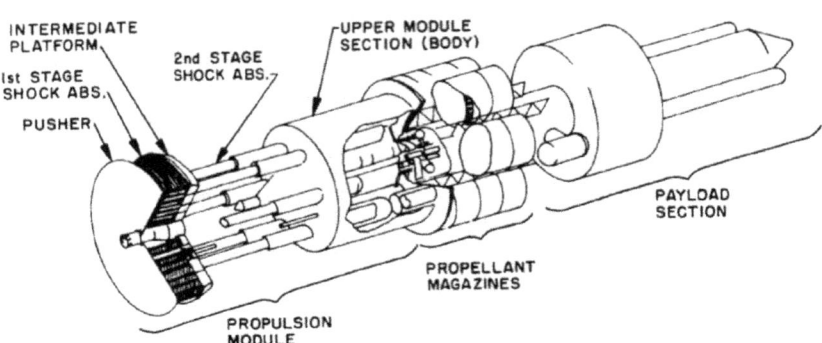

INTERMEDIATE
PLATFORM
1st STAGE
SHOCK ABS.
2nd STAGE
SHOCK ABS.
PUSHER
UPPER MODULE
SECTION (BODY)
PAYLOAD
SECTION
PROPELLANT
MAGAZINES
PROPULSION
MODULE

Fig. 8.6   The design of a pulse fission rocket produced as part of the NASA project Orion in the 1960s. The pusher plate at the left end has a diameter of about 10 m. The propulsion module has a mass of 100 t and was expected to deliver an effective exhaust velocity of about 20 km s^{-1}. Credit: NASA, public domain, from Wikimedia Commons.

the propellant but are then absorbed by the walls of the combustion chamber, heating that instead. This would require significant cooling. The cooling apparatus as well as the combustion chamber itself add significant mass. Theoretical studies indicated that the effective exhaust velocities achievable by such internal engines would be an order of magnitude lower than the external, pusher-plate arrangement.

The idea of nuclear pulse propulsion was explored experimentally in the USA by the Department of Defence and NASA between 1958 and 1965 under the name Project Orion (figure 8.6). It went through a series of designs, including ones launched from the Earth's surface, which would of course have produced significant amounts of nuclear fallout. Each bomb would release 0.01–0.1 kt TNT-equivalent of energy. One design for a crewed round-trip mission to Mars was to be started from Earth's orbit. It used a 100 t engine that provided an effective exhaust velocity of 20 km s^{-1} and a thrust of 3.5 MN. The total spacecraft mass was 1100 t, of which 150 t was payload. Using one to two thousand fission explosions, the travel time to Mars was to be 125 days. The spacecraft included shielding to reduce the radiation dose for the crew from the engines down to 500 mSv for the round-trip mission, the same as the additional dose expected from cosmic rays.[3]

One of the experiments undertaken for Project Orion involved exploding a 20 kt TNT-equivalent fission bomb 10 m away from two 1-m-diameter graphite spheres. Surprisingly they survived, suffering just the ablation of a fraction of a millimetre of their surfaces. Before testing nuclear bombs for propulsion itself, tests were done using chemical explosions. The most successful lifted an object 56 m into the air in 23 s using a series of seven explosions. Towards the end of the project, preliminary designs were being made for a spacecraft that would be assembled in Earth's orbit and launched from there. Ultimately, however, the project was cancelled in the wake of the 1963 nuclear test ban treaty, which banned the testing of nuclear weapons in the atmosphere, under water, or in space (underground tests were still permitted).

---

[3] The Sievert measures the biological effect of ionizing radiation, whereby 1 Sv = 1 J kg^{-1}. The average annual dose that a person receives on the Earth is of order 2.5 mSv, although it varies with location and activity. A chest X-ray delivers around 0.1 mSv and a whole body CT scan around 10 mSv, for example, while a 12-hour commercial flight delivers a dose of around 0.03 mSv. NASA and ESA limit an astronaut's lifetime exposure to 1000 mSv. The health effects of radiation also depend on how quickly the dose is received.

The idea of using small bombs to propel an interstellar spacecraft was not dead, however, as we shall see in section 8.6.

8.4 Magnetic nozzles

A key idea underlying several types of nuclear propulsion is to couple the charged products to the rocket using a magnetic field. How does this work?

Consider a particle moving at velocity v relative to a rocket. If the particle is not coupled to the rocket, it moves off at this velocity and the rocket remains at rest. If the particle is instead coupled to the rocket, it still moves at velocity v relative to the rocket, so there has been no momentum or energy transfer in the rocket's instantaneous rest frame. But due to conservation of momentum, the rocket is now moving in the initial rest frame, so in this frame the particle must be moving in the opposite direction at a velocity less than v. In this frame the particle has transferred momentum and kinetic energy to the rocket. This is of course the basis of the rocket equation (section 3.1). Conceptually it does not matter how this coupling takes place. It could be via the elastic reflection of a gas molecule, or via a magnetic field for a charged particle. Just as a magnetic field accelerates a charged particle, so a moving particle will accelerate whatever the magnetic field is attached to, in this case the rocket.

Immediately after a set of nuclear reactions, there are numerous charged particles moving in random directions. These have zero net momentum in their centre-of-mass frame, which is also the rocket's instantaneous rest frame. We want to convert these random motions – thermal energy – into directed kinetic energy. Recall that this is the purpose of the convergent–divergent nozzle in thermal rockets (chapter 5). By collimating the particle motions, more momentum is transferred along the axis of the rocket rather than perpendicular to it.

In chemical rockets, the molecules of the hot gas transfer momentum by bouncing off the inside walls of the nozzle. With nuclear reactions, the temperature of the products is so high it would melt the rocket. We can instead use a magnetic field to couple the charged particles to the spacecraft without them making physical contact. The magnetic field is designed to convert the radial motions (perpendicular to the axis) into axial ones, and just like the convergent–divergent nozzle this converts thermal energy into directed kinetic energy. This is the *magnetic nozzle*.

How strong does the magnetic field need to be? We can get an order-of-magnitude estimate by computing the magnetic field strength required to keep a particle moving in a circle of fixed radius. To do this we equate the Lorentz force, qvB, with the (relativistic) centripetal force, $\gamma m v^2/r$, to give

$$B = \frac{\gamma m v}{qr}.$$

(8.2)

The radius r is sometimes called the Larmor radius. In a uranium-235 fission fragment rocket (section 8.3.2), an example product nucleus is $^{139}_{56}$Ba with a kinetic energy of 112 MeV. This corresponds to $v = 0.042\,c$, $\gamma \simeq 1$ (not very relativistic), and so a field strength of $B = 0.33$ T is required to turn it in a circle of radius 1 m. This is a strong field that would need large massive magnets to collimate the propellant.

The actual field configuration required for the magnetic nozzle is much more complicated than this simple calculation implies. In particular, the field must diverge in the right way at the nozzle exit to ensure the plasma detaches from the rocket.

8.5 Nuclear fusion

When two nuclei with low atomic mass numbers combine, their product often has a higher binding energy per nucleon than the parent nuclei (figure 8.1). This combination, or fusion, releases the binding energy as the kinetic energy of the fusion product and other released particles.

The difficulty with achieving fusion is that nuclei are positively charged and so repel each other. For fusion to occur, the nuclei need to have sufficient kinetic energy to overcome this Coulomb barrier (or actually somewhat less, due to quantum tunnelling). This energy can be supplied by heating the fuel to high enough temperatures, of the order of tens or hundreds of millions of kelvins, which also turns it into a fully ionized plasma.

Fusion takes place in the centres of stars in the middle part of their lives and is the source of their luminosity. In lower-mass stars like the Sun (core temperature 1.6×10^7 K), the dominant reaction is the fusion of hydrogen (H, $Z = 1$) into helium (He, $Z = 2$). This takes place through a series of steps starting with[4]

$$p + p \rightarrow D + e^+ + \nu_e + 0.42 \, \text{MeV}. \tag{8.3}$$

The deuteron and another proton then undergo another fusion to produce a helium-3 nucleus plus a gamma photon. Two such helium-3 nuclei then combine in yet another fusion reaction to produce a helium-4 nucleus (an alpha particle) plus two protons. The overall reaction is

$$6p \rightarrow {}^4\text{He} + 2p + 2e^+ + \nu_e + 2\gamma + 24.68 \, \text{MeV}. \tag{8.4}$$

Each positron quickly annihilates with an electron in the plasma, in each case producing two gamma photons (with a combined energy of 1.02 MeV), so the total energy released by the reaction is 26.72 MeV. This appears as the kinetic energy of the protons and alpha particles as well as the gamma photons and neutrinos. The fraction of the initial rest energy released[5] is $26.72 \, \text{MeV}/(4m_p c^2) = 0.0071$, where m_p is the mass of the proton. Note that the initial rest energy is considered to be four protons, not six, because the two protons produced can be used again in this reaction.

The above process is one of three branches of a more complicated reaction known as the proton–proton (p–p) chain, other branches involving the intermediate production of heavier elements. The details are unimportant from the point of view of propulsion systems, however, for one simple reason: the reaction rate of two protons is extremely small, a consequence of it being mediated by the weak interaction (something we can conclude from the production of a neutrino). In the Sun, for example, the average time before a proton fuses with another

---

[4] I use the symbols D=${}_1^2$H and T=${}_1^3$H for deuterium and tritium, the two heavy forms of hydrogen, respectively. p and n indicate the proton (${}_1^1$H) and neutron (${}_0^1$n) respectively. e$^+$ is the positron and ν_e is the electron neutrino.

[5] Conventionally this fraction is denoted ϵ. In the context of rockets, where some of the particles produced cannot be used for propulsion, the energy useable for propulsion is smaller, and was denoted by the factor $\eta\epsilon$ in figure 7.9.

proton is nine billion years, which not coincidentally is close to the main sequence lifetime of the Sun.

We must turn to other fusion processes for propulsion. One of the most important from the point of view of future energy production on the Earth is the fusion of deuterium and tritium

$$D + T \rightarrow \ ^4He\ (3.52\,MeV) + n\ (14.07\,MeV). \tag{8.5}$$

The energy released equals the difference in the masses, 17.59 MeV. The fraction of the initial mass converted into energy is 0.0038. This goes entirely into the kinetic energies of the product nuclei, in a proportion that can be determined here by the conservation of momentum. The momenta of the two products in the initial centre-of-mass frame must be equal in magnitude, which in Newtonian mechanics is $p = \sqrt{2Km}$. Hence the kinetic energy K of each product is inversely proportional to its mass m. The helium-4 nucleus therefore receives 3.52 MeV kinetic energy and the neutron four times as much, 14.07 MeV. This is a Newtonian calculation, but as the velocity of the neutron in the centre-of-mass frame is only about $0.17\,c$, it is a reasonable approximation. In reactions where there are three or more products, conservation of energy and mass does not yield a unique solution to the energy partition, and a range is possible (as in equations 8.3 and 8.4).

The above D+T fusion reaction produces 3.5 MeV per nucleon, compared to 0.8 MeV per nucleon for the fission of uranium-235. This is because the D+T reaction produces a nucleus – the alpha particle – that is unusually tightly bound for its atomic number: Note the peak of the binding energy per nucleon at helium-4 in figure 8.1.

In order to overcome the Coulomb barrier and get a significant probability of fusion occurring, the nuclei require a combined energy of order 10 keV. This corresponds to a temperature of $T = 8 \times 10^7$ K, from $(3/2)kT = E$, where k is the Boltzmann constant. (Recall that the temperature only represents the average kinetic energy of the the plasma. Fusion can occur in a cooler plasma in the high-energy tail of the energy distribution.) This kinetic energy is nonetheless much smaller than the energy released, so can be neglected in the energy partition just mentioned.

The D+T reaction is being intensely pursued for large-scale energy production on the Earth. Deuterium is a stable, naturally occurring isotope of hydrogen. Although it only makes up a small proportion of hydrogen on the Earth, just 0.016% by number, hydrogen is widely abundant in water, and its extraction is routine. Deuterium has been used for many years as a moderator in fission reactors that use natural (non-enriched) uranium (section 8.2). Tritium, on the other hand, is a radioactive isotope with a half-life of just 12.3 years. It barely occurs naturally on the Earth (mostly via the interaction of cosmic rays with the atmosphere), and unnaturally as a result of nuclear weapons testing. Tritium must therefore be manufactured, usually via the reaction

$$^6Li + n \rightarrow \ ^4He\ (2.06\,MeV) + T\ (2.75\,MeV) \tag{8.6}$$

by lining a nuclear reactor with lithium (Li, $Z = 3$) to take advantage of the energetic neutrons produced therein. That is, the tritium fuel required for the fusion (equation 8.5) can be bred within the reactor itself. Lithium-6 is a naturally occurring stable isotope of lithium (1.9–7.8% by number).

The probability of a fusion reaction occurring is determined by the cross-section (its unit is area). A significant probability is only achieved once the temperature reaches tens of millions of kelvins, as can be seen in figure 8.7 for several different reactions. The plasma must also

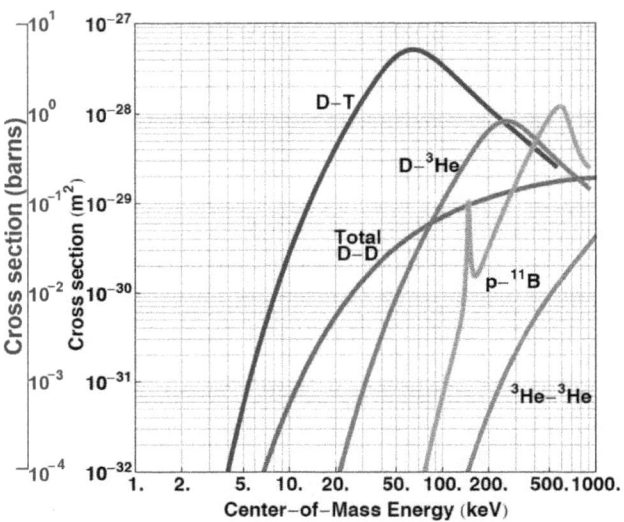

The cross-section for different fusion reactions as a function of the kinetic energy (temperature). An energy of 10 keV corresponds to a temperature of 8×10^7 K. Credit: WikiHelper2221, licensed under CC-BY-SA-4.0, obtained from Wikipedia Commons.

be dense enough to get a significant amount of fusion. Such a hot plasma will lose energy, the rate of which is measured by the inverse of the confinement time. In order for a fusion reaction to be self-sustaining, the amount of energy produced needs to exceed the energy losses so that the temperature can be maintained. These three aspects indicate that for significant fusion to occur, there is a minimum of the product of the ion number density n, the temperature T, and the confinement time τ. This triple product, sometimes also referred to as the *Lawson criterion*, is often used as a figure-of-merit of fusion systems. For D+T fusion to produce power in a reactor, a triple product $nT\tau$ in excess of a few $10^{21}\mathrm{m}^{-3}\,\mathrm{s\,keV}$ must be achieved.

No solid material can tolerate the high temperatures required for fusion, so much research has been done on two different approaches to contain the plasma within a small volume. The first approach is to heat the reactants – which ionize to become a plasma – with an electric current via Joule heating. Strong magnetic fields are then used to keep the nuclei at high enough densities for reactions to occur, giving this approach its name, *magnetic confinement fusion*. One type of reactor, known as a tokamak, is key to the International Thermonuclear Experimental Reactor (ITER). The plasma will be heated to 1–1.5×10^8 K and retained at this temperature for extended periods while the plasma is confined to a small volume using powerful superconducting magnets that produce a field strength of the order of 10 T. With a plasma density of around $10^{20}\,\mathrm{m}^{-3}$ and confinement times of a few seconds, this will enable a self-sustained net-power-producing reaction. ITER is a large reactor, several metres in radius with a plasma volume of nearly $1000\,\mathrm{m}^3$, intended to provide around 500 MW of thermal power. Most of the energy from the D+T reaction in equation 8.5 is in the neutron, so to generate electricity in a reactor (not planned for ITER), the neutrons as well as the alpha particles need to be absorbed to heat a working fluid. This can then be used to drive a high-pressure steam turbine to generate electricity, just like in other power stations.

The second approach to achieving fusion goes by the name of *inertial confinement fusion*. The approach here is to manufacture the fuel into mm-sized pellets and then to use high-power lasers to compress and heat these pellets to achieve fusion, releasing the energy in a single explosive reaction. One such experiment, the National Ignition Facility (NIF) in the USA, uses 192 lasers with a combined power of 500 TW, although these only operate in bursts lasting a few nanoseconds. The process is not very efficient, with only about 0.5% of the initial 2 MJ of energy from the laser light ending up in the pellets. However, this is enough to compress them to very high densities (at pressures of 10^{16} Pa) and heat them to about 10^8 K, causing the deuterium and tritium in the pellets to undergo fusion. Theoretically this could yield in the order of 100 MJ per pellet. By the end of 2022 only about 3 MJ could be produced, but this was above the 2 MJ of laser energy put in, making it the first fusion experiment to achieve an energy gain Q above 1.0. Although a significant milestone, this overlooks the 400 MJ of energy that was required to charge the laser. NIF is a science experiment. Clearly, to become a useful source of domestic power, or to form the basis of a propellant system, pellets must be fused in rapid succession, but currently the turnover time is many hours.

Magnetic confinement fusion needs large superconducting magnets, which are both massive and consume a lot of energy, making them rather unsuitable for spacecraft. ITER, for example, will have an external diameter of 20 m and a mass of around 5000 t. The lasers of inertial confinement fusion are likewise massive and energy-hungry (just the NIF target chamber is 10 m in diameter and weighs 130 t). Thus the facilities currently being developed for experiments and power plants cannot be used in practical propulsions systems. However, some of the ideas and results coming from this work have inspired the development of hybrid systems that may ultimately lead to practical propulsion methods, as we shall see in section 8.7.

8.5.1 Reactions for rockets

While the D+T reaction is suitable for energy generation on the Earth, it is not suitable as a basis for propulsion for one main reason: 80% of the kinetic energy is carried by the neutron. Being neutral, its momentum cannot be coupled to the spacecraft using a magnetic field (section 8.4), so this energy is essentially lost. With the D+T reaction we could only really make use of one-fifth of the energy released, namely the 3.52 MeV in the alpha particle.

In practice, some of the neutrons from the D+T reaction would collide with the spacecraft and be stopped by it, thereby transferring some of their kinetic energy to the spacecraft. But this is actually a problem for a number of reasons. First, the collisions are inelastic, so release a significant amount of energy that heats the spacecraft. The spacecraft then needs large radiators to remove this heat to avoid melting. Second, these high-energy neutrons can create collision cascades, thereby dislocating atoms from lattices to create point defects that degrade the spacecraft structure. Third, some neutrons would be captured by nuclei in the material of the spacecraft, creating excited, unstable nuclei. These then decay radioactively, releasing alpha particles, electrons, and/or gamma photons, or even undergoing fission, a process known as neutron activation. This is unlikely to be good for the structural integrity of the spacecraft and would demand the use of shielding to protect its instruments. We therefore look to other reactions.

A more promising reaction is the fusion of deuterium with helium-3:

$$D + {}^3He \;\rightarrow\; {}^4He \;(3.67\,MeV) + p \;(14.68\,MeV)\,. \tag{8.7}$$

This too produces a helium-4 nucleus and releases a similar amount of energy as the D+T reaction. But instead of a neutron, it produces a proton. Being positively charged, its energy can be coupled to the spacecraft electromagnetically, potentially allowing us to harness all of the energy released (but see below).

This reaction presents two difficulties compared to D+T, however. The first is that because there are two protons in helium-3 as opposed to just one in tritium, the Coulomb barrier is higher. Energies of order 60 keV must be attained in order to get significant fusion, several times higher than that required for D+T (see figure 8.7). The second difficulty is that helium-3 hardly exists on the Earth, even though it is a stable isotope. It is a primordial isotope, making up 1–10 ppm of the helium on the Earth (which itself is very rare; only 5 ppm of the atmosphere), but most is in the Earth's mantle and so inaccessible. Helium-3 does occur in natural gas sources, but at current rates of gas extraction only a few kg per year could be collected in this way. It is also produced in small quantities by some reactions, including the radioactive decay of tritium, itself produced by reaction 8.6, for example, meaning helium-3 is a trace product from nuclear power stations and weapons testing. Some is also produced by cosmic ray spallation. But these sources are far too small to be a source of fuel for future rockets.

There may be more helium-3 on the surface of the Moon, as the result of interactions of the solar wind with lunar regolith. Estimates are of a few parts per billion in lunar surface rock, a thousand times more than in the Earth's atmosphere. Helium-3 appears to be more abundant in the outer atmospheres of the gas giant planets. The Galileo probe, for example, measured the ${}^3He/{}^4He$ ratio in the atmosphere of Jupiter to be 1.7×10^{-4}. Taking the abundance of helium to be 0.14 by number, this corresponds to a helium-3 number density of around 20 ppm. Although mining helium-3 from there may seem like a stretch of the imagination, not least because of Jupiter's deep gravitational potential well, that may not be the hardest aspect of an interstellar mission.

Another reaction that could be suitable for rockets is the fusion of two deuterons. This can be achieved with a slightly lower temperature than D+^{3}He (figure 8.7), of the order of 30 keV, and has two branches that occur with equal probabilities:

$$D + D \;\rightarrow\; T \;(1.01\,MeV) + p \;(3.02\,MeV)\quad 50\% \tag{8.8a}$$

$$D + D \;\rightarrow\; {}^3He \;(0.82\,MeV) + n \;(2.45\,MeV)\quad 50\%. \tag{8.8b}$$

The first reaction produces two charged products. The second, in contrast, gives three-quarters of its energy to the neutron. These reactions produce less useable energy than the D+^{3}He reaction, with only 66% of the energy going into the charged particles. However, the products in both branches can undergo further fusion reactions. If the conditions are right, the tritium in the first branch can fuse with another deuteron via the D+T reaction (equation 8.5), and the helium-3 in the second branch can fuse with another deuteron via the D+^{3}He reaction (equation 8.7). The overall reaction is

$$3D \;\rightarrow\; {}^4He + p + n + 21.62\,MeV. \tag{8.9}$$

This is sometimes referred to as the catalyzed deuterium reaction. The energy partition among the products is different for the two branches, but overall 62% of the energy goes into the charged particles, and the rest into the neutrons.

The catalyzed deuterium reaction puts more energy in the charged particles per input nucleon than the D+^{3}He reaction (4.76 MeV vs 3.67 MeV), implying a better energy-to-fuel mass ratio, and it uses a readily available fuel, just deuterium. The D+^{3}He reaction appears to have the advantage that it does not produce any neutrons, which can damage the spacecraft through the process of neutron activation mentioned earlier. However, while this is true for the principal reaction, there are inevitably side reactions that do create neutrons, in particular equation 8.8b. Equation 8.8a creates tritium, which can either react with deuterium via equation 8.5 or react with itself, both reactions creating neutrons. Thus the production of neutrons, and the problem this creates, is more or less unavoidable. In practice, though, only about 5% of the kinetic energy would be lost to neutrons resulting from D+D. The lower neutron flux from D+^{3}He gives it a considerable advantage over the catalyzed deuterium reaction.

Several more side reactions occur when trying to fuse D+^{3}He (or indeed other nuclei), not only between D and D, D and ^{3}He, and ^{3}He and ^{3}He, but also between these and the products of those reactions, which include hydrogen, tritium, and small quantities of lithium-5 and beryllium-6. Some reactions give off photons too. Which reactions take place with what relative frequencies depends on the exact conditions of the fusion, in particular the temperature. Thus in practice the products of fusion, and the energy that can be extracted from them, is more complex than has been summarized here.

The fuel for any practical fusion reaction would be stored in atomic or molecular form, and this fully ionized into a plasma before the above fusion reactions occur. But the ionized electrons have not disappeared. Through collisions with the nuclei, the electrons become heated, gaining kinetic energy at the expense of the nuclei. These moving electrons are then decelerated by the electric field in the plasma (generated by the positive nuclei), and so lose some of their kinetic energy again in the form of X-ray photons, a process known as *bremsstrahlung*. If the plasma is not optically thick, these photons will remove energy from the plasma. Bremsstrahlung is an important source of energy loss from fusion reactions, especially for the D+^{3}He and D+D reactions. As this reduces the kinetic energy of the charged fusion products, it also reduces the effective exhaust velocity of a fusion-powered rocket.

Bremsstrahlung X-rays are furthermore problematic as a source of ionizing radiation that can damage the spacecraft structure and as a source of heat that can melt the spacecraft. It may be possible to absorb some of the bremsstrahlung in a fluid that is used in a closed-cycle heat engine to generate electricity. Better still is a direct conversion of the heat to electric power that avoids moving parts, for example via the thermoelectric effect. This could be used either to power systems onboard the spacecraft, or it could be used to drive an electric field that accelerates the charged fusion products further (as in an ion engine; chapter 6). In practice the spacecraft would still need large radiators to remove the excess heat, and these add significant additional mass.

There exist many other fusion reactions, but most are less practical because they have lower fusion cross-sections, require higher temperatures to initiate fusion, or involve nuclei with higher atomic numbers. The latter is problematic because the intensity of the bremsstrahlung is proportional to the square of the atomic charge.

One final reaction worth mentioning, however, is the interaction[6] of boron (B, $Z = 5$) with a proton to produce three alpha particles,

$$^{11}\text{B} + \text{p} \ \rightarrow \ 3\,^4\text{He} + 8.7\,\text{MeV}, \tag{8.10}$$

---

[6] This reaction is actually fission, not fusion, but it is usually considered among fusion reactions.

because this uses a readily available fuel. The energy released per nucleon is relatively low compared to the earlier reactions examined because boron-11 has a large binding energy per nucleon. It also needs a very high temperature to achieve a reasonable cross-section (several hundred keV; see figure 8.7). Although nominally aneutronic, side reactions such as boron reacting with an alpha particle to produce nitrogen and a neutron can remove some of the kinetic energy. However, only about 0.2% of the total energy yield is taken away by neutrons. It does, however, produce significant amounts of bremsstrahlung, because the fully ionized boron fuel contributes a lot of electrons.

8.5.2 Effective exhaust velocities

When the combined kinetic energy of the fusion products is K, and the sum of their masses is m, then classically we can write an equivalent velocity as $\sqrt{2K/m}$. Assuming no energy losses, this is the effective exhaust velocity of the propellant. The quantities K and m only include the particles that can be coupled to the rocket, so this usually excludes the neutrons. For the D+^{3}He reaction in equation 8.7, $K = 18.35\,\text{MeV}$ and $m = 5.009\,m_u$, so $v_e = 0.0887\,c$. This is well below the speed of light, so this classical calculation is a reasonable approximation. The temperature corresponding to the kinetic energy is $2E/3k = 1.4 \times 10^{11}\,\text{K}$, making it obvious that a physical nozzle could not be used to collimate this propellant.

An alternative (and relativistic) way to compute v_e is the approach in section 7.8. For this we need to compute the factors ϵ, η, and ζ. The energy released is the difference between the initial and final masses of the nuclei. For the D+^{3}He reaction, the sum of the initial masses is $2.013553 + 3.014932\,m_u$ or 4684.005 MeV, and the sum of the final masses is $4.001506 + 1.007276\,m_u$ or 4665.652 MeV, giving a difference of $0.0197\,m_u$ or 18.35 MeV. As the kinetic energy required by the parent nuclei to overcome the Coulomb barrier is only a few tens of kiloelectronvolts, it can be neglected. Assuming there are no losses ($\eta = 0$), then the factor ϵ is the fraction of the initial mass converted into energy, which here is $\epsilon = 18.35/4684.005 = 0.003918$. No photons are produced (matter rocket) so $\zeta = 1$. Equation 7.40 gives $v_e = \sqrt{\epsilon(2 - \epsilon)} = 0.0884\,c$. This is very close to what we just computed using classical mechanics in this case.

8.6 Fusion pulse rockets

Having established that fusion reactions produce charged particles with very large kinetic energies, the question is how to use these to propel a rocket. The direct approach is to use a magnetic nozzle to couple the charged fusion products to the rocket, as explained in section 8.4.

One way of achieving this is pulse fusion, in which fusion takes place in sequential pulses, each giving the rocket a small impulse. A well-developed concept for this was presented in the 1970s by a team associated with the British Interplanetary Society. The mission, based on the D+^{3}He reaction, was called Daedalus after the mythological Greek character who made artificial wings to enable him to fly. The goal of the modern Daedalus was to carry an uncrewed payload with a mass of 450 t to Barnard's Star in 50 years. At 6.0 ly distance, this is the second closest stellar system to the Sun (section 2.5), and was chosen because at the time there were claims of the detection of a planet around the star. This journey requires the spacecraft to achieve an average velocity of $0.12\,c$.

According to one plan, Daedalus is fuelled by deuterium and helium-3 mixed into small pellets of mass 3 g (first stage) and 0.3 g (second stage). These are compressed by electron beams or lasers to initiate fusion, and assumed to provide an effective exhaust velocity of $10^4\,\mathrm{km\,s^{-1}}$, or $0.033\,c$. This is a lot lower than the ideal value of $0.089\,c$ derived in section 8.5.2 because of the inevitable inefficiencies in extracting kinetic energy from the fusion.

The total mass of the spacecraft is 52 670 t (figure 8.8) of which 50 000 t is fuel and the remainder is the structure (engines, tanks, payload, etc.) This is split across two stages. The first has a structural mass of 1690 t and 46 000 t of fuel. The second has a structural mass of 980 t, of which 450 t is the payload, plus 4000 t of fuel. The mass ratio (initial/final) of the first stage is $R = 52\,670/(52\,670 - 46\,000) = 7.90$ and thus delivers $\Delta v = 2.1 \times 10^4\,\mathrm{km\,s^{-1}} = 0.069\,c$ (assuming $v_e = 10^4\,\mathrm{km\,s^{-1}}$; the actual design study used slightly different values for the two stages). The structure of the first stage is then ejected. The mass ratio of the second stage is $R = (4000 + 980)/980 = 5.1$ and thus $\Delta v = 1.6 \times 10^4\,\mathrm{km\,s^{-1}} = 0.054\,c$. The total Δv is (classically) the sum of these, $3.7 \times 10^4\,\mathrm{km\,s^{-1}}$ or $0.123\,c$. The actual mission design was more optimized than this, with the fuel in each stage stored in multiple tanks, each of which is discarded as soon as it is empty, which slightly increases the Δv achievable with each stage.

The engines are designed to burn for an appreciable fraction of the mission time. In the first stage, pellets of mass 3 g are fused at a rate of 250 per second, each pellet releasing 170 GJ of energy (over four orders of magnitude more than is currently achieved per pellet

An artist's impression of Daedalus, 190 m tall, shown in comparison to a Saturn V rocket (111 m tall). Credit: Nick Stevens, http://nick-stevens.com/.

Fig. 8.8

in the NIF), corresponding to a power of 44 TW. With 46 000 t of fuel this takes 2.05 years to exhaust. Lower-mass pellets are used in the second stage (each releasing 13 GJ), but fused at the same rate, taking 1.76 years to exhaust. A total of 30 billion pellets are consumed. From these mass expulsion rates we can also compute that the thrusts of the two stages are 7.1 MN and 0.72 MN, respectively.

The enormous propellant mass arises from the large payload mass of 450 t (via the rocket equation). This payload would permit a substantial amount of scientific equipment to be taken. However, this is a flyby mission with an encounter velocity of 0.9 au per hour, leaving little time to gather data near the target star. Daedalus addresses this by including 18 smaller spacecraft that are released well before the encounter. Although they would not decelerate in the nominal mission, they would each fly much closer to different targets of interest. They would send their data back to Earth via the main spacecraft, which repurposed its 40-m-diameter engine reaction chamber as a radio antenna.

It is historically interesting that the first planned scientific investigation to be done from Daedalus as it approached the target was to use onboard telescopes to look for planets. This was the late 1970s, and within 20 years the first exoplanets had been discovered around a nearby star. Given the continuing improvements in technology, and the strong focus on exoplanets by astronomers, we can reasonably expect all but the smallest planets around the target star to be discovered from the Earth (or nearby space telescopes) well before a mission would ever get to its target. But there would still be much to discover and explore from the spacecraft, as discussed in chapter 2.

Given the enormous mass of Daedalus, it was not intended for launch from the Earth's surface. Indeed, most of its mass is helium-3, which the study writers assumed would be mined from the Moon or from the atmospheres of the gas giant planets. The spacecraft would therefore be assembled in orbit about the source planet/Moon and launched from there. Exactly how helium-3 would be mined remained an open question.

We saw in section 8.3.3 that a lot of effort and money was investigated in the past into the use of nuclear fission bombs to power a rocket. Daedalus is similarly propelled by a series of explosions. One of the main differences, other than the magnetic coupling, is that with fusion there is no minimum mass for the reaction, so very small explosions can be used.

There are many other concept studies for interstellar spacecraft based on pulsed fusion engines, in particular using inertial confinement fusion. One set of studies followed up directly on Daedalus, going by the name of Icarus, who in Greek mythology is the son of Daedalus (although unlike his father, he did not survive their joint attempt to escape from Crete using artificial wings).[7] Long et al. (2011) optimized the Daedalus design for a fixed mass of propellant, investigating solutions with one to four stages, still as a flyby mission. They came to the preliminary conclusion that more stages were not necessarily better, because with a fixed propellant mass the extra stages add structural mass and thus decrease the payload mass for the given total mass.

Other Icarus designs focused on smaller payloads and so required much less propellant. One design has a total mass of 2700 t, of which 320 t is the engine, 150 t the payload, and 2000 t the D+^{3}He fuel. It is a single stage rocket (one engine with a jet power of 1.2 TW), but ejects fuel tanks once they are empty. After 2.2 years it would reach its cruise velocity of

---

[7] This should not be confused with a Soviet Daedalus-like project also called Icarus that appeared around the same time as Daedalus, but about which little has been published.

$0.045\,c$ and it would take a total of 98 years to travel to Alpha Cen. This relaxing of the travel time requirement (from 50 years) permits smaller engine powers and smaller mass ratios, and so makes the mission far more doable.

Another variation of the Icarus design called Firefly uses a different plasma confinement approach called the z-pinch. When electrical currents are passed in the same direction along parallel wires, they generate magnetic fields that interact with the currents, creating Lorentz forces that attract the wires to one another. In the same way, a current run through a plasma will compress or 'pinch' the plasma, until the Lorentz force is balanced by the gas pressure. If the compression is large enough, it could initiate fusion. The Firefly design requires a large current of 5×10^6 A and a huge power to sustain this. It uses readily available deuterium in a D+D reaction, thus avoiding the need to find helium-3, but pays the price in having to cope with a large neutron flux that would overheat the spacecraft. Thus a significant fraction of the spacecraft's dry mass has to be dedicated to cooling, in terms of coolant fluid (hundreds of tonnes of beryllium), coolant pumps, and large radiators.

There are many variations of the Icarus design using different fusion reactions and approaches to fusion. They, like the original Daedalus study, demonstrate the complexities and trade-offs in designing a mission. What they have in common is very massive spacecraft of thousands of tonnes with payloads of tens of tonnes getting to one of the nearest stars within a hundred years.

8.7 Other fusion propulsion schemes

Several other approaches have been proposed that use fusion for rocket propulsion. Like the fusion engines of the previous sections, none of these systems have been developed, nor have they achieved net fusion power output. Although some informative experiments have been done with plasmas, many of the detailed results rely on computer simulations. Here we take a brief look at a few different ideas.

The first idea is to use a sustained fusion reaction. By siphoning off some of the high-energy fusion products from the plasma, and directing them with a magnetic nozzle, we obtain thrust. This is analogous to the fission fragment engine (section 8.3.2). One way to implement this is with mirror magnetic confinement, in which the plasma is trapped between two magnetic fields, one of which leaks a small amount of plasma into the nozzle. Fuel must be continuously supplied to the engine, and a sufficiently high temperature and density must be maintained such that the energy siphoned off for propulsion does not stop the reaction (recall the triple product). The challenge with this approach is that the mass flow rate of the products is low, and so despite the high velocity of the fusion products, the thrust remains low.

Another idea is represented by the *direct fusion drive* (DFD), in which energy from the fusion is used to heat up a separate propellant, in analogy to the thermal fission rocket (section 8.3.1). The DFD uses the D+^{3}He reaction of equation 8.7 and is based on the field-reversed configuration. This is a way of confining the plasma in a torus with a magnetic field (the tokamak and stellarator are other examples). A rotating magnetic field heats the plasma using radio-frequency waves, aiming to achieve a temperature of around $100\,\text{keV}$ ($T \sim 10^9$ K) at a density of $5 \times 10^{20}\,\text{m}^{-3}$.

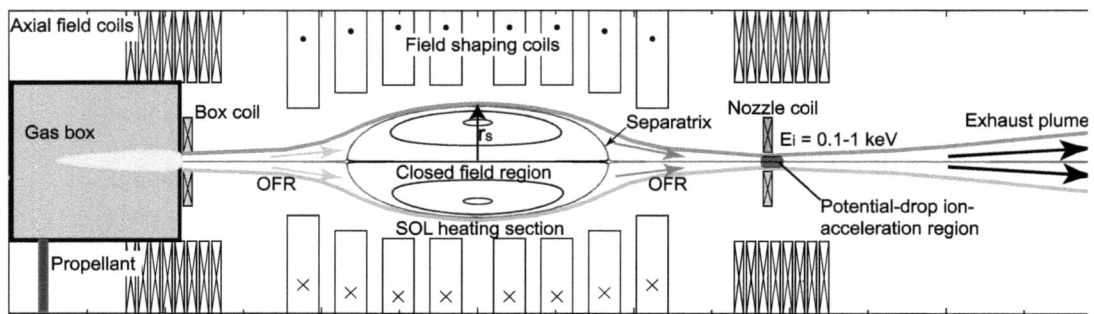

The direct fusion drive. Deuterium from the gas box is heated as it flows around the closed field region in which the fusion occurs. OFR = open field region; SOL = scrape-off layer. Credit: Used with permission of Elsevier Science & Technology Journals, from Razin et al. (2014), conveyed through Copyright Clearance Center, Inc.

The propellant, which is also deuterium, is injected into the magnetic field from the gas box where it is ionized (figure 8.9). This initially cold plasma (comprising deuteron ions and electrons) remains outside of the closed field region where the fusion is taking place. The plasma is heated by collisions with the high-energy fusion products as it flows along the axis and around the closed field region. However, as there is no physical separation, the propellant picks up some of the main fusion products (as well as products of side reactions, like tritium) from the surface of the closed field region, which is known as the 'scrape-off layer'. This now high-energy plasma propellant is directed out of the engine via a magnetic nozzle (section 8.4). More propellant can be added at this point to increase the thrust at the expense of a lower effective exhaust velocity (a technique we will look at in section 8.8.3).

The DFD is in principle scalable to various powers of 1–100 MW. Suggested uses range from Earth-orbit manoeuvres through solar system missions to interstellar travel. Cohen et al. (2019) outline a mission to Alpha Cen using a 100 MW engine. When configured to produce an effective exhaust velocity of $15\,000\,\text{km s}^{-1}$ ($0.05\,c$), this would ideally produce a thrust of $F = 2P/v_e = 13.3\,\text{N}$ (equation 3.23), although the authors assume a power-to-thrust efficiency of just 0.3, so this is lowered to 4 N. If a specific power of $25\,\text{kW kg}^{-1}$ can be achieved, they compute the spacecraft would take about 500 years to get to Alpha Cen. (The specific power is defined here as the power of the engine divided by the mass of the engine, thus excluding the mass of the fuel, propellant, and payload.) However, this specific power is far above the published targeted specific powers of the DFD, which are around $0.3–1.5\,\text{kW kg}^{-1}$. A current design with a plasma volume of $2.2\,\text{m}^3$ and total mass of 14.5 t is predicted to produce 14.5 MW fusion power, of which 10 MW could be utilized, giving $0.7\,\text{kW kg}^{-1}$.

The drawback of reactors currently being developed for Earth-based power generation is that they require massive structures – magnets or lasers – that are impractical for a rocket. Conventional magnetic confinement as done in a tokamak aims to hold a relatively low density plasma at high temperatures for a long period of time. Inertial confinement, in contrast, aims for extremely rapid compression to high densities to achieve fusion far from equilibrium. Some concepts for achieving sustained fusion reactions in a rocket seek to combine aspects of both confinement approaches, an approach known as *magnetized target fusion*.

A rocket concept that uses this combined approach is the *helicity drive*. This uses a relatively modest magnetic field strength (0.3–1 T) to confine a plasma at an intermediate density for an intermediate length of time. While in this state, the plasma is compressed in pulses. This inertial confinement part is done using a specially shaped magnetic field, rather than lasers. As the confinement takes place at higher densities, the engine should be more compact than a tokamak (or similar device), and as the compression can be done over a longer timescale, a less powerful pulsing system is needed. The whole system is cyclic, with pulses applied at a rate of about 1 Hz.

A key idea of the helicity drive is to use magnetic reconnection to heat the ions more efficiently than conventional approaches, by not transferring as much energy to the electrons and so reducing losses due to bremsstrahlung. One consequence of the approach is that the fusion triple product scales with the number of plasma sources N as $N^{3/2}$. This means that fusion can be achieved by simply delivering more plasma. Different fuels and fusion reactions are possible, but D+^{3}He and D+D are the favoured choices.

There have been several other investigations into fusion engines. The *fusion drive rocket* uses a convergent array of magnetic fields to compress a plasma to ignite fusion. This directly heats a solid lithium propellant. A solid propellant is favoured over a liquid or gas propellant because it saves mass on tanks. The goal is to achieve an effective exhaust velocity of order $30 \, \mathrm{km \, s^{-1}}$ in a 100 t spacecraft, making this an interesting application for interplanetary missions, for example travelling to Mars in 30 to 90 days.

Whether or when any of these ideas will see development even on modest scales for space missions is, as of mid 2025, an open question. Some startups, as well as more established companies and governments, are investing into fusion power as part of a long-term move away from fossil fuels. Some of these endeavours are more serious than others, and many will no doubt produce neither fusion power nor rockets. Yet as a long-term, sustainable energy source on the Earth, as well as a high-power source for interstellar travel, fusion power is one of the relatively few options that human civilization will eventually have to develop.

8.8 Antimatter propulsion

The fission of uranium-235 releases about 0.1% of its rest energy into the kinetic energy of its products. The D+^{3}He fusion reaction releases about 0.4%. Matter–antimatter annihilation can convert a much larger fraction of the mass into potentially useable energy.

8.8.1 Main reactions

For almost every type of particle there exists an antiparticle which has many of the same properties as the particle, such as mass, but also some opposite quantities, such as charge. A collision between a particle and antiparticle results in their mutual annihilation, with their masses released as photons or as particles with varying degrees of kinetic energy.

The antiparticle of the electron is the positron, the first antimatter particle to be predicted and discovered. When an electron and position collide at low energies, they annihilate to produce two or more photons:

$$\mathrm{e^+ + e^- \rightarrow 2\gamma \; (+ \; more \; \gamma)}. \tag{8.11}$$

The rest energy of an electron is 0.511 MeV, so if two photons are produced (by far the most likely outcome) then each will have a wavelength of $hc/E = 0.0024$ nm. Any initial kinetic energy that the particles have will result in shorter photon wavelengths. The two photons will be moving in opposite directions, so to use them as a means of propulsion in a photon rocket (section 7.7), they need to be collimated into a narrow beam by means of reflection, which is not easy at these very short wavelengths.

If the electron and positron collide at very large energies – in excess of 105 MeV – then other particles such as muons and antimuons can be produced. At even higher energies other exotic particles including bosons are produced, as is done in large particle colliders for research purposes. But such energies would not be available aboard a spacecraft, so such results can be ignored for our purposes.

The primary matter–antimatter reaction of interest for propulsion is the annihilation of protons by antiprotons. This is more complex than the above reaction because the proton is made up of three quarks (and the antiproton of three antiquarks). At low energies the most probable reaction is

$$\mathrm{p}^+ + \mathrm{p}^- \rightarrow n\pi^+ + n\pi^- + m\pi^0 \quad (2n + m \simeq 5). \tag{8.12}$$

This produces several pions (pi-mesons), which are hadrons comprising two quarks. Due to charge conservation, the same number of positive and negative pions must be produced, and there is often also a neutral pion. The actual number varies, but in what follows we will assume that on average there are two neutral pions and 1.5 of each of the charged pions ($m = 2, n = 1.5$).

Pions are unstable. The mean lifetimes (in their rest frames) are $\tau = 2.6 \times 10^{-8}$ s for the charged pions and $\tau = 10^{-16}$ s for the neutral pion. These decay as follows:[8]

$$\pi^0 \rightarrow 2\gamma, \tag{8.13a}$$

$$\pi^+ \rightarrow \mu^+ + \nu_\mu, \tag{8.13b}$$

$$\pi^- \rightarrow \mu^- + \overline{\nu}_\mu \tag{8.13c}$$

where μ^+ and μ^- denote the muon and antimuon (which are leptons), and ν_μ and $\overline{\nu}_\mu$ denote the (muon) neutrino and antineutrino. These are all fundamental particles, but whereas the neutrinos and the photons are stable, the muons are not. They have a mean lifetime of $\tau = 2.2\,\mu$s and decay as

$$\mu^+ \rightarrow \mathrm{e}^+ + \nu_\mathrm{e} + \overline{\nu}_\mu, \tag{8.14a}$$

$$\mu^- \rightarrow \mathrm{e}^- + \overline{\nu}_\mathrm{e} + \nu_\mu\,. \tag{8.14b}$$

Once all the decays have occurred the overall reaction of the proton and antiproton is

$$\mathrm{p}^+ + \mathrm{p}^- \rightarrow n(\mathrm{e}^+ + \mathrm{e}^-) + 2m\gamma + \text{ neutrinos}. \tag{8.15}$$

8.8.2 Use in a rocket

With the proton–antiproton reaction we have a few options of which products to use. Considering the initial reaction in equation 8.12, the neutral pion is hard to couple to the spacecraft. As we shall see below, this receives about a third of the kinetic energy. We could in principle

[8] The charged pions can also decay to positron/electron and electron neutrino/antineutrino, but this is much rarer.

make use of the photons it decays into (equation 8.13), but as these have energies of the order of hundreds of megaelectronvolts, and thus wavelengths of the order of 10^{-5} nm, they are very hard to direct. Thus about a third of the total energy from the initial proton–antiproton reaction is lost, and we focus instead on capturing the energy of the charged particles. We can either make use of the charged pions before they decay, or of their resulting muons after they decay. The former is preferable because after the decay, some of the pions' energy is given to the neutrinos, which barely interact with matter and so their kinetic energy would be lost.

For low-energy interactions, the total energy available from the initial proton–antiproton pair is the sum of their masses, which is $2 \times 938.3\,\mathrm{MeV} = 1876.6\,\mathrm{MeV}$. As the number of charged pions produced in the first reaction varies, so do their energies, so they have a broad spectrum of total energies. The mean is $E = 389.1\,\mathrm{MeV}$, of which $E_0 = 139.6\,\mathrm{MeV}$ is the rest energy. Assuming three charged pions ($2n = 3$ in equation 8.12) are produced with these energies, then 40% of the rest energy of the fuel goes into the kinetic energy of these pions. As $E = \gamma E_0$, the pions move with a speed of $\gamma = 2.8$ or $\beta = 0.93\,c$ on average. This is their speed relative to the centre-of-mass of the reactants, which is the spacecraft's rest frame. The distance they travel depends on their lifetime in the spacecraft's rest frame, which is $t = \gamma\tau$ from the Lorentz transformations (section 7.2). Thus the distance travelled in the spacecraft rest frame is

$$L = vt = \beta c\gamma\tau = c\tau\sqrt{\gamma^2 - 1} = c\tau\sqrt{\frac{E^2}{E_0^2} - 1} \qquad (8.16)$$

using the relation for $\gamma\beta$ from equation 7.4. We find that charged pions with the mean energy from the reaction travel 20.3 m before decaying. To gain momentum from them, the spacecraft must therefore couple to them within this distance. This would typically be done via a magnetic nozzle as shown in figure 8.10. From equation 8.2 we can estimate the field strength required to collimate them. Charged pions have the same charge as the proton and a mass of $0.15\,m_{\mathrm{u}}$. Thus to turn these pions in a radius of 1 m requires a field strength of $B = 1.2\,\mathrm{T}$. Because the pion energy distribution has a long tail to large energies, most of the pions have smaller energies than the mean value, and so don't move as far before they decay. So they need to be coupled over less than 20 m, although their lower energies also mean their turning radii are smaller.

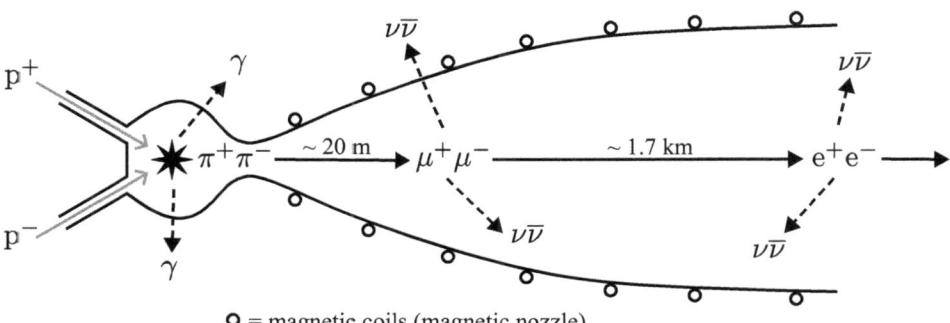

O = magnetic coils (magnetic nozzle)

An antimatter engine based on proton–antiproton annihilation to pions. The momentum of the pions can be coupled to the spacecraft by magnetic fields, which collimate the pions to create a magnetic nozzle. After about 20 m the pions decay to muons, which after another 1.7 km decay to electrons. Adapted from Morgan (1982) and Forward (1982).

Fig. 8.10

Table 8.2 Rest and kinetic energies for the hydrogen–antihydrogen antimatter reaction, given by equation 8.12 for the nucleons and equation 8.11 for the electron and positron. Data taken from Table 2 of Frisbee (2003a).

Particle	Rest energy [MeV]	Kinetic energy [MeV]
Fuel		
p^+	938.3	0
p^-	938.3	0
e^+	0.5	0
e^-	0.5	0
Reaction products		
$2\pi^0$	269.9	439.1
$1.5\pi^+$	209.4	374.3
$1.5\pi^-$	209.4	374.3
2γ	0	1.0

If we wait until the charged pions decay via equation 8.13, the resulting muons again have a range of energies, with an average of 298.0 MeV of which 105.7 MeV is the rest energy. The average neutrino energy is therefore $389.1 - 298.0 = 91.1$ MeV. Thus about 37% of the kinetic energy of the charged pion will be lost if we don't couple it before it decays. About 31% of the rest energy of the original proton–antiproton fuel ends up as the kinetic energy of the muons. The muons have $\gamma = 2.8$, $\beta = 0.93$, more or less the same as the pions, but as their lifetimes are longer the muons travel an average of 1.7 km before decaying.

If we wait even longer, the muons will decay to electrons, positrons, and neutrinos via equation 8.14. The neutrinos carry off about two-thirds of the energy. Many of these will annihilate via equation 8.11, leaving photons that we could then try to harness as propellant in a photon rocket.

Whether we use the pions or muons to propel the rocket, in both cases they are moving at about $0.93\,c$ relative to the rocket on average. We learned in section 7.8 that when a non-negligible amount of fuel mass is converted into energy, the effective exhaust velocity is not equal to the particle velocity. In the present case a significant fraction of the initial mass goes into neither propellant mass nor propellant energy. To be able to apply the relativistic rocket equation (equation 7.27) when we have losses, we need to compute the appropriate effective exhaust velocity.

Table 8.2 shows the typical energies of the particles involved in the reaction of hydrogen with antihydrogen to produce pions, adopting $n = 1.5$ and $m = 2$ in equation 8.12. Let us assume that magnetic fields are used to couple all of the momentum of the two charged pions to the spacecraft, but that the neutral pion is unusable. The initial energy is the sum of the initial rest and kinetic energies in the fuel, which is $2 \times (938.3 + 0.5) = 1877.6$ MeV (we now include the electrons and positrons too). Referring back to figure 7.9, a fraction $1 - \epsilon$ of this energy goes into the rest energies of the propellant particles, the charged pions. Thus

$$1 - \epsilon = \frac{209.4 + 209.4}{1877.6} = 0.2231 \quad \Rightarrow \quad \epsilon = 0.7769. \tag{8.17}$$

The fraction of the initial energy that goes into propelling these particles (and potentially any photons) is $\eta\epsilon$. This is equal to the kinetic energy of the charged pions, assuming that the gamma photons are not used to propel the spacecraft (so they will be counted as part of the losses). Thus

$$\eta\epsilon = \frac{374.3 + 374.3}{1877.6} = 0.3987, \tag{8.18}$$

from which it follows that $\eta = 0.5132$. We are assuming $\zeta = 1$ (no photons used for propulsion), so putting these three numbers into equation 7.39 we get $v_e = 0.5804\,c$.

The effective exhaust velocity is less than the velocity of the pions, calculated above to be $0.93\,c$. This is because of the energy lost to the neutral pions (and a tiny amount to the photons). These losses are a fraction $(1 - \eta)\epsilon = 0.3782$ of the initial energy, which is $710.1\,\text{MeV}$. This is the sum of the rest and kinetic energies of the neutral pions and the energy of the photons. If we could somehow couple also the neutral pions to the spacecraft, then we would achieve

$$\epsilon = 1 - \frac{209.4 + 209.4 + 269.9}{1877.6} = 0.6332, \tag{8.19}$$

and no losses (we ignore the small energy of the gamma), so $\eta = 1$, giving $v_e = 0.93$, which is indeed the velocity of the pions.

8.8.3 Thrust augmentation

We have found that exploiting the charged pions from hydrogen–antihydrogen reactions can theoretically deliver $v_e = 0.58\,c$. Using this to propel a rocket to a velocity of $0.1\,c$ so we can get to Alpha Cen within 50 years would need an initial-to-final mass ratio of only 1.19 from the relativistic rocket equation (the classical one gives virtually the same result). Thus just 16% of the total rocket mass needs to be fuel.

However, the dry spacecraft mass (everything minus fuel/propellant) will not be small on account of the engine (fuel containment, magnetic nozzle, etc.). So if the dry spacecraft mass were 100 t (surely optimistic), we would still need 19 t of fuel, which is 9.5 t of antihydrogen. This is 15 orders of magnitude more than the current annual production (see section 8.8.4).

Assuming that the amount of antihydrogen available is limited, we can achieve a larger thrust by mixing in more hydrogen and letting the high-energy pions heat the excess non-annihilated hydrogen through collisions. This will of course lower the effective exhaust velocity. But it turns out this can still produce a larger Δv overall.

This idea is quite general and not specific to antimatter or even nuclear reactions. To calculate the resulting Δv, we turn again to the general relativistic formulation for the mixed matter–photon rocket in section 7.8, now for the case of a matter rocket ($\zeta = 1$). Suppose the total mass of fuel is m_a. For the antimatter rocket this is half hydrogen and half antihydrogen. Let m_p be the mass of inert propellant that will be added to the exhaust; it does not undergo any reactions and does not release energy. The fraction of the initial mass that is ejected as propellant is

$$\frac{(1 - \epsilon)m_a + m_p}{m_a + m_p} = \frac{1 - \epsilon + x}{1 + x}, \tag{8.20}$$

where $x = m_p/m_a$. Using this in place of $(1 - \epsilon)$ in the conservation of momentum (equation 7.36), the effective exhaust velocity (equation 7.37b) becomes

$$v_e = \left(\frac{1 - \epsilon + x}{1 + x}\right) \gamma_m v_m. \tag{8.21}$$

In the conservation of energy equation (equation 7.38b), the mass fraction to be accelerated is $1 - \epsilon + x$ with the same amount of energy as before, and so the velocity of the particles is given by

$$\gamma_m = \frac{\eta\epsilon}{1 - \epsilon + x} + 1. \tag{8.22}$$

Continuing as before, we find that the effective exhaust velocity of the thrust augmented matter rocket is

$$\frac{v_e}{c} = \frac{\sqrt{\eta\epsilon(\eta\epsilon + 2[1 - \epsilon + x])}}{1 + x}. \tag{8.23}$$

When $x = 0$ we recover the non-augmented result. When $\epsilon \ll 1$ or $x \gg 1$ we get

$$\frac{v_e}{c} \simeq \sqrt{\frac{2\eta\epsilon}{1 + x}} \quad (\epsilon \ll 1 \text{ or } x \gg 1). \tag{8.24}$$

This is the same result that we get for the velocity of propellant particles when we do a classical calculation in which a fraction $\eta\epsilon$ of the rest energy of mass m_a provides the kinetic energy of the remaining mass, that is,

$$\eta\epsilon m_a c^2 = \frac{1}{2}([1 - \epsilon]m_a + m_p)v_e^2, \tag{8.25}$$

which, being classical, requires $\epsilon \ll 1$.

Let us use these results to compare how the Δv attained by a thrust augmented rocket compares with that of a non-augmented rocket. The mass ratio is $R = (m_d + m_a + m_p)/m_d$, where m_d is the dry mass of the rocket (without fuel or propellant), and for the non-augmented rocket $m_p = 0$. Using this and the appropriate values of v_e in the relativistic rocket equation (equation 7.27), we can compute Δv. Figure 8.11 shows the case for $m_d = 100$ t and $m_a = 1$ t (although we can use any units as only the mass ratios are relevant) for increasing R, that is, increasing m_p. We see that although augmenting the thrust decreases v_e, it significantly increases Δv. The non-augmented rocket has $v_e = 0.5804\,c$, $R = 1.01$, and $\Delta v = 0.0058\,c$. If we augment this with $m_p = 1$ t of inert propellant ($R = 1.02$), v_e only decreases slightly to $0.5325\,c$, but Δv nearly doubles to $0.0105\,c$. If we augment with $m_p = 393$ t ($R = 4.94$), we achieve $\Delta v = 0.0718\,c$ at $v_e = 0.0449\,c$. This is the optimal amount of inert propellant to add (the maximum of the curve). Adding more inert propellant then lowers Δv, because the increase in ejected mass is now being offset faster by the decrease in v_e.

We see that adding inert propellant increases Δv compared to just using the antimatter. In practice we need more mass for the propellant tanks and pumps, which would increase m_d, but a net increase in Δv should still be achievable. Adding inert propellant won't give a larger Δv than when burning that additional propellant as fuel, of course. But adding inert propellant becomes a useful option if the fuel is hard to acquire (as in the case of antimatter) or if we cannot react it fast enough to get a large thrust (as might be the case with fusion).

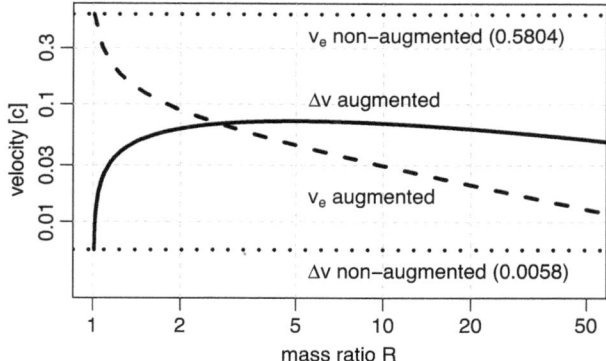

The thrust augmented hydrogen-antihydrogen rocket driven by the charged pions (equation 8.12, which gives $\epsilon = 0.7769$, $\eta = 0.5132$, $\zeta = 1$) with a dry mass of $m_d = 100$ and unit mass matter+antimatter fuel ($m_a = 1$; the units are arbitrary). The dashed line shows how the effective exhaust velocity (equation 8.23) varies as we increase the mass m_p of inert propellant added to the reactant, in terms of the rocket's mass ratio $R = (m_d + m_a + m_p)/m_d$. The solid line shows the corresponding Δv achieved by the rocket according to the relativistic rocket equation. The two horizontal dotted lines show the same quantities for the non-augmented rocket, where $m_p = 0$, which here are independent of R. Note the log scales on both axes.

Fig. 8.11

8.8.4 Practicalities

Use of antimatter as an energy source presents two significant challenges. The first is that it hardly exists in the local universe. Some is produced via interactions with cosmic rays, and positrons are produced in some beta decay processes. Antiprotons can be produced in particle accelerators by firing protons at very high energies into a dense target, such as tungsten. The resulting antiprotons have a very large spread in energies, so the beam must be cooled to be able to store them in a magnetic ring. Current production rates are extremely small. Accelerators produce of order 10^{16} antiprotons (10^{-11} kg) per year, although of course the experiments are not designed to produce them on industrial scales. There are ideas to increase annual worldwide production to around 0.02 kg, but this is still far below the amount needed to propel a fast spacecraft of any size.

The second problem with antimatter is its storage. As a spacecraft would be made from normal matter, storage cannot involve contact and would have to use electric or magnetic fields instead. Antiprotons are currently stored in motion in circular storage rings, but only for short periods (accelerated charge loses energy through synchrotron radiation), and these magnetic systems are likely to be too massive for a spacecraft. One idea for high-density storage is as molecular antihydrogen ice suspended in a magnetic field, which must be held at very low temperatures (1–2 K) to prevent evaporation. Small amounts could be extracted and channelled electromagnetically to the combustion chamber as required.

One advantage of antimatter propulsion is that the reaction occurs spontaneously. Unlike fusion reactors, we do not need massive equipment for inertial or magnetic confinement to initiate or sustain the reaction. This mass saving could offset some of the mass needed for the antimatter confinement.

Antimatter has an alternative potential use in propulsion as a means of triggering nuclear fission in subcritical masses. As an antiproton has a negative charge, it can penetrate the nucleus of a uranium atom. Here the antiproton annihilates with a proton (equation 8.12) and the resulting high energy pions can induce nuclear fission. This can even induce fission in uranium-238. The resulting energy and momentum can then propel a rocket in the same way as either a fission fragment rocket (section 8.3.2) or a fission pulse rocket (section 8.3.3). The advantage of the antimatter-induced fission, however, is that it no longer needs a critical mass, and so fission can be achieved with very small amounts of fuel. Only microgrammes of antihydrogen may be necessary to induce the fission. This approach is particularly attractive for the deceleration of spacecraft at its target destination.

8.9 Summary

Nuclear reactions release about 10^7 times more energy per unit mass than chemical reactions, and so in principle can deliver in the order of $10^{3.5} = 3000$ times larger effective exhaust velocities (as $v \sim \sqrt{K}$).

The products of nuclear reactions could be exhausted directly from the rocket. As the products have extremely high energies, they cannot be collimated with a conventional nozzle. The charged products can instead be coupled to the spacecraft using a magnetic nozzle.

One example of direct use is the fission fragment rocket, which can theoretically achieve $v_e = 0.039\,c$ with uranium-235 fuel. The thrusts tend to be low, however. The fusion pulse rocket instead extracts energy and momentum from a rapid sequence of small fusion explosions. Using the D+^{3}He reaction, this can theoretically deliver $v_e = 0.089\,c$. The Daedalus and Icarus projects have outlined concepts that could get to Alpha Cen in 50–100 years. A big drawback of this reaction, though, is the lack of helium-3 on Earth, which would have to be mined from the Moon or Jupiter.

The energy from a nuclear reaction could instead be used to heat an inert propellant, as in the thermal fission rocket or direct fusion drive. In the former case this lowers the effective exhaust velocity significantly, but can provide quite large thrusts.

Most nuclear reactions put a lot of energy into neutrons or produce bremsstrahlung, both of which will heat and damage the spacecraft. Shielding and massive large radiators for cooling will be essential.

Antimatter reactions release much more energy than fission or fusion. The annihilation of protons with antiprotons to produce pions, for example, can provide a v_e of up to $0.580\,c$. However, only tiny amounts of antimatter can currently be produced, so this remains a very far-fetched concept.

Perhaps counter-intuitively, the Δv of a rocket can be increased by mixing in an inert propellant, such as hydrogen, a process known as thrust augmentation. Although this lowers the effective exhaust velocity, it can increase the mass flow rate sufficiently to more than compensate for this and the extra propellant mass.

Nuclear fusion is one of the best options we have for accelerating spacecraft to 3–9% the speed of light, and so to get to Alpha Cen in 50–100 years. Such spacecraft would necessarily be massive (on the order of thousands of tonnes, or more), much of which would be fuel and propellant and probably cooling systems, but they could then carry fairly substantial payloads.

8.10 Exercises

1. Following the approach in section 8.3.2, work out the initial mass and propellant power of a fission fragment rocket that can get to Alpha Cen within 100 years. Assume the engine has an effective exhaust velocity $v_e = 0.017\,c$ and that the total initial mass is 200 t plus the mass of the propellant. Write code to compute the time taken, using the relevant equations from chapter 3. There is no unique answer – the spacecraft could accelerate rapidly then spend most of the journey cruising, or it could accelerate more leisurely for most of the journey – so it's probably easiest just to use trial and error. This exercise should give a feel for the trade-offs involves, and an appreciation for the magnitudes of the quantities.

2. Using both approaches described in section 8.5.2, compute the effective exhaust velocity for the catalyzed deuterium reaction of equation 8.9 assuming the neutron (a) can, and (b) cannot, be coupled to the rocket.

3. In section 8.8.3 we looked at thrust augmentation in an antimatter engine. The concept applies to a classical engine too, for example mixing inert cold hydrogen into a hot exhaust. Suppose a non-augmented engine exhausts mass m_a at rate $\dot{m}_a$ and velocity v_a. We now mix in an inert propellant of mass m_p at rate $\dot{m}_p$. (i) Show that the effective exhaust velocity becomes $v_a/\sqrt{1 + m_p/m_a}$. (ii) Show that this is the same as equation 8.24 when $\eta = 1$. (iii) Compute the thrust for the non-augmented and the augmented cases. Show that the thrust in the augmented case is larger when $\dot{m}_p/\dot{m}_a > \sqrt{1 + m_p/m_a} - 1$. (iv) Equation 5.10 predicts that adding a low-mass inert propellant – in order to lower the molar mass of the exhaust – *increases* the effective exhaust velocity. How can we reconcile this with the answer to part (i)?

Relativistic optical effects

Special relativity affects what an interstellar traveller observes. This includes the apparent direction, size, and brightness of a source, as well as the frequency of signals and the wavelength of light. We examine these phenomena – aberration and the Doppler effect – and see how they arise from the Lorentz transformations. General relativistic effects, due to light passing through gravitational fields, are not considered.

9.1 Source direction and size: aberration

Aberration is the effect by which the observed position of an object changes when its transverse velocity relative to the observer changes. This is an everyday experience. If rain is falling vertically to the ground, then when a person starts to run, they experience the rain arriving diagonally (figure 9.1). The rain appears to have acquired some horizontal velocity. Generally, if $\mathbf{u}_{rg}$ is the velocity of the rain relative to the ground, and $\mathbf{u}_{pg}$ is the velocity of the person relative to the ground, then $\mathbf{u}_{rp} = \mathbf{u}_{rg} - \mathbf{u}_{pg}$ is the velocity of the rain relative to the person.

The same thing occurs if we replace rain drops with photons. Consider a light source shining in the negative y' direction in its rest frame S' (figure 9.2). The velocity of the photons in this frame is $(u'_{x'}, u'_{y'}, u'_{z'}) = (0, -c, 0)$. This source moves at velocity v relative to the observer (at rest in frame S) in the x direction. Frame S' is like the ground in the previous example, but the person is now running to the left. Under a classical Galilean transformation, where special relativity is neglected, the observer measures velocities in their rest frame of $(u_x, u_y, u_z)_{\text{classical}} = (v, -c, 0)$. Thus the observer sees the photons arriving at angle

$$\theta_{\text{classical}} = \arctan\left(\frac{|u_x|}{|u_y|}\right) = \arctan \beta \tag{9.1}$$

where as usual $\beta = v/c$. I've taken the absolute values of the velocities because for now we only consider the magnitude of the aberration angle, not its sign. Using instead the relativistic Lorentz transformation (equation 7.13), the observer measures velocities in their rest frame of $(u_x, u_y, u_z)_{\text{relativistic}} = (v, -c/\gamma, 0)$. The observer then sees the photons arriving at an angle of

$$\theta_{\text{relativistic}} = \arctan\left(\frac{|u_x|}{|u_y|}\right) = \arctan \gamma\beta = \arcsin \beta. \tag{9.2}$$

For velocities small compared to the speed of light, the classical and relativistic results are very similar, as we can see from a Taylor expansion:

$$\theta_{\text{classical}} = \arctan \beta = \beta - \frac{\beta^3}{3} + \mathcal{O}(\beta^5) \tag{9.3a}$$

$$\theta_{\text{relativistic}} = \arcsin \beta = \beta + \frac{\beta^3}{2 \cdot 3} + \mathcal{O}(\beta^5). \tag{9.3b}$$

The concept of aberration. A person running experiences the rain as coming from a different direction ($\mathbf{u}_{rp}$, middle panel) compared to when standing still ($\mathbf{u}_{rg}$, left panel).

Fig. 9.1

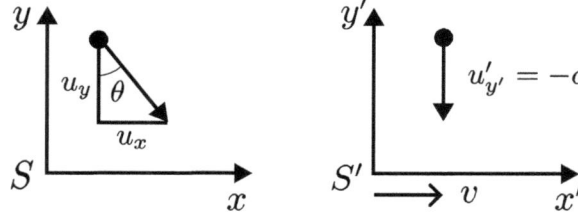

Aberration of light. In reference frame S', where the photon source is at rest, the photon moves vertically down. This frame moves at velocity v along the x-axis with respect to reference frame S. In frame S the photon appears to arrive at an angle θ from the vertical.

Fig. 9.2

Figure 9.3 shows the size of the aberration as a function of the relative velocity. For a spacecraft with a velocity $\beta = 0.01$, the aberration is $0.6°$, and the difference between the classical and relativistic predictions is $0.1''$, so becomes measurable.

At low velocities, aberration is small, but is still significant in astronomical observations. The Earth orbits the Sun in a near-circular orbit with a velocity of $30 \, \mathrm{km \, s^{-1}}$. The velocity of a star relative to an Earth-based observer therefore changes by $\pm 30 \, \mathrm{km \, s^{-1}}$ over the course of a year. This induces an observed change in position in an inertial reference frame of $\pm 20.5''$. The difference between the classical and relativistic predictions here is just $0.1 \, \mathrm{\mu as}$. The rotation of the Earth about its axis induces another change in position of up to $\pm 0.3''$ over the course of the day.

The magnitude of the velocity of the photon in S' is $(u_x^2 + u_y^2 + u_z^2)^{1/2}$. With the Lorentz transformation this is equal to c, as expected, but with the Galilean transformation it is greater than c.

Let us now consider the more general case in which a photon moves at an angle α' to the x'-axis in S', as shown in figure 9.4. S' is moving to the right relative to S as usual, but the origin of S' is now shown to the left of S to make the geometry a bit clearer. In S' the velocity of the photon along the x'-axis is $u'_{x'} = c \cos \alpha'$. In S the x-component of its velocity is $u_x = c \cos \alpha$. Substituting these into the Lorentz transformation for this velocity component (equation 7.13), we find that the photon is observed in S to be moving in the direction

$$\cos \alpha = \frac{\cos \alpha' + \beta}{1 + \beta \cos \alpha'} \, . \tag{9.4}$$

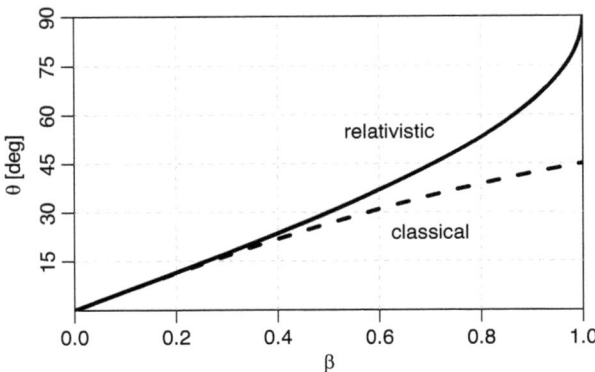

Fig. 9.3 The magnitude of the aberration (change in position) when a source moves with a transverse velocity relative to the observer of β, for the classical case (dashed line) and the relativistic case (solid line).

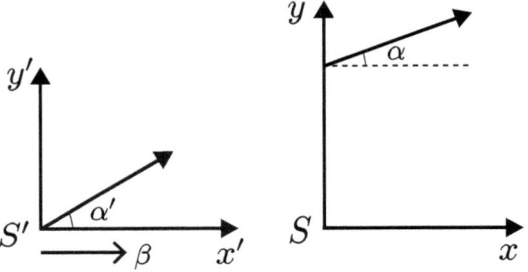

Fig. 9.4 Relativistic aberration. A light source at rest at the origin of frame S' emits photons in direction α' to the x'-axis. Frame S' moves with velocity β relative to frame S parallel to the x-axis. An observer at rest in S observes the photons to be moving in a direction α to the x-axis given by equation 9.4.

This function is shown in figure 9.5. Note that α is the angle at which the photons are travelling in S. The angle at which they appear to be coming from is therefore $\alpha + 180°$.[1] If a source is emitting photons only in direction α', then the observer in S will receive them only if the source is positioned at an angle $\alpha + 180°$ in S.

Consider a source emitting isotropically in its rest frame. For the spatial configuration shown in figure 9.4 with $\beta > 0$, the source is approaching the observer and the received angle α is smaller than the emitted angle α'. Once the source moves past the observer in S, $90° < \alpha < 180°$ and the source is receding; α is still less than α' (figure 9.5), but this corresponds to the received photons making a *larger* angle to the *negative* x-axis. Thus an approaching source always appears less inclined with respect to the line of sight than when it was at rest. For a receding source the opposite is true: it always appears more inclined with respect to the line of sight than when it was at rest. When observing from a spacecraft moving

---

[1] If we define the angles α and α' instead as the direction which the photons are coming from, then because $\cos(\alpha \pm \pi) = -\cos(\alpha)$, equation 9.4 becomes

$$\cos \alpha = \frac{\cos \alpha' - \beta}{1 - \beta \cos \alpha'}.$$

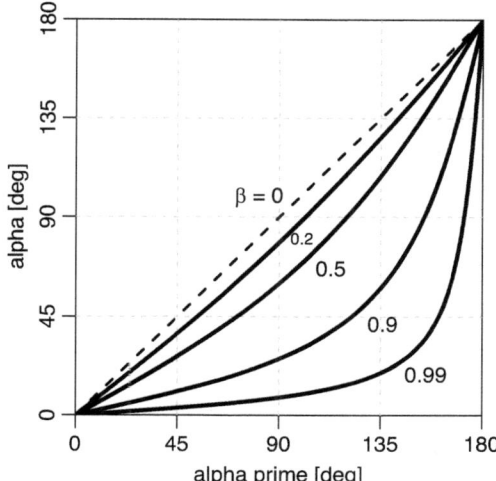

Fig. 9.5

The relation between the direction of travel of a photon α' in frame S' and the direction of travel α observed in frame S (configuration in figure 9.4) for various values of the velocity β of S' relative to S. This is given by equation 9.4. For negative values of β the curves are mirrored about the $\alpha = \alpha'$ ($\beta = 0$) line. Note that β gives the relative direction of motion of the two frames. Whether a specific source is observed to be approaching or receding depends also on its angular coordinate relative to the observer, that is, whether S is to the left or right of S' in figure 9.4.

at high velocity, stars in the forward direction will appear more bunched up in the direction of motion compared to when the spacecraft was at rest, because those stars are approaching the spacecraft. Conversely, stars behind the spacecraft will appear more spread out than they were when the spacecraft was at rest. This is illustrated in figure 9.6.

For small angles in both frames, we find that (see exercises)

$$\alpha \simeq \alpha' \sqrt{\frac{1 - \beta}{1 + \beta}} \quad (\alpha' \ll 1). \tag{9.5}$$

For an approaching source ($\beta > 0$), light is received at an angle closer to the direction of travel the faster it is moving.

Aberration is purely a result of the change in velocity of the observer relative to the source. It has nothing to do with their change in separation over time. Indeed, as we move away from a source of finite size at a finite distance, we would expect it to get smaller simply from the fact that we are getting further away from it. But our motion actually induces the opposite effect at any instant because of aberration (although in practice this is not noticeable).

Accounting for aberration is essential for navigation, as we shall see in chapter 14. Even for the low orbital velocity of the Earth mentioned above ($30\,\mathrm{km\,s^{-1}}$), the aberration of $\pm 20.5''$ for stars means that the pointing of telescopes has to be adjusted over the course of the year. Indeed, aberration was discovered and explained by James Bradley in 1727, in the course of trying to measure the parallax of nearby stars. Aberration is nearly 30 times as large as the largest stellar parallax, so it is not surprising that it was discovered nearly 100 years before the first stellar parallaxes were measured. Aberration, unlike parallax, does not depend on the distance to the stars. Furthermore, the displacement of a star due to aberration is 90° out of phase with respect to the displacement due to parallax.

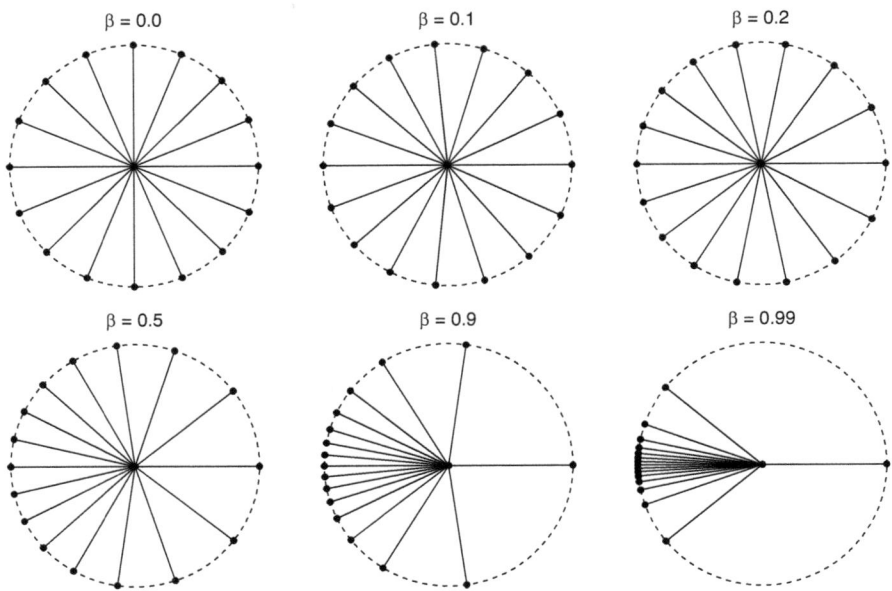

$\beta = 0.0$ $\beta = 0.1$ $\beta = 0.2$

$\beta = 0.5$ $\beta = 0.9$ $\beta = 0.99$

Fig. 9.6 The effect of aberration on the observed positions of objects, such as stars. The stars (points on the circle) are at rest with respect to each other. When the spacecraft (at the centre of the circle) is at rest with respect to the stars, the stars are regularly spaced around the spacecraft (the top-left panel with $\beta = 0$). The faster the spacecraft moves along the horizontal axis to the left, the closer to the direction of travel the stars appear to lie. In the context of figure 9.4, the stars (rest frame S') are moving to the right with respect to the spacecraft (frame S), so β is positive.

By changing the angular position of point sources, it follows that aberration also changes the angular size of extended sources. The solid angle element is $d\Omega = d\phi\, d\alpha \sin\alpha$, where ϕ is the azimuthal angle in spherical coordinates. Because ϕ is defined in terms of the lengths perpendicular to the direction of motion, that is, in terms of y and z in equation 7.10, and because these lengths are invariant between S and S', the angle ϕ is also invariant. Thus

$$\frac{d\Omega}{d\Omega'} = \frac{\sin\alpha\, d\alpha}{\sin\alpha'\, d\alpha'} = \frac{d(\cos\alpha)}{d(\cos\alpha')} = \gamma^2(1 - \beta\cos\alpha)^2. \tag{9.6}$$

In the last step I evaluated $d(\cos\alpha')/d(\cos\alpha)$ using the inverse version of equation 9.4 – with primes swapped and the sign of β changed – to obtain a result in terms of α rather than α'. As the quantity on the right of the above equation appears in many places in relativistic optics, I denote it with

$$\psi = \gamma(1 - \beta\cos\alpha). \tag{9.7}$$

(We will examine this function below in figure 9.9 when we discuss the Doppler effect.) For a moving observer, the solid angular size of an object is multiplied by ψ^2 compared to its size when the observer was at rest. When directly approaching a source, $\alpha = 0$ and $\beta > 0$ for the spatial configuration[2] shown in figure 9.4, so this factor is $\gamma^2(1 - \beta)^2 = (1-\beta)/(1+\beta)$ (equation 7.5), which is less then unity. Thus the approaching source appears

---

[2] The observer would likewise be approaching the source if the source S' lay to the right of the observer S, with $\alpha = 180°$ and $\beta < 0$.

smaller. Receding sources, conversely, appear larger. This is consistent with figure 9.6. We will use this result again in section 9.3 when we consider the brightness of extended sources.

There are many other interesting optical effects induced by relativistic motion. One goes by the name of the Terrell–Penrose rotation. If we take a photograph of an extended object, we observe the photons that arrive at the observer (camera) at the same time, but which may have left different parts of the object at different times due to the finite speed of light. This makes small moving objects appear rotated, and large moving objects will appear distorted. In practice this is usually dominated by other effects, so is hard to observe.

9.2 Source frequency and wavelength: the Doppler effect

When a light source emits regular pulses at frequency ν' in its rest frame, then the frequency at which those pulses are received by an observer is affected by their relative motion via two phenomena. First, the finite speed of light means that pulses arrive less frequently when the source is receding from the observer, because the source is further away at each successive pulse and so has to cover a larger distance at the same velocity. Second, there is a time dilation between the source's and the observer's rest frames. The first effect is a classical phenomenon, and is familiar as the lowering of pitch (frequency) of acoustic waves as a vehicle with a siren drives past and recedes into the distance. The second effect is purely relativistic.

Consider the source to be emitting photons isotropically in its rest frame S'. The source is moving at velocity β relative to an observer O, who in their rest frame S observes the photons moving in a direction α, as shown in figure 9.7. Suppose the period between pulses in S' is T'. In frame S, the source emits a pulse when it is at position x_1 at time t_1, and the next pulse when it has moved to position x_2 at time t_2. Due to time dilation, the time in frame S between the pulses is $t_2 - t_1 = \gamma T'$ (equation 7.11). In order to be received by the observer, the first pulse must travel a distance r_1 and the second pulse a distance r_2. The time that elapses between the arrival of the two pulses at the observer, which is the period in S, is

$$T = \left(t_2 + \frac{r_2}{c}\right) - \left(t_1 + \frac{r_1}{c}\right) = (t_2 - t_1) + \frac{r_2 - r_1}{c} = \gamma T' + \frac{r_2 - r_1}{c} . \qquad (9.8)$$

From the geometry in figure 9.7, if the distance travelled by the source between the pulses is small compared to the distance to the observer, then $(r_2 - r_1) \simeq -\Delta r = -(x_2 - x_1)\cos\alpha$. The distance travelled by the source is $x_2 - x_1 = (t_2 - t_1)\beta c = \gamma T'\beta c$. Substituting this into equation 9.8 gives

$$T = T'\gamma(1 - \beta\cos\alpha) . \qquad (9.9)$$

When using this equation, recall that the direction to the source from the observer is $\alpha + 180°$.[3]

As the frequency is the inverse of the period, we can also write

$$\nu = \frac{\nu'}{\gamma(1 - \beta\cos\alpha)} = \frac{\nu'}{\psi}, \qquad (9.10)$$

---

[3] If we instead define α as the direction to the source, then the minus sign in equation 9.9 becomes a positive sign.

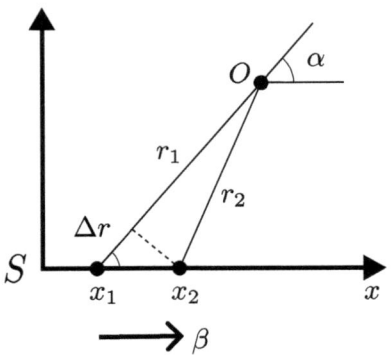

Fig. 9.7 The Doppler effect. A source emits light pulses at frequency ν' in its rest frame. It moves at velocity β relative to an observer O at rest in frame S. When the observer receives light pulses moving in direction α relative to the direction of travel, the observed frequency is given by equation 9.10. The points x_1 and x_2 are the positions of the source in S when it emits its pulses. This is the same configuration as in figure 9.4.

where we see again the ψ term appearing (equation 9.7). The above equation applies for any periodic signal, so could refer to the frequency of the photons, and therefore also to their energies, as $E = h\nu$. For photons specifically $c = \lambda\nu = \lambda'\nu'$, so the wavelength dependence is as for the period:

$$\lambda = \lambda'\gamma(1 - \beta\cos\alpha) = \lambda'\psi. \tag{9.11}$$

A source moving radially away from the observer emitting photons back to the source has $\alpha = 180°$, $\beta > 0$ (source to the right of the observer in figure 9.7). Using $\gamma(1 + \beta) = \sqrt{(1 + \beta)/(1 - \beta)}$ (equation 7.6) in equation 9.11 with $\alpha = 180°$ we get

$$\lambda = \lambda'\sqrt{\frac{1 + \beta}{1 - \beta}} \quad \text{(radial)}. \tag{9.12}$$

This is plotted in figure 9.8 as the line marked 'radial (receding)'. For a source receding at $\beta = 0.1$, $\lambda/\lambda' = 1.11$: the wavelength is increased, or redshifted. The quantity $\lambda/\lambda' - 1$ is called the redshift in astronomy. For an approaching source we could either have $\alpha = 0°$ and $\beta > 0$, or $\alpha = 180°$ and $\beta < 0$. Equation 9.12 uses $\alpha = 180°$, and with $\beta = -0.1$ we get $\lambda/\lambda' = 0.90$, meaning the wavelength is decreased, or blueshifted.

For a source moving tangentially with respect to the observer, $\alpha = 90°$, so

$$\lambda = \lambda'\gamma \quad \text{(transverse)} \tag{9.13}$$

shown in figure 9.8 as the line marked 'transverse'. Successive pulses travel the same distance, so there is now no Doppler effect arising from the different light travel times. Here the Doppler effect is caused only by the time dilation, so is a purely relativistic phenomenon. As $\gamma \geq 1$, this transverse Doppler effect always produces a redshift for the observer. Given that $\lambda \propto 1/\nu$, wavelength behaves like time under time dilation, and indeed equation 9.13 has the same form as equation 7.11.

The variation of the general Doppler shift (equation 9.11) with angle for various velocities is shown in figure 9.9. To interpret this, consider a large spherical surface covered with stars that are at rest with respect to one another (in frame S'), and a spacecraft at the centre of the sphere moving in some direction. Figure 9.9 shows how the Doppler shift, λ/λ', varies for

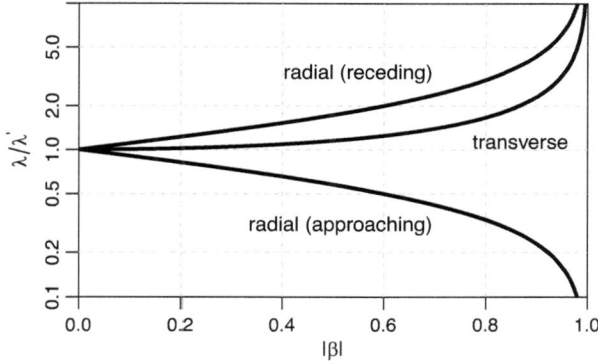

The radial and transverse Doppler effects. Note the logarithmic vertical scale.

Fig. 9.8

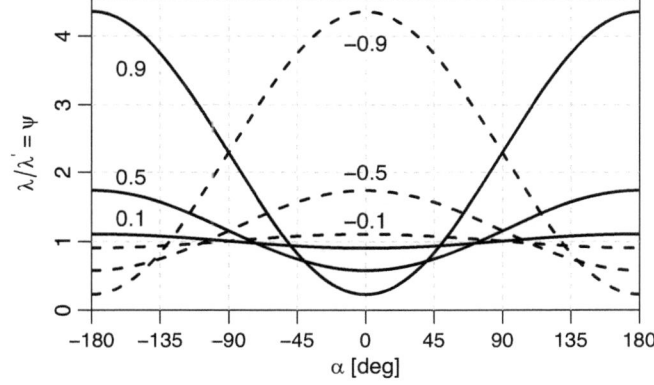

The variation of the Doppler effect (equation 9.11) with angle of travel of the photons for various velocities (β, labelled on the curves) for the configuration in figure 9.4. The curves for negative β are just phase shifted by 180° relative to the curves for positive β. A source moving radially away from the observer emitting photons back to the source has $\beta > 0$ and $\alpha = \pm 180°$ (or $\beta < 0$ and $\alpha = 0°$). The term $\psi = \gamma(1 - \beta \cos \alpha)$ appears in many expressions in relativistic optics.

Fig. 9.9

photons moving in direction α, which corresponds to light received from a star that lies at an angle $\alpha + 180°$ from the positive x-axis. Recall that β is the velocity of the stars relative to the spacecraft (figure 9.4), so a positive β corresponds to the spacecraft moving to the left in the rest frame of the stars.

Consider the line with $\beta = 0.5$ in figure 9.9. A value of $\alpha = 0°$ corresponds to light received from stars lying further off to the left directly in front of the spacecraft. These stars are approaching the spacecraft and their light is blueshifted ($\lambda/\lambda' < 1$). Increasing α, that is, examining the light that is increasingly off-axis in the direction of travel, we see that λ/λ' increases to the point that at sufficiently large angles the light becomes redshifted ($\lambda/\lambda' > 1$). For larger β the curve has a similar shape, but the variation of the Doppler shift is larger.

For any velocity there is an angle α at which the signal crosses from blueshift to redshift. This is found by setting $\lambda/\lambda' = 1$ in equation 9.11; the result is plotted in figure 9.10. For small positive velocities, stars lying in the forward direction of the spacecraft (photons moving in directions $-90° < \alpha < 90°$) are blueshifted, and all those in the backward direction are redshifted. As the velocity increases (β more positive), the opening angle of the

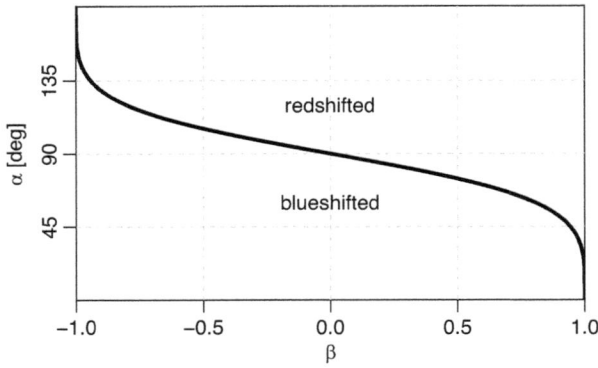

Fig. 9.10 The variation of the angle at which there is no Doppler shift ($\lambda = \lambda'$ in equation 9.11) as a function of β, for the configuration in figure 9.4.

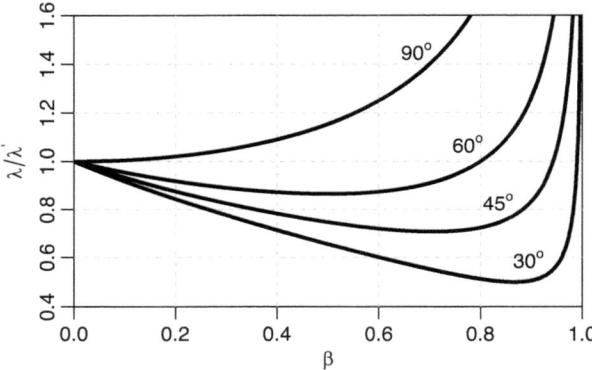

Fig. 9.11 The variation of the Doppler shift with velocity for various angles (α, labelled on the curves) for the configuration in figure 9.4 as given by equation 9.11.

cone over which we see stars blueshifted gets smaller. At $\beta = 0.5$, for example, only stars lying within 74.5° of the direction of travel (photons travelling in directions with $|\alpha| < 74.5°$) will be redshifted.

What is less evident in figure 9.9 is that the Doppler shift does not vary monotonically with velocity at a given angle. This variation is shown in figure 9.11. Take for example sources lying 60° from the direction of travel, and so have photons travelling at $\alpha = 60°$. As the speed increases from zero to more positive β, these sources initially become more blueshifted, reach a maximum blueshift (minimum λ/λ'), and then become increasingly redshifted.

9.3 Source brightness

The brightness of a source is also affected by relative motion. The spectral radiance is the power per unit area per unit solid angle per unit frequency, with units $\mathrm{W\,m^{-2}\,str^{-1}\,Hz^{-1}}$. (See section 1.5 for a summary of radiation terms.) The 'unit area' in this definition is the area at the receiver at some arbitrary distance from the source. In general, the spectral radiance can vary with direction and with the frequency of the radiation. Let J_ν be the spectral radiance received by an observer from a source that is moving at a velocity β in the usual configuration (figure 9.4). Let $J'_{\nu'}$ be the spectral radiance the observer would receive from the same position when at rest relative to the source. Using the expressions for time dilation, aberration, and Doppler shift (which applies to the energies of the photons), we find that the spectral radiance transforms as

$$J_\nu = J'_{\nu'} \frac{1}{\psi^3}. \tag{9.14}$$

Recall that ψ (equation 9.7) is independent of wavelength, so the spectral radiance transforms at all wavelengths in the same way. For an approaching source, $\psi < 1$ (see figure 9.9), and so the source will appear brighter than when it was at rest. Receding sources, in contrast, appear fainter. It may be useful to observe that, because J_ν and ν^3 both transform from S to S' with a factor ψ^3, the quantity J_ν / ν^3 is invariant between reference frames.

Because frequency and wavelength are not linearly related, the spectral radiance of a source defined per unit frequency (J_ν) is not the same as its spectral radiance defined per unit wavelength (J_λ). From $|J_\nu d\nu| = |J_\lambda d\lambda|$ and $d\lambda = -(\lambda^2/c)\, d\nu$ we get $J_\nu = J_\lambda \lambda^2 / c$ and likewise for the primed quantities. Using this and the expression for the Doppler shift, $\lambda / \lambda' = \psi$, in equation 9.14, we see that the transformation for J_λ is

$$J_\lambda = J'_{\lambda'} \frac{1}{\psi^5}. \tag{9.15}$$

This expression (and the one for frequency) refers to monochromatic light. By integrating spectral radiance over a range of wavelengths (or frequencies) we get the radiance (units $\mathrm{W\,m^{-2}\,str^{-1}}$). Using $d\lambda = \psi\, d\lambda'$ we see that the relativistic transformation of the radiance is

$$J = \int J_\lambda d\lambda = \frac{1}{\psi^4} \int J'_{\lambda'}\, d\lambda' = \frac{1}{\psi^4} J'. \tag{9.16}$$

We of course get the same result if we replace wavelength with frequency. For the case of a source moving radially with respect to the observer, the radiance at the receiver is, using equation 7.6,

$$J = J' \left(\frac{1 - \beta}{1 + \beta} \right)^2 \tag{9.17}$$

where $\beta > 0$ means the source is moving radially away from the observer. An extended source receding at 10% the speed of light appears just two-thirds as bright (0.67) compared to when at rest.

In going through the derivation leading to the ψ^3 term in equation 9.14, we find that a factor of ψ^2 arises from the transformation of the solid angles, and another factor of ψ arises through a factor of γ from the time dilation (rate of reception of the photons) and a factor of $(1 - \beta \cos\alpha)$ from the aberration (of the projected area). The transformation for the photon frequencies cancels, because $\delta\nu/\nu$ is invariant. If we had a point source, as opposed to an extended source, the factor from the solid angle transformation would not appear, and so the spectral intensity (power per unit area per unit wavelength) would transform as $I_\lambda = I'_{\lambda'}/\psi^3$. The wavelength integral of this, which is just the intensity (power per unit area), would transform as $I = I'/\psi^2$.

9.4 Summary

Special relativity affects the appearance of objects that are moving at large velocities relative to the observer.

Aberration describes the change in apparent direction to a source on account of a change in its velocity relative to the observer. It is a result of vectorial velocity addition, and so has a classical equivalent (e.g. a person running through the rain). One consequence of aberration is that for a spacecraft moving at relativistic velocities, stars appear to bunch towards the direction of travel, whereas the backward direction appears sparser (figure 9.6). An extended object in the forward direction appears smaller and also brighter (larger spectral radiance). Aberration is easy to measure astronomically: the directions to stars seen from the Earth vary by $\pm 20.5''$ due to the Earth's orbit around the Sun ($\pm 30\,\mathrm{km\,s^{-1}}$).

The Doppler effect describes the change in frequency of a pulsating signal, such as the frequency (and therefore wavelength) of light. Pulses from an object moving towards an observer are received at higher frequency. Thus for a spacecraft moving at relativistic velocities, stars in the forward direction are blueshifted. Stars in the backward direction are receding and are redshifted. Even an object moving transversely relative to an observer appears redshifted, as a consequence of time dilation.

These optical effects are relevant to interstellar travel in a number of ways. In communicating back to the Earth, the spacecraft's signal will be redshifted, so a receiver will have to be correctly tuned. To point instruments in the right direction during a fast flyby, aberration must be taken into account. In chapter 14, we will see how we can exploit aberration and the Doppler effect for navigating by the stars.

9.5 Exercises

1. Prove equation 9.5 for the aberration. Hint: Consider the Lorentz transformation for the y-component of the velocity.
2. The cosmic microwave background is the highly redshifted radiation from the hot early universe described by the Planck function (equation 15.9), with a temperature observed at our current epoch of about 2.7 K. Explain why its spectral radiance should be a Planck function for an observer at any other redshift.
3. How fast would a spacecraft have to move so that the peak of the cosmic microwave background radiation in the direction of travel appears blue (400 nm)? This will produce a concentric rainbow pattern around the direction of travel. At what angle to the direction of travel is the red circle (700 nm)?
4. How does figure 9.11 look if you extend the range to larger angles, and to negative velocities?

Solar sails

A solar sail is propelled by the pressure of photons emitted by the Sun or other star. Unlike a rocket, a sail needs to carry neither propellant nor an energy source. Although photons are emitted radially away from the Sun, tilting a sail produces a continuous non-radial force, and thus permits non-Keplerian orbits. In this chapter we explore the basic properties of solar sails and their orbits. Photon pressure is low, meaning sails must be light, so we also look at the demanding requirements on sail materials. Solar sails have already been deployed and tested in space, although not yet used to achieve significant propulsion. We look briefly at some past and future missions.

10.1 Radiation pressure

Consider a single photon incident normally on a flat surface. The energy E and momentum p of the photon relative to the surface are related by $E = pc$, where c is the speed of light. When the photon is absorbed by the surface, then from Newton's second law the force imparted by the photon on the surface is

$$F = \dot{p} = \frac{\dot{E}}{c} = \frac{IA}{c} \quad \text{(photons absorbed)}, \tag{10.1}$$

where in the last step we have written the rate of energy absorption in terms of the intensity I (units W m^{-2}) and cross-sectional area A. The expression above tells us that the force per unit power of an absorbed photon is very small, just $1/c = 3.3\,\text{nN W}^{-1}$. Equation 10.1 also applies to an ensemble of photons of different energies incident at different rates, with total intensity I.

As the pressure P is force per unit area, it follows from equation 10.1 that

$$P = \frac{I}{c} \quad \text{(photons absorbed)}. \tag{10.2}$$

When the photons are perfectly reflected by the surface, their change of momentum is twice that given in equation 10.1, and so

$$P = 2\frac{I}{c} \quad \text{(photons reflected)}. \tag{10.3}$$

At 1 au from the Sun, the solar intensity is $I = 1361\,\text{W m}^{-2}$, and so the pressure imparted when photons across all wavelengths are reflected is $P = 9.1\,\mu\text{Pa}$.

Perfect reflection means that the reflected photons have the same energy, and therefore same magnitude of momentum, as the incoming photons. But a reflection changes the direction of the photons, so due to conservation of momentum they must have imparted some momentum to the surface. Perfect reflection can only occur if the surface has infinite mass. For a finite mass sail, the reflected photons must actually lose some of their energy, as we will see in section 11.1.1 when we consider laser sails. But for the small velocity changes of solar sails we can ignore this, and in practice we can get very close to perfect reflection. Later in this chapter we'll anyway generalize some expressions to the case of imperfect reflection.

10.2   Forces on a sail

So far we considered photons incident normal to a surface. The more general case is shown in figure 10.1 for a flat sail of area A. We can think of the incoming photons as being absorbed by the sail and the reflected photons as being emitted by the sail (which at a microphysical level is closer to what really happens). The incoming photons exert a force $\mathbf{F}_i$ on the sail in direction $\hat{\mathbf{u}}_i$ (hats denote unit vectors). If the intensity of the incoming photons is I, then their pressure is I/c (equation 10.2). The projected area of the sail is $A \cos \alpha$. Hence the force exerted by the incoming photons on the sail is

$$\mathbf{F}_i = \frac{IA}{c} \cos \alpha \, \hat{\mathbf{u}}_i. \tag{10.4}$$

By a similar argument the force on the sail due to the reflected photons is

$$\mathbf{F}_r = -\frac{IA}{c} \cos \alpha \, \hat{\mathbf{u}}_r. \tag{10.5}$$

From trigonometry we see that $\hat{\mathbf{u}}_i - \hat{\mathbf{u}}_r = 2 \cos \alpha \, \hat{\mathbf{n}}$, where $\hat{\mathbf{n}}$ is the direction normal to the sail. Thus the net force exerted by photons reflecting perfectly off the sail is

$$\mathbf{F} = \mathbf{F}_i + \mathbf{F}_r = \frac{2IA}{c} \cos^2 \alpha \, \hat{\mathbf{n}}. \tag{10.6}$$

Note the $\cos^2 \alpha$ factor. We might have expected simply a $\cos \alpha$ factor on the basis of this being the projection of the forces along the normal, but there is an extra factor due to the projected area of the sail. There are also components of force along the surface of the sail (in the plane of the page) but these cancel out. In the limit of normal incidence, $\alpha = 0$, equation 10.6 reproduces equation 10.3 (as $F = PA$). The main thing to notice here is that the net force on the sail is in the direction $\hat{\mathbf{n}}$, which is not parallel to the incident radiation. This means that if

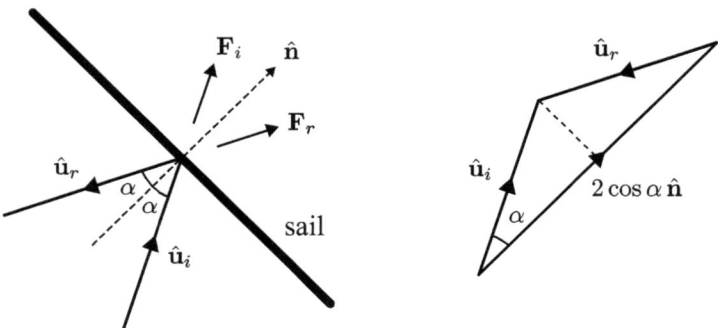

Fig. 10.1 Forces on a solar sail. Left: A photon wavefront with normal $\hat{\mathbf{u}}_i$ at an angle α to the normal $\hat{\mathbf{n}}$ of the sail is perfectly reflected into the direction $\hat{\mathbf{u}}_r$. Due to conservation of momentum, the incident and reflected wavefronts impart forces of $\mathbf{F}_i$ and $\mathbf{F}_r$, respectively, on the sail. Right: Vector triangle $\hat{\mathbf{u}}_i - \hat{\mathbf{u}}_r = 2 \cos \alpha \, \hat{\mathbf{n}}$. Left panel adapted from McInnes (1999) with permission of Springer Nature BV.

we tilt the normal of the sail away from the radial direction to the Sun, there is a non-central force on the sail. This can produce some interesting non-Keplerian orbits, which we will examine in section 10.3.2.

For a sail of mass m, the magnitude of its acceleration from the reflected photons is

$$a = \frac{2I \cos^2\alpha}{c\sigma} \tag{10.7}$$

where $\sigma = m/A$ is the mass per unit area of the sail, also known as the *loading parameter*. When the sail is 1 au from the Sun, $I = 1361\,\mathrm{W\,m^{-2}}$, and so the maximum acceleration (when $\alpha = 0$) is

$$a_{\mathrm{rad}} = \frac{2I}{c\sigma} = \frac{9.1 \times 10^{-6}}{\sigma}\,\mathrm{m\,s^{-2}}, \tag{10.8}$$

with σ in $\mathrm{kg\,m^{-2}}$. This compares to the acceleration of the sail due to the Sun's gravity at the same distance of $a_{\mathrm{grav}} = GM_\odot/r^2 = 5.9 \times 10^{-3}\,\mathrm{m\,s^{-2}}$. As the Sun's gravity and radiation pressure are the two main forces acting on an deep space solar sail, a useful parameter is the ratio of these two accelerations. The Sun's intensity can be written as

$$I = \frac{L_\odot}{4\pi r^2}, \tag{10.9}$$

where $L_\odot$ is the solar luminosity (units W). The ratio of the accelerations is therefore

$$\lambda = \frac{a_{\mathrm{rad}}}{a_{\mathrm{grav}}} = \frac{L_\odot}{2\pi G c M_\odot}\frac{1}{\sigma} = \frac{1.53 \times 10^{-3}}{\sigma}. \tag{10.10}$$

This ratio λ is known as the *lightness number*. It is independent of distance because both the solar radiation pressure and the Sun's gravity vary as $1/r^2$. Indeed, for a given star – $L_\odot$ and $M_\odot$ – the lightness number depends only on the mass per unit area of the sail, σ. The larger λ, the larger the acceleration from radiation compared to gravity, and so the 'lighter' the sail.

From equation 10.10 we see that the loading parameter required to balance gravity in the solar system (i.e. to achieve $\lambda = 1$) is $\sigma = 1.53 \times 10^{-3}\,\mathrm{kg\,m^{-2}}$, or 1.53 grammes per square metre. This is small. For comparison, plastic food wrap has a volume density of about $1000\,\mathrm{kg\,m^{-3}}$ (similar to water) and a thickness of about $10\,\mu\mathrm{m}$, and thus an area density of around $10^{-2}\,\mathrm{kg\,m^{-2}}$, around seven times larger. The sail material should also be highly reflective. Furthermore, σ is the average loading parameter of the entire spacecraft, including the supporting structures and payload. This makes it clear that a solar sail spacecraft will require both extremely light materials and very large sails to compensate for the payload and structural mass. We will examine possible materials in section 10.5.

It is evident from equation 10.10 that a given sail acquires a larger net acceleration when orbiting a star of larger luminosity and/or lower mass than the Sun. For stars on the main sequence, which produce energy in their cores by the fusion of hydrogen to helium, the luminosity is related to the mass approximately as $L \propto M^{3.9}$ for lower-mass stars. Equation 10.10 then tells us that $\lambda \propto M^{2.9}/\sigma$ for such stars. Thus a given sail orbiting a main sequence star that is twice as massive as the Sun would have a lightness number 7.4 times larger. Solar sailing would be much more favourable around such stars. Most stars are less massive than the Sun, and for the lowest-mass stars the luminosity drops very rapidly. For example, Proxima Cen has $L/M = 0.0017/0.12 = 0.014$ in solar units, and so the lightness number of a given sail is about 70 times lower at this star than it is at the Sun.

10.3 Solar sail orbits

We can now write down the equation of motion for a solar sail of mass m in the gravitational and radiation field of a star of mass M. The gravitational force is $-(GMm/r^2)\,\hat{\mathbf{r}}$, where $\hat{\mathbf{r}}$ is the unit vector in the radial direction away from the star. Using the definition of the lightness number λ from above, the radiation force is $(GMm/r^2)\lambda\cos^2\alpha\,\hat{\mathbf{r}}$. From Newton's second law, the vectorial equation of motion of the sail relative to the star is

$$\ddot{\mathbf{r}} = -\frac{\mu}{r^2}\hat{\mathbf{r}} + \frac{\mu}{r^2}\lambda(\hat{\mathbf{r}}\cdot\hat{\mathbf{n}})^2\hat{\mathbf{n}}\quad(\lambda\geq 0),\tag{10.11}$$

where I have used $\cos\alpha = \hat{\mathbf{r}}\cdot\hat{\mathbf{n}}$ and $\mu = G(M+m) \simeq GM$. In general, $\hat{\mathbf{n}} \neq \hat{\mathbf{r}}$, so the acceleration is not in the radial direction, and the orbits will not be Keplerian.

In all of this we are implicitly assuming that the Sun is a point source of radiation. In reality the Sun has a finite extent, so the radiation pressure is not exactly radial, with a consequence that the inverse square law for the radiation pressure is only an approximation. This becomes more significant the closer the sail is to the Sun. We will ignore this here, but for accurate or practical calculations this must be taken into account.

10.3.1 The sail pointed at the Sun: Keplerian orbits

To solve the equation of motion, let us first consider the simpler case in which the sail is kept pointed at the Sun at all times, $\hat{\mathbf{n}} = \hat{\mathbf{r}}$. Equation 10.11 then becomes

$$\ddot{\mathbf{r}} = -\frac{\mu(1-\lambda)}{r^2}\hat{\mathbf{r}}\quad(\lambda\geq 0).\tag{10.12}$$

This is the same as the equation of motion for Keplerian orbits in which the mass of the central body has been decreased by a factor of $(1-\lambda)$. The sail acts to effectively decrease the gravity, and the lightness number introduces a new degree of freedom into the classification of orbits. We can simply replace μ with $\mu(1-\lambda)$ in some of the equations for Keplerian orbits. For a circular orbit, for example, the orbital speed is $\sqrt{\mu(1-\lambda)/r}$, and so will be smaller for a sail than for a non-sail body at the same distance. Whether or not a particular spacecraft is actually on a circular orbit or not depends also on the initial conditions, of course. When $\lambda = 1$ the net force is zero, so the sail moves on a straight line or remains at rest relative to the star. When $\lambda > 1$ the radiation force is larger than gravity, so the net force points outwards, and the sail moves either radially outwards (if starting at rest) or on a hyperbolic orbit.

We can understand the orbits of a solar sail better if we consider how we might use the sail to achieve interstellar travel. Consider a spacecraft in a circular orbit of radius r about a star. The orbital speed is $\sqrt{\mu/r}$, and the escape velocity is $\sqrt{2\mu/r}$ (section 4.1.4). If this spacecraft now unfurls a solar sail such that the sail plus spacecraft has a lightness number of λ, and points this sail at the Sun, the speed of the spacecraft does not change in that instant, but its escape velocity has been reduced to $\sqrt{2\mu(1-\lambda)/r}$. So if $\lambda = 1/2$, then the spacecraft's velocity is now equal to its escape velocity and the spacecraft will begin to move on a parabolic orbit. If $\lambda < 1/2$ the spacecraft will remain bound to the star on an elliptical orbit. For $\lambda > 1/2$ the spacecraft will move on a hyperbolic orbit, and will reach infinity with a non-zero velocity, paving the way for interstellar travel.

For this last case, how fast will the spacecraft be when it gets very far from the Sun? We can address this using conservation of energy. Suppose the spacecraft is initially a distance r_i from the Sun moving with velocity v_i. The specific potential energy once the sail is open and pointed at the Sun is $-\mu(1-\lambda)/r$ and the specific kinetic energy is $v_i^2/2$. At infinity the potential energy is zero, and the spacecraft is moving with velocity v_∞. Equating the initial and final energies we get

$$v_\infty^2 = v_i^2 + \frac{2\mu(\lambda - 1)}{r_i} . \tag{10.13}$$

When $\lambda > 1$, then the closer the spacecraft is to the Sun initially, the faster the spacecraft will be at infinity. Even though a spacecraft closer to the Sun has to overcome more gravity to escape, the higher radiation pressure more than compensates in this case. Even when $\lambda < 1$, a spacecraft initially on a Keplerian orbit that is closer to the Sun also has a larger velocity v_i, and so v_∞ could still be larger. For the specific case of a spacecraft initially on a circular orbit, we have $v_i^2 = \mu/r_i$ before it opens its sail, and so the above equation becomes

$$v_\infty^2 = \frac{2\mu}{r_i}\left(\lambda - \frac{1}{2}\right) \quad \text{(initial circular orbit).} \tag{10.14}$$

This only gives a real value – the sail only reaches infinity – if $\lambda \geq 1/2$, but then a smaller r_i gives a larger v_∞. For $\lambda = 1$, equation 10.14 yields $v_\infty = \sqrt{\mu/r_i}$, which for $r_i = 1$ au from the Sun gives $v_\infty = 29.8\,\mathrm{km\,s^{-1}}$. But if we started on a smaller orbit of $r_i = 0.5$ au, then the sail achieves $v_\infty = 42.1\,\mathrm{km\,s^{-1}}$. For a general initial Keplerian orbit we can use the vis-viva equation (equation 4.3) to substitute for v_i in equation 10.13 to get

$$v_\infty^2 = \mu\left(\frac{2\lambda}{r_i} - \frac{1}{a}\right) . \tag{10.15}$$

This motivates the idea of the sundiver, which we will examine in section 10.7.

As λ is defined as the ratio of the solar photon force to the gravitational force, we might think that $\lambda > 1$ is required for the spacecraft to escape to infinity. But this is only the case if the spacecraft is initially at rest. If it is initially moving, then it already has kinetic energy, so less solar photon energy is required for escape, which can therefore be achieved with a lower lightness number.

10.3.2 The sail inclined: non-Keplerian orbits

When the sail is not pointed at the Sun, the net force on the sail is not radial and so its orbit is not Keplerian. Equation 10.11 then has an extra degree of freedom, namely how the angle α varies over time, and so the equation of motion generally does not have a closed-form solution. A special case is when α is held constant. The orbit is then described by

$$r(\theta) = r_0 \exp(\theta \tan \gamma), \tag{10.16}$$

where γ is the angle between the direction of travel and the tangential direction at position (r, θ) (figure 10.2). The angle γ is a function of α and λ and must be solved for numerically. This orbit is called a 'logarithmic spiral' orbit. An example is shown in figure 10.3 for $\lambda = 0.3$ for $\alpha = \pm 10°$.

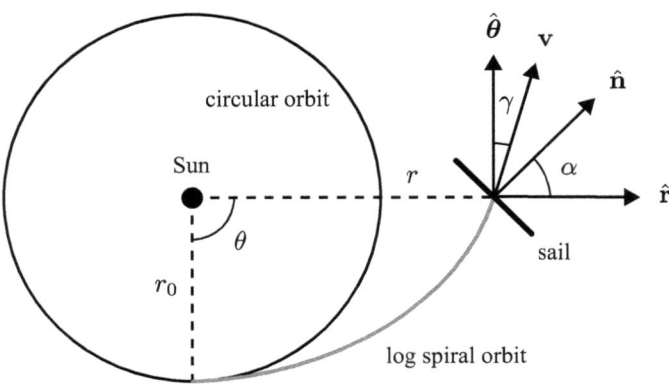

Fig. 10.2 The configuration of a solar sail in a logarithmic spiral orbit as described by equations 10.16 and 10.17. Definitions: $\hat{\mathbf{r}}$ and $\hat{\boldsymbol{\theta}}$ are the radial and tangential unit vectors respectively; $\hat{\mathbf{n}}$ is the normal to the sail; $\mathbf{v}$ is the direction of travel of the sail. The sail moves at an angle γ from $\hat{\boldsymbol{\theta}}$. The configuration shown here has a positive inclination angle α of $\hat{\mathbf{n}}$ relative to $\hat{\mathbf{r}}$, so the sail spirals outwards. Adapted from McInnes (1999) with permission of Springer Nature BV.

Denoting the radial and tangential components of the sail's velocity with v_r and v_θ respectively, then

$$v_r = v_{\text{tot}} \sin \gamma \,, \tag{10.17a}$$

$$v_\theta = v_{\text{tot}} \cos \gamma, \quad \text{where} \tag{10.17b}$$

$$v_{\text{tot}} = k \sqrt{\frac{\mu}{r}} \quad \text{and} \tag{10.17c}$$

$$k = \sqrt{1 - \lambda \cos^2 \alpha (\cos \alpha - \sin \alpha \tan \gamma)}. \tag{10.17d}$$

Note that k is a positive constant depending only on λ and α. For positive values of α, a component of the solar radiation pressure acts to increase the speed of the sail in the direction of travel and the sail moves outwards. For negative values of α the opposite happens. The total velocity, as well as the radial and tangential components, are monotonically decreasing functions of the radius in both cases, as we see in figure 10.3.

As the force is no longer central, the potential energy of the sail is a more complicated function than for Keplerian orbits. Furthermore, the total energy is no longer a constant of the orbit, although it still only depends on r and not θ. When the sail moves outwards its total energy increases.

A particular property of the logarithmic spiral orbit is that when $\alpha > 0$, the sail will spiral outwards indefinitely. In the limit of infinite time it will reach infinite distance, albeit with zero velocity. Note, however, that the velocity of the logarithmic spiral orbit is not equal to the Keplerian velocity at that radius, neither in magnitude nor direction. So to move a spacecraft onto a logarithmic spiral orbit from a Keplerian one, we would need to apply additional Δv boosts.

10.4 Imperfectly reflecting sails

So far we have considered perfectly reflecting sails. In practice, some of the photons can be absorbed by the sail or, if it is thin enough, transmitted through the sail. Let $\tilde{r}$, $\tilde{a}$, and $\tilde{t}$ be the reflectance, absorptance, and transmittance of the sail, respectively, where

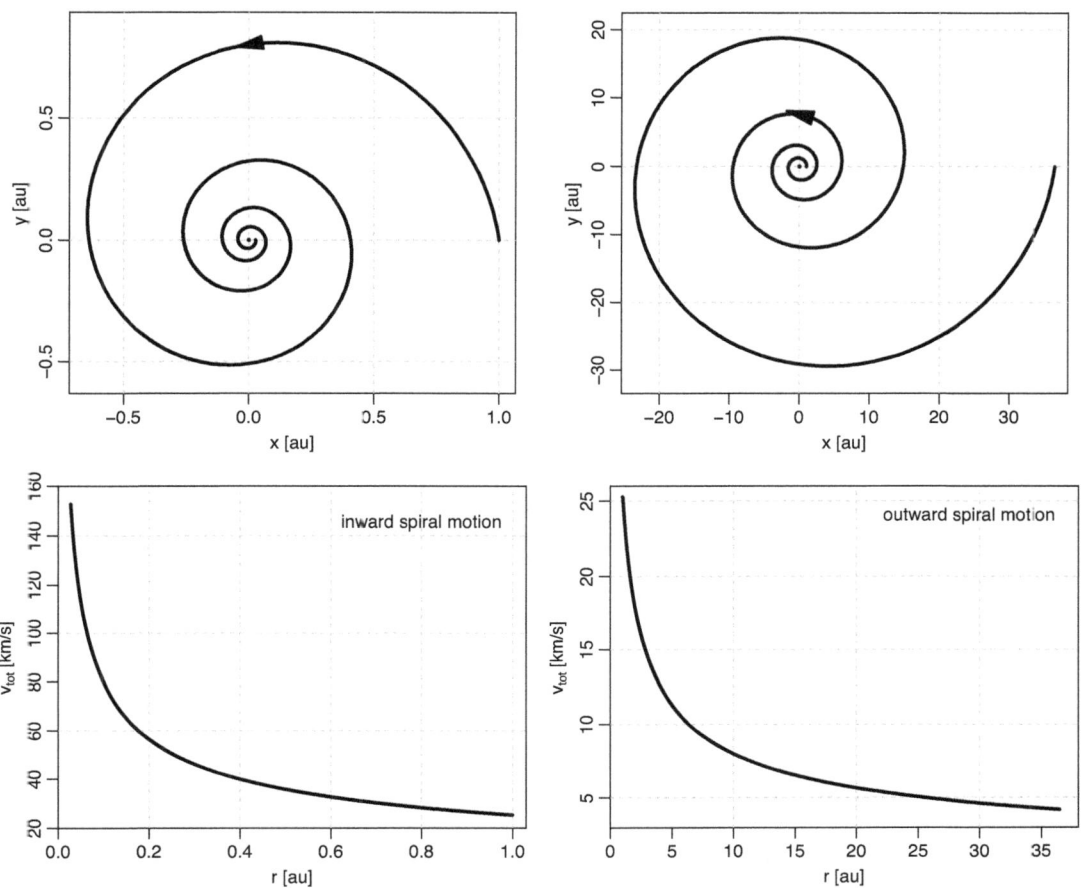

Logarithmic spiral orbits of a solar sail with $\lambda = 0.3$ for $\alpha = -10°$ ($\gamma = -8.14°$, left), and $\alpha = +10°$ ($\gamma = +8.14°$, right) starting at $(r, \theta) = (1\,\text{au}, 0°)$ and computed over four revolutions ($\Delta\theta = 4 \times 360°$). Both cases have $k = 0.840$ (equation 10.17). The top panels show the orbit in Cartesian coordinates; both plots have 1:1 aspect ratios but different scales. The bottom panels show how the total velocity varies with radial distance.

Fig. 10.3

$$\tilde{r} + \tilde{a} + \tilde{t} = 1. \tag{10.18}$$

I use the tilde symbols for several quantities in this chapter to avoid duplication of notation. If we return to equation 10.4 for the force imparted by the incident photons, then as a fraction $\tilde{t}$ now pass through the sail, we need to include a factor $1 - \tilde{t}$ on the right side of the equation for the incident force. Similarly, only a fraction $\tilde{r}$ are reflected, so we need to include a factor $\tilde{r}$ on the right side of equation 10.5 for the reflected force. In the case of normal incidence ($\alpha = 0$), the magnitude of the pressure on the sail becomes

$$P = (1 - \tilde{t} + \tilde{r})\frac{I}{c} = (2\tilde{r} + \tilde{a})\frac{I}{c}. \tag{10.19}$$

This generalizes equations 10.2 and 10.3.

Equation 10.19 tells us that imperfect reflection reduces the radiation pressure. Furthermore, in the general case of non-normal incidence shown in figure 10.1, now that the magnitudes of the forces in the $\hat{\mathbf{u}}_i$ and $\hat{\mathbf{u}}_r$ directions are different, the net force is no longer parallel to $\hat{\mathbf{n}}$. In other words, when a tilted sail reflects imperfectly, the sail does not accelerate along its normal.

The above generalization is a bit simplified, because the photons that are absorbed by the sail will, once the sail reaches thermal equilibrium, be re-emitted. In the above, I implicitly assumed that the emissivities of the sail are the same on both sides of the sail. If they are not, perhaps because the sail has a coating, this changes not only the force normal to the sail, but also the force tangential to the sail, so it also changes the direction of the net force. A further generalization of the material properties considers different types of reflection. I have assumed specular reflection (like a mirror), but some materials also exhibit diffuse reflection, like from a matte surface (Lambertian reflectance), and/or back reflection, whereby some photons are reflected directly back to the source.

With imperfect reflection (but ignoring unequal emissivity and assuming just specular reflection again) we need to redefine the lightness number. In equation 10.8 the factor of 2 is replaced with $2\tilde{r} + \tilde{a}$ and equation 10.10 becomes

$$\lambda = \frac{2\tilde{r} + \tilde{a}}{\sigma} \frac{\mathrm{L}_\odot}{4\pi G c \mathrm{M}_\odot} = \frac{2\tilde{r} + \tilde{a}}{2} \frac{\sigma_c}{\sigma}, \tag{10.20}$$

where $\sigma_c = 1.53 \times 10^{-3}\,\mathrm{kg\,m^{-2}}$.

10.5 Sail materials

The acceleration for photons incident normally on an imperfectly reflecting sail is, from equation 10.19,

$$a = \frac{P}{\sigma} = (2\tilde{r} + \tilde{a})\frac{I}{c\sigma}. \tag{10.21}$$

To maximize this we should maximize the reflectance $\tilde{r}$ and minimize the mass per unit area σ of the sail material. The reflectance of materials varies with wavelength, because how a photon interacts with the free or bound electrons in the material depends on the energy of the photon. As most of the Sun's flux is in the optical part of the spectrum, we want materials that have large optical reflectance, such as metals. Aluminium has a high reflectivity[1] across a broad part of the optical spectrum. Silver has one of the highest reflectivities, even higher than aluminium, and for this reason is sometimes used in reflecting telescopes, although its reflectivity is poorer in the blue and ultraviolet, so would be less suitable in a solar sail for reflecting the broad solar spectrum. Gold is high too, but only in the red part of the spectrum (and it is sometimes used for mirror coatings in infrared telescopes).

To achieve a small mass per unit area, we should use the thinnest layer of reflecting material possible. There are at least two practical limits to this. First, if arbitrarily thin, the sail will be mechanically weak. It needs to be kept in a specific shape and must be attached to the

---

[1] The *reflectance* of a material measures how much light is reflected from a material of a given thickness. The *reflectivity*, in contrast, measures how much light is reflected off the surface of the material. They are equal for infinitely thick materials, but for thin materials, multiple internal reflections can result in variable reflectance for a given reflectivity.

payload, so the sail will have to tolerate some tension. In practice one can attach a very thin layer of reflecting material to a stronger substrate, although that then contributes mass.

The second limit to the sail thickness is that very thin conductors become increasingly transparent. This can be understood when we consider the behaviour of electromagnetic waves incident on a thin film of conducting material. By solving the wave equation, we find that the amplitude of the electric and magnetic fields that penetrate a distance d through the material is proportional to $\exp(-d/L)$, so the degree of transmission is determined by the value of d/L, where L is a property of the material. Low transmittance – and so high reflectance – is achieved when $d \gg L$, and so L is a characteristic minimum thickness of the material known as the *skin depth* (or penetration depth).

The value of the skin depth depends on the conducting properties of the material. A good electrical conductor has $\tilde{\sigma} \gg \tilde{\epsilon}\omega$, where $\tilde{\sigma}$ is its electrical conductivity, $\tilde{\epsilon}$ is its permittivity, and ω is the angular frequency of the light. For good conductors it turns out that

$$L = \sqrt{\frac{\tilde{\lambda}}{\pi c \tilde{\mu} \tilde{\sigma}}}, \tag{10.22}$$

where $\tilde{\lambda}$ is the wavelength of the light ($\omega = 2\pi/\tilde{\lambda}$) and $\tilde{\mu}$ is the magnetic permeability of the material. The latter is often listed in terms of the dimensionless relative permeability $\tilde{\mu}_r$, where $\tilde{\mu} = \tilde{\mu}_r \mu_0$ and $\mu_0 \simeq 1.26 \times 10^{-6} \, \text{NA}^{-2}$ (newtons per square ampere) is the vacuum permeability. We see that the longer the wavelength, the thicker a given material must be to keep the reflectance high.

To minimize the mass of the sail, L should be small. Thus for a given wavelength, the larger the conductivity and permeability, the thinner the sail can be. Non-ferromagnetic metals like copper, aluminium, silver, and gold have relative permeabilities near to 1.0. Aluminium has a conductivity of $\tilde{\sigma} = 3.9 \times 10^7 \, \text{S m}^{-1}$ at room temperature (S is siemens, the inverse of the ohm); at $\tilde{\lambda} = 550 \, \text{nm}$, the peak of the solar spectrum, it has a skin depth of $L = 3.5 \, \text{nm}$.[2] This skin depth is much less than the wavelength of the electromagnetic radiation, which is typical for conductors. Copper, silver, and gold have slightly higher conductivities than aluminium of 4.5–$6.3 \times 10^7 \, \text{S m}^{-1}$, but much higher densities (3–7 times as large). Iron has a very high relative permeability – over 100 at 99% purity, and up to tens or hundreds of thousands at very high purities – as well as a reasonably high conductivity ($\tilde{\sigma} = 1 \times 10^7 \, \text{Sm}^{-1}$), and so a much smaller skin depth than the non-ferromagnetic metals just mentioned. As it is only three times denser than aluminium, iron appears to be a good candidate sail material. But it is not considered because it absorbs in the bluer part of the spectrum and so would heat up. Being magnetic is also undesirable, given the presence of magnetic fields in space, plus it could interfere with instruments.

When a sail is made of a thin conducting material, there can be an optimal thickness that provides maximum acceleration. A large thickness increases the reflectance, but it also increases the mass. As the power of an electromagnetic wave is proportional to the square of its field amplitude, the intensity of the radiation that is reflected back from the surface is $1 - \exp(-2d/L)$, which is plotted in the left panel of figure 10.4. For a material with volume density ρ, then from equation 10.21 with $\sigma = \rho d$ the acceleration is

---

[2] Electrical conductivity generally decreases with increasing temperature. As the sail will inevitably absorb some radiation, it will heat up, thereby increasing the skin depth.

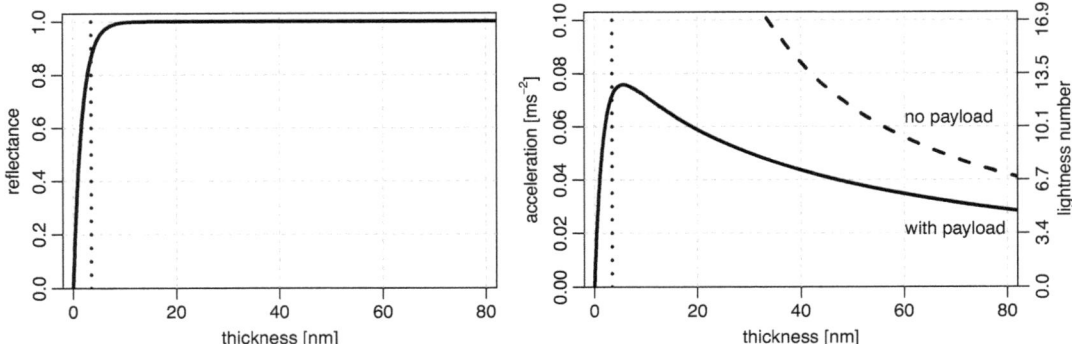

Fig. 10.4 Left: The variation of reflectance of a sheet of aluminium under normal incidence as a function of its thickness. Right: The corresponding variation of acceleration of a thin aluminium solar sail 1 au from the Sun at normal incidence for two cases. The dashed line shows just the sail (no payload), according to equation 10.23. This curve increases towards infinity as $d \rightarrow 0$ and drops to zero as $d \rightarrow \infty$. The solid line shows the case for a sail of 10 m^2 area carrying a payload of 1 g, as given by equation 10.24. The properties adopted for the aluminium in both cases are density $\rho = 2700$ kg m^{-3}, conductivity $\tilde{\sigma} = 3.9 \times 10^7$ S m^{-1}, and relative permeability $\tilde{\mu}_r = 1.0$. The intensity of the solar radiation is $I = 1361$ W m^{-2} and is assumed for simplicity to be entirely at a wavelength of 550 nm, so that we can compute the skin depth to be $L = 3.5$ nm (the vertical dotted line).

$$a = \frac{2I}{c\rho d}[1 - \exp(-2d/L)] . \tag{10.23}$$

In the limit $d \rightarrow 0$ we get $a \rightarrow \infty$: Despite the trade-off between reflectance and mass with thickness, the decreasing mass wins, and so the thinner the material, the higher the acceleration. But this limit is useless in practice because the sail needs a finite thickness for structural integrity, plus it must carry some kind of payload. Let us assume the sail carries an additional non-reflecting and non-absorbing mass m, which for conciseness I refer to as 'payload'. We must now replace the mass per unit area σ with $\sigma + m/A$, where A is the area of the sail, so equation 10.23 becomes

$$a = \frac{2I}{c}\left[\frac{1 - \exp(-2d/L)}{\rho d + m/A}\right] . \tag{10.24}$$

This expression has a maximum at an intermediate value of d. This is plotted as the solid line in the right panel of figure 10.4 for a 10 m^2 aluminium sail carrying a 1 g payload. The maximum acceleration is achieved when the sail is just a few nm thick.

The above example is given to show how there is a trade-off in principle. As we will see below, current solar sails are made by depositing a thin film of reflecting metal on a sturdier substrate, often a polymer. These substrates are much thicker than the metal film, so we can make the metal film thick enough to maximize reflectance without adding much additional mass. When the substrate has a mass per unit area of 10^{-4} kg m^{-2}, for example a polymer around 100 nm thick, and the aluminium is 20 nm thick, then the substrate is almost twice as massive as the aluminium. A 10 m^2 sail out of this composite material has a total mass of 1.54 g (1 g for the susbtrate, 0.54 g for the aluminium). When carrying a 1 g payload and assuming unit reflectance, this sail has a lightness number of 6.0 and would achieve an acceleration at 1 au of 0.036 m s^{-2}. This adopted substrate area density is much lower than is currently achievable, however. Current sails use polymers such as polyethylene

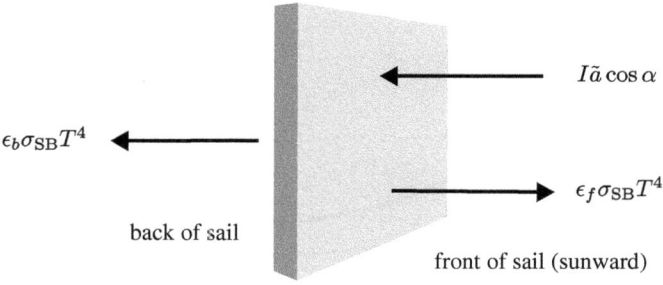

A schematic of the radiation balance of a solar sail at equilibrium temperature T.

Fig. 10.5

terephthalate (PET; density of $1400 \, \text{kg m}^{-3}$) 2500 nm thick, which corresponds to an area density of $35 \times 10^{-4} \, \text{kg m}^{-2}$. This lowers the lightness number in the above scenario to 0.42. The total mass of the substrate is 35 g, so is much more than the payload too. Clearly, in order to use a metallic film for a solar sail, we need to reduce the thickness or density of the substrate as much as possible.

Several other factors beyond thickness affect a sail's reflectance. One example is the flatness of the metal film, at both a microscopic and macroscopic level. For the sail that has already been made for the Solar Cruiser spacecraft, for example (see section 10.6), the non-flatness is expected to reduce the overall reflectance by around 5%.

The above discussion is a simplified treatment. Real metals absorb some of the incoming radiation, so the reflectance is not unity even for large thicknesses. Furthermore, this absorption will heat the sail, which could eventually melt the sail. The sail will also radiate the heat, which can accelerate the sail.

The equilibrium temperature of the sail can be calculated assuming it to be a very thin black body with thermal emissivity ϵ_f on the Sun-facing side (front) and ϵ_b on the other side (figure 10.5). From the Stefan–Boltzmann law, the total power emitted per unit area from a black body of emissivity ϵ and temperature T is $\epsilon \sigma_{\text{SB}} T^4$, where σ_{SB} is the Stefan–Boltzmann constant. In thermal equilibrium, the energy radiated by the sail equals the energy it absorbs. When the sail is inclined at an angle α to the Sun, then the incident intensity is $I \cos \alpha$, and so

$$I \tilde{a} \cos \alpha = (\epsilon_f + \epsilon_b) \sigma_{\text{SB}} T^4. \tag{10.25a}$$

$$T = \left[\frac{I \tilde{a} \cos \alpha}{(\epsilon_f + \epsilon_b) \sigma_{\text{SB}}} \right]^{1/4}. \tag{10.25b}$$

For a sail coated on one side with aluminium, we can adopt $\tilde{a} = 0.1$ and $\epsilon_f = 0.1$ as reasonable averages for our purposes. Ideally the back side of the sail (the substrate) has $\epsilon_b \simeq 0$, because any photons emitted in this direction would decelerate the sail. Adopting this, then for a sail 1 au from the Sun pointing at it, the above expression gives an equilibrium temperature of 394 K. Figure 10.6 shows how this temperature varies with distance from the Sun (solid line). Once this sail approaches within 0.18 au of the Sun, its temperature reaches the melting point of aluminium (933 K). Thus if we want to operate close to the Sun (for example with the sundiver – section 10.7), we need a material with a lower absorptance, a higher emissivity, or a higher melting temperature.

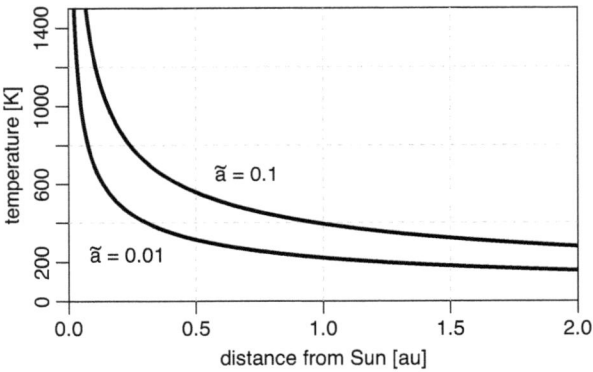

Fig. 10.6   The variation of the equilibrium temperature of a sail with distance from the Sun for two values of the absorptance $\tilde{a}$ according to equation 10.25b. The sail has emissivities $\epsilon_f = 0.1$ and $\epsilon_b = 0.0$ and is pointed at the Sun ($\alpha = 0$).

If $\epsilon_f \neq \epsilon_b$, then the rate of emission of photons from the two sides of the sail is not equal. This generates a net force on the sail. This was not taken into account in section 10.4, where the pressure in equation 10.19 implicitly assumed that any absorbed photons (i.e. when $\tilde{a} > 0$) were emitted isotropically.

The above treatment of the sail has been entirely classical. One quantum mechanical effect of some relevance is the photoelectric effect. Photons of high enough energies eject free electrons from a metallic sail. The photon energy required is given by the work function of the material, and is of order a few electronvolts. For aluminium it is 4.3 eV, which corresponds to photons with wavelength below 290 nm. Not only does the ejection of electrons represent lost energy (used to overcome the work potential), but it also leads to charging the sail. As the solar spectrum has a non-negligible intensity in the ultraviolet, this is unavoidable.

We have only considered the use of good conductors – metals – as the primary sail material. Poor conductors – dielectrics – generally have lower reflectances, so at first sight appear to be less interesting for solar sail applications. However, when it comes to the much higher intensity light from laser sails, non-conductors are interesting on account of their potentially higher melting temperatures (see section 11.4.1). They can also be made very thin and yet strong, and this reduced mass can compensate for their lower reflectances, also as a solar sail.

A solar sail cannot launch from the Earth's surface, of course. The sail material therefore has to be packaged into a rocket's fairing for launching. It must be strong enough to withstand folding or rolling and the stresses of launch. Once in space it must be deployed. Various methods have been developed and tested for doing this, including using motors to pull it along rigid booms, and rotating the spacecraft to open it via centrifugal forces. As we shall see in the next section, sails built so far have thicknesses no less than 2.5 μm. Thinner sails are at risk of tearing during fabrication, launch, or deployment. If sails can be manufactured in space, then thinner sails may be possible with current materials and technology.

In summary, we have the following requirements for a sail material: high reflectance; low mass per unit area (thin or low density); high thermal emissivity or melting point; foldable; sufficient tensile strength so as not to tear. These requirements cannot be optimized independently, and so the choice of material will be a compromise between different factors.

10.6 Missions

Several solar sail missions have already flown. Several more were planned and partially developed but never flown. The German space agency DLR, for example, produced and deployed several sail materials in the late 1990s, and had a technology programme through the first decade of this century.

Of those missions that have successfully flown (as of mid 2025), all were tests and technology demonstrators. NASA, for example, has mapped out a series of missions of increasing size and sophistication, leading also into laser sails that we will discuss in the next chapter. Not all of the missions were successful, though. Nanosail-D2 (a replacement of Nanosail-D that was lost due to rocket failure) was a small mission primarily intended to test the deployment of its $10\,\text{m}^2$ sail, which it did successfully in 2010. NASA's NEA Scout was a cubesat with a deployed sail area of $85\,\text{m}^2$ of $2.5\,\mu\text{m}$-thick aluminized polyimide. It was meant to last for two years and encounter a small asteroid. It was launched in November 2022 on Artemis 1, but its sail failed to deploy, and communication was lost just after launch. The similar-sized Advanced Composite Solar Sail System (ACS3), launched in April 2024, did deploy successfully, but started to tumble in October of the same year, possibly after one of the sail support arms was bent. Solar Cruiser is a much larger sailcraft, comprising four sails each with an area of $440\,\text{m}^2$ made from a $2.5\,\mu\text{m}$-thick metal-coated polyimide. Sail deployment has been successfully tested on the Earth. Funding permitting, the spacecraft hopes to launch in 2029 to study the Sun.

A non-governmental organization, The Planetary Society, built and launched two solar sails into low Earth orbit in 2015 and 2019, called LightSail 1 and 2 (figure 10.7). Both were miniature cubesats, in which the 5 kg sail was folded for launch into a rectangular volume

LightSail 2 in orbit above the Red Sea. The Nile is clearly visible. The sail appears curved due to the wide angle lens. Credit: The Planetary Society, reproduced with permission.

Fig. 10.7

of just 10 litres. Once deployed, the square sail had an area of $32\,\mathrm{m}^2$. This was made out of a thin polyester (boPET) film and an aluminium coating with a total thickness of $4.6\,\mu\mathrm{m}$, and providing a reflectance of around 0.99. Together these data imply a lightness number of 0.01. Solar panels provided an average power of $8.5\,\mathrm{W}$. After sail deployment, the sail was reoriented several times between facing the Sun and being edge-on to the Sun, which could be used to determine the acceleration due to the solar radiation pressure.

Several other organizations have developed solar sails, for example the French company GAMA, which launched a small sail into Earth orbit in January 2023. It was able to adjust its orbit using solar radiation pressure.

We now look in a bit more detail at one past successful mission, and one proposed interstellar mission.

10.6.1 IKAROS

One of the first solar sail missions, and so far the only interplanetary one, was IKAROS (Interplanetary Kite-craft Accelerated by Radiation Of the Sun), launched by the Japanese space agency JAXA in 2010. The sail itself had an area of $200\,\mathrm{m}^2$ and was made from a poly-imide substrate $7.5\,\mu\mathrm{m}$ thick with an $80\,\mathrm{nm}$-thick coating of aluminium. This had a mass per unit area of around $0.01\,\mathrm{kg\,m}^{-2}$, corresponding to a lightness number of 0.14 from equation 10.10. However, extra supports and tensioning masses were required, increasing the total mass from $2.1\,\mathrm{kg}$ to $16\,\mathrm{kg}$, and so reducing its mass per unit area to $0.08\,\mathrm{kg\,m}^{-2}$. More significant, though, was the payload mass of $55\,\mathrm{kg}$, which lowered the mass per unit area to $0.4\,\mathrm{kg\,m}^{-2}$, to give an overall lightness number of 0.0043.

The square sail was deployed by rotating at a low rate, several revolutions per minute (rpm). It continued to rotate at around $1\,\mathrm{rpm}$ after deployment to help stabilize the sail. IKAROS was not designed as a high-performance sail to achieve maximum acceleration, but rather as a technology demonstrator and test of interplanetary solar sailing, testing not only propulsion, but also power generation, guidance, navigation, and control.

For its power, IKAROS was equipped with thin-film photovoltaic silicon cells just $25\,\mu\mathrm{m}$ thick. These covered 5% of the sail area (i.e. $10\,\mathrm{m}^2$) and produced $500\,\mathrm{W}$ of power. Successfully generating electricity from these was a major mission goal. The instruments onboard included a number of cameras, a polarimeter to observe gamma-ray bursts, and a dust detector.

Attitude control is an important task for any spacecraft. As sailcraft must be light, it would be counter-productive to equip them with either traditional cold gas thrusters or massive gyroscopes. Instead, IKAROS used a number of liquid crystal panels on its sails to control its orientation. When a voltage is applied to the liquid crystals, they align and reflect light in a specular fashion, like the rest of the sail. When the voltage is removed, the crystals no longer point in a common direction and then reflect diffusely, which lowers their net reflectance. When this is done for panels on just one side of the sail, the solar radiation exerts a torque on the sail, thus changing its orientation. In this way the attitude of the sail can be controlled without requiring propellant, but just a small amount of energy.

IKAROS was launched in May 2010 and spent about two weeks carefully deploying the sail. A thrust of $1.12\,\mathrm{mN}$ due to solar radiation pressure was measured from the Earth by Doppler radio. For a total mass of $71\,\mathrm{kg}$ this implies an acceleration of $1.7 \times 10^{-5}\,\mathrm{m\,s}^{-2}$, or a lightness number of 0.0027. This is only slightly less than the (simplified) expected value based on the sail area and spacecraft mass computed above. Over the next six months IKAROS cruised to Venus, and during this period the sail provided a total Δv of about $100\,\mathrm{m\,s}^{-1}$. Clearly this Δv was not enough to get the spacecraft from Earth to Venus; most of the thrust was provided in advance by chemical rockets. IKAROS passed Venus at a distance of $80\,000\,\mathrm{km}$ on 8 December 2010, and since 2015 has been in hibernation mode.

10.6.2 Interstellar mission concepts

The the late 1990s, NASA/JPL studied a concept for an interstellar probe propelled by a solar sail. The nominal design is a circular boPET/aluminium sail 400 m in diameter with a mass of 100 kg. Combined with a payload mass of 150 kg, this corresponds to $\sigma = 2.0 \times 10^{-3}\,\mathrm{kg\,m}^{-2}$ and thus a lightness number of 0.77. When this sail is opened from an initial 1 au circular orbit about the Sun, it would escape from the solar system, achieving an asymptotic velocity of $22\,\mathrm{km\,s}^{-1}$ (equation 10.14). In practice the mission design foresaw the use of chemical rockets, planetary gravity assists, and/or initially diving closer to the Sun (see section 10.7) to achieve a total Δv of $70\,\mathrm{km\,s}^{-1}$. The mission is designed to reach 200 au 15 years after launch (an average velocity of $63\,\mathrm{km\,s}^{-1}$), a distance that is well beyond the orbit of Neptune. Its main science objectives are to study the outer solar system and the interstellar medium and their interaction. The sail would be ejected once 5 au from the Sun, as the benefit of the small additional acceleration beyond this distance was considered inferior to the requirement to avoid interference of the sail with onboard instruments.

One of the big challenges of such a deep space mission is the data rate. Using a 2.7 m radio dish on the spacecraft with signals picked up by NASA's Deep Space Network, an average data rate of just $30\,\mathrm{b\,s}^{-1}$ from a distance of 200 au was predicted. We shall examine the challenge of communication in chapter 15.

There have been several other concepts for using solar sails to reach either the outer solar system or interstellar space. An ESA study considered a $60\,000\,\mathrm{m}^2$ sail to take a payload out to 200 au within 20 years in order to study the heliopause; this is similar in size and scope to the JPL concept just mentioned. Several authors have proposed solar sails as a fast way to get to Mars or to an asteroid, or to chase high-speed interstellar objects (ISOs) that pass through the solar system on hyperbolic orbits (see section 12.3.2). With velocities of many tens of kilometres per second in the inner solar system, ISOs are hard to catch. One concept therefore involves using solar sails with a lightness number close to unity that can hover, waiting, before tilting their sails to follow a particular target. Some solar system missions also envisage use of the sundiver to help the sail get to the outer solar system more quickly.

10.7 The sundiver

We saw earlier from equation 10.15 that provided λ is large enough, then the closer a sailcraft is to the Sun when it opens its sail, the larger its velocity at infinity. For fast interstellar travel it therefore makes sense to manoeuvre a sailcraft as close to the Sun as possible before starting the outward journey. How can we achieve this?

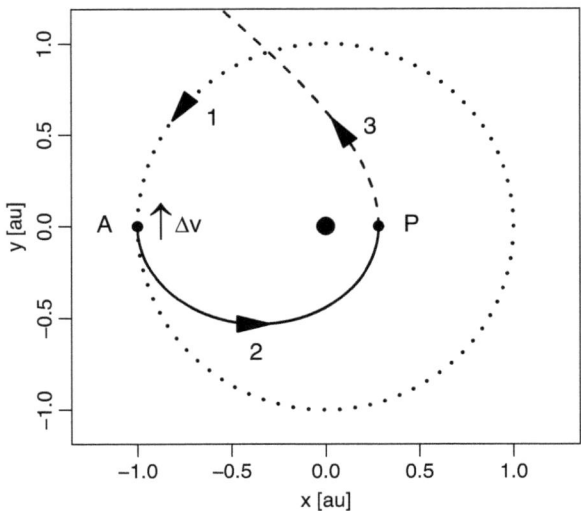

A simple sundiver. A spacecraft initially on a circular orbit (1) about the Sun at $(0, 0)$ performs a retrograde boost at point A to reduce its orbital velocity by Δv. This puts it on an elliptical orbit (2) with aphelion A and perihelion P. At perihelion the spacecraft jettisons its rocket and deploys its solar sail, keeping it pointed at the Sun. The solar radiation pressure accelerates the spacecraft onto a hyperbolic orbit (3), which achieves a velocity at infinity given by equation 10.26. The plot is drawn to scale with $r_a = 1$ au, $r_p = 0.28$ au (achieved using $\Delta v = 10$ km s^{-1}), and $\lambda = 0.3$. The velocity at infinity is 22.3 km s^{-1}. Adapted from figure 1 of Bailer-Jones (2021b).

Suppose our spacecraft has been launched from the Earth and has escaped Earth's gravity, and is initially on the Earth's orbit about the Sun (orbit 1 in figure 10.8). The most obvious way to drop close to the Sun is to use a rocket to apply a retrograde boost (at A in the figure) almost equal to the orbital velocity. We of course do not want to drop into the centre of the Sun, and let us assume the spacecraft cannot approach closer than a distance r_p to the Sun's centre for thermal reasons. After this boost, the spacecraft is on an elliptical orbit (orbit 2) with perihelion r_p and aphelion r_a, which in this case is 1 au. At perihelion (point P), the spacecraft ejects its engines and fuel tanks, opens its sail and points it at the Sun. If λ is large enough, the sailcraft will move on a hyperbolic orbit to infinity (orbit 3) where it will have a velocity given by equation 10.15 with $r_i = r_p$. The semi-major axis of the initial elliptical orbit is $a = (r_a + r_p)/2$, so we can write the velocity equation as

$$v_\infty^2 = \frac{2\mu}{r_p} \left(\lambda - \frac{r_p}{r_a + r_p} \right). \tag{10.26}$$

Adopting $r_p = 10\,\mathrm{R}_\odot = 0.047$ au (the closest approach of the Parker Solar Probe), the perihelion velocity at the moment of opening the sails is $v_p = 191$ km s^{-1}(from the vis-viva equation, equation 4.3). With $\lambda = 1/2$ the velocity at infinity is $v_\infty = 132$ km s^{-1}. For $\lambda = 1$ we get $v_\infty = v_p$ (from equation 10.13).

We found earlier that a sailcraft on an initial circular orbit can only reach infinity if $\lambda \geq 1/2$. For an initial elliptical orbit, a smaller λ is permissible. By setting $v_\infty = 0$ in equation 10.26, we see that the sail can reach infinity if $\lambda \geq 1/(1 + r_a/r_p)$. For example, with

$r_a/r_p = 2$ the minimum λ is $1/3$. Once λ is large enough for the sailcraft to reach infinity, a lower r_p for given r_a increases v_∞.

For the spacecraft to dive down to 10 solar radii with a single retrograde boost, we need a large Δv. The velocity at aphelion for this transfer orbit can be found from the vis-viva equation (or via conservation of angular momentum), and the difference between this and the circular velocity is the required Δv. In the above case with $r_a = 1$ au and $r_p = 10\,R_\odot$ this is $20.9\,\mathrm{km\,s^{-1}}$. We saw in section 4.2.2, however, that we can dive just as close with a lower total Δv by using a bielliptic Hohmann transfer orbit. This involves first boosting to a larger aphelion distance and then applying the retrograde boost (figure 4.7). In this case the perihelion velocity will be larger (for the same perihelion distance), and so the final velocity at infinity will also be larger. This gain is very small, however. If we now adopt $r_a = 2$ au, then the perihelion velocity increases by just $2.2\,\mathrm{km\,s^{-1}}$, to $193\,\mathrm{km\,s^{-1}}$. For $\lambda = 1$ this will increase the velocity at infinity by the same amount, just 1%.

Assuming one has a rocket engine to boost a sailcraft, an interesting question is whether, before opening the sails, it is better to use the available Δv to make a retrograde boost to drop down to the Sun, or to use it instead to make a prograde boost from the initial orbit. After all, having dropped to the Sun to gain kinetic energy (an application of the Oberth effect; section 4.2.3), the sailcraft would then have to use more of this energy to climb out of its now deeper gravitational potential well. It turns out that the strategy that maximizes the final velocity depends on the lightness number of the sail and the ratio of Δv to the initial velocity. For a given velocity ratio, when the lightness number is large, it is better to use a retrograde boost to dive to the Sun. But when the lightness number is low, it is better to use a prograde boost and avoid the dive. For $\lambda > 1/2$ it is always better to dive. Curiously, it is never optimal to use a combination of retrograde and prograde boosts. This is explored in detail in Bailer-Jones (2021b).

Instead of using a rocket to bring the sailcraft close to the Sun, we could use the sail itself by tilting it to spiral into the Sun, as outlined in section 10.3.2. This would take longer, perhaps a lot longer, but as we see from the bottom left panel of figure 10.3, the sail can still achieve quite large velocities close to the Sun.

In section 11.4.1 we will discuss carbon-based materials such as graphene and aerographite with extremely low densities, which, while not very reflective, are absorptive and so could act as solar sail materials. They will therefore heat up, but they have high enough emissivities and melting points that they could survive a very close encounter with the Sun. A sail made only from graphene could theoretically have a lightness number as high as 500, albeit without any payload. If this is released by a spacecraft that has dived from the Earth's orbit to a perihelion distance of 10 solar radii, then it will achieve a velocity at infinity of $4370\,\mathrm{km\,s^{-1}}$, or 1.5% the speed of light, and so gets to Proxima Cen in 290 years. If it could dive to five solar radii (four from the Sun's surface), then the velocity at infinity would increase to $6180\,\mathrm{km\,s^{-1}}$, a 210-year journey. Naturally, any payload would significantly lower the lightness number of the sail. If we could achieve a high but more modest lightness of 20, and then dive to within two solar radii (one from the Sun's surface), it would take 650 years to get to Proxima Cen. This is optimistic, but it shows that if we can design a very light sail to get very close to the Sun, and are patient enough for travel to the stars to take hundreds, rather than tens, of years, then a solar sail might become an option for interstellar travel.

10.8 Summary

A solar sail is a spacecraft propelled by solar radiation pressure. As this is low, the sail must be light. A mass density of $1.53 \times 10^{-3}\,\mathrm{kg\,m^{-2}}$ is required for a perfectly reflecting material to balance the force of gravity. The ratio of the radiation to gravitation force is called the lightness number λ (equation 10.10). As both the Sun's gravity and luminosity vary as $1/r^2$, λ is independent of the distance from the Sun.

The net force on a sail kept pointed at the Sun is radial, so it still moves on a Keplerian orbit, but effectively with a lower-mass star (for $\lambda < 1$). When the sail is tilted, the net force is no longer radial, allowing for a wide set of non-Keplerian orbits. For $\lambda > 1$, the net force from the Sun is outwards.

Very thin metallic sails become transparent, so there is a limit to how light a metal sail can be. In current sails, a thin layer (tens of nanometres) of highly reflective metallic foil is deposited on a supporting substrate; the latter usually dominates the mass budget.

A solar sail with a sufficiently large lightness number can escape the solar system, depending on its initial orbit. By first bringing the sailcraft close to the Sun, more kinetic energy can be transferred to the sail, thereby raising its velocity at infinity. This is known as the sundiver. If the sailcraft can be brought to within a few solar radii of the Sun's surface, a velocity at infinity of order $1000\,\mathrm{km\,s^{-1}}$ can be achieved with a large lightness number of 10 (equation 10.26). This optimistic scenario still results in a long journey time to the nearest stars – over a thousand years to Proxima Cen – so a solar sail is probably not a viable solution for fast interstellar travel.

Solar sails have already been successfully tested and deployed in space, although none have yet achieved significant propulsion. Within the coming decades they are likely to be used for interplanetary missions, and for exploring the outer reaches of the solar system.

10.9 Exercises

1. What is an appropriate figure-of-merit for a solar sail? Is a single mathematical expression in terms of the sail's properties appropriate or sufficient?
2. In the Table 10.1, fill in the type of orbit a solar sail will move on for the two different initial conditions specified in the columns, for the various values of the lightness number in the rows. The sail is always kept pointed at the Sun.

Table 10.1		
λ	Initially at rest	Initially on a circular orbit
0		
0 to 1/2		
1/2		
1/2 to 1		
1		
> 1		

The velocity that can be attained by a solar sail is limited by the Sun's intensity. In this chapter we examine replacing the Sun with a much more intense artificial radiation source, such as a laser, to achieve a higher velocity. We look more closely at the mechanics of photons incident on sails, taking into account relativistic effects and the finite travel time of light. Diffraction of the directed radiation source is a major issue, because it limits the amount of light intercepted by the sail, requires large optics, or demands high laser power, or some combination of these. We look into achieving the kilometre-scale optical systems required using aperture synthesis and adaptive optics. Given the large laser powers involved (tens to hundreds of gigawatts), sail heating becomes a limiting factor, and the choice of materials is critical. Having looked at the theoretical issues, I summarize a few of the numerous laser sail designs that have been published that bring together many of the issues discussed.

11.1 Relativistic kinematics of a laser sail

We start with the fundamental kinematics of a photon-propelled sail. Analyzing the momentum and energy transfer at the sail, initially for perfect reflection, we derive an expression for the velocity of the sail as a function of the laser energy. We then derive the equation of motion of the sail allowing for imperfect reflection, and solve this for a constant power laser. In this section we neglect diffraction loses.

11.1.1 Reflection of photons off a relativistic sail

Consider a sail receding directly away from a laser at velocity $\beta > 0$, as shown in figure 11.1. The laser emits a single photon of frequency ν_e in the laser rest frame. In the sail rest frame, the laser appears to be receding, so when received, this photon is Doppler shifted to a lower frequency of $\nu'_e = \nu_e\sqrt{(1-\beta)/(1+\beta)}$ (see section 9.2).[1] If we assume the photon is perfectly reflected by the sail, then in the sail rest frame the reflected photon still has a frequency of ν'_e. But when viewed from the laser rest frame, which sees a receding sail, this photon is Doppler shifted again, and by the same factor if the sail speed is unchanged. Thus the frequency of the reflected photon in the laser rest frame is

$$\nu_r = \nu_e\left(\frac{1-\beta}{1+\beta}\right). \tag{11.1}$$

The reflected photon has a lower energy. This means it must have given some of its energy to the sail. However, we assumed that the reflected photon had the same frequency as the

---

[1] The primed and unprimed labels are swapped with respect to the notation in chapter 9, because now the photons are emitted from the laser rest frame S.

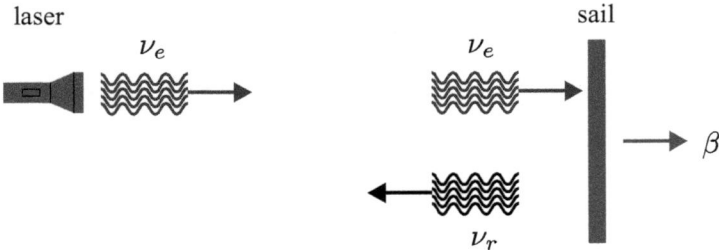

laser   sail

ν_e ν_e

β

ν_r

Fig. 11.1 The view from the laser rest frame of the reflection of photons off a sail receding at velocity β.

incident photon (in the sail rest frame), and so did not lose energy. Clearly this assumption is incompatible with the assumption that the speed (and thus kinetic energy) of the sail did not change. This is similar to the classical problem of a ball bouncing elastically off a wall, in which the ball returns with the same speed. This can only happen if the wall is infinitely massive, otherwise the ball will give energy to the wall, and so return more slowly.

We will deal with this dilemma correctly below, but in practice the sail has a rest mass–energy that is much larger than the energy of a single photon. In that case, equation 11.1 is a very good approximation. It tells us that the larger the sail velocity, the larger the redshift, and so the larger the fraction of energy that has been transferred to the sail. In other words, the efficiency of energy transfer from laser to sail increases with increasing velocity.

We might stop at this point and ask how energy conservation works with the Doppler effect, before we even get to the problem of reflection. Specifically, if the emitted photon has energy $h\nu_e$ in the laser rest frame, but energy $h\nu'_e$ in the sail rest frame, where has the energy gone? Energy, like velocity and momentum, is a frame-dependent quantity, so it is not conserved between different rest frames. This is true in classical systems also: An object at rest has zero kinetic energy, but it has a non-zero kinetic energy in a reference frame that it moving relative to it. The velocity of a photon is invariant across all reference frames, but its energy (and momentum) is not.

Let us now consider more rigorously the problem of a single photon reflecting off a sail, from the viewpoint of the laser rest frame. The sail has mass m and initial velocity β_i. A photon of frequency ν_e approaches the sail, then is reflected in the opposite direction with frequency ν_r. This exchange of momentum and energy results in the sail changing its velocity to β_f (its mass is assumed unchanged). Recall that for photons $E = h\nu = pc$, where p is the momentum. Equating the initial and final momentum and energy, we get two conservation equations in two unknowns, ν_r and β_f:

$$\frac{h}{c}\nu_e + \gamma_i mc\beta_i = -\frac{h}{c}\nu_r + \gamma_f mc\beta_f \quad \text{(momentum)}, \qquad (11.2a)$$

$$h\nu_e + \gamma_i mc^2 = h\nu_r + \gamma_f mc^2 \quad \text{(energy)}. \qquad (11.2b)$$

We rearrange each of these equations into something more convenient:

$$\frac{h}{mc^2}(\nu_e + \nu_r) = \gamma_f \beta_f - \gamma_i \beta_i \quad \text{(momentum)}, \qquad (11.3a)$$

$$\frac{h}{mc^2}(\nu_e - \nu_r) = \gamma_f - \gamma_i \quad \text{(energy)}, \qquad (11.3b)$$

and then sum them to eliminate ν_r:

$$\frac{2h\nu_e}{mc^2} = \gamma_f(1 + \beta_f) - \gamma_i(1 + \beta_i). \qquad (11.4)$$

Using the identity in equation 7.6 and defining $q = h\nu_e/mc^2$, the ratio of the photon energy to the sail rest mass–energy, this becomes

$$2q = \sqrt{\frac{1 + \beta_f}{1 - \beta_f}} - \sqrt{\frac{1 + \beta_i}{1 - \beta_i}}. \tag{11.5}$$

Rearranging this we arrive at an expression for the final velocity of the sail:

$$\beta_f = \frac{(2q + \phi_i)^2 - 1}{(2q + \phi_i)^2 + 1}, \quad \text{where} \quad \phi_i = \sqrt{\frac{1 + \beta_i}{1 - \beta_i}}. \tag{11.6}$$

When the sail starts at rest, $\phi_i = 1$, and we get (with $E_l = h\nu_e$)

$$\beta_f = \frac{(2q + 1)^2 - 1}{(2q + 1)^2 + 1}, \quad \text{where} \quad q = \frac{E_l}{mc^2}. \tag{11.7}$$

When $q = 1$, where the photon energy supplied equals the rest energy of the sail, the sail attains $\beta_f = 4/5$. For $q \ll 1$, equation 11.7 becomes $\beta_f \simeq 2q$. A sail of mass $10\,\text{g}$ at rest reflecting a $1\,\mu\text{m}$ photon has $q = h/\lambda mc = 2.2 \times 10^{-34}$ and so would acquire a velocity of order $10^{-34}\,c$. Clearly, the approximation we used to arrive at equation 11.1 earlier is valid in all practical applications. In the derivation of equation 11.7, we have assumed perfect reflection in the sense that no photons are absorbed by or transmitted through the sail, but we have *not* assumed that the photons are reflected with the same energy. Note that equation 11.7 is also the equation we would derive in the initial rest frame of the sail, with q then defined as $n\nu_e'/mc^2$ (see exercises).

The above derivation also applies if, instead of a single photon hitting the sail, we had a set of N photons. In this case $E_l = Nh\nu_e$ is the total energy (in the laser rest frame) emitted by the laser that arrives at the sail. It turns out, as will be shown below, that the equation for β_f still applies if the photons arrive at different times when the sail has a range of velocities. That is, the final velocity depends only on the total amount of energy received, and not on how it is received over time. The variation of the sail velocity with q is shown in figure 11.2. A $10\,\text{g}$ sail (rest energy $9.0 \times 10^{14}\,\text{J}$) propelled by a $100\,\text{GW}$ laser for $1000\,\text{s}$ ($10^{14}\,\text{J}$) attains a final velocity of $0.20\,c$, as does a $100\,\text{g}$ sail powered for 10 times as long.

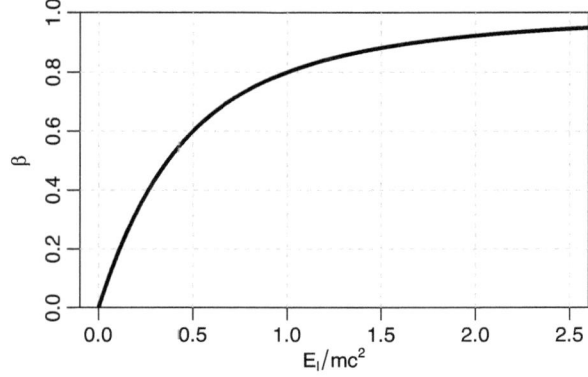

The dependence of the velocity acquired by a sail of mass m when accelerated from rest by photons of total energy E_l (equation 11.7). The curve asymptotes to $\beta = 1$ as $q = E_l/mc^2$ increases to infinity. This assumes all photons reach the sail and are reflected (zero transmission through, or absorption by, the sail).

Fig. 11.2

11.1.2 Energy efficiencies for perfect reflection

How much of the laser energy is put into the kinetic energy of the sail? The energy provided by the laser (in the laser rest frame) and received by the sail is $E_l = qmc^2$. From equation 11.5 with $\beta_i = 0$, and using the identity in equation 7.6, we can write this in terms of the velocity as

$$E_l = \frac{1}{2}[\gamma(1+\beta) - 1]mc^2. \tag{11.8}$$

As the kinetic energy (KE) of the sail is $(\gamma - 1)mc^2$, the overall energy efficiency is the ratio of these:

$$\text{overall efficiency} = \frac{\text{KE}_{\text{sail}}}{E_l} = \frac{2(\gamma - 1)}{\gamma(1+\beta) - 1}. \tag{11.9}$$

This is plotted in figure 11.3. We see that the efficiency increases monotonically with increasing velocity. At very low velocities, the efficiency is near zero. Most of the energy is reflected off the sail. This is consistent with equation 11.1, which tells us that at low velocity the reflected photon has a very similar frequency, and thus energy, to the incident photon.[2] As the velocity increases, more energy is put into the sail, and correspondingly the reflected photon is redshifted more.

It is interesting at this point to compare the laser sail with the photon rocket, discussed in section 7.7. The energy E_r that the photon rocket has to expend to get its velocity to β is given by equation 7.35. This is twice the amount that the laser has to deliver (E_l in equation 11.8) to give the sail the same kinetic energy. Although we can in principle achieve near unit energy efficiency with a laser sail, with a photon rocket the maximum efficiency is only $1/2$. This is in keeping with the disadvantage of a rocket having to carry its energy/propellant source. Observe that whereas E_l is measured in the laser rest frame, E_r was measured in the photon rocket's (accelerating, instantaneous) rest frame. But in both cases the amount of

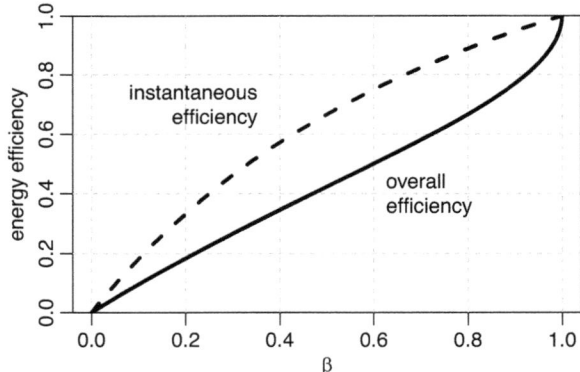

Fig. 11.3
The variation of the energy transfer efficiencies of a laser sail as a function of the sail's velocity. The solid line shows the overall efficiency, equation 11.9, and the dashed line the instantaneous efficiency, equation 11.10.

---

[2] Recall that β is the velocity of the sail after the first photon has hit it, so the energy transfer and efficiency are not exactly zero for the first photon.

energy expended to achieve the motion is measured in the rest frame in which the energy is expended.

This factor of two difference also occurs when we consider the momentum. The photon rocket is propelled by photons emitted from it, whereas the laser sail is propelled by photons reflected off it, which, for perfect reflection, give twice the change in momentum as a reflected photon.

As we are primarily interested in accelerating sails to speeds of 0.1–0.2 c, figure 11.3 shows that most of the energy sent by the laser will be reflected back to the laser. When the laser power is being sent by a large telescope (discussed below), we could simply reflect the (redshifted) photons back to the spacecraft, where they can again transfer some of their kinetic energy to the spacecraft. This can in principle be repeated more than once, as long as the spacecraft has not attained its final velocity. But as each reflected beam will be significantly broadened by diffraction in practice, this 'photon recycling' may not help much.

The efficiency in equation 11.9 is the overall efficiency of energy transfer when accelerating the sail from rest to β. Another efficiency is that of a single photon transfer (or at least a small number of photons) when the sail is moving at velocity β. If the incident photon has energy E_e, and the reflected photon energy E_r (both in the laser rest frame), then the energy transmitted to the sail is $E_e - E_r$ and the 'instantaneous efficiency' is $(E_e - E_r)/E_e$. With $E = h\nu$ and using equation 11.1, we get

$$\text{instantaneous efficiency} = 1 - \frac{E_r}{E_e} = \frac{2\beta}{1+\beta}. \tag{11.10}$$

This is plotted as the dashed line in figure 11.3.

11.1.3 The equation of motion

So far we have considered how the final velocity of the sail varies with the total amount of energy it receives, but not how its velocity varies with time or distance travelled. To determine these we need to take into account that the photons require a finite amount of time to reach the sail. Consider time measured in the laser rest frame. Because of the photon delay, the sail will achieve the velocity derived above at a time later than when the last photons were emitted by the laser. This is illustrated in figure 11.4. Let t denote the time in the laser rest frame when a photon arrives at the sail. This photon left the laser at an earlier time $T = t - x/c$, where $x \geq 0$ is the distance of the sail from the laser when the photon arrives at the sail. The distance x is also measured in the laser rest frame. The laser is switched on and the first photons are emitted at time T_A. These arrive at the sail at event A, when the time is t_A. The laser is switched off at time T_B, and the final photons arrive at the sail at event B, when the time is t_B. Even though the 'coordinate' time t and the 'retarded' time T are measured in the same reference frame, the interval $t_B - t_A$ is longer than the interval $T_A - T_B$ because of the distance travelled by the sail over that duration. Moreover, because t depends on x – the distance of the sail from the laser – these intervals will vary as the sail accelerates, as we will see below. It also follows from figure 11.4 that in the laser rest frame, the rate of photon arrival at the sail is lower than the rate of emission from the laser, because each successive photon takes longer to arrive. This is a purely classical effect resulting from the finite speed of light.

Following this discussion, we now recognize that equation 11.7 gives the velocity of the sail at time t in the laser rest frame corresponding to the energy E_l emitted by the laser up to

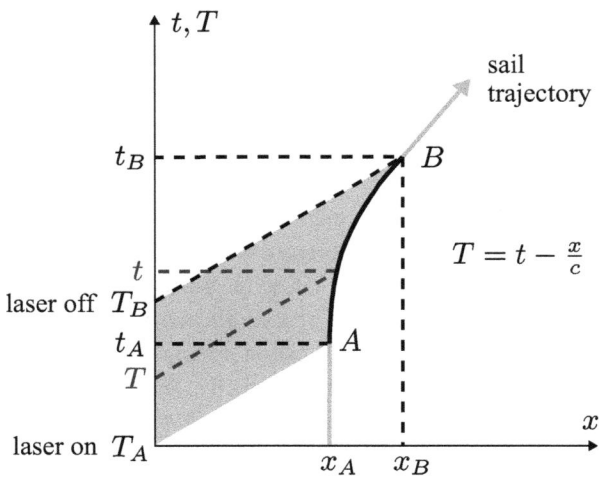

Fig. 11.4 A spacetime diagram of a laser sail, where x is the distance of the sail from the laser in the laser's rest frame. Both t and T are times in the laser rest frame. Photons emitted from the laser at time T arrive at the sail at time t, where $T = t - x/c$ is referred to as the 'retarded' time. The sail starts at $x = x_A$ and remains at rest until the first photons hit it at event A. The sail then accelerates (the curved part of its trajectory) until the last photons hit it at event B; after that the sail moves with constant velocity. The grey area shows the photons travelling from the laser to the sail.

retarded time T. If the laser continues to emit photons up until time T_B, then the final velocity of the sail – once all the photons have caught up with it – at time t_B is given by equation 11.7 with a larger value of E_l.

With this distinction between the times in mind, we now derive an equation of motion for the sail.

To evaluate the net force on the sail, we consider the reflection, transmission, absorption, and emission of photons for a relativistically moving sail. All photon momenta (equivalently frequencies, as $p = h\nu/c$) and photon rates $\dot{n}$ are defined in the laser rest frame, and all are positive.

The laser emits photons of momentum p_l at a rate $\dot{n}_l$. Because the sail is receding from the laser at a speed β, the photons arrive at the sail at a lower rate of

$$\dot{n}_s = \dot{n}_l(1 - \beta). \tag{11.11}$$

This is a purely classical effect that we considered as part of the Doppler effect in section 9.2. The momentum (frequency) of the photons is unchanged because we are always considering the laser rest frame. These incident photons may be reflected, absorbed (and then re-emitted), or transmitted through the sail, as illustrated in figure 11.5. We now consider the effect of each of these on the sail as they enter or leave the sail. As in chapter 10, the absorptance and reflectance of the sail are denoted $\tilde{a}$ and $\tilde{r}$, respectively; the transmittance is $1 - \tilde{r} - \tilde{a}$:

- The incident photons of momentum p_l, rate $\dot{n}_s$, act to increase the momentum of the sail.
- Photons of momentum p_r are reflected at rate $\dot{n}_r = \tilde{r}\dot{n}_s$ and act to increase the momentum of the sail. These photons undergo a double Doppler shift (equation 11.1) and so

$$p_r = p_l \left(\frac{1 - \beta}{1 + \beta} \right). \tag{11.12}$$

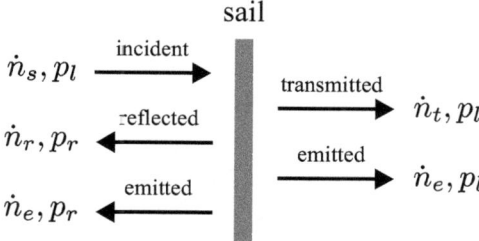

Fig. 11.5

A sketch of the various contributions of the photon rates and momenta to the motion of a light sail. Quantities are defined in the laser rest frame.

- Photons of (unchanged) momentum p_l are transmitted through the sail at rate $\dot{n}_t = (1 - \tilde{r} - \tilde{a})\dot{n}_s$, and so are a source of loss of momentum of the sail. (We could instead think of these as being subtracted from the incident beam.)

- Photons are absorbed at rate $\dot{n}_a = \tilde{a}\dot{n}_s$. Assuming the sail is in thermal equilibrium, these will be re-emitted at the same rate. We can consider half to be re-emitted in the forward direction as though they were transmitted (so at momentum p_l) and the other half to be re-emitted in the backward direction as though they were reflected (so at momentum p_r). In that case $\dot{n}_e = \dot{n}_a/2$.

Bringing all of the above points together, the rate of change of momentum of the sail is

$$\frac{dp}{dt} = \dot{n}_s p_l + \dot{n}_r p_r - \dot{n}_t p_l + \dot{n}_e p_r - \dot{n}_e p_l \tag{11.13a}$$

$$= \dot{n}_s p_l + \dot{n}_s \tilde{r} p_r - \dot{n}_s (1 - \tilde{r} - \tilde{a}) p_l + \dot{n}_s \frac{\tilde{a}}{2}(p_r - p_l) \tag{11.13b}$$

$$= \frac{1}{2}\dot{n}_s(2\tilde{r} + \tilde{a})(p_r + p_l). \tag{11.13c}$$

Using equations 11.11 and 11.12 we can eliminate $\dot{n}_s$ and p_r to get

$$\frac{dp}{dt} = p_l \dot{n}_l (2\tilde{r} + \tilde{a}) \left(\frac{1 - \beta}{1 + \beta} \right). \tag{11.14}$$

As the power of the beam in the laser rest frame is $P_l = c p_l \dot{n}_l$, we can write the expression for the force on the sail due to the laser beam as

$$F = \frac{dp}{dt} = \frac{P_l}{c}(2\tilde{r} + \tilde{a}) \left(\frac{1 - \beta}{1 + \beta} \right). \tag{11.15}$$

Using the relativistic version of Newton's second law (section 7.4, equation 7.25) to relate this force to the coordinate acceleration $a = c\, d\beta/dt$, and neglecting any other forces, we get the equation of motion for the sail of mass m:

$$\frac{d\beta}{dt} = \frac{P_l}{mc^2} \frac{1}{\gamma^3} \left(\frac{1 - \beta}{1 + \beta} \right)(2\tilde{r} + \tilde{a}) = \frac{P_l}{mc^2} \frac{(1 - \beta)^2}{\gamma}(2\tilde{r} + \tilde{a}). \tag{11.16}$$

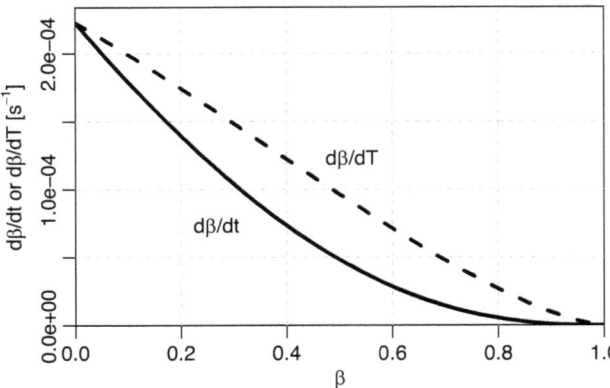

Fig. 11.6 The variation of the sail acceleration as a function of velocity ($\beta c = dx/dt$), for the acceleration defined in terms of both the coordinate time $d\beta/dt$ (solid line; equation 11.16) and the retarded time $d\beta/dT$ (dashed line; equation 11.25). This is illustrated for a 10 g sail propelled by a laser power of 100 GW, assuming perfect reflection ($\tilde{r} = 1$).

This gives the acceleration of the sail at coordinate time t when its velocity is β, where P_l is the power of the photons that have just arrived at the sail, and so were emitted from the laser at the earlier retarded time T (see figure 11.4). The variation of this acceleration with velocity is shown as the solid line in figure 11.6 for a 10 g sail accelerated with a 100 GW laser.[3] The acceleration decreases monotonically with increasing velocity. Thus even though at velocities near the speed of light the energy transfer efficiency is very high (figure 11.3), the acceleration drops to zero.

In the non-relativistic limit of $\beta \ll 1$, equation 11.16 becomes, with $v = \beta c$,

$$\frac{dv}{dt} = \frac{P_l}{mc}(2\tilde{r} + \tilde{a}) \quad \text{(non-relativistic limit)}. \tag{11.17}$$

In this case the acceleration is independent of velocity, as we would expect. This is of course the same equation that we would derive from equation 10.19 with pressure equal to $(1/A)mdv/dt$ and intensity equal to P_l/A, where A is the area of the sail.

11.1.4 The solution of the equation of motion

The integration of equation 11.16 is

$$\int \frac{1}{2\tilde{r} + \tilde{a}} \gamma^3 \left(\frac{1+\beta}{1-\beta} \right) d\beta = \int \frac{P_l}{mc^2} dt. \tag{11.18}$$

Let us now assume that both the sail mass and the laser power are constant, so $\int_{t_A}^{t} P_l \, dt = P_l(t - t_A)$. Assume also that the reflectance and absorptance are independent of wavelength over the range that it varies due to the increasing redshift (range of β). Integrating the left side and setting $\beta = 0$ at $t = t_A$ we get

$$t - t_A = \frac{mc^2}{3P_l} \frac{1}{2\tilde{r} + \tilde{a}} \left[\frac{(1+\beta)(2-\beta)\gamma}{(1-\beta)} - 2 \right]. \tag{11.19}$$

This is the (coordinate) time that has elapsed in the laser rest frame when the spacecraft reaches velocity β.

[3] For comparison, the acceleration due to gravity from the Sun at 1 au is $d\beta/dt = 2 \times 10^{-11}\,\mathrm{s}^{-2}$.

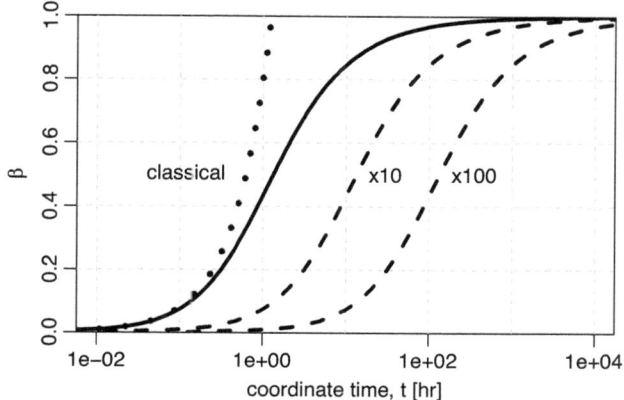

The variation of the velocity with coordinate time t for a 10 g sail spacecraft propelled by a laser power of 100 GW assuming perfect reflection (solid line; equation 11.19). The two dashed lines show an increase in the mass (or decrease in the power) by factors of 10 and 100. The dotted line shows the classical, non-relativistic result (equation 11.20) for the nominal case.

Fig. 11.7

As t is the time when a photon arrives at the sail, $P_l(t - t_A)$ is the energy that has arrived at the sail by this time. As we can see from figure 11.4, this is the energy that was emitted by the laser in the time period T_A to T. Thus $P_l(t - t_A)$ is *not* the energy that has been emitted by the laser by time t. By this time the laser has emitted more photons (those after time T), but these photons have not yet reached the sail.

Equation 11.19 does not have a closed-form solution for β in terms of t, but we can still plot the motion. This is shown in figure 11.7 for three different masses, or equivalently, three different values of the velocity-independent factor in equation 11.19. A 10 g spacecraft powered by a 100 GW laser will reach a velocity of $0.1\,c$ after 8.3 minutes and $0.2\,c$ after 19 minutes. Due to the linear dependence of the time on mass and power, a spacecraft 10 times more massive will take 10 times as long, for example.

In the non-relativistic limit of $\beta \ll 1$, the term in square brackets in equation 11.19 becomes 3β, and so we get

$$t - t_A = \frac{mc^2}{P_l} \frac{1}{2\tilde{r} + \tilde{a}} \beta \quad \text{(non-relativistic limit)}. \tag{11.20}$$

This is shown as the line labelled 'classical' in figure 11.7.

How far does the sail travel before it achieves a certain velocity? Using

$$\frac{d\beta}{dt} = \frac{d\beta}{dx} \frac{dx}{dt} = \frac{d\beta}{dx} \beta c \tag{11.21}$$

we can rewrite equation 11.16 in terms of $d\beta/dx$ and then integrate. With $x = x_A$ when $\beta = 0$ we get

$$x - x_A = \frac{mc^3}{3P_l} \frac{1}{2\tilde{r} + \tilde{a}} \left[1 + \frac{(2\beta - 1)}{\gamma(1 - \beta)^2} \right]. \tag{11.22}$$

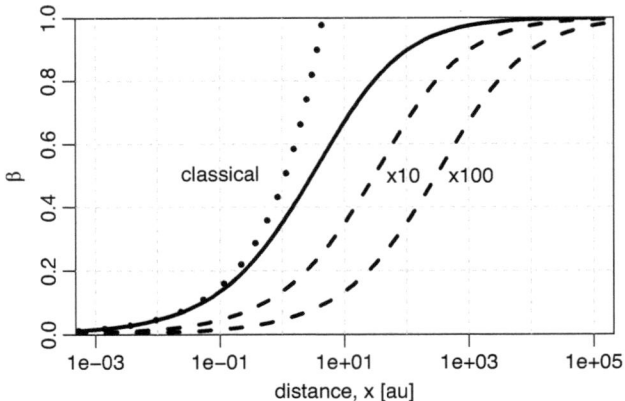

Fig. 11.8   As figure 11.7 but now showing the variation of the velocity with distance travelled x (equation 11.19).

This is plotted in figure 11.8, again for a 10 g sail spacecraft powered by a 100 GW laser. By the time the sail has reached a velocity of $0.1\,c$, the sail has travelled 7.8 million km (0.05 au), and when it has reached $0.2\,c$, it is 37 million km (0.24 au) away.

In the non-relativistic limit of $\beta \ll 1$, the term in square brackets in equation 11.22 becomes $3\beta^2/2$, and so we get

$$x - x_A = \frac{mc^3}{2P_l}\frac{1}{2\tilde{r}+\tilde{a}}\,\beta^2 \quad \text{(non-relativistic limit)}. \tag{11.23}$$

This is shown as the line labelled 'classical' figure 11.8.

We also need to know for how long we must run the laser in order to achieve a certain velocity. To do this we need to write equation 11.16 in terms of the retarded time T. As $T = t - x/c$, $dT = dt - dx/c$. With $\beta c = dx/dt$ this is

$$dT = dt(1-\beta). \tag{11.24}$$

As was already apparent from figure 11.4, time intervals between a pair of events at the sail, such as events A and B, are shorter in retarded time than in coordinate time because of the acceleration of the sail, and it therefore takes longer for the later photons to catch up with the sail. In terms of dT, equation 11.16 is

$$\frac{d\beta}{dT} = \frac{P_l}{mc^2}\frac{(1-\beta)}{\gamma}(2\tilde{r}+\tilde{a}). \tag{11.25}$$

This is shown as the dashed line in figure 11.6.

Integrating equation 11.25 with $\beta = 0$ at $T = T_A$, we find the velocity of the sail at retarded time T to be

$$\beta = \frac{(2q+1)^2 - 1}{(2q+1)^2 + 1}, \quad \text{where} \quad q = \frac{2\tilde{r}+\tilde{a}}{2mc^2}\int_{T_A}^{T} P_l(T^*)\,dT^*. \tag{11.26}$$

The integral is the total energy emitted by the laser between retarded times T_A and T, and therefore is the energy that will arrive at the sail between coordinate times t and t_A. This is the same as equation 11.7, but now with the energy replaced by a time integral of the power (over retarded time) and accounting for imperfect reflection. So we have generalized the case analyzed in section 11.1.1 of photons arriving simultaneously to photons arriving at arbitrary times for constant or variable laser power. The integrated power is the total energy E_l provided by the laser; its variation with sail velocity for the perfect reflection case was shown in figure 11.2. If the power is constant, then $E_l = P_l(T - T_A)$, and so this figure also shows how the velocity of the sail varies with retarded time (that is, the velocity it will eventually achieve once all the photons have arrived).

As the sail recedes from the laser, its angular size will get smaller, so in reality an increasing amount of the transmitted laser power will be lost around the edge of the sail due to diffraction from a finite-sized transmitter. We will examine this in section 11.2.1. We could accommodate this through an appropriate variation of the power with distance.

11.1.5 The power received by the sail

Consider now the power P'_s received by the sail in its rest frame. How does this relate to the power P_e provided by the laser in its rest frame? In the laser rest frame, photons of frequency v_e are emitted at constant rate $\dot{n}_l$, so $P_e = \dot{n}_l h v_e$. The power received by the sail in its rest frame can likewise be written $P'_s = \dot{n}'_s h v'_e$ (all primed quantities are measured in the sail rest frame). The photons hit the sail when it is moving at velocity β, so the sail sees them Doppler shifted to frequency $v'_e = v_e / \psi$, with ψ defined in equation 9.7. The primed and unprimed labels are swapped with respect to the notation in chapter 9, because now we consider the unprimed frame as the source of the photons. As the laser is moving away from the sail, we have $\alpha = \pi$ and $\beta > 0$ so $\psi = \gamma(1 + \beta)$, which is greater than 1; the photons are redshifted. The rate at which photons are received is also a frequency and so we also have $\dot{n}'_s = \dot{n}_l / \psi$. Putting these two terms together we get $P'_s = P_e / \psi^2$. Using equation 7.6 we can write this as

$$P'_s = P_e \left(\frac{1 - \beta}{1 + \beta} \right). \tag{11.27}$$

This is the power received by the sail in its rest frame when it is moving at velocity β. The factor on the right decreases monotonically from one to zero as β increases from zero to one. Thus the power received by the sail is less than the power emitted by the beam, and the larger the velocity, the less power is received.

Where has this power gone? It is not in the reflected photons, because we are not yet considering the reflection process; we are just looking at the power that arrives in the rest frame of the sail. It is also not due to any practical losses, because we are dealing with an idealized situation. Part of the explanation is that equation 11.27 gives the instantaneous power, not the total energy. Not all of the energy emitted per unit time in the laser rest frame arrives in the same interval time in the sail rest frame. Because of the motion of the sail, some of the energy is still in photons that have not yet arrived at the sail. This can be seen by referring once again to figure 11.4. This is a purely classical effect. On top of this comes the relativistic effects of shifting between rest frames. An exercise at the end of the chapter examines this a bit more.

Even once all of the photons have arrived at the sail, not all of the energy emitted by the laser as measured in the *laser* rest frame is received at the sail as measured in the *sail* rest frame. This is because energy is not conserved between different rest frames, as noted earlier in the context of the Doppler effect (section 11.1.1).

When we previously considered the energy exchange from within the laser rest frame, we saw that at larger velocities a larger fraction of the laser energy is put into the kinetic energy of the sail (figure 11.3). In the present section, we considered the energy exchange from within the sail rest frame, and found that the larger the velocities, the lower the rate of energy transfer to the sail. This apparent contradiction is resolved by the fact that previously we were comparing an integrated quantity (energy) in a common rest frame, whereas here we are comparing a differential one (power) between two different rest frames.

11.2 Optics and arrays

11.2.1 Diffraction

Solar sails are propelled by an isotropically emitting source, namely the Sun or other star. The wavefront from a star expands spherically, and when it reaches the spacecraft, the wavefront is close to flat with uniform intensity. But when we propel the sail with a directed source like a laser, we must consider diffraction. Even if we create a perfectly collimated beam, the fact that it passes through an aperture of finite size leads to an unavoidable spreading of the beam due to diffraction. Consequently, some of the laser power may pass around the sides of the sail. This is still the case when we use a lens or mirror to focus the beam to a point on the sail. What in geometric optics would have been a point will be extended. This can be understood in Kirchhoff's diffraction theory as the result of interference from a continuum of sources in the plane of the emitting aperture.

Consider a circular aperture – the laser source – of diameter d_t transmitting monochromatic light of wavelength λ (figure 11.9). The interference pattern on a surface (the sail) a distance r from the aperture can be described by Fraunhofer diffraction (far-field diffraction). For a circular aperture, the variation in intensity (power per unit area) on a plane perpendicular to the direction of propagation is

$$I(u) = I_0 \left(\frac{2J_1(u)}{u} \right)^2, \quad \text{where} \quad u = \pi \alpha \frac{d_t}{\lambda}, \tag{11.28}$$

α is the angle from the direction of propagation, I_0 is the intensity at the centre, and $J_1(u)$ is a first-order Bessel function. This is shown in figure 11.10. For this function, 84% of the total power is contained between between the centre and the first minimum, a region known as the *Airy disk*. The first minimum is at $u = 1.22\pi$, so the angular radius of the Airy disk is $1.22\lambda/d_t$. A total of 91% of the power is contained within the second minimum at $u = 2.23\pi$, and 94% within the third minimum at $u = 3.24\pi$. The intensity at the centre of the Airy disk is

$$I_0 = \frac{P_t A_t}{\lambda^2 r^2} \tag{11.29}$$

where P_t is the power emerging from the aperture, $A_t = \pi d_t^2/4$ is the area of the aperture, and r is the distance from the aperture. (We'll see in section 15.2 how the dependencies in equation 11.29 arise.)

The Airy disk is conventionally taken as the size of the beam at the receiver. For a 1 m telescope transmitting light at 1 μm for example, we get $\alpha = 1.22$ μrad, or $0.25''$. For other shaped apertures, the angular size of the region containing a fixed fraction of the total power

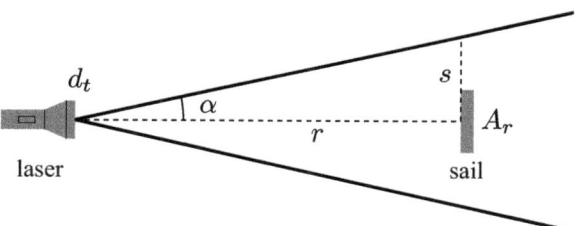

The diffraction of a light source through an aperture of diameter d_t. From diffraction theory, $\alpha \sim \lambda/d_t$, where λ is the wavelength of the light, so at the sail (of area A_r) the beam has a characteristic radius of $s \sim \lambda r/d_t$.

Fig. 11.9

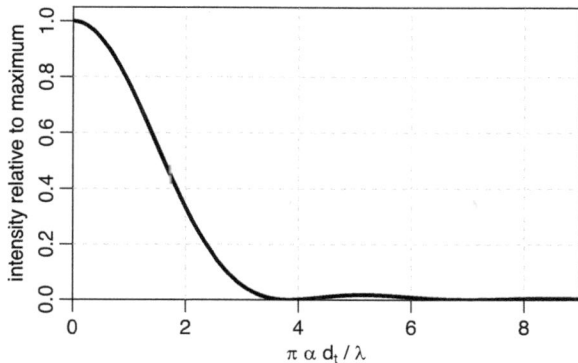

Diffraction pattern far from a circular aperture of diameter d_t with light of wavelength λ. This shows how the intensity varies as a function of angular distance α from the axis in the Fraunhofer diffraction regime.

Fig. 11.10

still varies as λ/d_t but with a different numerical constant. Writing this constant as a, the *diffraction angle* can be written more generally as

$$\alpha = a\,\frac{\lambda}{d_t},\tag{11.30}$$

with $a = 1.22$ for the first minimum of the Airy disk. The physical radius of the beam at a distance r from the source is $s = \alpha r$ from geometry when $\alpha \ll 1$. As an example, for a 10 m diameter telescope transmitting at $\lambda = 1\,\mu$m, then at a distance of $r = 10^6$ km the Airy disk on the sail has a radius of 122 m, and thus a diameter of 244 m. In the previous section, we saw that when accelerating a 10 g sail with a 100 GW laser, it will travel 7.8 million km by the time is reaches a velocity of 0.1 c. At this distance the Airy disk has a diameter of 1.9 km. This is much larger than the sail, which at this mass is unlikely to be much larger than a few metres. Thus almost all of the light would be lost around the edge of the sail by diffraction, meaning the sail would never have attained this velocity by this distance in the first place. To reduce these diffraction losses, we need to use a much larger telescope aperture. When the aperture is 1 km in size, for example, then $\alpha = 0.25$ mas and the Airy disk at 7.8 million km has a diameter of only 19 m, although this is probably still much larger than the sail.

The above discussion has assumed that the sail is stationary relative to the laser/telescope. When in motion, the diffraction angle is increased by aberration, as given by equation 9.5.

When $\beta = 0.1$, the diffraction angle is 11% larger and the solid angle 22% larger, and this will spread out the laser power even more.

It is clear that we need very large transmitting optics to propel laser sails, and in the next section we'll look at how to achieve this. Even then, the transmitting aperture and sail will likely not be large enough for the sail to intercept all of the Airy disk at all distances during the acceleration. Limits on laser power mean the sail won't reach its final velocity before there is significant loss of power around the edge of the sail. This needs to be taken into account in a more detailed treatment of the sail kinematics (see exercises).

We discussed in the previous section that much of the laser power incident on the sail is reflected back to the laser. This could be reflected back again to the sail, thereby increasing the power or saving energy. The problem with this photon recycling is that the beam reflected by the sail is also diffracted due to the finite size of the sail. Because the sail is much smaller than the laser aperture, this diffraction is very large, so relatively little power returns to the laser, and even less makes it back to the sail again. This can be repeated, but every reflection dilutes the beam significantly.

11.2.2 Aperture synthesis

Building a kilometre-sized optical telescope would be extremely challenging, and almost impossible on the Earth's surface given the need to support and steer it. However, it is not necessary to build an individual telescope structure in order to achieve a large effective aperture with a narrow diffraction pattern. If we take a large telescope mirror and remove small parts of it, this changes the detailed shape of the diffraction pattern, but it does not significantly change the width of the central part of the pattern that contains most of the energy. A large telescope mirror with bits missing is equivalent to a number of smaller telescopes with the appropriate surface shapes and relative positions. This is the idea of aperture synthesis, shown in the left panel of figure 11.11. This allows a large effective aperture to be achieved without having to build a large single telescope. The 'baseline' is the separation between any two individual telescopes, and the largest baseline is equal to the diameter of the original monolithic telescope. The transmitting area (or collecting area, if it is being used as a receiver) of this effective aperture is reduced, but for many applications this is not a limitation.

Telescope arrays are commonly used in astronomy, so let us first consider using the array as a receiver to observe a distant object (incoming plane wavefront). The array shown in figure 11.11 (left panel) has the individual telescopes in the physical locations they would have in the filled aperture. This is impractical for a large array on the Earth's surface, because to point it in different directions we would have to lift individual telescopes far above the ground. However, the purpose of the precise shape of a mirror is to ensure that all of the wavefront reflected off the different parts of the mirror travel the same distance and so maintain a common phase at the focus. We can instead place the individual telescopes in a common plane, and then (somehow) remove the phase shifts that arise because different portions of the wavefront that went to different telescopes had to travel different distances (figure 11.11, right panel). These phase shifts are called the 'geometric delay'. By varying the phase shifts, the effective aperture is sensitive to different directions, equivalent to pointing it in different directions. This is the idea of an astronomical interferometer. The resulting diffraction pattern (also for a filled aperture) is often referred to as the *point spread function* (PSF).

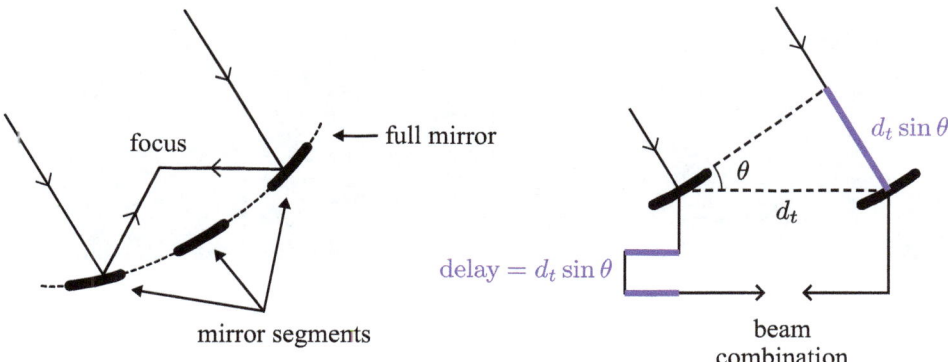

Aperture synthesis. Left: Parts of the the primary mirror of a telescope can be removed to leave just some mirror segments. This reduces the amount of light gathered and changes the diffraction pattern, but the telescope still has an effective aperture diameter equal to the original complete telescope. Right: This same-sized aperture can be synthesized with the mirror segments in a common plane (e.g. the surface of the Earth) by introducing appropriate optical path delays between the wavefronts received at the different segments to ensure that the wavefronts are in phase when combined.

Fig. 11.11

How can we introduce the phase shifts? Astronomical interferometry was first developed most extensively at radio wavelengths, because at these longer wavelengths we can detect both the amplitude and the phase of electromagnetic waves and record them electronically. This allows the signals from the different individual telescopes to be combined electronically too. As the signals are stored on a computer, the combination does not even have to be done in real time. This has enabled the combination of signals from radio telescopes separated by thousands of kilometres and thus the synthesis of apertures on this scale, in a process known as very long baseline interferometry (VLBI). It was this technology that allowed the first images of black holes to be obtained with the Event Horizon Telescope in 2019, for example.

At optical wavelengths, the direct electronic detection of phase is currently impossible, so astronomers instead remove the phase shifts by making the light that arrives at the first telescopes travel a longer path, before optically combining the light from all the telescopes, as shown in the right panel of figure 11.11. By varying the length of these delay lines – at the same precision required for a telescope mirror to keep wavefronts in phase – the effective aperture can be pointed in different directions. (The individual telescopes would be pointed too.)

In principle, this same process can be used backwards to transmit power from a powerful light source through an optical telescope array. This would produce an outgoing wavefront with a common phase that acts as though it were a beam focused from a single large telescope. In practice this would be very difficult, on account of the extremely high light intensity that has to be transmitted through the delay line optics. It is also not possible to produce an individual laser with anything close to the required power. We would instead need to feed the telescopes with separate lasers, and somehow ensure that the lasers have the appropriate phase offsets needed to synthesize the larger aperture. We look at how to do this in section 11.3 below.

With a phase-corrected array of telescopes, the size of the central part of the PSF it produces is of order λ/d_t, where d_t is the largest baseline in the array. However, the amount of power in that region is less than in the case of an Airy disk produced by a filled aperture. This

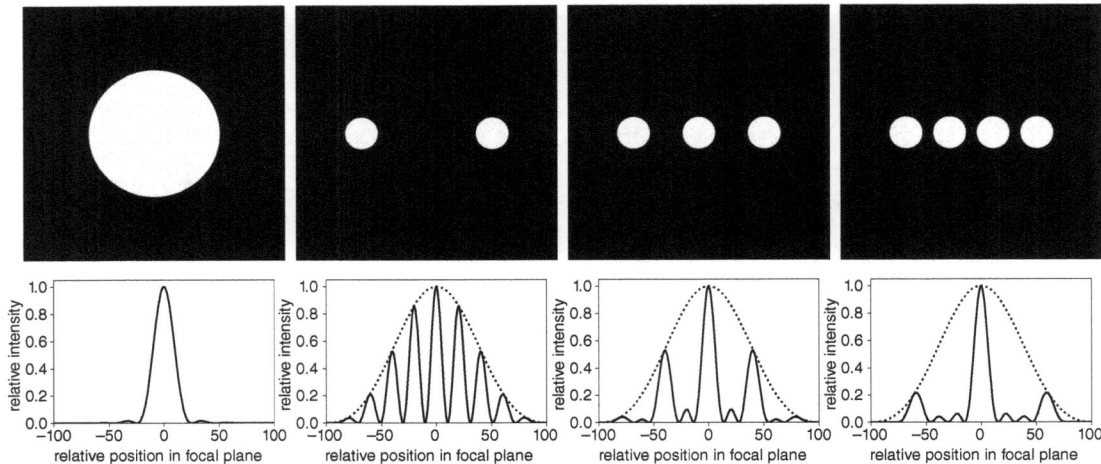

Fig. 11.12 Four different apertures (top row) and their corresponding PSF (bottom row). The white circles in the top row show the size and location of individual telescope apertures that are used to synthesize an aperture. The largest baseline is the same in all cases (equal to the diameter of the single large aperture in the left panel). The bottom rows shows the variation in intensity of a horizontal cut through the middle of the PSFs. The dashed line in these panels shows the PSF of a single small aperture for comparison. The scale on the horizontal axis is common to all panels, but in arbitrary units, and their vertical axes have been scaled to the maximum intensity in each case.

can be seen in figure 11.12, which shows four different apertures that have the same largest baseline d_t, but a different number of telescopes and thus a different filling fraction of the synthesized aperture. (In practice the array would extend in the other direction too; a linear array is shown for simplicity.) The first PSF on the left is for the filled aperture of diameter d_t. This is the Airy pattern mentioned earlier. The second PSF comprises two smaller apertures. The central peak of this PSF is narrower than the Airy disk for the filled aperture, but a larger fraction of the power is now in the so-called side lobes. As we increase the number of same-sized apertures to three and then four at constant baseline, we see that a larger proportion of the power (total area under the curve) is concentrated into the central peak of the PSF. The dashed line is the PSF for a single small aperture, which shows that the envelope of the synthesized PSF is determined by the individual aperture size.[4]

For most situations it turns out that the array can deliver more power to a sail when we use a compact array, with individual telescopes as close to each other as possible, rather than a sparse array with a larger effective aperture (so narrower central peak) but more power put into the side lobes.

11.2.3 Atmospheric distortion and adaptive optics

A significant complicating factor for an optical array on the Earth's surface is the atmosphere through which the light must pass. Spatial and temporal variations in the refractive index of the air – caused by temperature variations, wind, and turbulence – distort the wavefront. This introduces phase differences across the wavefront on a range of length and timescales.

---

[4] More precisely, the PSF of the synthesized aperture is the product of the Fourier transform of an individual telescope aperture with the Fourier transform of the array (represented as a set of delta functions).

In astronomy, this wavefront distortion is referred to as 'seeing'. Without correction, seeing limits the size of the PSF in the optical and near-infrared to be no less than about $1''$ (to within factors of a few). This is a thousand times larger than the diffraction limit we want to achieve.

Seeing not only limits the angular resolution, but also spreads faint light from a point source across a larger area, mixing it with diffuse background light, thereby reducing the sensitivity of the observations. The seeing varies with wavelength (better at shorter optical wavelengths) and altitude (better on mountains). But with a seeing of $1''$, then at $1\,\mu m$, a telescope with an aperture larger than 0.25 m is seeing-limited rather than diffraction-limited. This applies equally well to an aperture synthesis telescope. If we cannot overcome the seeing, the spot size of laser power sent through the array will still be of the order of $1''$, and not reduced by the large extent of the array.

The problem of seeing is addressed in astronomy by measuring the distortion of the wavefront and using this information to adjust the shape of the mirrors to undo the distortion. This is known as adaptive optics, the principle of which is summarized in figure 11.13. The wavefront can be modelled as a contiguous set of small patches that have become tilted with respect to one another by the atmosphere. By taking some of the light from the observed source, the tilt of each patch can be determined. This can be done via intensity measurements. Instead of the target source itself – which may be too extended or too faint to provide enough light for wavefront monitoring – we can instead use a bright star that lies a small angular distance from the target, a so-called natural guide star. In either case, these tilt measurements provide a map of the distortions across the pupil (aperture) of the telescope. A deformable mirror in the pupil plane that has subapertures corresponding to the measured patches is then adjusted to undo the tilts. The result is that the wavefront reflected off this mirror is restored to being flat.

For this correction procedure to work, each patch must be small enough such that the wavefront distortion can be approximated by a simple tilt about two axes. This patch size is known as the Fried parameter r_0, and is of the order of a few tens of centimetres, depending on wavelength and atmospheric conditions. The wavefront must also be measured, and the mirror adjusted, on a timescale that is shorter than the variation timescale of the wavefront. As the turbulence in the atmosphere evolves relatively slowly, this timescale is set primarily by the time it takes air to cross the patch, and so is of order r_0/v, where v is the wind speed. Timescales are typically a few milliseconds, so mirror adjustments need to be made several hundred times per second.

Adaptive optics for near-infrared wavelengths is routinely employed on large ground-based telescopes, such the four 8 m telescopes of the ESO's VLT in Chile. Building such a large primary mirror that can be adjusted at high frequencies would be challenging. But as large telescopes invariably employ smaller mirrors to help focus the light, one of these is usually the deformable one.[5] The VLT secondary mirrors have a diameter of 1.12 m, and one of them has 1170 actuators that adapt its shape at high frequency.

When individual telescopes are connected to make an interferometric array (as can be done at the VLT Interferometer), adaptive optics can be applied to each telescope independently. With the wavefront from each telescope beam corrected in this way, and the appropriate

---

[5] The primary mirror is often adjustable too, but at a much lower frequency, foremost to compensate for the distorting effects of gravity when this large thin mirror is tilted.

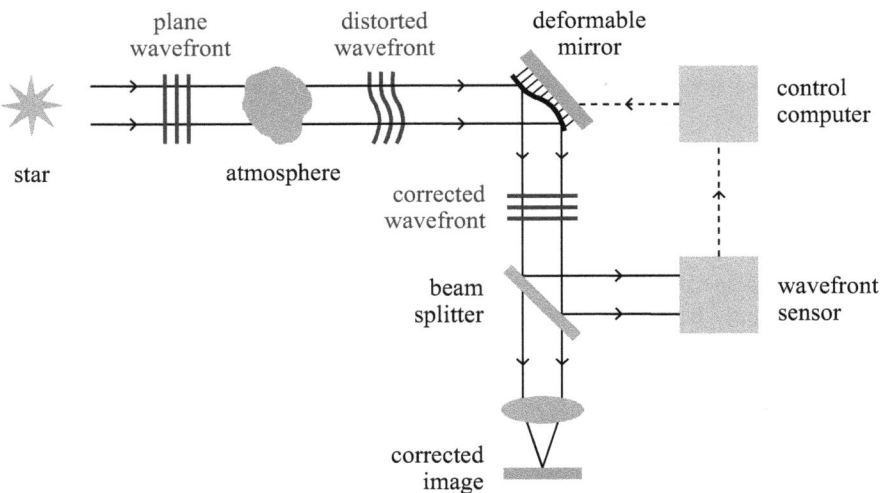

Fig. 11.13 A schematic outline of an astronomical adaptive optics system. The wavefront from a star, distorted by the atmosphere, enters the telescope and reflects off a deformable mirror. The wavefront is then divided, with a portion of the light going to a wavefront sensor that measures the distortions, and the other portion going to the image plane. The information from the wavefront sensor is used by a control computer to adjust the form of the deformable mirror to remove the wavefront distortions, so that the corrected wavefront can be imaged in a camera. The sensing and mirror correction typically takes places a few hundred times per second.

optical delays introduced for each beam to synthesize an aperture (figure 11.11), the array can achieve a PSF size that is determined by the diffraction limit of the array, rather than by the seeing or the individual mirrors.

The same issues apply to transmitting light from a laser array to a sail. An outgoing laser beam will have its wavefront distorted by the atmosphere just as incoming starlight does. If we want the laser array to be diffraction-limited to achieve the narrowest possible beam on the sail, then we need to correct these distortions. This could again be done with deformable mirrors. To measure the distortions we again need a coherent light source above the atmosphere that is imaged by the adaptive optics system. Natural guide stars could probably not be used because they would be much fainter than any backscattered laser light (and also fainter than the scattered sunlight if the laser is used during the day).

As the laser light leaving the telescope is designed to be coherent, one solution might be to use laser light reflected off the sail (recycled photons) for the adaptive optics corrections. The laser is powerful, and although the reflected photon beam is strongly diffracted by the small size of the sail, there would be enough photons for the corrections. Unfortunately, though, by the time the photons have travelled up from the telescope to the sail and back again, the atmosphere will have changed. If the atmospheric timescale is 1 ms, then the maximum distance to the sail permitted for this approach would work is $c \times 10^{-3}/2 = 150\,\mathrm{km}$. This is barely above the atmosphere, yet we saw in section 11.1.4 that the laser needs to propel the sail over millions of kilometres. So we can't use this reflected light for the corrections either.

Some of the outgoing laser light will be reflected back to the telescope by the atmosphere, and this could potentially be used for the wavefront correction. This is rather complicated, however, as the reflections occur across a range of altitudes, so at best only a partial correction could be achieved.

A better solution to this problem is to place a satellite above the atmosphere, close to the path of the laser, that shines a laser back at the telescope array, thus acting as a bright artificial guide star. This has already been explored as a possibility for use by astronomical telescopes. Such a beacon satellite needs to be in a relatively high orbit so that its downward beam – which is conically expanding due to diffraction – passes through the same portion of the atmosphere as the upward-propagating beam from the telescope array (which is conically expanding in the other direction). Without this, the atmospheric distortion cannot be properly corrected, a problem with laser guide stars in astronomy known as focal anisoplanatism, or the cone effect. A satellite altitude of at least 160 000 km has been suggested, which is over four times the geostationary altitude. The beacon's orbit has to be set according to the location of the telescope array and the direction to the target (star or sail). As the beacon needs to be very close to the line of sight between telescope and target, yet is moving, it should be placed in a highly eccentric orbit and used near apogee, where it is moving the slowest. Even then it would be usable only for short time periods. A beacon in a non-Keplerian orbit, achievable with an ion engine, or the deployment of multiple beacons, could be considered.

11.3 Lasers

11.3.1 How much power?

To accelerate a 10 g sail to 0.1 c requires that it receive 4.5×10^{12} J or 1.3 GWh of energy. As we have seen, the laser has to provide more energy than this because most of the energy remains in the reflected photons and is not transferred to the sail. From figure 11.3 we see that the overall efficiency to reach 0.1 c is only 10%, so in this case the laser must provide 13 GWh of energy. If this energy is transmitted at 1 μm by a telescope array with an effective diameter of 1 km, then a laser spot focused on the sail when it is 0.01 au from the array has spread by diffraction to an Airy disk diameter of 3.7 m (a bit more when we include aberration). This is the size of a 10 g sail with a lightness number of 1.6 (equation 10.10), which is large but plausible with newly identified materials (discussed below). Thus in order to avoid a significant loss of laser energy due to its beam spilling out around the sides of the sail, we must transfer all of the energy to the sail when it lies within 0.01 au. Assuming a constant acceleration up to 0.1 c over this distance, this means we must provide the energy within about 0.01 au/(0.05 c) = 100 s (actually about 5% less due to the light travel time, but we are just doing an order of magnitude calculation here). This implies we need a laser power of 13 GWh/100 s, or 500 GW.

Compared to currently available lasers (discussed below), 500 GW is an enormous power for a laser that must operate for this duration. Furthermore, it corresponds to an intensity on the sail that surpasses what current materials can withstand (see section 11.4.1 below). With a laser of only half the power we could still achieve the same final velocity for the same sail if we illuminate it for twice as long, which will also lower the intensity. However, to keep the same diffraction losses we would have to double the size of the telescope aperture, in this case to 2 km.

In practice, both laser power and array size will be limited, so we will have little choice but to accept a longer acceleration time, and therefore diffraction losses, and therefore an even longer acceleration time. The energy efficiency will go down, and storing enough energy – and then delivering it fast enough – is also a major challenge.

The sail mass and size are critical to the scaling of the problem. If the sail is twice as massive, it requires twice as much energy to get to the same final speed. Assuming fixed mass per unit area, then with twice the mass, the diameter is $\sqrt{2}$ times larger. So using the same telescope array (same PSF size), the sail can now travel $\sqrt{2}$ times as far before suffering the same diffraction losses, and thus has a travel time that is $\sqrt{2}$ longer. But as power is energy divided by time, this requires $2/\sqrt{2} = \sqrt{2}$ times as much power. In other words, for a given telescope array, a more massive sail also requires more laser power, not just more energy. This favours small sails. However, the sail must also carry a payload that adds mass but does not contribute to the propulsion, so a minimum-sized sail is required for a given payload mass. In practice there is a complicated trade-off between array size, sail size, laser power, and diffraction losses, as well as total energy required.

11.3.2 Arrays of lasers

Lasers of tens or hundreds of gigawatts of power that can run for extended periods of time do not yet exist. Some of the most powerful lasers available are those used in inertial confinement fusion, discussed in section 8.5. Each of the lasers in the National Ignition Facility, for example, has a power of about 2.5 TW, but delivers this for only a few nanoseconds. Perhaps the most powerful long-duration laser was the military MIRACL laser, able to deliver 1 MW for about a minute. Others have powers closer to tens of kilowatts. The input power for such lasers is many times larger than their laser output, and power is additionally required for water cooling. In any case, a large number of individual lasers will have to be combined to achieve tens or hundreds of gigawatts.

There are many different laser technologies operating at a wide range of wavelengths. An Nd:YAG laser, for example, is typically used to lase at 1064 nm. There are other transitions that can be used, and frequency doubling would lower the wavelength, which is beneficial for reducing the diffraction. Solid state diode lasers also offer many possibilities.

As discussed earlier, we need to synthesize a large transmitting aperture using multiple smaller telescopes (section 11.2.2). To achieve this, we could split the light from each laser across all of the telescopes, such that the entire array provides a coherent beam for each laser separately, and these beams then overlap incoherently on the sail. But this is impractical, because it would require a lot of optics in order to split and geometrically delay the light from each of the lasers (tens of thousands) into each of the telescopes (thousands).

A more practical approach is for each laser to feed only one telescope. In this case we need to ensure that all the lasers are in phase, and we have to introduce the correct geometric delay for all the different lasers that feed a given telescope. All the beams from the different individual telescopes are then coherent with one another and the outgoing wavefront has the diffraction limit of the array. This approach poses significant challenges too. Getting and keeping all the different lasers in phase is difficult, because the frequency and phase of a laser can drift over time due to numerous effects, including temperature variations, acoustic vibrations, and noise in the electric current driving the laser. The different lasers must therefore be locked to the same frequency and phase. This can be done using a common external reference, together with one of several active methods that rely on feedback from the laser to monitor changes and make corrections.

In all of this we have been assuming that the laser is ultraviolet, visual, or near-infrared. Longer wavelength sources, such as masers, bring simplifications in the aperture synthesis. Specifically, coherence could then be achieved electronically instead of using optical delay lines, in a so-called phased array antenna, which we will examine more in section 15.8.3. However, to keep diffraction losses within reasonable limits, the apertures would need to be two or three orders of magnitude larger, which seems to rule them out. Phased arrays in the optical are possible by integrating the delay lines in solid state without any moving parts. If these could be scaled up over very large areas to deliver the huge amounts of laser power required, this could do away with the need for large individual telescopes.

11.4 Sail structure

We have looked at the energy and power requirements of a sail and at how we might get this power to the sail by means of a laser and telescope array. We turn now to the properties of the sail itself.

11.4.1 Materials

In the context of solar sails (section 10.5), we identified that the ideal sail material has low mass per unit surface area, high reflectance (equation 10.21), and low absorptance and/or high melting temperature (equation 10.25b). These requirements cannot be optimized independently of one another. We saw that a very thin metallic sail will eventually become transparent, for example. A sail propelled by a laser will be exposed to much larger intensities than a solar sail. In section 11.3.1 we considered a sail of 3.7 m diameter, and a laser power of around 500 GW. Even if we consider a more modest 100 GW, this still corresponds to an intensity on the sail of $9 \, \mathrm{GW \, m^{-2}}$. In comparison, at a distance of 10 solar radii, a solar sail is exposed to an intensity of less than $1 \, \mathrm{MW \, m^{-2}}$. Laser sailing must pay much closer attention to the thermal properties of the spacecraft.

The Stefan–Boltzmann law in equation 10.25b tells us that the equilibrium temperature of a sail increases only slowly with the incident intensity, namely to the one-quarter power. Nonetheless, with a laser intensity of $9 \, \mathrm{GW \, m^{-2}}$, and a sail absorptance of 0.01 and emissivity of 0.1, this equation predicts an equilibrium temperature of 9400 K. This is well above the highest melting temperature of any known material, which is around 4000 K (diamond 3800 K; tantalum hafnium carbide alloy 4200 K). Even with unit emissivity, the absorptance would have to be below 0.003 to get down to these temperatures (with any factor decrease in the emissivity having to be offset by the same factor decrease in the absorptance). In practice, the maximum operating temperature would have to be lower, because absorptance tends to increase with temperature, and materials may weaken well before they melt. In any case, the absorptance will never be zero, so the sail should have a high reflectance at the wavelength of the incident laser beam – optical or near-infrared – yet a high emissivity at longer wavelengths corresponding to its equilibrium temperature. By choosing the right material(s), the reflection and emission can be optimized for these different wavelengths.

Although metals are highly reflective, they have neither very low absorptances nor high melting temperatures. Metallized films like those considered for solar sails are therefore inappropriate. For laser sail materials we need to turn to semiconductors, dielectrics, and composite materials. Zirconium dioxide (ZrO_2), for example, has a high emissivity of 0.95 and melting temperature of 2990 K. Its refractive index n at 1 μm is around 2.13, and so a thick piece of material has a reflectance of $[(n - 1)/(n + 1)]^2 = 0.13$. Germanium has a higher refractive index of around 4.1, and therefore a higher reflectance of 0.37.

The reflectance for monochromatic light of wavelength λ can be increased by using a thin film of dielectric material with thickness $\lambda/4n$, a result of the constructive interference of the beams reflected from the front and the inside back of the film. For such a 'quarter-wave plate' the reflectance increases to $[(n^2-1)/(n^2+1)]^2$, which for ZrO_2 is 0.41. Assuming zero transmittance, this implies an absorptance of 0.59. Using equation 10.25b, we conclude that a quarter-wave plate of ZrO_2 could tolerate an intensity of about 7 MW m^{-2} (with $\epsilon_b = 0.0$, so emission is only from the front). This is still far below the intensities considered in the previous section.

By using many layers of material, we can increase the reflectivity further, albeit at the cost of increased mass. Hein et al. (2017) cite several examples of multiple layers of dielectrics on metallized plastic films. A set of 15 layers of silicon dioxide (SiO_2) and titanium dioxide (TiO_2) dielectrics on a copper film achieves a reflectance of 0.9999294 (for light of wavelength 1.06 μm) and emissivity of 0.07. This can tolerate an intensity of nearly 0.2 GW m^{-2} at its maximum temperature of 1360 K.

The problem with using quarter-wave plates in this way, however, is that they are tuned to a single wavelength, whereas the sail will see an increasingly redshifted laser beam as it accelerates. One solution to this might be to vary the wavelength of the laser during the acceleration (difficult in practice), or to accept a lower average reflectivity due to imperfect tuning of the sail to the range of incident wavelengths.

Two other materials with high reflectance and low absorptance that could be suitable sail materials are molybdenum disulphide (MoS_2) and silicon nitride (Si_3N_4). The former has a high refractive index of around 3.7, and single layer samples of this have been measured to have negligible absorptance at visible wavelengths. Their densities are not that low, however (5060 kg m^{-3} for MoS_2).

One way to reduce the mass of the material is to fabricate it as a two-dimensional photonic crystal. This is a layer of material with a regular grid of micron-sized holes that have a diameter not much less than their separation, so most of the structure is holes (figure 11.14). Not only does this reduce the mass, but because the length scale of the repeating structure is similar to the wavelength of the light, this actually increases the reflectance. This is due to the interference produced by a large number of holes, similar in concept to a diffraction grating. So far it has been possible to manufacture photonic crystals of MoS_2 many centimetres in size just 80 nm thick, with holes 1 μm in diameter separated by 1.14 μm, which achieves a reflectance of 0.90. The thermal emissivity is increased by adding a 5 nm layer of Si_3N_4 to each side.

Counter-intuitively, some materials with very low reflectance may be of interest if they remain strong enough even when extremely thin. One such material is graphene, which comprises a single layer of carbon atoms arranged in a hexagonal structure. It is therefore extremely light, with an area density of just 7.4×10^{-7} kg m^{-2}. Graphene normally has a very low reflectance and an absorptance of just a few per cent or less (and thus a high

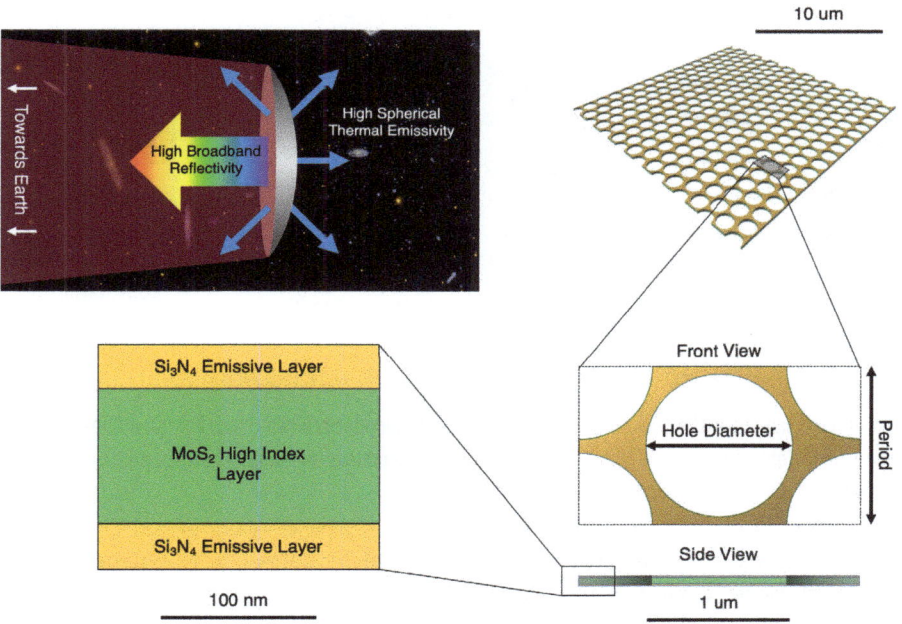

A two-dimensional photonic crystal. Credit: Reprinted with permission from Brewer et al. (2022), copyright 2022 American Chemical Society.

Fig. 11.14

transmittance), although doping it with impurities might raise the absorptance to 0.40. With emissivities of $\epsilon_b = \epsilon_f = 0.95$ and adopting a maximum temperature of 3000 K, this would allow the sail to operate at intensities of up to 20 MW m^{-2}. Even with zero reflectance, a sail from this material would have a lightness number (equation 10.20) of 410, but both the doping and the payload would lower this value considerably.

Another form of carbon that is promising for laser sails is aerographite. This is a mesh of carbon tubes with an incredibly low density, just 0.2 kg m^{-3}, which is 14 000 times lighter than aluminium and six times lighter than air. If it can be made into a sail, it would provide an extremely high (pre-payload) lightness number. Its absorptance is near unity and it has a melting temperature around 3900 K. With unit emissivity, the maximum intensity it can tolerate is about 25 MW m^{-2}.

The maximum tolerable intensities for all these materials is much less than the GW m^{-2} scale intensities on a sail we computed at the beginning of this section, so major technological developments are still required. However, with very light materials, the mass and thus the lightness number of the spacecraft will in practice be determined by the mass of the payload and the size of the sail; the mass of the sail itself will hardly play a role. We can therefore make the sail much larger. This will permit a larger diffraction angle and thus a smaller telescope array, for a given power and illumination duration, which will lower the intensity on the sail. Alternatively, a longer illumination duration could be used for a given array size, thereby allowing the laser power to be decreased.

11.4.2 Shape and stability

So far we have implicitly been assuming that the sail is flat and circular, corresponding to the circular symmetry of the laser beam's diffraction pattern. A flat, specularly reflecting sail is

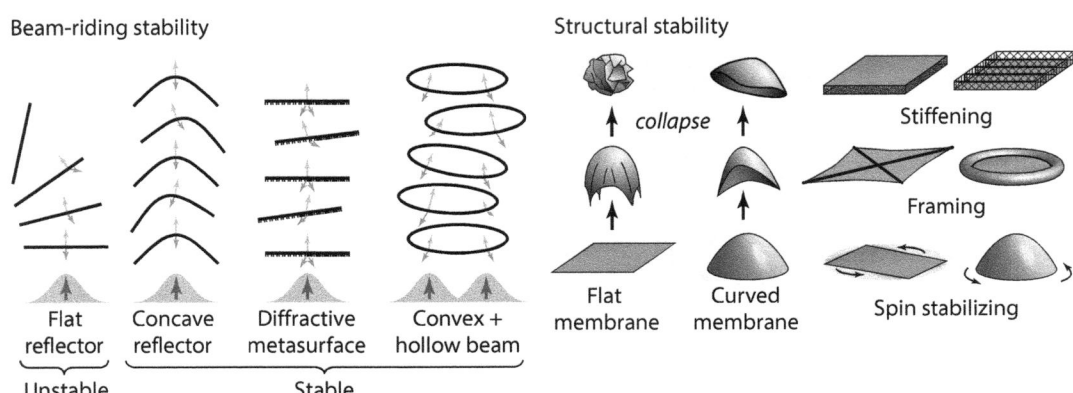

unstable, however. If the laser beam's intensity profile is not perfectly symmetric, or if the beam is not perfectly centred on the sail all the time, the resulting uneven photon pressure creates a torque on the sail and tilts it. This tilt not only changes the direction of travel of the sail, but would likely increase the tilt further and flip the sail out of control (figure 11.15). Some concave reflectors, in contrast, are stable to small variations in the beam intensity or centring. This includes conical, parabolic, and hyperbolic shaped structures. Diffusely reflecting surfaces can also be designed to give stability, even for flat surfaces. Convex structures, like a gas-inflated sail, can be stable if a 'hollow' light beam is used, that is, one in which the PSF is a circular ring with a minimum in the centre, rather than a Gaussian-like bump with a maximum in the centre. When the ring is similar in size to the sail, with the sail in the central intensity minimum, it provides stability against perturbations that try to shift or tilt the sail laterally.

Stability could also be achieved through feedback, but this would require additional sensing and control mechanisms acting at high frequency. Any feedback loop between the laser and the sail that adjusts the profile of the former in response to the orientation of the latter is infeasible given the large light travel time.

The above refers to the 'beam-riding' stability of the sail, namely its ability to remain on the beam and moving parallel to it. Another type of stability is structural, so that the sail retains its shape and does not tear. A flat sail would likely crumple when exposed to a Gaussian-like beam because the intensity would be much larger in the centre than at the edges (figure 11.15). To be stable against collapse, a flat sail would have to be able to resist very large stresses. A curved sail, in contrast, experiences much lower stresses. This is similar to the way in which parachutes or boat sails billow in the wind to minimize wind-induced stress. If I is the laser intensity on the sail, then for a spherical sail, the maximum tensile stress in the sail scales as IR/b, where R is the radius of curvature of the sail and b is its thickness (figure 11.16). Thus to minimize the stress for a sail material of a given thickness, we should minimize R, by having a maximally curved sail. However, to avoid the light incident on the sail's perimeter from reflecting back onto another part of the sail, we need $R > D/\sqrt{2}$, where D is the projected diameter of the curved sail. (Recall that the focus of a spherical reflector is at a distance $R/2$ from the surface of the reflector on its axis.) If we set $R = D$, then the

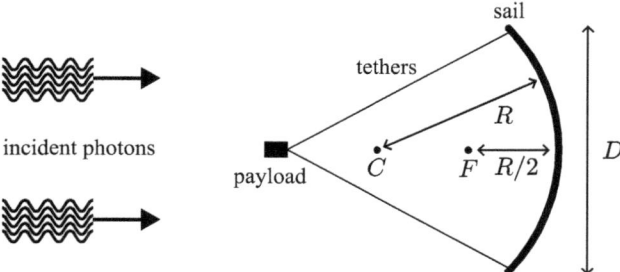

A spherical laser sail of diameter D and curvature radius R. C is the centre of curvature and F is the focus. To be stable the payload must be located at least a distance $R/2$ from the sail surface along the axis. The payload is shown here at $3R/2$.

Fig. 11.16

surface area, and thus mass, is only 7% larger than a flat sail. Such concave sails gain stability at the cost of only a small mass penalty.

The sail must also carry a payload. From the point of view of mechanical stability, for a concave sail it is best to put the payload on the front (laser) side of the sail, attached to the sail's perimeter by tethers. For a spherical sail this is stable once the payload is beyond the focal point, that is, further than $R/2$ from the surface of the sail along the axis (figure 11.16). However, the payload is then fully exposed to the high-intensity laser beam. Another possibility might be to distribute the payload across the backside of the sail, but this would affect the stability.

A flat laser sail can be stable if it is made from a material that has anisotropic reflective properties. The idea here is to use nanophotonic elements to control the phase of light that is reflected from different points on the surface. It is possible to design this in such a way that if the sail tilts away from its nominal orientation, the reflections produce a restoring torque that acts to return the sail back to its nominal orientation. Sails that reflect light using diffraction gratings can likewise be designed to be self-stabilizing.

11.5 Lasers in space

From a construction perspective, it makes sense to place a large, powerful laser array on the surface of the Earth. This has several drawbacks, however. We have already looked at the problem of distortion of the wavefront by the atmosphere, which makes it difficult to produce a coherent beam from the array of apertures (section 11.2.3). The atmosphere will also absorb some of the laser energy, and this absorption itself can exacerbate the atmospheric turbulence. Absorption increases with the air mass through which the beam has to travel, which in a plane-parallel approximation varies as $\sec\theta$, where θ is the zenith angle. So at $\theta = 60°$ the absorption is twice that at the zenith. Aircraft, birds, and orbiting satellites may also be damaged or destroyed by passing through the high-intensity beam.

Another disadvantage of an Earth-bound laser is that it can access only half of the sky at any one time, and less than a quarter at low zenith angles. Even over the course of a year, a laser placed anywhere on the Earth other than on the equator never sees all of the sky. As the laser, sail, and target star must be more or less colinear, this limits the sets of stars that can be targeted with a single laser. The Earth's rotation also places a limit on how long the laser can be used to propel a given sail. If the required illumination time is more than a few hours, then either there have to be daily breaks in the transmission, or additional lasers have to be placed at different longitudes around the Earth. Target stars nearer to the pole, such as Alpha Cen (declination $-63°$), offer the possibility of continuous viewing from a high latitude.

Placing the laser array in space gives more flexible access to the sky. This also overcomes the problem of atmospheric turbulence, obviating the need for adaptive optics (although we still need to phase lock the different lasers to achieve a coherent synthesized beam). We would want to place the array in an orbit far from the Earth's surface, because in a low Earth orbit the Earth itself still blocks half the sky, and there would still be many satellites in higher orbits that might cross the beam.

The big drawback of lasers in space is how to power them. If the laser power is as low as a few gigawatts, nuclear fission reactors may be feasible. Alternatively, the power could be beamed from the Earth's surface at either visible or microwave frequencies. This beam would of course suffer from diffraction, but the distance would only be thousands of kilometres, and so the diffraction minimal. Power could only be transmitted for the time that the array is above the horizon as seen from the power source, a duration that depends on the altitude and thus the orbital period of the array.

An alternative approach is to use the power of the Sun. With an intensity of $1.36\,\text{kW}\,\text{m}^{-2}$ at the Earth's distance, we require a collecting area of $0.74\,\text{km}^2$ for each gigawatt of input power required, neglecting conversion losses. When covered with photovoltaics, the sunlight could be converted to electricity and this used to power a laser.

A major non-technical problem with a space-based laser is that it could be repurposed as a weapon to hit either satellites or targets on the Earth's surface. This dual use could make it very hard to get political approval for laser-based propulsion. An Earth-based laser is almost as problematic, because it could be used to destroy satellites, or even target locations on the Earth's surface when used in conjunction with reflectors in space. One solution to this objection is to place the laser on the dark side of the Moon, where it never sees the Earth. The Moon's lack of atmosphere and slow rotation overcome many of the problems with an Earth-based laser, and we could imagine having very large photovoltaic collectors to power it.

A laser can be avoided entirely by using a large mirror in space to collect sunlight and focus it directly onto the sail to propel it. The big challenge of this approach is that in order to transmit a focused collimated beam towards the sail – and not just reflect light diffusely – the mirror has to have a precise surface over its entire area, which is $74\,\text{km}^2$ for $100\,\text{GW}$. This may be very hard to achieve.

The nearer we bring such a big mirror to the Sun, the smaller it can be for a given power. It would need a total area of 'only' $0.74\,\text{km}$ when operating 0.1 au from the Sun (for $100\,\text{GW}$), and could even be segmented. The mirror could be made of very thin, reflective material, and in fact could be quite similar to a solar sail. Indeed, if we make its lightness number large enough, we could even use it as a sail, and by tilting, get it to spiral closer to the Sun without having to use rockets (see section 10.3.2). Naturally, when we then use it as a reflector to propel our sailcraft, the mirror itself will be pushed away from the Sun, although as it is significantly more massive that the sailcraft, its motion would be much less. Once we have finished propelling the sailcraft, we can again use the mirror as a tilted solar sail to return it to its initial position close to the Sun, ready to propel another sailcraft.

11.6 Laser sail missions

Although sails propelled by solar photons have been successfully tested in space, laser sails have not yet been deployed. A number of concepts have been developed, however, both for interplanetary and interstellar missions.

There are many trade-offs and optimizations to make when designing a laser sail mission. For example, to accelerate a given sail to a specified velocity, we need a certain amount of energy. This could be provided either by a high-power laser over a short period of time, or by a lower-power laser over a longer period of time. A short period of time means the sail is nearer to the laser source on average during the acceleration, and so diffraction losses are kept small. But the larger power means a larger intensity on the sail, and so a higher temperature that is more demanding of the sail materials. Higher-power lasers are also more difficult and expensive to build, and to retain cool during operation. The longer period of acceleration required for a low-power laser, on the other hand, means either there are larger diffraction losses (and so more energy in total is required), or we need a larger laser array to reduce diffraction, which also comes with an expense.

With these kinds of trade-off in mind, let us look at a few specific mission designs.

11.6.1 High-mass payloads

Forward (1984) looked into using very large sails to carry payloads of many tonnes. One of these concepts is explored in detail by Frisbee (2004), who proposes a sail with a diameter of 32 km and a mass of 78 t carrying a payload of 100 t. This corresponds to an average areal mass density of the spacecraft of $2.3 \times 10^{-4}\,\mathrm{kg\,m^{-2}}$. It uses an aluminium layer just 16 nm thick and yet with a reflectance of 0.82 and absorptance of 0.135 (compare to figure 10.4). This would give the overall spacecraft a lightness number of 6.0 (or 13.5 just for the sail). Presumably additional supporting structures would be required in practice, here considered as part of the payload. This sail is driven by a variable power laser, which can deliver a peak power of 58 TW. This exceeds the rate of energy use by all humans on the Earth (for all purposes), which was about 20 TW on average in the year 2022.

The optical array used to transmit the laser power (at $1\,\mu\mathrm{m}$ in the laser rest frame) has an enormous diameter of 1000 km. One motivation for such large optics is to use the laser also to brake the sail at the target star (see section 11.6.4). The Airy disk of this diffraction pattern has a diameter of only 100 km at Proxima Cen. When all of the peak laser power is intercepted by the sail, the average intensity on the sail's surface is $72\,\mathrm{kW\,m^{-2}}$, which is only about 50 times the intensity of the Sun at the top of the Earth's atmosphere. When the sail is nearer to the laser, the diffraction spot is smaller and so the intensity would be much larger, heating the sail to unacceptable temperatures. For this reason, the laser power is lower in the earlier parts of the journey.

The laser is used to accelerate the sail to a velocity of $0.5\,c$ over a period of 5.5 years – that's a lot of laser energy – at which point the spacecraft is 1.4 ly from the Earth. A further 5.9 years cruising results in a total travel time to Proxima Cen as measured on an Earth clock of 11.4 years. This is laser sailing on a grand scale, to say the least.

11.6.2 Low-mass payloads

More recently there has been a trend towards designing extremely lightweight sails in order to bring the laser power and optic sizes closer to the range of the plausible. An example of such a study undertaken as part of the Breakthrough Starshot initiative is by Parkin (2018). This uses a 4.1 m diameter sail with an areal mass density of 2.0×10^{-4} kg m^{-2}, corresponding to a mass of 2.6 g. The payload is tiny, just 1 g. The sail is assumed to have a reflectance of 0.7 and an absorptance of just 10^{-8}, so the overall lightness number of the sailcraft is 3.9. It is propelled by a laser beam with a maximum power of 200 GW, although the power is initially lower and then increased as the spacecraft recedes in order to compensate for diffraction losses. The laser power is channelled through an array of telescopes with an effective diameter of 2.7 km. The peak intensity on the sail is 8.7 GW m^{-2}, which is why such a small absorptance is necessary. This intensity is considerably higher than that considered in the high-mass payload design above, which used used two orders of magnitude more laser power, but eight orders of magnitude more sail area. Despite this enormous intensity, the pressure on the sail is only 41 Pa (equation 10.19), yielding a total force of 540 N, yet the sail achieves an acceleration of 1.5×10^{5} m s^{-2}, or 15 000 times standard gravity g. A bullet in a rifle is accelerated by a similar amount, but only for a few milliseconds. The sailcraft must tolerate this acceleration for several minutes.

Figure 11.17 shows how various properties of the laser and sail vary with time (measured in the laser rest frame). Initially, the diffraction pattern on the sail is smaller than the sail, and so the laser is not run at full power to avoid overheating. The acceleration profile is designed so that the sail temperature never exceeds 625 K. As the sail recedes, the diffraction spot gets larger, and the output power is increased. After three minutes, the angular size of the sail is too small to fill most of the beam and the number of photons lost around the sides becomes significant. After five minutes the maximum laser power is attained. The laser remains on for a total of eight minutes in the laser rest frame, but this corresponds to the sail receiving photons and being accelerated for nine minutes (also in the laser rest frame) due to the difference between coordinate time and retarded time (see figure 11.4). During this acceleration period the sail has travelled 67 light seconds, or 0.13 au, and has acquired its final velocity of $0.2\,c$. This corresponds to a kinetic energy (relative to the laser rest frame) of 1.9 GWh. As the laser has provided a total of 63 GWh of energy, the overall energy efficiency is just 3%. This is much lower than the value given by equation 11.9 (18%), because much of the energy never makes it to the sail, due in particular to diffraction losses around the side of the sail, but also absorption by the Earth's atmosphere.

11.6.3 Intermediate-mass payloads

Various laser sail designs were put forward in response to the Project Dragonfly competition workshop, organized by the Initiative for Interstellar Studies (i4is) in 2014. This required a sail propelled by a laser with 100 GW power or less to reach the Alpha Cen system within a century.

One solution was put forward by Häfner et al. (2019). They consider a sail made of a graphene monolayer of volume density 2210 kg m^{-3} with a thickness of 0.5 nm, and thus an areal density of 1.1×10^{-6} kg m^{-2}. Adopting a reflectance of 0.05 and absorptance of 0.40, this corresponds to a lightness number of 345 for the sail alone. A sail 29.4 km in

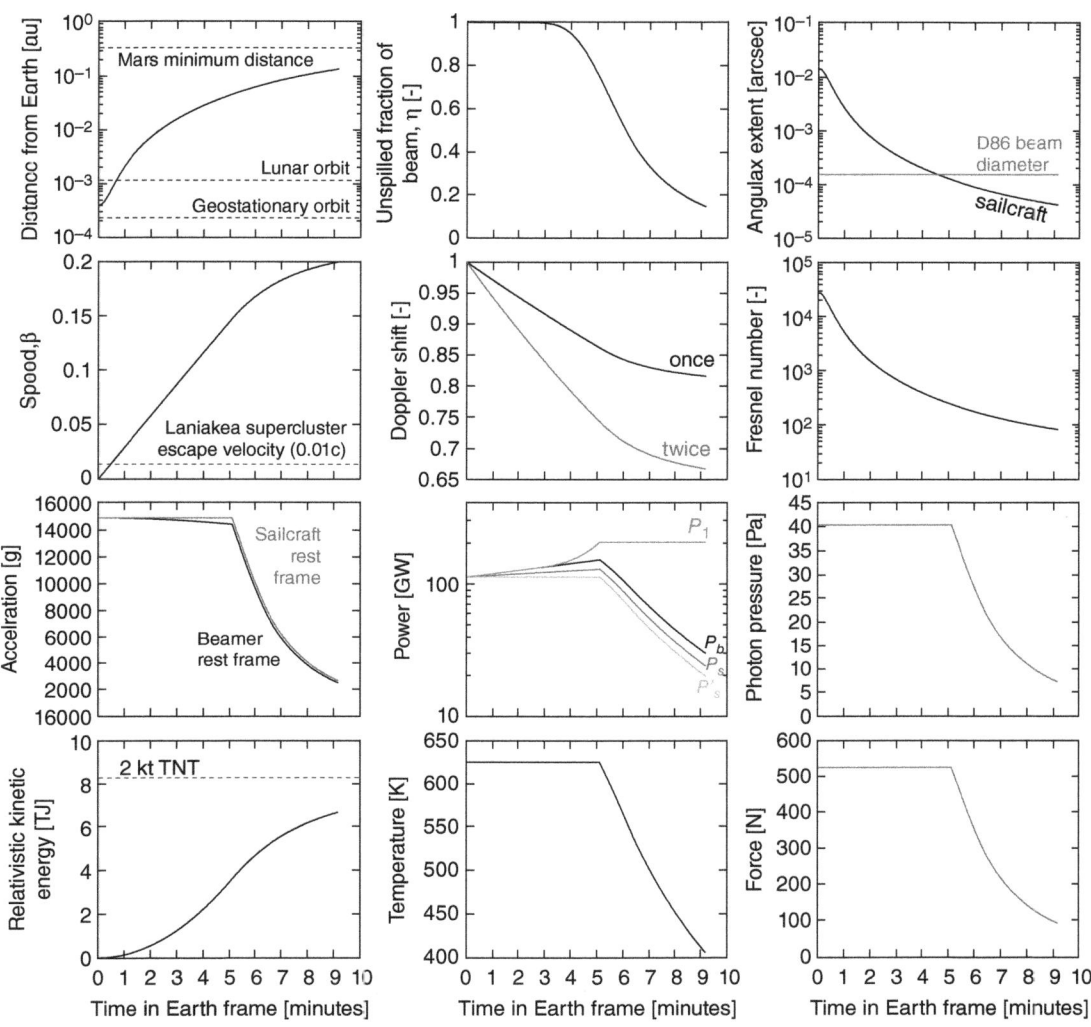

The variation of various properties of a laser sail system as a function of time during the acceleration of a sail. In the power panel, P_1 is the power emitted by the laser in its rest frame, P_b is the power received by the sail in the laser rest frame (which is lower than P_1 due to diffraction losses), and P_s' is the power received by the sail in the sail rest frame (due to the double Doppler shift, equation 11.27). The timescale for P_1 is not correctly represented here as it does not take into account the photon travel time from laser to sail. Credit: From Parkin (2018), used with permission of Elsevier (conveyed through Copyright Clearance Center, Inc.) and the author.

Fig. 11.17

diameter has a mass of 750 kg, and they adopt a total spacecraft mass of 2750 kg, of which 1000 kg is the payload and another 1000 kg is a magnetic sail used to brake the spacecraft at its destination, Alpha Cen. (We examine magnetic sails in chapter 13.) The overall lightness number is therefore 95.

A 100 GW laser with an effective aperture of 29.4 km (equal to the sail size) accelerates the sail at a nominal 0.061 m s^{-2} (equation 11.17) to 0.05 c over 7.8 years, after which the spacecraft is 12 000 au (0.19 ly) from the Earth. At this distance the sail has an angular radius of just 0.0082 nrad, or 1.7 µas. They use a 400 nm laser, so its diffraction angle is 0.017 nrad. The laser has to be pointed with a higher accuracy than this, which is very challenging.[6]

At its cruising speed of $0.05\,c$ the spacecraft will arrive at Alpha Cen A after another 85 years. However, when the spacecraft is 7500 au (0.12 ly) from its destination, it jettisons the laser sail, powers up a current through its magnetic sail, and uses this to brake, first using the ions in the interstellar medium then using the higher density ions in the heliosphere (from the stellar wind) of Alpha Cen A. Adopting a very large current density of $J_s = 1.5 \times 10^{11}\,\mathrm{A\,m^{-2}}$, they compute that the magnetic sail can decelerate the spacecraft to a velocity of just $250\,\mathrm{km\,s^{-1}}$ by the time it reaches the star, taking around 21 years.

Dragonfly is a much more massive spacecraft than the one considered by Parkin (2018) above, and although they use a similar laser power, Dragonfly must use it for nearly eight years as opposed to eight minutes, and correspondingly consumes far more energy. Although the Breakthrough Starshot approach uses extremely light spacecraft with limited payloads (and essentially no braking ability), it could launch many more spacecraft within the same time period. Dragonfly, in contrast, permits a larger payload, and if it could go into orbit about the target star (or potentially a planet around it), it could perform significantly more science, potentially even dropping a small lander. The larger mass would also permit the spacecraft to carry more onboard power and carry larger communication systems for a higher capacity data link to the Earth.

11.6.4   Using a light sail to brake at the target system

By leaving the means of propulsion behind, a sailcraft overcomes the restrictions of the rocket equation. However, lacking an internal propulsion system limits the options for braking the spacecraft at the target system, whether to prolong the flyby or to go into orbit. We saw in the previous section that a massive sailcraft could be equipped with a magnetic sail to decelerate, but this is not compatible with a low-mass sailcraft, which could not carry the heavy equipment needed to generate a magnetic field.

A sailcraft of any size could use the light from the target star to decelerate. We can compute the change in velocity using equation 10.13, where v_∞ is the velocity of the sailcraft far from the star, and v_i is its velocity once it has approached to a distance r_i from the star. A 4 m diameter sail with a mass of 1 g and unit reflectance has a lightness number at the Sun of 19. When inbound to a Sun-like star with $v_\infty = 0.1\,c$, then by the time it reaches $r_i = 1$ au, its velocity has barely decreased: by only $0.5\,\mathrm{km\,s^{-1}}$. This might seem odd, given that this sailcraft starting at rest 1 au from the Sun would be accelerated to $v_\infty = 180\,\mathrm{km\,s^{-1}}$ (equation 10.13 with $v_i = 0$). But this is just a consequence of the quadratic dependence of kinetic energy on velocity: subtracting $180\,\mathrm{km\,s^{-1}}$ in quadrature from $0.1\,c$ makes hardly any difference.

We can only start to achieve non-negligible decelerations from the star if the sail (a) has a much higher lightness number, (b) has a smaller incoming velocity, and (c) approaches

---

[6] Häfner et al. (2019) adopt a larger acceleration of $0.127\,g$, which appears to be based on their erroneous equation 1, which has the term $1 + \tilde{r}$ in place of $2\tilde{r} + \tilde{a}$ in equation 10.19. They therefore quote an acceleration distance and time that are about half the values I obtain using the constant acceleration formulae (speeds are barely relativistic, so we don't need to use equations 11.19 and 11.22). On the other hand, there will have been significant diffraction losses later in the journey, which have not been taken into account.

much closer to the star. Adopting a lightness number of 410 (e.g. zero-reflectance graphene with absorptance of 0.4), an initial velocity of $0.05\,c$, and getting it to approach to five solar radii, then the spacecraft can be decelerated by $1080\,\mathrm{km\,s^{-1}}$ to $0.046\,c$, which is still not a significant decrease in speed.

The above calculations assume the target star is the Sun. If the star has a larger luminosity-to-mass ratio, then it will provide a larger deceleration. Alpha Cen A (section 2.5.2) has a luminosity-to-mass ratio about 1.4 times that of the Sun, and therefore sails have a lightness number at Alpha Cen A that are also larger by this factor (equation 10.10). Taking advantage of this, and also using an even lighter sail and a slightly lower initial velocity, it may be possible to decrease the sail velocity at Alpha Cen A to around half its initial value. Furthermore, the sail's path can then be deflected in a gravitational – actually, a photo-gravitational – assist toward Alpha Cen B, where it is decelerated again, and then redirected to Proxima Cen. Provided the incoming velocity is below some maximum value, it should be possible to put the spacecraft into a range of desired orbits around Proxima Cen.

An ingenious if somewhat impractical idea for braking the sailcraft using the laser itself was outlined by Forward (1984) and is summarized in figure 11.18. The sail has a diameter of 100 km and a mass of 785 t. The sailcraft is accelerated continuously over most of the distance to the target star (Alpha Cen). This reduces the power required compared to a shorter acceleration distance, but it also requires a large focusing array to avoid diffraction loses. For this purpose, a 1000 km Fresnel lens is put into space near to the Sun to focus the laser beam. Operating at $1\,\mu\mathrm{m}$, this lens produces a diffraction pattern with 100 km diameter at the distance of Alpha Cen. The laser power is nonetheless enormous at 7.2 TW (a third of the world's power in 2022). After 40 years of illumination at a leisurely acceleration of

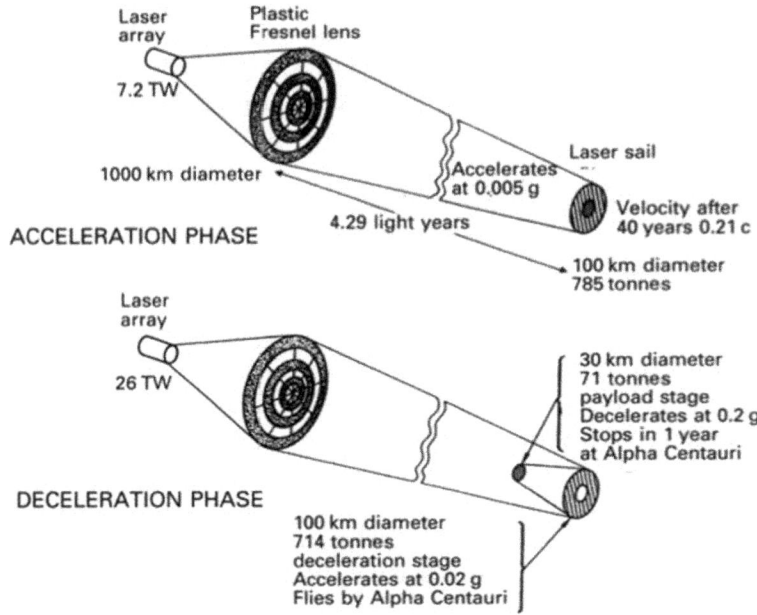

A two-part laser sail that uses the Earth-based laser to decelerate at the destination. Credit: Used with permission of Springer Nature BV, from Turner (2009); permission conveyed through Copyright Clearance Center, Inc. Based on figures in Forward (1984).

Fig. 11.18

0.005 g, the sail reaches a velocity of 0.21 c and has travelled 4.29 ly, and so is nearly at its destination. At this point the sail splits into a central disc of diameter 30 km (71 t), and an outer annulus (714 t). The disc – which carries the payload – is then turned around so that its highly reflecting side faces the target star. Its non-reflecting side faces the laser, and as it has a higher areal density than the rest of the sail, its acceleration (from absorption of the laser light) is now much lower. The annulus continues to be accelerated by the laser as before and so moves away from the disc. The annulus now reflects laser light back onto the disc – it sees the reflecting side – thus decelerating the disc. The laser power is increased to 26 TW and after one year the disc has decelerated to zero velocity at Alpha Cen. The annulus continues past Alpha Cen and is lost.

This ambitious idea obviously has many challenges, not least the enormous size of the optics, the huge laser power, the need to run the laser continuously for over 40 years, and the heating of the payload. But it could even be configured to not only stop the sailcraft at the target star, but even bring it back to Earth.

11.7 Summary

Laser-propelled sails offer an alternative to nuclear rockets for high-speed interstellar travel. They are conceptually similar to solar sails, but operate at much larger radiation intensities. Laser power is transmitted by large telescope arrays on the Earth (or Moon or orbit). Even with a very large array (kilometres), diffraction of the beam results in significant amounts of light spilling around the edge of the sail once it exceeds some distance. Most of the laser light should therefore be delivered within a limited distance, and therefore time, which demands a very large laser power. To accelerate even an ultralight sail a few grammes in mass and a few metres in diameter to 0.1–0.2 c requires hundreds of gigawatts of power, ideally delivered within several minutes. The energy efficiency is of the order of a few per cent, with most of the energy lost either to diffraction spillage or unavoidably in the reflected photons (figure 11.3).

Due to the much higher intensities, laser sail materials face much larger thermal demands than solar sails. Metallic films are too absorptive with low melting temperatures. Dielectric materials – such as molybdenum disulphide and silicon nitride – are preferable. Surprisingly, low-reflectance, high-absorptance materials may perform well if they are light enough and have high emissivities and melting temperatures. One such example is graphene. Another approach to materials engineering is to design nanophotonic structures to achieve high reflectance and low mass.

As the laser array must be of the order of kilometres in size, a single filled aperture is implausible. It must instead be synthesized as a compact array of many smaller apertures (e.g. tens of thousands each of 10 m diameter). These need to deliver a coherent wavefront, so their phases must be carefully controlled. If placed on the surface of the Earth, adaptive optics is required to correct the effects of atmospheric turbulence that would otherwise break the coherence.

Numerous laser sail mission concepts have been developed, some more plausible than others. Those originating in the 1980s tended to involve massive sails and payloads, on the scale of hundreds of tonnes and many kilometres in diameter. They required tens of terrawatts of power maintained for years, propelled by arrays hundreds of kilometres in size.

More recent designs have focused on ultralight sailcraft. Their low mass (grammes) severely limits the scientific instruments they can carry. However, each sail is accelerated only for a few minutes or hours. As the marginal cost of a sail is small compared to the expensive optical array, it makes sense to launch many sailcraft, perhaps hundreds or thousands over the course of a year. In this way, the scientific instruments and objectives can be spread over many spacecraft. We will look at this distributed science concept more in section 16.3.

11.8 Exercises

1. Show that equation 11.7 gives the velocity of the sail relative to its initial rest frame when $q = h\nu'_e/mc^2$.
2. Figure 11.3 shows that the overall efficiency of energy transfer from the laser to the sail increases with sail velocity β. To reach a large velocity where the efficiency is high, the sail had to pass through a series of lower velocities where the overall efficiencies were all lower. How can we reconcile this?
3. Show that equations 11.20 and 11.23 follow directly from equation 10.19 in the non-relativistic limit.
4. In section 11.1.5 we used the Doppler expression to compute the rate of arrival of photons at the sail in the sail rest frame $\dot{n}'_s$, in terms of the rate of emission of photons from the laser in the laser rest frame $\dot{n}_l$. Derive this instead in two steps, first using equation 11.11, which is purely classical, and then using the time dilation (to relate $\dot{n}_s$ to $\dot{n}'_s$). Plot the rate ratios vs β for these two steps, as well as the rate ratio in the final expression (i.e. $\dot{n}'_s/\dot{n}_l$)
5. The diffraction angle of the laser beam observed on the moving sail is larger than the diffraction angle at the laser source due to aberration, and can be described by equation 9.5. Show that this expression can also be obtained from equation 11.30 using the Lorentz transformations and/or Doppler effect.
6. Modify the derivation of the equation of motion in section 11.1.3 to accommodate the diffraction losses discussed in section 11.2.1. To keep it analytic, assume all the power is distributed uniformly over the Airy disk.

The interstellar medium

Before a spacecraft encounters its target star, it must pass through our solar system and cross the vast extent of interstellar space. At a few hundred astronomical units from the Sun, the spacecraft will cross the edge of the heliosphere, where the interstellar medium (ISM) begins to dominate the solar wind emitted by the Sun. The ISM is not empty, and the spacecraft will encounter gas and dust particles at high speeds. We examine the impact of such particles on the spacecraft, quantify the damage they would make, and look at mitigating actions. We will also see how the concept of the Bussard ramjet could, in principle, exploit the ISM particles as a source of fuel and propellant. When arriving at its target star, the spacecraft can study its heliosphere, and look for analogues to the minor bodies in our own solar system.

12.1 The solar system

What will an interstellar spacecraft encounter as it leaves our solar system?

To acquire its large velocity, a spacecraft may first dive very close to the Sun, either as a rocket to take advantage of the Oberth effect (section 4.2.3), or as a solar sail to benefit from the higher photon intensity (section 10.7). The spacecraft could even pass close to a planet to perform a gravity assist (section 4.3), primarily to change its direction of travel because this manoeuvre would not add significantly to the otherwise very large velocity we are aiming for.

In many cases, though, our interstellar spacecraft will avoid the Sun and the planets. The spacecraft may pass through the asteroid belt. This has such a low number density that the probability of a collision with an unmapped object is negligible.

Beyond the orbit of Neptune there is a large population of minor planets and comets organized into various classes according to their orbits and properties. One of these classes is the Kuiper belt, objects lying mostly at distances of 30–50 au, the best and longest-known member of which is Pluto. Some of these objects, including Pluto, are in resonant orbits with Neptune. Another class of minor planets is the scattered disk objects, which have eccentric and inclined non-resonant orbits. The largest of these is Eris, which is currently the most massive trans-Neptunian object known ($0.0027 \, M_\oplus$), with aphelion and perihelion distances of 98 au, and 38 au, respectively. In total there are of the order of 10^5 trans-Neptunian objects larger than 100 km, and probably of the order of 10^{11} larger than 1 km. Yet their total mass is perhaps only $0.1 \, M_\oplus$. Our interstellar spacecraft could possibly be steered to examine one of these, perhaps as a test of how to do science observations during a high-speed encounter.

Beyond the scattered disk there lies a large population of comets with aphelia extending up to 0.25 pc or more, the Oort cloud. No Oort cloud objects have been directly observed in situ, but gravitational interactions with passing stars (or with the stellar population as a whole in the form of the Galactic tide), can push them into the inner parts of the solar system, where their orbits can be further perturbed by Jupiter. This is believed to be the source of long-period comets (longer than 200 yr). Estimates of the size of the Oort cloud are highly uncertain, but there are perhaps 10^{12} objects of radius 1 km, which at a typical comet density of $500\,\mathrm{kg\,m^{-3}}$ implies a mass of $2 \times 10^{24}\,\mathrm{kg}$ or $0.35\,\mathrm{M_\oplus}$, but this could easily be out by an order of magnitude. If their masses are distributed like other solar system bodies, then most of the mass will be in the few largest objects. In any case, given the vast volume they are spread over, the probability of a transiting spacecraft colliding with any of them is negligible (see exercises). Correspondingly, the closest expected flyby distance to any one of them is a few astronomical units (if they all had the same size). Given the extreme faintness of Oort cloud objects both in reflected sunlight and in the thermal infrared, it is unlikely a single spacecraft would have the opportunity to observe one as it flies through.

12.2 The solar wind and its interaction with the interstellar medium

The outermost part of the Sun's atmosphere is the corona, visible to the naked eye during a solar eclipse as a set of jagged structures extending beyond the otherwise near-spherical surface of the Sun. The corona has a low density but is heated to $1\text{–}2\times10^{6}\,\mathrm{K}$ by magnetic fields and is therefore completely ionized. Some of this plasma has enough energy to overcome the Sun's gravity and streams away as the solar wind.

The solar wind comprises mostly protons (hydrogen ions) and electrons with some helium nuclei (2–4% by number) and heavier ions. Overall the solar wind is neutral. As a charged fluid that supports pressure, its individual particles move not only under the force of gravity. A simple hydrodynamical model of the solar wind shows that it initially accelerates away from the Sun. Considering just the protons (as these dominate the mass), this model shows that the solar wind becomes supersonic after a few solar radii. The wind has a range of temperatures (and thus energies), but for wind at $10^{6}\,\mathrm{K}$, it continues to accelerate out to about $40\,\mathrm{R_\odot}$ (0.19 au), reaching a velocity of $400\text{–}500\,\mathrm{km\,s^{-1}}$, after which it maintains a more or less constant velocity out to large distances.

Observations show that the particles in the solar wind can be divided into two distinct populations of different energies and densities. At 1 au from the Sun, the so-called fast solar wind has a velocity between $400\,\mathrm{km\,s^{-1}}$ and $800\,\mathrm{km\,s^{-1}}$ and an ion flux density of $2 \times 10^{12}\,\mathrm{m^{-2}\,s^{-1}}$, corresponding to a ion number density of $2.5\text{–}5\times10^{6}\,\mathrm{m^{-3}}$. The proton temperature is about $2 \times 10^{5}\,\mathrm{K}$; the electron temperature is half this. The fast solar wind comes primarily from coronal holes, which are cooler regions that emit less ultraviolet and X-ray radiation. The so-called slow solar wind originates mostly from the equatorial regions of the Sun. This component has velocities of $250\text{–}400\,\mathrm{km\,s^{-1}}$, about twice as dense as the fast solar wind, and it has a lower proton temperature, about $3 \times 10^{4}\,\mathrm{K}$. The total solar wind (fast and slow) ion density varies with time, but at 1 au it is around $10^{7}\,\mathrm{m^{-3}}$ on average. This corresponds to a mass loss rate from the Sun of a few $10^{-14}\,\mathrm{M_\odot\,yr^{-1}}$.

The dynamic pressure, or ram pressure, exerted by the solar wind is $(1/2)\rho v^{2}$, where ρ is its mass density and v its velocity. Adopting a particle number density of $1 \times 10^{7}\,\mathrm{m^{-3}}$ and

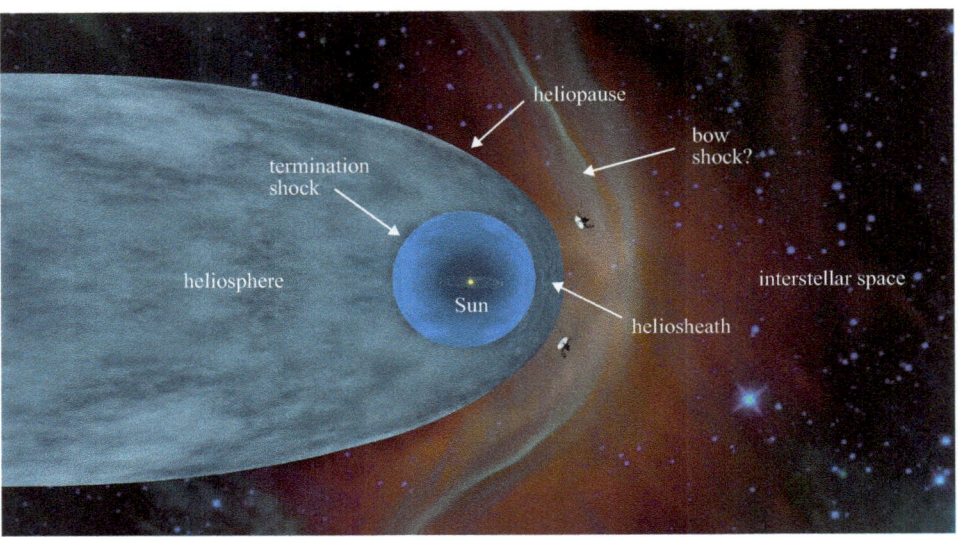

Fig. 12.1 The heliosphere and the edge(s) of the solar system. Credit: Courtesy NASA/JPL-Caltech, annotated by the author.

assuming the solar wind is all protons moving at $400\,\mathrm{km\,s^{-1}}$, the ram pressure is $1 \times 10^{-9}\,\mathrm{Pa}$. This is several thousand times less than the solar photon pressure at 1 au (section 10.1).

The *heliosphere* is that region of space that is occupied by the solar wind and can be considered as a very low-density extension of the solar atmosphere. As the solar wind pressure decreases with distance from the Sun, there comes a point at which this pressure drops to the pressure of the gas in the ISM. The surface at which this occurs is the *heliopause*, the boundary of the heliosphere with interstellar space. If the Sun where stationary with respect to the ISM, the heliopause would be roughly spherical (although not exactly because the solar wind is neither isotropic nor constant in time). But due to the motion of the Sun with respect to the ISM, the heliosphere is expected to be compressed in the 'upwind' direction (the solar apex) and have an extended tail in the 'downwind' direction (solar antapex), as shown in figure 12.1, although some observational evidence suggests it is in fact more symmetric.

Voyager 1 became the first spacecraft to cross the heliopause in August 2012 when it detected a significant decrease in the ambient particle density. It was 121 au from the Sun at the time, and was travelling in the direction $\mathrm{RA} = 263°$, $\mathrm{Dec.} = 12°$, which is close to the solar apex ($\mathrm{RA} = 275°$, $\mathrm{Dec.} = 30°$). Voyager 2 crossed the heliopause in November 2018 at the same distance, travelling in nearly a perpendicular direction ($\mathrm{RA} = 316°$, $\mathrm{Dec.} = -68°$). So 120 au is a reasonable estimate for the distance to the heliopause at the solar apex. It possibly extends to tens of thousands of astronomical units in the antapex direction.

The heliopause is often taken as the definition of the edge of the solar system, although the Voyager spacecraft passed another edge some years earlier. This was the *termination shock*, defined as the boundary where the solar wind speed drops from supersonic to subsonic due to interactions with the ISM. Voyager 1 crossed the termination shock 94 au from the Sun,

Voyager 2 at 84 au. The location of the termination shock is probably variable in time too, as Voyager 2 seemed to cross it several times over the course of several hours.

Between the termination shock and the heliopause is the *heliosheath* (figure 12.1). This is the region of the heliosphere that is influenced by the ISM. The plasma here has a higher temperature and density than the solar wind internal to the termination shock. Observations from Voyager indicate that this region is not homogenous.

Finally, beyond the heliopause, there may exist a bow shock. This is where the velocity of the ISM relative to the solar wind drops from supersonic to subsonic (i.e. drops below the ISM speed of sound) due to the interaction with the heliosphere, creating turbulence. Bow shocks have been observed around other stars. It has been hypothesized that the solar bow shock is about 230 au from the Sun in the apex direction. The International Boundary Explorer (IBEX), which observes the heliosphere from an orbit near to the Earth using energetic neutral atoms, measured the velocity of the local ISM relative to the Sun to be $23.2 \, \text{km s}^{-1}$. This implies the pressure from the ISM is too low to produce a bow shock, and that there is instead an unshocked region of higher density. A suitably equipped interstellar spacecraft could shed some light on this issue (see chapter 16).

12.3 Composition of the interstellar medium

The space between the stars is not empty. In the ISM there are photons, neutral atoms, ions, electrons, molecules, dust and larger particles, as well as electric and magnetic fields generated by the charged particles. Evidence for matter in the ISM comes from both broad band absorption and reddening of starlight due to dust, and also from atomic and ionic spectroscopic absorption lines at characteristic wavelengths. Particle densities vary across space, and are not precisely determined, so some of the numbers given below may not be entirely self-consistent.

12.3.1 Atoms, ions, and electrons

The early universe consisted mostly of hydrogen and helium (from big bang nucleosynthesis), but after the first stars formed, nuclear fusion created heavier elements from these building blocks.[1] These elements are returned to the ISM through various processes, including stellar winds and more energetic events such as core-collapse (type II) supernovae or the merger of a white dwarf with another star in a binary system (type Ia supernovae).

The density and composition of the ISM varies by orders of magnitudes depending on location in the Galaxy, in particular the proximity to star forming regions. Interstellar material collapsing under gravity to form new stars can reach such high densities that they form clouds that are optically thick, with the protostars themselves barely visible (in the optical, at least). The nearest star-forming regions lie a few hundred light years from the Sun, so are unlikely to be the first targets of interstellar travel.

---

[1] This includes both fusion in the cores of stars to produce elements up to the iron/nickel peak (see figure 8.1), as well as proton and neutron capture processes in giant stars and core-collapse supernovae that create heavier elements.

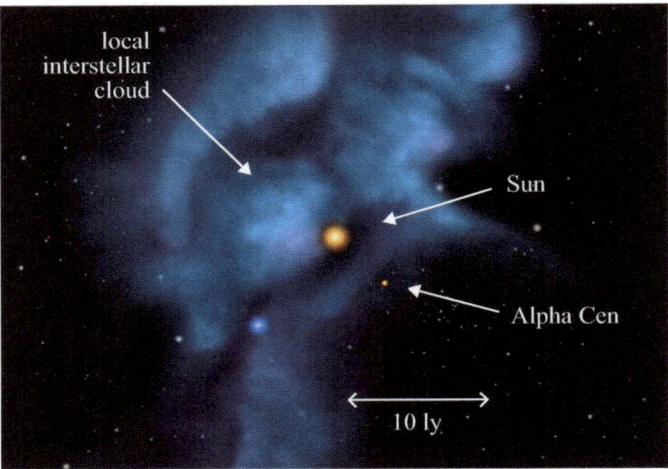

Fig. 12.2 The local ISM. Credit: NASA/Goddard/Adler/U. Chicago/Wesleyan, cropped and re-annotated by the author.

In regions far from stars and star-forming regions, the typical number density of the ISM is of the order of 10^6 gas atoms per cubic metre, or one per cubic centimetre. Most of these atoms are hydrogen. By number, hydrogen makes up about 90% of the atoms, most of the rest being helium, with about 0.1% heavier elements, mostly carbon, nitrogen, and oxygen. By mass, these correspond to 70% hydrogen, 28% helium, and 2% for the rest (which astronomers refer to collectively and confusingly as 'metals'). With this composition, a number density of 10^6 m^{-3} corresponds to a mass density of about 2×10^{-21} kg m^{-3}. The density can be four orders of magnitude larger in denser clouds.

Interestingly, the Sun is in or at the edge of a lower than average density region known as the *local interstellar cloud*, a region roughly 10 pc (30 ly) across with a gas atom number density of about 0.3×10^6 m^{-3} (figure 12.2). This cloud is embedded within the even lower density *local bubble*, an irregularly shaped region extending 50–200 pc from the Sun with a gas atom number density of just 0.05×10^6 m^{-3}. The local bubble was probably carved out by one or many supernova over the past several million years.

Not all of the atoms in the ISM are neutral, as some atoms are ionized by short wavelength radiation from stars, in particular hot OB stars. The ionization fraction in the ISM therefore depends on the proximity to such stars. At the edge of the solar system, about 25% of the hydrogen atoms are ionized. About 40% of helium atoms are ionized, with a very small fraction (less than 1%) doubly ionized. Given that helium has a higher first ionization energy than hydrogen – 24.6 eV vs 13.6 eV – it might seem surprising that helium shows a higher ionization fraction. The reason appears to be that whereas the hydrogen is ionized mostly by continuum starlight emission (at wavelengths below 91 nm), the helium is ionized primarily by very short wavelength emission lines (below 50 nm) coming from the hot gas phase of the ISM (see below).

As there are ions in the ISM, there must also be electrons. Overall the ISM is neutral, so the average electron number density must be similar to the ion density (accounting for multiple ionizations). Absorption line data combined with photoionization models of the very nearby ISM indicate that the electron number density is about 0.06×10^6 m^{-3}. Like the ion density, this too varies significantly. The low density of the ions and electrons means collisions between them are rare, so they form a free-moving plasma. The electrons can be excited by electromagnetic waves of frequencies at or below the electron plasma frequency $v_p = \sqrt{n_e e^2 / 4\pi^2 \epsilon_0 m_e}$, where m_e is the electron mass and n_e is the electron number density. This can be written more conveniently as $v_p = 9\sqrt{n_e}$ Hz. With $n_e = 10^6 \, \text{m}^{-3}$ we get a plasma frequency of $v_p = 9 \, \text{kHz}$. Waves with a frequency below this will be strongly dispersed by the ISM, making it effectively opaque. As an aside, the Earth's ionsphere has an electron density of around $n_e = 10^{12} \, \text{m}^{-3}$, so is opaque to frequencies below 9 MHz. Thus radio waves at lower frequencies coming from astronomical objects (or interstellar spacecraft) are not observable from the ground. As the ions in the plasma are much more massive than the electrons, their plasma frequency is much lower.

12.3.2 Molecules, dust, and rocks

The hydrogen in the ISM will bind as molecular hydrogen (H_2) if the temperature is low enough and the density high enough. These conditions occur in molecular clouds, which is where new stars form out of the ISM. Under the right conditions, in particular in the absence of strong ultraviolet radiation, molecules of other elements can also form, including the hydroxyl radical (OH), carbon monoxide, water, ammonia (NH_3), and formaldehyde (H_2CO). Over 100 different molecules have been discovered in the ISM, including quite complex organic molecules comprising many atoms.

The ISM is filled also with dust particles. This is both dust that was formed in the ISM through the conglomeration of atoms, and dust that was formed around stars and then blown out by stellar radiation pressure. Interstellar dust particles have a broad range of masses with a mean around 10^{-16} kg. As the Sun is moving through the ISM, interstellar dust can be detected within the solar system, and has been measured by various interplanetary spacecraft, in particular Cassini. This found that dust is made of a range of elements, including carbon, magnesium, calcium, silicon, oxygen, and iron often in the form of silicates. Adopting a characteristic density of $2500 \, \text{kg m}^{-3}$, a spherical dust particle of mass 10^{-16} kg has a diameter of $0.4 \, \mu\text{m}$.

Dust particle masses extend down to at least 10^{-18} kg, probably further, and extend up to 10^{-12} kg, corresponding to a diameter of $10 \, \mu\text{m}$, although there are proportionally fewer large particles. The mass distribution has a small tail towards even larger masses. There is no clear definition of where dust stops and larger solid particles start, but 1 mm is sometimes taken as a cut-off. The number density of dust particles in the ISM ranges from $10^{-8} \, \text{m}^{-3}$ to $10^{-5} \, \text{m}^{-3}$. Taking an average number density of $10^{-7} \, \text{m}^{-3}$ and an average mass per particle of 10^{-16} kg, the mass density of dust in the ISM is of order $10^{-23} \, \text{kg m}^{-3}$, a hundred times less than the atomic density of $10^{-21} \, \text{kg m}^{-3}$ given earlier. This is at best an order of magnitude estimate, and both atom and dust number densities vary considerably across space. But broadly speaking, dust makes up about 1% of the microscopic mass in the ISM (a fraction widely adopted in astronomy). The number and mass densities of atoms and dust are summarized in Table 12.1.

Moving up in size, interstellar rocks of centimetre or metre sizes presumably exist, but have never been detected. Once we get to objects ranging from tens of metres to kilometres in size, we are in the realm of interstellar meteoroids, asteroids, or comets, which can be individually imaged if they come close enough to the Sun. As of mid 2025, only three

Table 12.1 Number and mass densities of gas atoms (all ionization states) and dust particles in different parts of the ISM. Gas number densities are measured. Gas mass densities have been derived from these values, adopting an average atom mass of 2.1×10^{-27} kg (1.26 protons). 'Typical ISM' refers to regions far from stars and star-forming regions. The measured number densities vary considerably, by three orders of magnitude for dust, so the gas density listed here is an approximate average. The gas density in the local bubble and local interstellar cloud are better determined. The dust number and mass densities have been derived from the gas densities assuming a gas-to-dust mass ratio of 100 and an average dust particle mass of 10^{-16} kg

Region	Particle	Number density $[\mathrm{m}^{-3}]$	Mass density $[\times 10^{-24} \, \mathrm{kg\,m}^{-3}]$
Typical ISM	Gas	1×10^6	2000
	Dust	20×10^{-8}	20
Local bubble	Gas	0.05×10^6	100
	Dust	1×10^{-8}	1
Local interstellar cloud	Gas	0.3×10^6	600
	Dust	6×10^{-8}	6

such interstellar objects (ISOs) have been found (in 2017, 2019, and 2025): the very flat-tened object 1I/'Omuamua[2], with a disk diameter of around 100 m, and the interstellar comet 2I/Borisov, with a nucleus size of around 500 m. The larger comet 3I/ATLAS was only just discovered at the time of writing. All three were observed when relatively near to the Sun. They could be identified as interstellar in origin from their hyperbolic orbits, and from hav-ing ruled out the possibility that a close encounter with a planet modified the orbit of what was previously a bound solar system object. Based on just three detections obtained from different surveys, we can only estimate the space density of such objects to within an order of magnitude. But taking into account the selection effects of the survey that found 'Omua-mua, one study concluded that there are of the order of $0.2 \, \mathrm{au}^{-3}$ objects of 'Omuamua size or larger near the Sun at any time.

The neutral atoms, ions, and dust particles in the ISM are moving with respect to one another, and so have a non-zero temperature (which, from the kinetic theory of gases, is proportional to the root mean square velocity dispersion). They absorb energy from starlight and re-emit it at various wavelengths (dust thermally; atoms through line emission). The ISM is often described as comprising several phases at different temperatures. These are relevant to astrophysical processes in the ISM. The warm neutral ISM, for example, has a temperature of around 8000 K, which corresponds to a velocity dispersion of $14 \, \mathrm{km\,s}^{-1}$ for hydrogen, from $(1/2)m\overline{v^2} = (3/2)kT$. This velocity dispersion is much smaller than the velocity of a fast interstellar spacecraft relative to the ISM. It is therefore this relative motion, rather than the thermal motion of the gas, that will generate a high energy flux of particles on the spacecraft. We consider the effects of this in section 12.4 below.

---

[2] Despite some excitement and inevitable speculation at the time, there was no good evidence to suggest this is an alien spacecraft as opposed to a natural object.

12.3.3 Cosmic rays and magnetic fields

The final material component of the ISM is Galactic cosmic rays. These are mostly ions with very high energies, ranging from 1 GeV to 10^{11} GeV or more (kinetic plus rest energy). They arise from a range of high-energy astronomical phenomena, in particular supernovae and black hole accretion disks. The distribution of cosmic rays is roughly isotropic in the Galaxy's rest frame, probably due to the action of interstellar magnetic fields. Cosmic rays are mostly protons and helium nuclei, but around 1% are heavier nuclei (ions). Their flux is largest at 1 GeV, where it is of the order of $2 \times 10^4 \, \text{s}^{-1} \, \text{m}^{-2} \, \text{str}^{-1} \, \text{GeV}^{-1}$. The number per unit energy decreases with increasing energy E as approximately $E^{-2.7}$. A 1 GeV proton has a velocity of $0.26 \, c$, so unlike the rest of the ISM constituents, cosmic rays will hit even a fast spacecraft from all directions.

As interstellar space is full of ions and electrons, we would also expect it to be permeated with magnetic fields. The strength of magnetic fields can be measured in the radio by Faraday rotation. There is a large variation, but a reasonable average field strength is 5×10^{-10} T. In the local interstellar cloud, calculations suggest the field strength is $2–5 \times 10^{-10}$ T. These compare to the Earth's magnetic field at the surface of around 5×10^{-5} T.

12.4 Impacts on a spacecraft

12.4.1 Collision rates

A spacecraft moving at velocity v relative to the ISM experiences a flux of neutral atoms, ions, and dust particles with a range of kinetic energies. With cross-sectional area A, the spacecraft passes through a volume Avt in time t. When the volume number density of particles is n, the spacecraft will encounter $Anvt$ particles in this time.[3] Thus the flux of particles – the number per unit cross-sectional area per second – is nv. With $n = 0.3 \times 10^6 \, \text{m}^{-3}$ – typical of gas atoms in the local interstellar cloud from table 12.1 – and $v = 0.1 \, c$, the flux is $9 \times 10^{12} \, \text{m}^{-2} \, \text{s}^{-1}$, corresponding to a mass flux of the order of $2 \times 10^{-14} \, \text{kg} \, \text{m}^{-2} \, \text{s}^{-1}$. Over a distance x, the total number of particle impacts per unit area is nx. For the 4.25 ly journey to Proxima Cen, this gives a total number of impacted gas atoms of $1 \times 10^{22} \, \text{m}^{-2}$, which corresponds to a gas mass of $2 \times 10^{-5} \, \text{kg} \, \text{m}^{-2}$. The number of dust particles impacting the spacecraft over this journey is $2 \times 10^9 \, \text{m}^{-2}$ (adopting $n = 6 \times 10^{-8} \, \text{m}^{-3}$) corresponding to a mass of $2 \times 10^{-7} \, \text{kg} \, \text{m}^{-2}$, 100 times less than gas.

The above estimate applies when the number of particles encountered is much larger than one. This is not the case for very large objects, such as asteroids in the asteroid belt, Oort cloud comets, or ISOs. The average number of objects encountered over a path of length x is Anx. The probability of there being k encounters is described by the Poisson distribution:

$$P(k) = \frac{e^{-Anx}(Anx)^k}{k!} \, . \tag{12.1}$$

---

[3] This assumes that the velocity dispersion of the particles is much less than v, so we can consider the spacecraft moving at velocity v relative to every particle.

The average distance travelled between encounters, the mean free path, is $1/An$. The probability of $k = 0$ encounters is $\exp(-Anx)$, so the probability of one or more encounters is $1 - \exp(-Anx)$. For $Anx \ll 1$ we can use a Taylor expansion to see that this probability is[4]

$$P(k > 0) \simeq Anx \quad (Anx \ll 1) . \tag{12.2}$$

Consider ISOs with a space density of $0.2\,\mathrm{au}^{-3}$. Assume these are larger than our spacecraft and so dominate the collisional cross-sectional area, which I take as $\pi \times 500^2\,\mathrm{m}^2$; this is a rough guess so our result is only good to within a couple of orders of magnitude. Over a journey to Proxima Cen, the probability of one or more collisions with an ISO is then of order 10^{-12}, which is not worth worrying about. The probability of hitting an Oort cloud comet when leaving the solar system is likely to be even lower (see exercises). The probability of hitting an asteroid in the solar system's asteroid belt if flying through it is higher, perhaps of the order of 10^{-9}, but is still tiny.

Although we can neglect the possibility of a collision with such large objects, we do need to consider the damage that microscopic dust particles and atoms could do to the spacecraft. As the spacecraft will be moving much faster than the velocity dispersion within the ISM, essentially all of the high-energy impacts will take place on the surface in the direction of travel. The exception to this is cosmic rays.

12.4.2 Damage by atoms and ions

Incident neutral atoms or ions from the ISM will interact with the atoms in the solid material of the spacecraft. This occurs at the atomic level in the form of electromagnetic interactions between the incoming nucleus (plus its electrons if it is not fully ionized) and the the nuclei and electrons of the material. At low energies, the incoming atom interacts with both the nucleus and its bound electrons, and – if the material is a conductor – the free electrons. At kinetic energies above about $1\,\mathrm{MeV}$, the cross-section for interactions of the incoming atom with a nucleus is much smaller than its cross-section for interactions with the material's electrons, so we need only consider the latter. As a hydrogen ion moving at $0.1\,c$ relative to the spacecraft has a kinetic energy of $4.7\,\mathrm{MeV}$, this is the regime that dominates for high-speed interstellar travel. Furthermore, at energies above $1\,\mathrm{MeV}$, an incoming ISM atom will be ionized completely by its initial interactions with the material (see figure 12.3). The electrons so stripped off will also impact the material, but as the electrons have masses and thus energies three to five orders of magnitude less than ions of the same initial speed, their impact effects can mostly be ignored, with the possible exception of bremsstrahlung, covered later. Thus at the typical velocities of the order of $0.1\,c$ for interstellar spacecraft, we consider mostly the interactions between incoming ions and the electrons in the spacecraft material.

The incident ion transfers energy to the atoms in the impacted material. This excites or ionizes the atoms. In the latter case, the secondary electrons transfer energy to neighbouring atoms. When these electrons recombine, they also release energy in the form of photons. In this way the energy of the impacting ion is dissipated through the material, primarily along a long thin track in the direction of travel. Most of this energy heats the material, and this may be enough to cause local melting. When the melted material later cools and solidifies, it likely does so into a different configuration, resulting in permanent damage to the structure.

---

[4] The probability of exactly one encounter is more or less identical when $Anx \ll 1$.

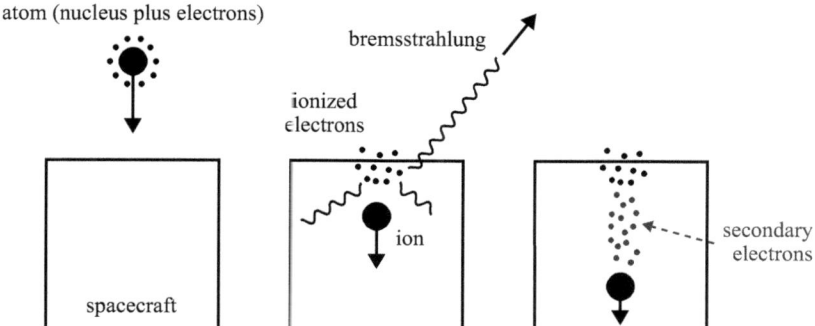

Fig. 12.3

A schematic of a collision of a high-velocity atom with a solid. The atom becomes fully ionized before it penetrates deeply into the solid; some of its electrons may be ejected from the surface. These decelerating electrons radiate their kinetic energy as bremsstrahlung. The ion penetrates more deeply into the solid, interacting primarily with the atomic electrons in the material (and free electrons if it is a conductor) and loses kinetic energy. This energy excites the electrons, heating the material, which may locally melt and lead to permanent track damage. Some atomic electrons may even become unbound (secondary electrons).

The energy may even be enough to vaporize the surface, leading to the loss of material. A penetrating particle can also displace an atom or ion, setting off a cascade of collisions. If these reach the surface, atoms can be ejected from the surface, a phenomenon known as sputtering.

When an ion penetrates a solid material, the rate at which the ion loses kinetic energy can be described by a relativistic stopping formula. The stopping power is defined as the energy E lost per unit depth x of penetration. For an ion with charge number q_i travelling at velocity $v\ (=\ \beta c)$ through a homogenous material with electron number density n_e, the stopping power is

$$S = -\frac{dE}{dx} = \frac{e^4}{4\pi \epsilon_0^2 m_e c^2} \frac{n_e q_i^2}{\beta^2} \left(\ln\left[\frac{2m_e c^2 \beta^2}{I(1-\beta^2)} \right] - \beta^2 \right) \tag{12.3a}$$

$$\simeq \frac{e^4}{4\pi \epsilon_0^2 m_e} \frac{n_e q_i^2}{v^2} \ln\left[\frac{2m_e v^2}{I} \right] \quad \text{when} \quad \beta \ll 1, \tag{12.3b}$$

where I is the ionization energy averaged over the energy levels of the target atom, m_e is the electron mass and ϵ_0 is the vacuum permittivity. Note that q_i refers to the ionization state (number of electrons removed) of the incident ion. If the target material has a mass density ρ, atomic number Z, and relative atomic mass A, then $n_e = \rho Z/m_u A$, where m_u is the atomic mass unit. As we might expect, the stopping power increases with the electron number density of the target material and with the charge of the incoming ion. It is also higher for a smaller ionization potential of the target material, as those secondary electrons can then carry energy away. The stopping power does not, however, depend directly on the mass of the incoming ion.

The dependence of the stopping power on velocity for a hydrogen ion incident on three different materials is shown in figure 12.4. For the mean ionization energy I of the target material, I have adopted a common approximation in which I depends only on the atomic

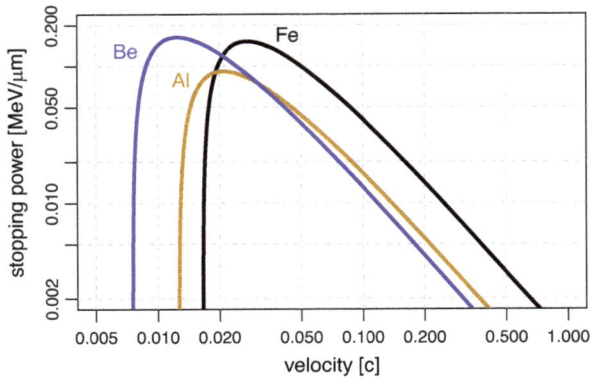

The stopping power as given by equation 12.3a as a function of velocity, for a hydrogen ion ($q_i = 1$) incident on three different materials: iron (Fe, $Z = 26$), aluminium (Al, $Z = 13$) and beryllium (Be, $Z = 4$). For comparison, the energy of a hydrogen ion moving at 0.1 c is 4.7 MeV.

number Z of the target material. Here I use $I = 58\,\text{eV}$ for beryllium, $I = 163\,\text{eV}$ for aluminium, and $I = 541\,\text{eV}$ for iron. In all cases we see that the stopping power is maximum at intermediate velocities, but smaller for low and high velocities.

Equation 12.3a and thus figure 12.4 is an idealized approximation. In real materials we do not see the sharp drop in stopping power at low energies. Furthermore, the target material properties depend on its temperature: at high temperatures, the stopping power is lower and so the penetration depth larger.

Hoang et al. (2017) used a more sophisticated numerical model to investigate the impact of high-energy ions on materials. They find that light ions like hydrogen and helium ions travelling at 0.1 c transfer about 13 eV to each atom in a target material like quartz or graphite. This is enough to eject a few of the outermost electrons. While these secondary electrons will heat the material, they are insufficient to cause permanent damage. Heavier incident ions like iron ions, in contrast, transfer about 1 keV per atom, so can cause permanent damage along the length of their tracks. The threshold stopping power above which such damage occurs depends on the material, but for silicon it is about $1.5\,\text{MeV}\mu\text{m}^{-1}$ and for graphite about $5.1\,\text{MeV}\mu\text{m}^{-1}$. For very good conductors like copper or diamond, the threshold is very high, possibly infinite. The width of the damage track depends on a number of factors, including the conductivity of the material and the velocity. Good conductors transport heat away more quickly, resulting in a narrower track. Larger velocities generally produce narrower tracks.

By integrating equation 12.3a – or a more accurate expression for the stopping power – we can compute the distance over which an ion is stopped as

$$R = \int_{E_i}^{0} \left(\frac{dE}{dx}\right)^{-1} dE \qquad (12.4)$$

where E_i is the kinetic energy of the incident ion. The results of some such more accurate calculations are shown in figure 12.5. At $v = 0.1\,c$, the stopping distance ranges from 0.2 mm for hydrogen ions down to 0.04 mm for heavy ions like iron. The reason heavy ions are stopped in a shorter distance, despite having more energy, is because the stopping power depends on the charge of the incident ion: $S \propto q_i^2$ in equation 12.3a, and we are considering

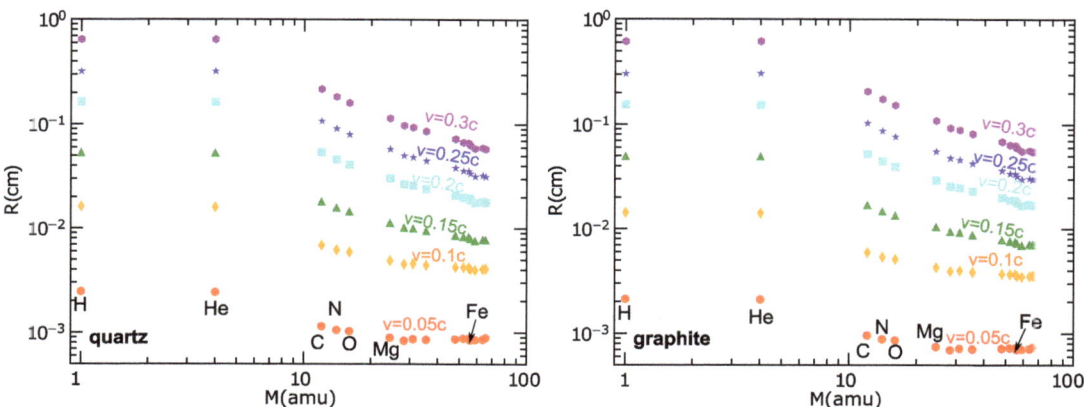

Stopping distances of high-energy ions of different masses (horizontal axis) and velocities (colours) incident on quartz (left) and graphite (right). Credit: From Hoang et al. (2017) © AAS. Reproduced with permission.

Fig. 12.5

complete ionization. At higher velocities, the stopping distance is larger, but is still below 2 mm at $v = 0.2\,c$.

Although impacting electrons do not cause as much direct damage to the spacecraft material as do impacting ions, the electrons are decelerated by interactions with the material, and as a result radiate their kinetic energy as bremsstrahlung. At $v = 0.1\,c$, an electron has a kinetic energy of 2.6 keV, so this is radiated at a range of X-ray wavelengths. These X-rays can damage the spacecraft by ejecting electrons from the atoms. The total X-ray flux from this process is expected to be relatively low, however, and as the electrons will penetrate only to depths of a few nanometres, the X-rays could be shielded without too much difficulty. Decelerating ions also produce bremsstrahlung, but the rate of energy loss by bremsstrahlung scales inversely with the mass, so the X-ray power is much lower. However, ions penetrate more deeply into the material (up to a few millimetres, as we just saw), so its X-rays may be harder to shield.

Although the ISM is overall neutral, a spacecraft can nonetheless acquire a net charge over time. The interactions just described tend to eject more electrons than ions from the spacecraft material, leaving the spacecraft positively charged. Ultraviolet starlight can also lead to the spacecraft becoming positively charged via the photoelectric effect. Spacecraft charging is a problem because it can damage onboard electronics and interfere with scientific measurements. A charged spacecraft would also interact with the ISM's or target star's magnetic field, which could modify the spacecraft's trajectory.

12.4.3 Damage by dust

Figure 12.6 illustrates the impact of a dust particle on a spacecraft. Initially, the particle has enough kinetic energy to unbind the atoms from the surface, that is, to evaporate them. As the particle loses energy through these interactions, it at some point no longer has enough energy to eject any more atoms, but it still has enough energy to break molecular bonds and thus melt the material.

Figure 12.7 shows the result of an experimental impact of a small aluminium sphere with a block of aluminium at an initial relative velocity of 6.8 km s^{-1}. The sphere is much larger

dust particle

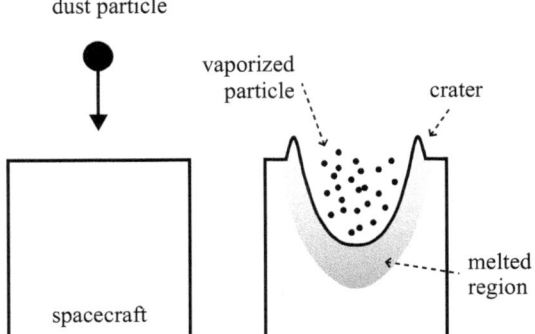

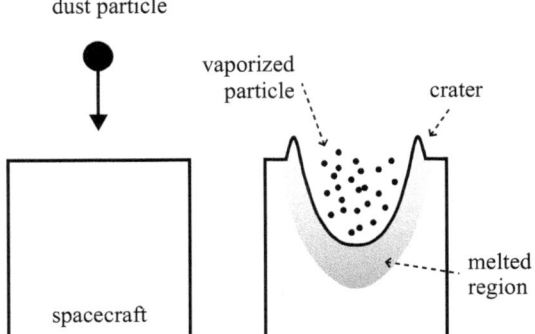

Fig. 12.6   A schematic of a collision of a high-velocity dust particle with a solid. The particle is completely vaporized by the collision and the resulting ions (and electrons, not shown) can continue to penetrate into the solid, heating it and causing subsurface melting.

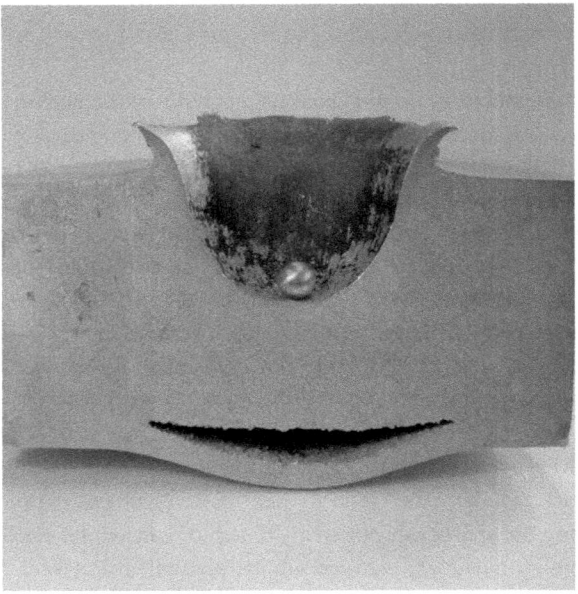

Fig. 12.7   The result of the impact of a 1.2 cm diameter sphere of aluminium (mass 2.4 g) with an aluminium block at a velocity of 6.8 km s^{-1}. The impact vaporizes a significant amount of the block, and also melts some of the surrounding material, which later solidifies. The sphere itself is also vaporized by the impact; the one shown is just for reference. Credit: European Space Agency, ESA Standard Licence.

than a dust particle, and the impact much slower than would occur on a fast interstellar spacecraft. But the photograph illustrates the general principle of what happens.

At large velocities, the kinetic energy per atom of the incident dust particle is larger than the binding energy that holds these atoms together as a particle. The impact can therefore be considered as an impact of many atoms simultaneously. As with individual atom impacts discussed in section 12.4.2, each atom is quickly ionized by the impact. The main difference to before is that now a much larger amount of kinetic energy is delivered at the same time.

We saw in section 12.4.1 that the total ISM dust mass impacting a spacecraft on a journey to Proxima Cen is about $2 \times 10^{-7} \, \mathrm{kg \, m^{-2}}$, so at a relative velocity of $0.1 \, c$ this dust would deliver $0.1 \, \mathrm{GJ \, m^{-2}}$ of kinetic energy. How much of this energy goes in ionizing, vaporizing, melting, or just heating the target material is a complex process that requires numerical simulation. The amount of material melted, for example, depends on both its thermal conductivity and its melting temperature. The higher the conductivity, the more quickly heat can be dissipated into the rest of the structure, and so the lower the temperature around the point of impact. If the material is melted, this can lead to permanent changes in the material, which would obviously destroy critical structures like electronics, but could also compromise other components. All other things being equal, conductors therefore generally make better shields than insulators, as will materials with higher melting temperatures, provided they can radiate the heat efficiently from the spacecraft. There is some trade-off here, however, because materials with low atomic numbers tend to have higher melting temperatures, whereas materials with higher atomic numbers tend to provide shorter stopping distances and have better heat dissipation.

Figure 12.8 shows the result of one calculation to determine how deeply the surface of a spacecraft is damaged by melting or evaporation by dust impacts. This is shown as a function of the total number of gas particles encountered per unit area over the journey (the fluence). We notice a number of things. First, the depth of damage caused by melting is larger than that caused by evaporation. Second, more damage is done at larger velocities, because the particles have more energy. Third, the damage tracks caused by gas ions are shallower than those of the dust, despite the higher total mass of gas. Finally, graphite, on the right, is less damaged than quartz because it is a better conductor. In either case, this figure shows that a shield of depth of order 1 cm would be enough to prevent damage to components below it during a mission to our nearest star.

Figure 12.8 does not show what fraction of the surface area is eroded or melted by the impacts to this depth. For the typical dust fluence we have been assuming for a journey to

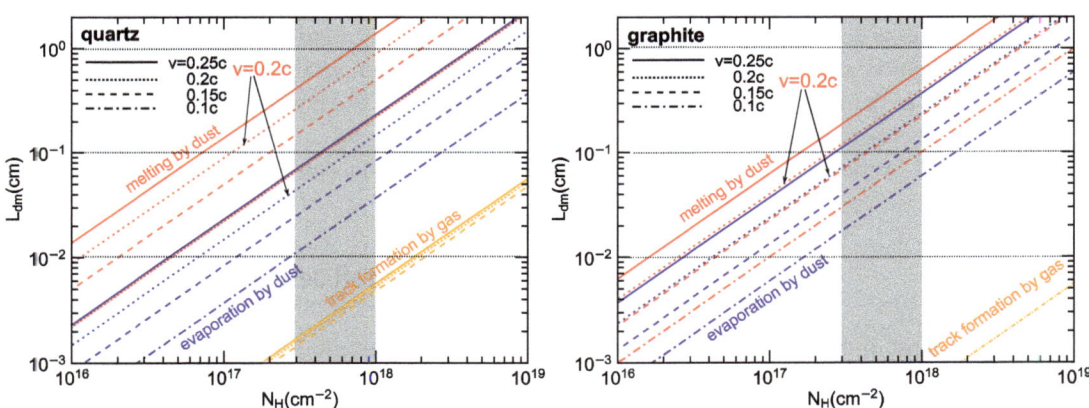

The total depth removed from the surface (L_{dm}) of a quartz (left) or graphite (right) shield as a function of the total number of gas particles encountered per unit area (N_H), assuming a gas-to-dust mass ratio of 100. The red lines indicate melting by dust, the blue lines evaporation by dust, and the orange lines damage by gas, for various velocities (indicated by the line types). The grey region shows the range of the total gas number considered by the authors of the study to be likely encountered on a journey to Alpha Cen. Credit: From Hoang et al. (2017), © AAS. Reproduced with permission.

Fig. 12.8

Proxima Cen, it turns out that 10% to 100% of the surface area would be vaporized, and 100% of it melted. So the entire area of the shield would be damaged to the depths reported. This implies that we may not want to deploy large unprotected structures, like antenna or telescope mirrors, until close to the destination.

Most of the damage done by dust would probably be done by the largest dust particles, even though these are rarer than small ones. A study by London and Early (2018), for example, showed that a single $10\,\mu$m dust particle moving at $v = 0.2\,c$ would require about 1 cm of silicon to stop it, and this would melt 1.6 g of the silicon in the process. Such an impact would be enough to destroy a very small spacecraft of the size considered for an ultralight laser sailcraft (see chapter 11). A $0.3\,\mu$m particle would only penetrate to 1.2 mm at $v = 0.2\,c$, but this is reduced to 0.25 mm at $v = 0.1\,c$. The amount of material melted is proportional to the kinetic energy of the dust particle, so a slower spacecraft ($E_k \sim v^2$) and thus longer mission time greatly reduces the damage done.

As the dust particles deliver a lot of kinetic energy to the spacecraft, they will raise the equilibrium temperature of the surface layer. The study of Hoang et al. (2017) shows that at a dust number density of $10^{-7}\,\mathrm{m}^{-3}$ and velocity of $0.1\,c$, the temperature should remain under 100 K. Even at 10 times this density and twice the velocity, the temperature remains under 300 K.

We saw in the previous section that individual ions will be stopped over a distance of 0.04 mm (for heavy ions) or more (for hydrogen ions). Thus a high-velocity ion or dust particle will pass straight through a very thin layer such as a sail (or perhaps a mirror). In such cases, not all of the energy from the particles will be deposited into the material, so the above calculations of melting and vaporization do not directly apply. But a dust particle would leave a hole in the spacecraft with a size equal to the particle's cross-sectional area. Adopting $d = 0.4\,\mu$m as the typical diameter of a dust particle (section 12.3.2), and $n = 2 \times 10^9\,\mathrm{m}^{-2}$ as the number of dust impacts per unit area over a journey to Proxima Cen (section 12.4.1), then the fraction of the surface area perforated by dust particles is of the order of[5] $n\pi d^2/4 = 3\times 10^{-4}$. Even if the average dust number density has been underestimated somewhat, a thin sail, for example, should make the journey mostly intact. A small fraction of holes would barely reduce its performance, although the torques of large dust particles might be a problem for navigation and control (see exercises). On the other hand, a sail used only for acceleration early in the mission would not need to be retained once the cruise velocity is attained, and could either be ejected or folded away.

All of the impacting particles – gas and dust – also induce drag on the spacecraft, a topic we examine in section 12.5.

12.4.4   Implications for spacecraft design

As most of the damage from the ISM comes from dust particles in the direction of travel, we should construct a spacecraft with the smallest possible cross-sectional area in that direction. In front of this we can place shielding material with a depth of several millimetres. Ideally this would be a good conductor with a high melting temperature and low density (to keep the mass down), although as noted above, there is a trade-off between high atomic mass solids with larger stopping powers and heat dissipation, and low atomic mass with higher melting

---

[5] If this number is close to unity, we need a more sophisticated calculation that allows for overlaps.

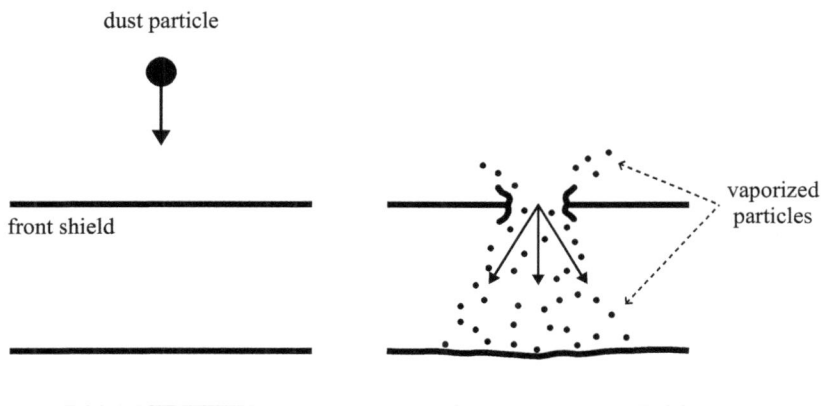

A Whipple shield. Dust particles are disintegrated by their impact on the front shield. Although the remnants penetrate this and still have significant energy, they are spread over a larger area and are more easily stopped by the next shield layer.

Fig. 12.9

temperatures. Behind this shield we can place critical components, such as electronics. Damage to passive structures like a sail or antenna may be less critical – a few holes would not make much difference – and if these can be made very thin, then dust particles could pass through them without depositing all of their energy.

Some spacecraft that have visited comets, such as ESA's Giotto and NASA's Stardust, as well the International Space Station in Earth orbit, have used Whipple shields to protect themselves against large dust fluxes. Instead of using one thick piece of material, a Whipple shield uses two or more layers of thinner shields, as shown schematically in figure 12.9. The incident particles penetrate the first shield, but in the process are disintegrated into many smaller pieces. These are then decelerated and eventually absorbed by the next layer, a process that can be repeated with multiple layers to stop particularly large or energetic particles. If the first shield is placed sufficiently far from the rest of the spacecraft, the disintegrated remnants may even spread out enough that some may miss the next layer altogether. In this way the total mass required for shielding can be reduced, although at the expense of a larger volume.

Incoming ions and charged dust particles could be deflected by magnetic or electric fields. Given the larger damage caused by dust particles, deflecting the ions only makes sense if the dust particles can also be deflected. Larger dust particles may be charged, so could in principle be deflected, but the apparatus to do so would presumably be more massive than some extra shielding. However, in chapter 13 we will consider the use of electric and magnetic sails for accelerating spacecraft, and these could also be configured to provide some shielding.

One thing we have not yet mentioned is that very high energy particles, in particular cosmic rays, can produce neutrons when they impact the spacecraft by the process of nuclear spallation. These neutrons can then go on to damage the spacecraft structure, and neutron capture by atoms in the spacecraft can result in unstable isotopes that then decay radioactively. This is the problem of neutron activation that we also encountered with nuclear rockets in section 8.5.1.

12.5 Drag on a spacecraft

The ISM particles that collide with a spacecraft have momentum. This drag will cause the spacecraft to decelerate.

Consider a spacecraft of mass M with cross-sectional area A travelling at a velocity v relative to the ISM of mass density ρ. Over distance dx (measured in the ISM rest frame) the spacecraft collides with mass $dm = \rho A dx$. Assuming these particles are stopped by the spacecraft (rather than pass through it or are elastically reflected), they impart momentum $dp = -\gamma v dm$ (equation 7.8). The force on the spacecraft in the ISM rest frame is therefore $F = dp/dt = -\rho A \gamma v^2$, as $v = dx/dt$. Assuming further that the mass of the spacecraft remains constant (the particles are brought to a halt but do not stick), then from equation 7.25 the proper acceleration of the spacecraft due to the drag is

$$a' = -\frac{\rho A}{M}\gamma v^2 . \tag{12.5}$$

Recall that the proper acceleration is what the spacecraft experiences in its instantaneous rest frame. Figure 12.10 shows how the velocity-dependent term in the drag varies with velocity. Adopting round values of $\rho = 10^{-21}\,\mathrm{kg\,m^{-3}}$ (Table 12.1), $A = 1\,\mathrm{m^2}$, $M = 1\,\mathrm{kg}$, and $v = 0.2\,\mathrm{c}$, we find that $a' = -3.7 \times 10^{-6}\,\mathrm{m\,s^{-2}}$. The coordinate acceleration of the spacecraft in the ISM rest frame is obtained via the Lorentz transformation as a'/γ^3 (equation 7.15). On account of the much larger density of gas than dust in the ISM, gas rather dominates the drag (whereas dust dominates the damage).

Integrating the coordinate acceleration we can determine how the velocity changes with distance due to the drag. For $\beta < 0.2$ we have $\gamma \simeq 1$, which makes the calculation easier. Separating the variables we get

$$\int \frac{dv}{v} = -\int \frac{\rho A}{M}dx, \tag{12.6a}$$

$$\frac{v}{v_0} = \exp\left(-\frac{\rho A}{M}x\right), \tag{12.6b}$$

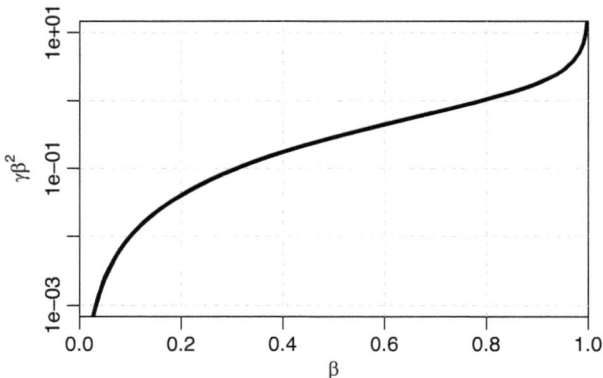

Fig. 12.10   The drag on a spacecraft, its proper acceleration, is $-\frac{\rho A}{M}c^2\gamma\beta^2$.

where v_0 is the initial velocity and x is the distance travelled. Using the above values and $x = 4.25$ ly, the term in the exponential is -4×10^{-5} and so $v/v_0 \simeq 1$. The velocity is hardly changed at all. If we instead had a sailcraft with a lightness number $\lambda = 1$, and thus a loading parameter $\sigma = M/A = 1.53 \times 10^{-3}$ kg m^{-2} (section 10.2), we have $\rho(A/M)x = -0.027$, and thus $v/v_0 = 0.973$, a very modest reduction in the velocity. In conclusion, drag from the ISM will not significantly slow down any moderately fast spacecraft over the distances to the nearest stars. However, even a small increase in travel time, and thus a later arrival time, caused by the drag has a significant impact on navigation, as we will see in chapter 14.

12.6 The Bussard ramjet

The big drawback of a rocket is that it has to carry its fuel. This limits its velocity to a few times its effective exhaust velocity in practice. Solar and laser sails overcome this by using externally provided photons as the source of momentum. Solar-electric ion engines partly overcome this by getting energy from an external source, but they still take their propellant with them. Are there other external sources we could use for both energy and momentum?

The ISM contains gas at a very low mass density, predominantly hydrogen atoms and ions. If we could collect these and force them to undergo fusion, this would release energy that accelerates the fusion products, as in a nuclear rocket. This is the idea of the Bussard ramjet. How the fusion might be achieved will be discussed later. Let us first look at the overall principle.[6]

12.6.1 Principle

Figure 12.11 sketches the idea of a Bussard ramjet. When a spacecraft moves at velocity β relative to the ISM, it sweeps up matter using a large collector of cross-sectional area A. With matter density ρ, then over distance dx the mass collected is $dm = \rho A dx$. Note that dx is measured in the ISM rest frame. In terms of the time dt in the ISM rest frame we have $dx = c\beta dt$. In order to make some of the expressions below more compact, in this section I use the (dimensionless) proper velocity $\omega = \gamma\beta$ (see equation 7.20). As we shall derive an expression for the proper acceleration, it will be convenient to express dm in terms of the proper time interval dt' on board the spacecraft. The relation between the two times is the time dilation, $dt = \gamma dt'$ (equation 7.11). The mass swept up by the ramjet in time dt' is therefore

$$dm = \rho A c \omega dt' . \tag{12.7}$$

Suppose the ramjet brings dm to rest relative to the spacecraft and converts a fraction ϵ of this mass into energy, which is used to propel the spacecraft. In general this propellant would be a combination of matter particles and photons, and so we have a situation rather similar to the mixed matter–photon rocket discussed in section 7.8, although we will start here from

---

[6] A conventional ramjet is a type of air-breathing jet engine that uses the high velocity of the aircraft to compress air, which is then heated by fuel combustion to produce the propelling jet. This contrasts with turbojet and turbofan engines on most modern aircraft that use a gas turbine to compress the air. A ramjet only works once the aircraft is flying fast enough that an additional compression is not required, typically at supersonic speeds.

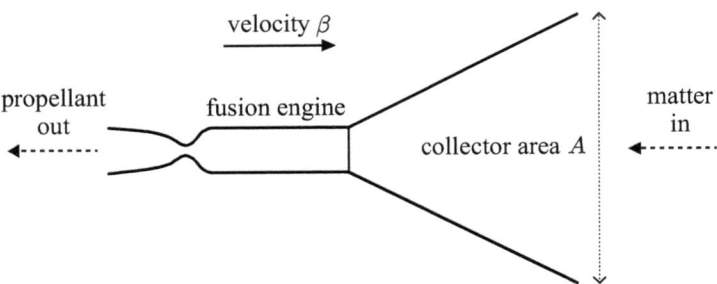

Fig. 12.11 A Bussard ramjet moving at velocity β relative to the ISM. The collector of cross-sectional area A collects mass. This is made to undergo nuclear fusion, and the high-velocity propellant is ejected out of the back.

first principles. We assume the mass of the spacecraft remains constant at M, and that there are no energy losses ($\eta = 1$ in the context of section 7.8).

In the ISM rest frame, the increase in the (kinetic) energy of the spacecraft is the energy provided by matter conversion minus the energy E_j that goes into the propellant jet:

$$dE = \epsilon c^2 dm - E_j . \tag{12.8}$$

The change in the momentum of the spacecraft dp is the momentum gained by ejecting the propellant, which is equal in magnitude and opposite in direction to the momentum of the propellant p_j, minus the momentum of the absorbed ISM matter, which is $\gamma v dm$. This latter term is the drag we discussed in section 12.5. Using the proper velocity we get

$$dp = p_j - \omega c dm . \tag{12.9}$$

Assuming dp is initially positive, equation 12.9 tells us that the spacecraft will accelerate, ω increases, and so the spacecraft will collect more matter per unit time (equation 12.7). Will it continue to accelerate indefinitely? Yes, but by ever decreasing amounts, because assuming p_j is constant (i.e. we extract the same momentum per unit mass), then as ω increases, dp decreases, eventually to zero. The spacecraft approaches a maximum velocity asymptotically due to the drag.

To proceed we consider the two limiting cases, first a photon-only propellant jet, and then a matter-only propellant jet.

Stopping matter: photon jet

Here the mass ϵdm is used to produce photons of energy E_j (relative to the rocket) that are collimated into a jet out of the back of the spacecraft, as with the photon rocket. The remaining mass, $(1 - \epsilon)dm$, is dumped overboard at zero velocity relative to the spacecraft. To find the asymptotic proper velocity ω_∞, we set $dE = 0$ and $dp = 0$ in equations 12.8 and 12.9. As $E_j = cp_j$ for photons, we get

$$\omega_\infty = \epsilon \quad \text{(photon jet)}. \tag{12.10}$$

The asymptotic proper velocity equals the fraction of the collected mass converted into photon energy. One of the highest values of ϵ for a fusion reaction is the proton–proton (p–p) chain, which converts hydrogen to helium (equation 8.4) and has $\epsilon = 0.007$. Although this reaction does not put much of the energy released into the photons, if this energy could nonetheless be converted into a photon jet, the ramjet would reach an asymptotic velocity of $\beta = 0.007$. To achieve a higher ϵ, we would have to use antimatter (section 8.8.1), but this hardly exists in the ISM.

Stopping matter: mass jet

Let us instead assume that the converted mass–energy $\epsilon c^2 dm$ is used to accelerate the remaining mass $(1 - \epsilon)dm$ as propellant (instead of dumping it overboard). No photons are produced. The energy and momentum of a mass jet are

$$E_j = (1 - \epsilon)(\gamma_j - 1)c^2 dm \quad \text{and} \tag{12.11a}$$

$$p_j = (1 - \epsilon)c\omega_j dm, \tag{12.11b}$$

where γ_j and ω_j refer to the velocity of the jet relative to the ISM. Proceeding as before, we find that the asymptotic velocity is

$$\omega_\infty = \sqrt{\epsilon(2 - \epsilon)} \quad \text{(mass jet).} \tag{12.12}$$

With $\epsilon = 0.007$, the asymptotic velocity is $\beta = 0.117$, considerably higher than for the photon jet. The variation of the asymptotic velocity with ϵ is shown in figure 12.12. For both the photon jet and the mass jet, the maximum asymptotic proper velocity, achieved when $\epsilon = 1$, is $\omega_\infty = 1$, which corresponds to $\beta = 1/\sqrt{2} = 0.71$. This could not be achieved using known matter available in the ISM, however.

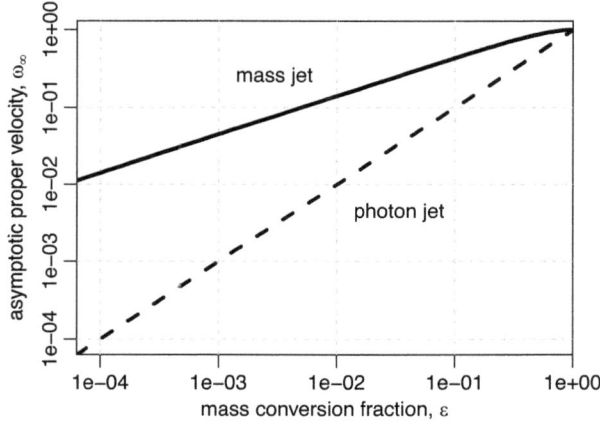

The asymptotic proper velocity of a Bussard ramjet when stopping the incoming matter in the spacecraft as a function of the fraction of mass converted to energy, for a photon jet (dashed line, equation 12.10) and a mass jet (solid line, equation 12.12).

Fig. 12.12

Acceleration

So far we have assumed that the matter must be stopped in the spacecraft for fusion to occur. If this could somehow be avoided (see below), then we can drop the drag term from equation 12.9. As the mass of the spacecraft is constant, the expressions for the change of energy and momentum in the ISM rest frame (equations 12.8 and 12.9) can now be written

$$dE = d\gamma Mc^2 = \epsilon c^2 dm - E_j \quad \text{and} \tag{12.13a}$$

$$dp = d\omega Mc = p_j. \tag{12.13b}$$

As there is no drag, it's clear from the second equation above that the proper velocity is no longer limited, and so the spacecraft will asymptotically approach the speed of light. We can find how quickly this happens by computing the proper acceleration. From equation 7.26 this is

$$a' = \frac{c}{\gamma} \frac{d\omega}{dt'}. \tag{12.14}$$

Let us compute this for the photon jet and mass jet cases. In doing this we use the relation $\gamma^2 = 1 + \omega^2$, from which we also get $\gamma d\gamma = \omega d\omega$.

Not stopping matter: photon jet

For photons, $E_j = p_j c$. Using this to combine equations 12.13b we get

$$M \left(\frac{\omega}{\gamma} + 1 \right) d\omega = \epsilon \, dm . \tag{12.15}$$

Using equation 12.7 for dm, the proper acceleration is

$$a'_{\text{no drag}} = \frac{\rho A c^2}{M} \frac{\epsilon \omega}{(\omega + \gamma)} = \frac{\rho A c^2}{M} \frac{\epsilon \beta}{(1 + \beta)} \quad \text{(photon jet).} \tag{12.16}$$

Figure 12.13 illustrates how this acceleration varies with the velocity for the cases $\epsilon = 0.007$ and $\epsilon = 1$. In this log–log plot, we see a near-linear increase up to an asymptotic non-zero acceleration of $\epsilon \rho A c^2 / 2M$. Thus, as long as the spacecraft sweeps up and fuses matter, its asymptotic velocity is the speed of light.

Had we retained the drag term $(-\omega c dm)$ in this derivation, we would get $\epsilon - \omega$ in place of ϵ in equation 12.16. This acts to decelerate the spacecraft when $\omega > \epsilon$, and gives zero acceleration when $\omega = \epsilon$, which is just the asymptotic velocity we found before (equation 12.10).

Not stopping matter: mass jet

For a mass jet, we use equations 12.11b for E_j and p_j in equations 12.13b. The maths is slightly more complicated, and results in a quadratic equation for the acceleration. The solution is

$$a'_{\text{no drag}} = \frac{\rho A c^2}{M} \omega \left[\sqrt{\omega^2 + \epsilon(2 - \epsilon)} - \omega \right] \quad \text{(mass jet).} \tag{12.17}$$

This is plotted as a function of velocity for $\epsilon = 0.007$ as the solid line in figure 12.13. The behaviour is qualitatively similar to the photon jet, but the mass jet reaches a higher asymptotic acceleration than the photon jet, and reaches this already at a lower velocity.

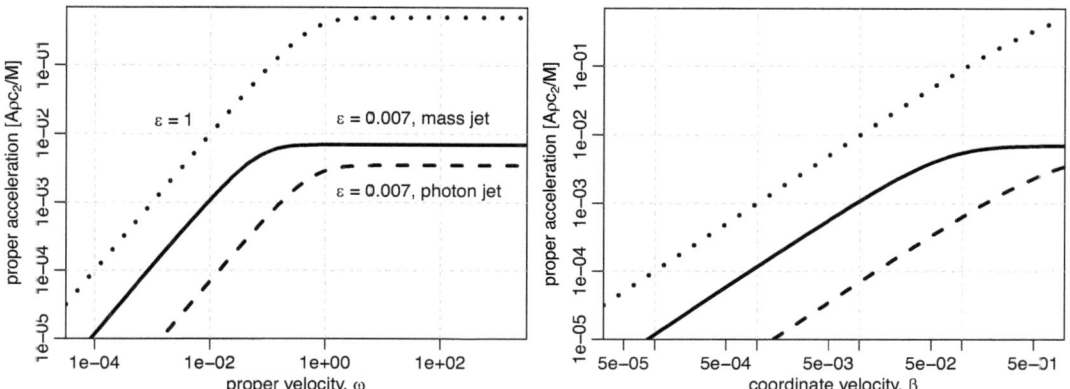

The variation of the proper acceleration of the Bussard ramjet with proper velocity (left) and coordinate velocity (right) when the incoming matter it not stopped. The dashed and solid lines are for a photon ramjet (equation 12.16) and a mass-ejecting ramjet (equation 12.17), respectively, both for $\epsilon = 0.007$. The dotted line is for 100% conversion of mass to energy ($\epsilon = 1$) for the photon ramjet, which is also the limit of the mass-ejecting ramjet.

Fig. 12.13

When $\epsilon = 1$, equations 12.16 and 12.17 are identical, although the mass-ejecting ramjet is then converting all of the matter into energy, leaving nothing to eject, so this is only a theoretical limiting case.

12.6.2 A ramjet's journey

From the expressions for the acceleration in terms of the proper velocity derived above, we can integrate equation 12.14 to determine the proper time taken to accelerate from from ω_i to ω_f:

$$t'_f = c \int_{\omega_i}^{\omega_f} \frac{d\omega}{\gamma a'} \ . \tag{12.18}$$

This has a closed-form solution, but it's rather complex, so here I just integrate numerically for a range of ω_f values to get a corresponding set of t'_f values. This gives us the variation of velocity with time as the spacecraft accelerates.

Using the definition of proper velocity we can integrate again to get the distance travelled in a given time

$$x_f = c \int_0^{t'_f} \omega dt' \ . \tag{12.19}$$

We can compute this by numerical integration too. Given a sufficiently dense list of (t'_f, ω_f) values from the first integral, we can compute x_f for each of these using the trapezium rule or Simpson's rule.

Let us compute the flight profile for a drag-free Bussard ramjet, for both the photon jet and mass jet scenarios. We assume the ramjet sweeps up and fuses hydrogen via the p–p chain, so $\epsilon = 0.007$. The spacecraft travels through the local interstellar cloud (table 12.1), which we take to have a gas number density of $0.3 \times 10^6 \ \mathrm{m}^{-3}$. Most of this is hydrogen, so we adopt a density of fusionable matter of $\rho = 5 \times 10^{-22} \ \mathrm{kg\,m}^{-3}$ (and ignore the drag of the non-fusionable material).

The mass of the spacecraft comprises the mass of the collector and the mass M_p of everything else ('payload', for brevity). The collector has constant cross-sectional areal mass density σ, and so its mass is σA. The acceleration scales as $\rho A/M$, so if the collector dominated the spacecraft's mass budget, the acceleration would scale as ρ/σ. In general we have $M = \sigma A + M_p$, so the acceleration scales instead as $\rho/(\sigma + M_p/A)$. As M_p must include a fusion engine, it is probably not going to be small. No such engine exists, so I will make a wild (and probably very optimistic) guess and assume this has a mass of 10^3 kg. The collector would presumably be some kind of wire mesh that electromagnetically captures hydrogen ions. I will somewhat arbitrarily assume $\sigma = 1.53 \times 10^{-4}$ kg m^{-3}, which if it were a solar sail corresponds to a lightness number of 10 (equation 10.10). Adopting a circular collector of radius 3 km, we have $M_p/A = 3.5 \times 10^{-5}$ kg m^{-3}, which is four times less than σ, and so the payload is not the limiting factor. A much larger collector therefore brings little benefit, because the extra collecting area would serve mostly to accelerate the extra mass of this collecting area. A larger collector presumably also means we need a larger engine to process the higher flux of protons, and thus a higher mass payload.

With these assumptions, we have $\rho A/M = 2.6 \times 10^{-18}$ m^{-1}. The acceleration scales with this quantity, so we can trade off the individual quantities to achieve the same acceleration. For example, if the ISM density is higher by some factor, we could increase the spacecraft mass or lower the collector area by the same factor.

A ramjet only works once it is moving relative to the ISM. We assume our spacecraft starts in Earth's orbit about the Sun with an initial velocity of 30 km s^{-1} relative to the ISM.[7] We now have enough information to compute the above integrations. The resulting variation of the acceleration, velocity, and distance with the proper time for both the mass jet and the photon jet variations are shown in figure 12.14, as well as a constant acceleration case for comparison. The photon jet shows far inferior performance to the mass jet, so we will only discuss the results for the mass jet.

To get to Proxima Cen, our ramjet requires 1650 years of proper time (spacecraft time). At this point, the spacecraft is moving with proper velocity $\omega = 0.012$, which corresponds to coordinate velocity $0.012\,c$ or 3610 km s^{-1}. These are almost equal because the spacecraft is barely relativistic.

With the same spacecraft, increasing the initial velocity increases the rate of matter consumption and therefore the acceleration. Suppose we used the Oberth effect in a close dive to the Sun (section 4.2.3) to increase the initial velocity to 200 km s^{-1}. In that case the time to Proxima Cen decreases to 1025 years, but the velocity at encounter is very similar (3760 km s^{-1}).

To increase the acceleration for given ϵ and ρ, we have to reduce the collector areal density σ. Reducing the payload mass M_p or increasing the area A of the collector do not help much in the adopted scenario, because the collector already dominates the mass. If we could reduce σ so far that $\rho A/M$ increases by a factor of 50, we find that the time taken to Proxima Cen decreases to 72 years. The spacecraft is moving at $0.24\,c$ as it passes the star.

In the original scenario, our spacecraft takes a long time to reach Proxima Cen and is not travelling particularly fast when it does. But it will keep on going, and because the acceleration never decreases to zero, it can eventually reach very high speeds. We see this in

---

[7] The Sun has a non-zero velocity relative the mean of the ISM, so the initial velocity would vary according to the initial direction the spacecraft travels.

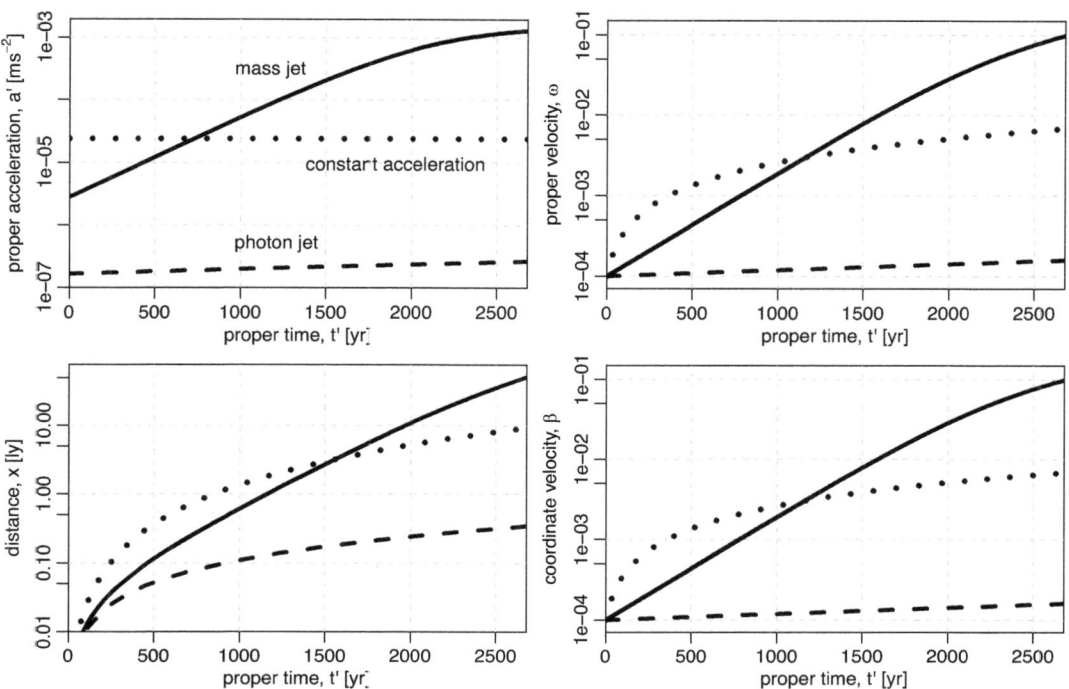

The variation of the acceleration, proper and coordinate velocities, and distance travelled as a function of proper time for a Bussard ramjet using either a photon jet (equation 12.16, dashed line) or a mass jet (equation 12.17, solid line) when the swept-up matter is not stopped (no drag). The proper and coordinate velocities are almost identical because the spacecraft barely reaches relativistic velocities. This simulation adopts $\rho A c^2 / M = 0.24$ m s^{-2} (the numerical term in the acceleration equations) and $\omega_i = 10^{-4}$. For comparison, the dotted line shows the case of a constant proper acceleration of 2.4×10^{-5} m s^{-2}.

Fig. 12.14

figure 12.15, where the integration is now extended to much longer times. The acceleration of the mass jet gets close to its asymptotic value after about 3000 years, when it is travelling at nearly half the speed of light (coordinate velocity) and is 100 ly from the Earth. Now that the velocities are relativistic, we see a big difference between the proper velocity and the coordinate velocity (note also the different axis scales in figure 12.15). After 36 000 years of spacecraft time, the spacecraft has covered one million light years and is moving at nearly the speed of light: $\beta = 0.999985$ or $\omega = 180$. This is the regime where the proper velocity is a more convenient measure than the coordinate velocity.

The problem with the Bussard ramjet is how to stop, or at least to decelerate. In principle this could be done by turning the mass (or photon) jet into the direction of travel. Recall that in the above scenario we assumed the matter is not stopped before undergoing fusion, that is, there is no drag from the ISM, so by additionally stopping the matter we could achieve a larger deceleration. The spacecraft could also also be steered through regions of higher ISM density, such as star-forming regions, to achieve a larger deceleration (or acceleration).[8]

---

[8] The Bussard ramjet is the centrepiece of the novel *Tau Zero* by Poul Anderson. The spacecraft carries a large human crew and is designed to reach an asymptotic acceleration of 1g. Things get even more interesting after something breaks, and the crew have no way of stopping the acceleration ...

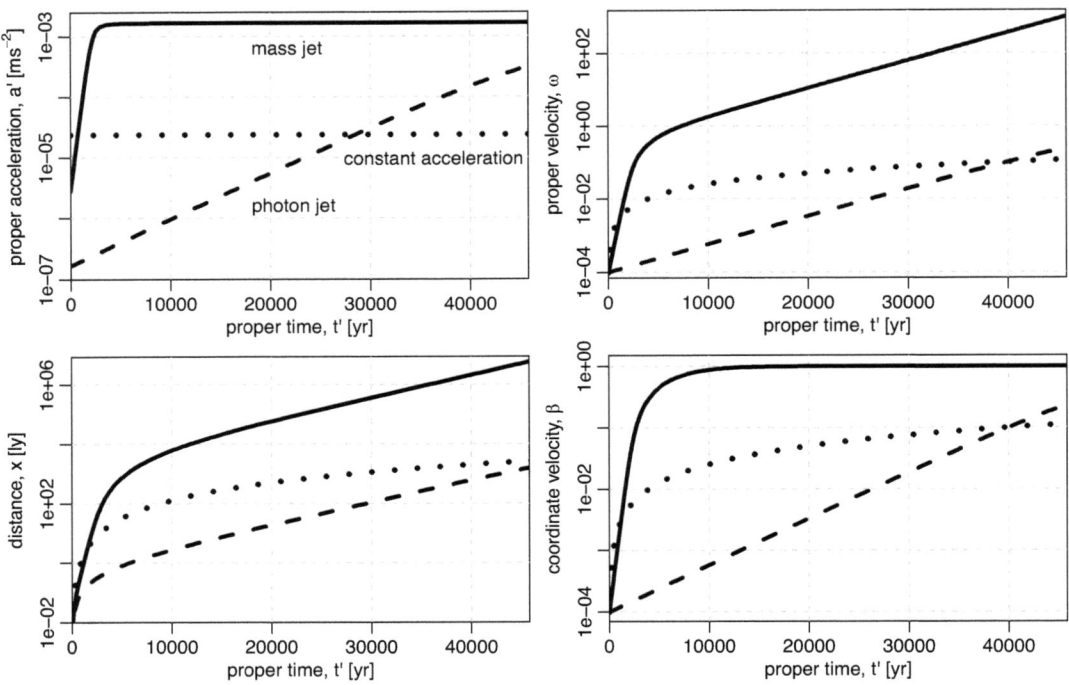

Fig. 12.15 As Fig 12.14 but extended to longer times.

12.6.3 Challenges

We have skipped some important details of ramjets, not least how we actually achieve fusion, especially when not stopping the protons. The value of $\epsilon = 0.007$ adopted implies use of the p–p chain outlined in section 8.5. This is an extremely slow reaction, whereas we have been implicitly assuming fusion would occur as fast as required. An alternative is to use the carbon–nitrogen–oxygen (CNO) cycle. This is the process that converts hydrogen to helium in higher-mass stars. It is much faster than the p–p chain, but it requires a higher temperature of around 15 million kelvins as opposed to about 4 million kelvins for the p–p chain. Furthermore, it requires carbon to get the cycle going, although as the carbon itself is not used up, it is a catalyst in this reaction. Thus the carbon could be taken along by the spacecraft rather than collected from the ISM.

An alternative to hydrogen burning it to use deuterium (^{2}H) fusion, for example in the D+D reaction we considered for a fusion rocket (equation 8.8). Relatively speaking, this is easier to achieve than the CNO cycle. But the deuterium abundance in the ISM is around a factor of 10^5 lower than the hydrogen (^{1}H) abundance, which lowers ρ by this factor.

Another significant challenge is how to collect the interstellar material. If this is done with a magnetic field, then this will only collect the ionized particles in the ISM (around 25% by number near to the Sun). One study, under the name of the Fishback ramjet, examined in detail how this collection could work. It found not only that the collector would have to be enormous (thousands of kilometres) but also that the asymptotic velocity of the ramjet (with drag) would be significantly lower than the simpler, idealized calculations shown above.

One way to overcome the low-density fuel problem is to send a precursor mission to drop much higher-density fuel pellets along the intended path of the Bussard ramjet, like a trail of breadcrumbs. The obvious problem with this, however, is that we need an additional mission to achieve it. If the path length over which the fuel must be dropped is large, then we already need a fast interstellar mission (not using a ramjet) to achieve another fast interstellar mission (with a ramjet). It would further require that the positions and velocities of the pellets are all carefully coordinated so that they are in the right place at the right time for the intended ramjet trajectory.

The results presented in the previous section assumed that the ISM matter does not have to be stopped to be fused. Current approaches to fusion require high plasma temperatures, densities, and/or confinement times (the Lawson criterion), and it's unclear how we could force fusion in matter travelling at a large velocity relative to the spacecraft. One idea is to use an electric field to decelerate the ionized incoming matter to rest with respect to the spacecraft, and to store this kinetic energy as electric energy in a capacitor. This energy is then given back to the fusion products to accelerate them to higher exhaust velocities than achieved just by the released fusion energy. This effectively removes the ion drag by returning the momentum absorbed to the exhausted propellant.

12.7 Summary

The ISM is made up of neutral atoms, ions, electrons, molecules, dust particles, and macroscopic-sized interstellar objects. Their space densities vary by orders of magnitudes across the Galaxy, but close to the Sun the gas number density is of order $0.3 \times 10^6 \, \mathrm{m}^{-3}$. This is mostly hydrogen (90% by number) and helium (9.9%), which corresponds to an overall mass density of $6 \times 10^{-22} \, \mathrm{kg \, m}^{-3}$. About 25% of the hydrogen is ionized. Dust makes up about a hundredth of the mass of the gas, with particle sizes spanning orders of magnitudes.

An interstellar spacecraft will encounter ISM particles at very high velocities. These can cause serious damage by melting or evaporating the surface and through the creation of secondary energetic particles. Although massive atoms like iron are far less abundant than light ones like hydrogen or helium, the massive ones do comparatively more damage. Dust particles are the most serious problem overall on account of their much larger masses, despite being even rarer.

Fortunately, it appears that several millimetres of shielding material, which will be eroded during the journey, is enough to protect a spacecraft against the most likely dust particles encountered. Above some size, a dust particle will be too large to shield and could destroy the spacecraft. As the particle size distribution is poorly known, it is difficult to estimate how likely this is, but it appears unlikely. To reduce the risk, the spacecraft should have the smallest cross-sectional area possible in the direction of travel. Large, thin structures like sails will inevitably be perforated by larger dust particles, but only a small fraction of the surface would be affected on a journey to Proxima Cen.

The solar system is surrounded by a weakly bound, roughly spherical distribution of icy minor bodies with sizes between metres and kilometres. This Oort cloud may extend to a light year from the Sun. Similar structures presumably surround other stars. The space density of the Oort cloud is far too low for it present a hazard to a spacecraft, however.

Finally, the hydrogen in the ISM could in principle be swept up by a spacecraft and used in a fusion engine as a source of energy and propellant. This so-called Bussard ramjet requires a collector many kilometres in size (perhaps thousands) to collect enough matter to produce adequate acceleration for a sub-century journey to Proxima Cen. How the spacecraft would collect the matter and then fuse it efficiently remains undetermined, so this remains a very far-fetched idea.

12.8 Exercises

1. Using equation 12.1 and the data on the Oort cloud in section 12.1, estimate the probability that an interstellar spacecraft will collide with an Oort cloud comet on its journey outwards. For simplicity, assume that the comets are stationary. Is that valid? How, qualitatively, would the estimate change if this assumption were not valid? If a spacecraft has to pass within one million kilometres of an Oort cloud object to observe it, what is the probability that it would observe at least one during its journey?
2. Figure 12.4 shows that the stopping power of a spacecraft's shield to incident ions has a maximum at intermediate velocities. To get to 0.1–$0.2\,c$, a spacecraft must first accelerate through lower velocities where the stopping power is lower. Might it be that ions cause more damage in this phase than when the spacecraft is moving faster?
3. In section 12.5 we found that the coordinate deceleration due to particle drag on a moving spacecraft is proportional to β^2/γ^2. Plot this against β. Is anything unexpected? From this, derive how the spacecraft velocity varies with distance travelled x, that is, find the relativistic version of equation 12.6b.
4. A flat circular laser sail 2 m in diameter of mass 0.01 kg is hit on its edge by a proton travelling normally to the surface at speed $0.1\,c$. This exerts a torque that makes the sail rotate. What angular velocity does the sail acquire, assuming the proton sticks to the sail? Is this a problem for navigation and control? What about when it is hit by a dust particle of mass 10^{-16} kg instead?
5. Derive the expressions for the asymptotic velocity of the Bussard ramjet in the photon jet case (equation 12.10) and mass jet case (equation 12.12).
6. Derive equation 12.17.
7. Extend the derivation of the distance travelled and time required to reach a given velocity in section 12.6.2 to the case of having to stop the incoming ISM matter. You first need to write down the expression for the acceleration of the ramjet when including the drag term, and then perform the numerical integrations.

We saw in the previous chapter that the Sun emits a wind of charged particles. If the momentum of these charged particles can be coupled to a spacecraft, it could be used as a means of propulsion, either to accelerate an outgoing spacecraft or to decelerate an incoming one. In a similar way, the charged particles of the ISM could be used to decelerate a spacecraft that is moving relative to them. In this chapter we look at two distinct methods of using charged particles in this way, namely magnetic sails and electric sails.

13.1 Magnetic sails

13.1.1 Concept

The idea of magnetic sail propulsion is to use a magnetic field to generate a magnetosphere around the spacecraft. This is conceptually the same as the magnetosphere around the Earth generated by its magnetic field. When such a spacecraft is within the solar system, the solar wind will in effect reflect off the surface of the magnetosphere, imparting its momentum onto the magnetosphere, and therefore onto the spacecraft from which the field is generated. The size of the acceleration depends on the velocity of the wind particles relative to the spacecraft. The same principle applies when the spacecraft is moving through the ISM, as this also contains charged particles.

To create a magnetosphere, a spacecraft deploys a large conducting loop. The loop is orientated so that its axis is perpendicular to the direction towards the Sun. Running a current through the loop generates a dipole magnetic field. In a vacuum, such a field would be symmetric about the loop's axis. But when impacted by a moving plasma, the shape of this field is distorted by the interaction of the charged particles with the field, as shown in figure 13.1. The field is compressed in the upwind direction and extended in the downwind direction. The boundary of the magnetosphere is the magnetopause, defined as the surface where the magnetic pressure from the field, $B^2/2\mu_0$, equals the ram pressure of the plasma, $(1/2)\rho_p v^2$, where ρ_p is its density and v is its velocity normal to the magnetopause. B is the magnetic field strength at the boundary and μ_0 is the vacuum magnetic permeability (approximately $1.26 \times 10^{-6}\,\mathrm{N\,A^{-2}}$). As the magnetic field strength and the velocity and density of the plasma vary spatially due to the interaction, the shape of the magnetopause needs to be found via an iterative numerical method. Few plasma particles penetrate deeply into the magnetosphere, so this is a region where the plasma density is low but the magnetic field strength is high. Outside the magnetosphere the plasma density is higher, but the magnetic field strength from the loop is low.

Charged particles that encounter the magnetopause with a velocity $\mathbf{v}$ are deflected by the magnetic field $\mathbf{B}$ according to the Lorentz force, $\mathbf{F} = q\mathbf{v} \times \mathbf{B}$. Charges of different sign are

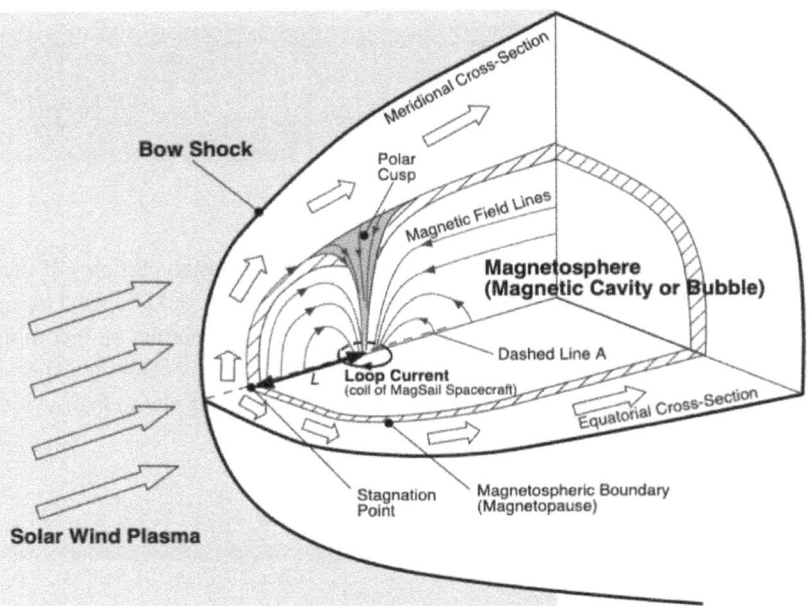

The magnetic sail concept. A current through the loop creates a magnetosphere that then deflects a moving plasma, shown by the large arrows. Credit: Funaki and Nakayama (2012), licensed under CC-BY-3.0, obtained from IntechOpen.

deflected in different directions (and by different amounts when their velocities differ). This generates a current in the magnetopause, and this current interacts with the magnetic field to create a Lorentz force on this field, and therefore on the spacecraft. The net force acts in the initial direction of the solar wind, which is radially away from Sun. The magnetopause can be considered as a magnetic sail against which plasma particles are deflected and thereby exert a force on it and, by extension, on the loop (spacecraft) that created the magnetic field.

We can compute the size of this magnetic force analytically to within numerical factors. From the Biot–Savart law, a loop of radius a carrying a current I produces a magnetic field at a large distance r ($r \gg a$) of magnitude[1]:

$$B \sim \frac{\mu_0 I a^2}{r^3}. \tag{13.1}$$

This is only valid in the absence of the plasma, of course, because that distorts the field. At the magnetopause, the magnetic pressure $B^2/2\mu_0$ equals the plasma ram pressure $(1/2)\rho_p v^2$. Equating these and inserting for B from the above equation gives us the approximate distance of the magnetopause from the loop,

$$r_m \sim \left(\frac{\mu_0 I^2 a^4}{\rho_p v^2} \right)^{1/6}, \tag{13.2}$$

to within numerical factors. To within an order of magnitude, the upwind cross-sectional profile of the magnetopause that exchanges momentum with the plasma is a circle of area

---

[1] There is a numerical factor missing here that depends on the position with respect to the loop axis, and which component of the vector field we are considering.

πr_m^2, and so the force from the plasma is $(1/2)\pi r_m^2 \rho_p v^2$. This is analogous to the drag force in fluid mechanics. The force on the magnetopause and thus on the spacecraft is therefore

$$F = k\,(\mu_0 I^2 a^4 \rho_p^2 v^4)^{1/3} \tag{13.3}$$

for some dimensionless constant k. This force acts in the direction that the plasma moves relative to the spacecraft. Evaluation of the constant k requires a proper integration of the magnetic pressure over the magnetopause, and depends on the approximations used in the field calculation to make the integrals tractable. Freeland (2015) gives a value of $0.354\pi = 1.11$, which I adopt here. Other authors also arrive at equation 13.3, but with different values of k. The values of the exponents in equation 13.3 are not intuitive, but the proportional dependence (as opposed to an inverse) of the force on all four physical terms on the right side of equation 13.3 makes sense. Note that ρ_p is the density of the charged particles in the plasma, so excluding any neutral particles. The solar wind can be taken as completely ionized, but the ISM is not (section 12.3).

Considering the plasma to be the solar wind, the force on the sail points radially away from the Sun when the spacecraft is moving radially. As long as the spacecraft is moving outward slower than the wind, this force will accelerate the spacecraft outward. Once the spacecraft reaches the velocity of the wind, their relative velocity is zero, so the force is zero, and the spacecraft is no longer accelerated. When the spacecraft is instead moving towards the Sun, or indeed another star, the oncoming solar wind will act to decelerate the spacecraft. In that case, the faster the spacecraft, the larger its velocity relative to the wind, and so the larger the deceleration. For this reason a magnetic sail is often considered as a means for braking at a target star instead of accelerating out of the solar system.

In a similar way, a high-velocity spacecraft moving through the ISM will also be decelerated. This can be used to decelerate the spacecraft over much larger distances than with the solar wind and so potentially achieve a larger change in velocity.

13.1.2 Application

We first use the above theory to calculate the acceleration of a magnetic sail by the solar wind. For this purpose I adopt a plasma density of $\rho_p = 8.35 \times 10^{-21}\,\mathrm{kg\,m^{-3}}$ (five protons per cubic centimetre), and a wind velocity of $v = 400\,\mathrm{km\,s^{-1}}$, values that are appropriate for the solar wind 1 au from the Sun. For the sail I adopt $I = 50\,\mathrm{kA}$ and $a = 31.6\,\mathrm{km}$. Such a large loop (compared to a solar sail, for example) is required because of the low ram pressure of the solar wind, and because the magnetic field of a loop falls rapidly with distance (equation 13.1). From equation 13.2, we find that the magnetopause is 115 km from the loop's centre, so the condition $r_m \gg a$ for equation 13.1 to be valid is only just satisfied (a larger current may be required in practice). Equation 13.3 tells us that the resulting force on the sail is 20 N.

The acceleration this force produces depends on the mass of the spacecraft, which is likely to be dominated by the mass of the loop. To be able to maintain such a large current for a long period of time without a heavy power supply, the loop needs to be superconducting. Superconductors are materials that conduct without resistance, so once the current has been set up, no power source is required to maintain it (although impacts from charged particles may complicate this). Only certain materials are superconducting, and then only if they are below a certain critical temperature. For niobium–titanium, for example, the critical temperature is about 10 K. High-temperature superconductors have a critical temperature above 77 K (but still well below room temperature), one class of which is based on copper oxides. Materials are only superconducting if the magnetic field strength remains below some critical value, which corresponds to a maximum current per unit cross-sectional area of the conductor, J_s. This depends on the material, but is of order 10^{10} A m^{-2}. Adopting this value as the upper limit, and assuming the superconductor material has density $\rho_s = 5000$ kg m^{-3} (approximately that of copper oxide), we can compute the minimum cross-sectional area – and hence minimum mass – of the loop required to carry a given current. The mass of a loop of radius a made from wire of constant cross-sectional area A is $m = 2\pi a A \rho_s$. The maximum current is $I = J_s A$, in which case the minimum mass loop is $2\pi a I \rho_s / J_s$.

Using these two expressions for the mass of the loop and equation 13.3 for the force on the loop, the magnitude of the acceleration is

$$\frac{dv}{dt} = \frac{F}{m} = 3bv^{4/3} \quad \text{where} \quad b = \begin{cases} \dfrac{k}{6\pi}\dfrac{1}{A\rho_s}(\mu_0 I^2 a \rho_p^2)^{1/3} & \text{(in general)} \\[2ex] \dfrac{k}{6\pi}\dfrac{J_s}{\rho_s}\left(\dfrac{\mu_0 a \rho_p^2}{I}\right)^{1/3} & \text{(minimum mass loop).} \end{cases}$$

(13.4)

Note the inverse dependence of the acceleration on the (maximum) current for the minimum mass loop. This suggests that for a given superconductor we should make the current as small as possible, because this will minimize the thickness and therefore the mass of the loop. There is a lower limit to the current we need, however, because a minimum magnetic field strength is required to set up the magnetosphere in the first place. With the above choices for the current and loop size and materials, the magnetic field strength at the centre of the loop is $B = \mu_0 I / 2a = 10^{-6}$ T and the minimum radius of the wire is $\sqrt{I/\pi J_s} = 1.3$ mm. This gives a mass of five tonnes for the loop. Equation 13.4 gives the acceleration of this minimum mass loop to be 0.004 m s^{-2}.

To put this value into context, the acceleration of a solar sail of lightness number $\lambda = 1$ at 1 au from the Sun is $GM_\odot/r^2 = 0.006$ m s^{-2} (section 10.2). This is quite an extreme solar sail by current standards, but so is a superconducting wire 200 km long with a cross-sectional radius of 1.3 mm. This wire must also have enough tensile strength to support the current, and to transfer the acceleration to the rest of the spacecraft.

The magnetic sail could be used to accelerate a spacecraft away from the Sun. By tilting the sail, the spacecraft receives a non-radial force, and thus can be made to move on a non-Keplerian orbit, just as can be done with solar sails (see section 10.3.2). This could be used to manoeuvre to the planets in the solar system, although the variability of the solar wind would make this a challenge in practice.

By integrating equation 13.4 we can compute the velocity of the loop as a function of time. This is the velocity of the loop relative to the solar wind. To find the velocity of the loop relative to the Sun (or other star) we need to know how the velocity of the solar wind varies with distance from the star. Zubrin and Andrews (1991) investigate using a magnetic sail for both an interplanetary mission and for escaping the solar system. For the latter they find that a sail could reach a terminal velocity of 95 km s^{-1}, reaching 1000 au in 50 years.

As a second application, we can use the magnetic sail to decelerate a fast moving spacecraft, one which has previously been accelerated by a nuclear rocket or laser sail, for example. In this case we can consider the charged particles in the wind or ISM as approximately stationary relative to the star. We neglect gravity. The acceleration is $-3bv^{4/3}$ from equation 13.4. Integrating this from an initial velocity v_0 at time $t = 0$ gives the velocity at some later time t to be

$$v = v_0(1 + bv_0^{1/3}t)^{-3} . \tag{13.5}$$

Integrating again we get the distance travelled after time t,

$$s = \frac{v_0^{2/3}}{2b}\left[1 - (1 + bv_0^{1/3}t)^{-2}\right] . \tag{13.6}$$

To get a sense of the magnitude of the plasma-induced deceleration, let us first compute the deceleration due to the solar wind at 1 au from the Sun for the previously discussed minimum mass loop (parameter values given above). When the initial velocity is $0.1\,c$, the deceleration is $1.3\,\mathrm{m\,s^{-2}}$. If the plasma density remains constant over 10 au, the velocity decreases by just 0.2%.[2] Assuming we cannot target a star with a much denser stellar wind, this result implies that to achieve a significant deceleration, the spacecraft needs to decelerate over a much larger distance. This in turn means it must brake using not the stellar wind, but the charged particles in the ISM.

Figure 13.2 shows a scenario of braking with the charged particles in the ISM, computed using equations 13.5 and 13.6. The spacecraft is initially moving at a velocity of $0.1\,c$. It then unfurls the same loop as used above: $I = 50\,\mathrm{kA}$, $a = 31.6\,\mathrm{km}$, $J_s = 10^{10}\,\mathrm{A\,m^{-2}}$, $\rho_s = 5000\,\mathrm{kg\,m^{-3}}$, which has a loop mass of five tonnes. The payload is assumed – somewhat unrealistically – to have negligible additional mass. The middle line in figure 13.2 shows how the spacecraft is braked by the ISM when the ion mass density is $150 \times 10^{-24}\,\mathrm{kg\,m^{-3}}$. This corresponds to the density of the local interstellar cloud with an ionization fraction of 0.25 (see Table 12.1). The figure shows that it takes many years for a significant deceleration to be achieved. When the loop is deployed 1 ly from the destination star, the spacecraft's velocity is reduced by a factor of four, to $0.024\,c$, by the time it arrives at the star. This deceleration takes 20 years, so adds 10 years to the travel time compared to the non-braked spacecraft continuing at $0.1\,c$.

The lower line in figure 13.2 shows the deceleration with four times the local interstellar cloud density. This gives a reduction in the velocity by a factor of 16 after 20 years. As the size of the magnetosphere is inversely proportional to the density of the plasma (equation 13.2), this scenario has the smallest magnetosphere of the examples shown, with 42 km radius. The upper line in the figure is for a density one-sixth that of the local interstellar cloud, characteristic of the local bubble. It gives much less reduction in velocity, a factor of only 1.7 after 20 years.

The density of the ISM varies and cannot be accurately measured at high spatial resolution from the Earth, so future mission planners are unlikely to be able to accurately predict what braking can be achieved with a magnetic sail. Note also that the masses involved here are large: five tonnes of loop in order to achieve the braking shown in figure 13.2. To accelerate such a mass to $0.1\,c$ in the first place requires a lot of energy. In practice, the superconducting coil needs to be shielded against erosion from collisions with the ISM (section 12.4), and this would add mass, more than the wire itself. This would decrease the deceleration and thus

---

[2] For this configuration, equation 13.2 gives a characteristic radius of the magnetosphere of $r_m = 27\,\mathrm{km}$. Although this is not an exact value, it is smaller than the loop radius, so the expressions used may no longer be a good approximation.

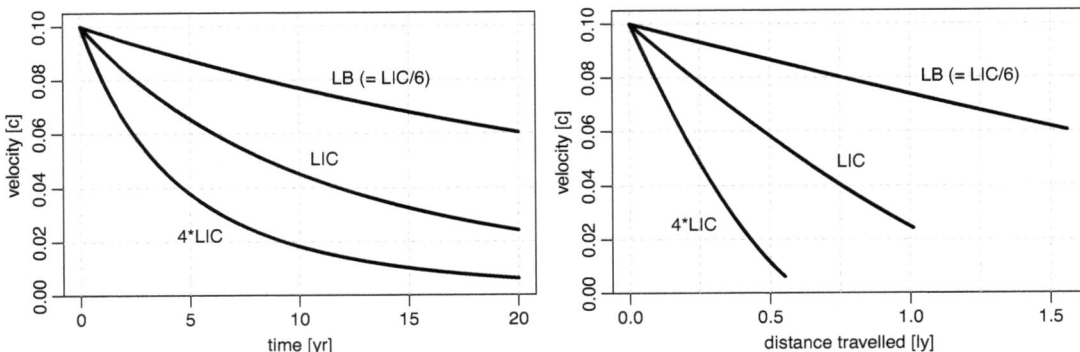

Braking with a magnetic sail using the drag of charged particles in the ISM. The panels show how the velocity of the spacecraft changes as a function of time (left) and distance travelled (right) over a period of 20 years. The middle line is for an ISM density typical of the local interstellar cloud (LIC) assuming an ionization fraction of 0.25. The lower line is for four times that density, and the upper line is for the typical density of the local bubble (LB), which is a sixth the LIC.

extend the total mission time significantly. Clearly, such a system cannot be used to brake a small lightweight spacecraft like a laser sail and so can only be considered for spacecraft with masses of many tonnes, which more plausibly would have been accelerated by nuclear rockets.

13.1.3 Other magnetic sail concepts

The big drawback of the magnetic sail described above is the large mass of the superconducting loop (plus shielding) required to produce the magnetosphere, resulting in a modest acceleration. This could be addressed using a much smaller coil – diameter of centimetres to metres – to produce a plasma that is injected into the space around the spacecraft, which then generates the magnetosphere. For this reason, the type of sail discussed in the previous section is sometimes called plasma-free magnetospheric propulsion. Whereas the magnetic field of a dipole decreases with distance r from the dipole as r^{-3} (equation 13.1), the field from the plasma decreases as r^{-1}, meaning a comparatively small coil could be used to generate the plasma. Essentially, the plasma inflates the coil-generated magnetic field.

This concept was originally introduced under the name mini-magnetic plasma propulsion (M2P2) by Winglee et al. (2000). They outline a scenario using a loop of 10 cm radius with over 1000 turns (total length 700 m) producing a magnetic field strength at the centre of the loop of 0.1 T. This ionizes a gas that is injected into the loop to produce a plasma with a temperature of around 5 eV. This plasma expands out of the loop to make a magnetosphere. When interacting with the solar wind, this produces a force on the artificial magnetosphere of a few newtons. They adopt a total mass of the spacecraft of 100 kg, of which 30 kg is the gas used to produce the plasma. Starting at the Earth's orbit and accelerating away from the Sun, they find that within a year the spacecraft achieves $\Delta v = 75\,\mathrm{km\,s^{-1}}$ and reaches 14 au, having used 20 kg of its gas to maintain the plasma, as this slowly leaks from the magnetosphere. This velocity is considerably larger than what has been achieved to date with chemical rockets, but is still far too slow for a desirable interstellar mission, requiring 17 thousand years to reach Alpha Cen.

There have been several criticisms of the original M2P2 concept, however. Khazanov et al. (2005), for example, argue that magnetohydrodynamics (MHD) is not an appropriate description and that the M2P2 design would deliver a much smaller thrust than claimed. Others have pointed out that much of the momentum of the solar wind imparted onto the plasma would not be transferred to the spacecraft, thus resulting in a much smaller acceleration. A variation of M2P2 goes by the name magnetoplasma sail (MPS), as reported by Funaki and Nakayama (2012). Numerical simulations show that a plasma produced by a coil can create a large magnetosphere and that it could achieve an acceleration comparable to the dipole-based design, but using a much smaller and less massive loop/coil.

13.2 Electric sails

An electric sail, like a magnetic sail, also uses the momentum of charged particles to accelerate, but is conceptually simpler. The electric sail is a large, positively charged body, for example a flat mesh of metal wires (see figure 13.3). When positively charged ions (from the solar wind or ISM) approach this, they are repelled and deflected by the Coulomb force. This force does work, decelerating the protons and passing their kinetic energy and momentum on to the sail.

The solar wind and ISM are neutral overall, so the solar sail also experiences a Coulomb attraction from the electrons. However, as the electrons have much smaller masses, this produces a much smaller acceleration. These electrons will tend to recombine with the positive charges in the sail, though, which over time would make it neutral. The sail therefore needs to be fitted with a gun to remove the electrons and keep the sail positively charged.

Just as with the magnetic sail, the electric sail could be used for accelerating away from a star on the solar wind or for decelerating against the ISM.

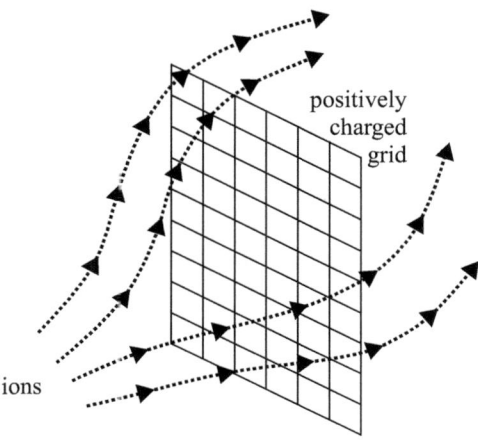

positively
charged
grid

ions

A schematic of an electric sail.   Fig. 13.3

To determine the force that charged particles exert on an electric sail, we consider first the electrostatic potential due to an infinitely long line of charges with charge per unit length λ. This corresponds to a long charged wire. The electric potential at distance r from the centre of the wire in a vacuum is found from Gauss's law to be

$$\phi(r) = \frac{\lambda}{2\pi\epsilon_0} \ln(r_0/r), \tag{13.7}$$

where r_0 is the distance at which the potential is zero. For a wire with a circular cross-section of radius r_w, the potential at the surface of the wire is

$$\phi_0 = \frac{\lambda}{2\pi\epsilon_0} \ln(r_0/r_w), \tag{13.8}$$

so we can write equation 13.7 as

$$\phi(r) = \phi_0 \frac{\ln(r_0/r)}{\ln(r_0/r_w)}. \tag{13.9}$$

When the wire is embedded in a proton/electron plasma, where each particle has number density n, the electrons in the plasma shield the potential and make it drop faster. From comparison with plasma simulations, Janhunen and Sandroos (2007) find that a good approximation for the potential in this case is

$$\phi(r) = \frac{\phi_0}{2} \frac{\ln[1 + (r_0/r)^2]}{\ln(r_0/r_w)} \tag{13.10}$$

for $r_w \ll r_0$. The distance r_0 is now determined by the plasma electron temperature T_e and charge e as

$$r_0 = 2\sqrt{\frac{\epsilon_0 k T_e}{ne^2}}, \tag{13.11}$$

where ϵ_0 is the vacuum permittivity. The length r_0 is twice the Debye length, which is the characteristic distance over which electric fields are screened by a plasma due to its charge separation. At the Earth's distance from the Sun, the solar wind has $T_e = 10^5$ K (about 9 eV) and a density of $n = 5 \times 10^6$ m^{-3}, giving $r_0 = 20$ m.

The wire presents an electrostatic barrier of some effective width r_b perpendicular to its length against which the plasma pushes. The force per unit length acting on the wire is this width times the plasma ram pressure $m_p n v^2$, where m_p is the mass of the proton and v is the asymptotic velocity of the plasma relative to the wire. Thus the force per unit length is

$$\frac{F}{L} = m_p n v^2 r_b. \tag{13.12}$$

We now assume that the barrier width r_b is proportional to the proton stopping distance r_s in the potential. Even though these two distances are perpendicular to one another, the cylindrical symmetry of the wire's potential suggests that these two lengths scale in the same way, so we can write $r_b = b r_s$, where b is a dimensionless constant. Using numerical simulations, Janhunen and Sandroos (2007) find $b = 3.09$ gives a good fit.

The stopping distance r_s is the distance protons travel as their kinetic energy is converted into electrostatic potential energy by the field. Solving the energy balance equation $e\phi(r_s) = (1/2)m_p v^2$, with $\phi(r = r_s)$ given by equation 13.10, we find

$$r_s = \frac{r_0}{\sqrt{\exp\left(\frac{m_p v^2}{e\phi_0} \ln\left[\frac{r_0}{r_w}\right]\right) - 1}}. \tag{13.13}$$

This stopping distance decreases monotonically and rapidly with increasing velocity. Inserting this into equation 13.12 gives the force acting on the wire,

$$F = \frac{bm_p n v^2 L r_0}{\sqrt{\exp\left(\frac{m_p v^2}{e\phi_0} \ln\left[\frac{r_0}{r_w}\right]\right) - 1}}.$$ (13.14)

We can use the above equation to calculate the acceleration of a sail away from the Sun under the action of the solar wind. As the solar wind ram pressure drops with distance from the Sun as $d^{-7/6}$, rather than d^{-2} for solar radiation pressure, electric sails appear favourable for exploring the outer solar. Studies have suggested that velocities of 40 km s^{-1} could be achieved in this way, which is an improvement over chemical rockets, but is insufficient for interstellar travel.

An alternative use of the electric sail is as a brake, because a sail moving at high velocity through the ISM experiences a drag force. Equation 13.14 shows that the force increases with increasing electric potential. It also increases with increasing wire radius, although the dependence is weak. Because the wire mass increases with the radius squared, we probably want very thin wires in practice.

The dependence of the force on velocity is more complex. It is shown in figure 13.4 for a wire of radius $r_w = 5\,\mu$m with potential $\phi_0 = 10^6$ V and electron temperature $T_e = 8000$ K, which is characteristic of the local ISM. The middle line assumes an electron number density of $n = 0.075\times10^6$ m^{-3}, which is appropriate for the local interstellar cloud with an ionization fraction of 0.25 (table 12.1). The force has its maximum value at around 0.01 c. This force is a product of the plasma ram pressure and the effective area of the barrier presented by the sail. At low velocities the force initially increases because the solar protons then have more momentum (pressure $\sim v^2$). But the stopping distance r_s decreases monotonically and rapidly with increasing velocity, meaning the effective barrier gets smaller. Above some velocity this effect dominates, and then the force decreases with increasing velocity, despite the increasing ram pressure.

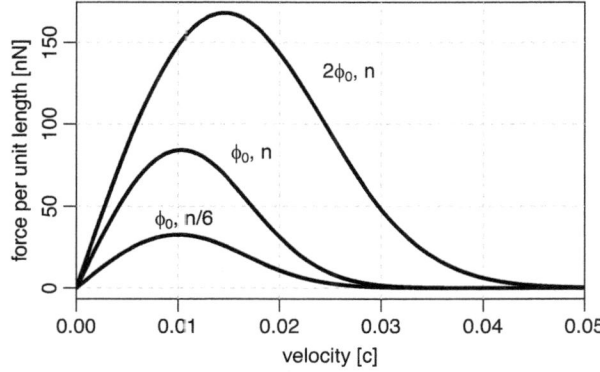

The variation of the force per unit length on an electric sail as a function of its velocity relative to the ISM (equation 13.14). The middle line assumes the sail is held at a potential $\phi_0 = 10^6$ V and that the electron number density in the ISM is characteristic of the local interstellar cloud with an ionization fraction of 0.25. The lower line is for a smaller plasma density (appropriate for the local bubble), and the upper line for twice the potential.

Fig. 13.4

The lower line in figure 13.4 shows the smaller force from a less dense ISM, and the upper line shows the larger force from a larger potential. For velocities greater than $0.05\,c$, the force becomes tiny in all cases shown. In order to shift the peak of the curve towards $0.1\,c$, and so to be able to decelerate the sail from this speed, we would need to increase the potential by two orders of magnitude, a tall order.

To determine how the velocity of a spacecraft of mass M varies with time due to the drag force, we use Newton's second law:

$$M\frac{dv}{dt} = -F(v) .$$

(13.15)

Integrating this from an initial spacecraft velocity v_0 at time $t = 0$, the time taken for the velocity to decrease to v is

$$t = -\frac{M}{bLr_0 m_p n} \int_{v_0}^{v} \frac{1}{v^2} \sqrt{\exp\left(\frac{m_p v^2}{e\phi_0} \ln\left[\frac{r_0}{r_w}\right]\right) - 1}\; dv .$$

(13.16)

This integral has no closed-form solution, but it can be solved numerically.

By way of illustration, we adopt the following parameters: $T_e = 8000\,\text{K}$, $n = 0.075 \times 10^6\,\text{m}^{-3}$, $r_w = 5\,\mu\text{m}$, and $\phi_0 = 2 \times 10^6\,\text{V}$, which corresponds to the upper line in figure 13.4. For the spacecraft we adopt $L = 50 \times 10^6\,\text{km}$ and $M = 10^4\,\text{kg}$. This is a huge length of wire. With a density of $1500\,\text{kg}\,\text{m}^{-3}$ its mass is $5890\,\text{kg}$. The remaining $4110\,\text{kg}$ is for wire shielding, current generation, and payload.

With initial velocity $v_0 = 0.05\,c$, I carry out a series of integrations at equally spaced velocity steps down to a final velocity of $30\,\text{km}\,\text{s}^{-1}$. The results are shown as an interpolated curve in figure 13.5. The initial deceleration is small. This is because at the large initial velocity, the drag force is small, as can be seen from figure 13.4. But after about 18 years, the velocity drops to a value where the drag is larger and the velocity drops much more rapidly. In total it takes 19.6 years to decelerate the spacecraft to $30\,\text{km}\,\text{s}^{-1}$, but it takes the first 17.6 years to decelerate to $0.04\,c$. Had the spacecraft started at $0.04\,c$, it would only have taken 2.0 years to decelerate to $30\,\text{km}\,\text{s}^{-1}$.

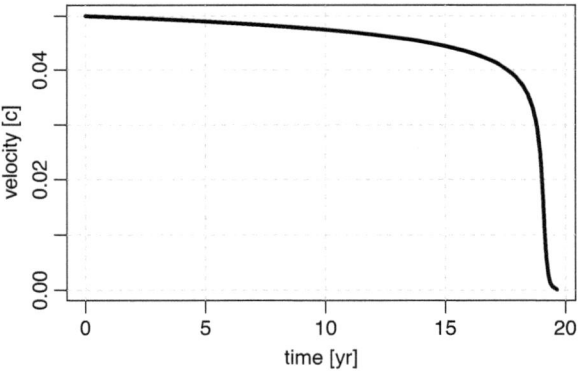

Fig. 13.5 The variation of the velocity of an electric sail with time due to the drag of the ionized ISM (equation 13.16) for $T_e = 8000\,\text{K}$, $n = 0.075 \times 10^6\,\text{m}^{-3}$, $r_w = 5\,\mu\text{m}$, $\phi_0 = 2 \times 10^6\,\text{V}$, $L = 50 \times 10^6\,\text{km}$, and $M = 10^4\,\text{kg}$.

The deceleration also depends strongly on the potential. When the potential is half as large, 1 MV, the deceleration from $0.05\,c$ takes 101 thousand years. On the other hand, when the potential is 3 MV, it takes just 1.9 years.

The deceleration time is linearly proportional to M/L. When the density of the wires is ρ_s, their mass is $\pi r_w^2 L \rho_s$. If the wires dominate the total mass of the spacecraft, the M/L term in equation 13.16 can be replaced with $\pi r_w^2 \rho_s$, in which case the deceleration time no longer depends on the wire length. But in reality a lot of extra mass is required for wire shielding, for the power supply for the electron gun to maintain the potential, and for the payload. So in practice there is a minimum wire length.

The sail would not be made of a single straight wire, as this may be broken by a dust impact, plus it would be rather unwieldy. We may instead arrange it as a set of many wires fanning out from the spacecraft in a plane at all angles, or in a flat grid as sketched in figure 13.3. When the wires are separated by the barrier width or less, the plasma essentially encounters a continuous potential barrier which it cannot pass through. We don't want the wires to be much closer than the barrier width, however, otherwise their barriers overlap and they will 'overly stop' the plasma, leading to a decrease in the amount of wire that acts as an effective barrier and so unnecessary wire mass.

As the width of the barrier increases rapidly with decreasing velocity, we need very large structures in order for all of the wire to be used effectively when decelerating at lower velocities. For the above configuration at $v = 0.01\,c$, the barrier width is $r_b = 130\,\text{m}$. If we arrange the wires on a square mesh with a separation equal to this, it would have a side length of $\sqrt{Lr_b/2} = 1800\,\text{km}$. Electric sails are big.

In conclusion, an electric sail can be used to decelerate a spacecraft from large velocities ($0.05\,c$) on a reasonable timescale (years). But to do so, we need an enormous length of wire (millions of kilometres) weighing many tonnes, together with large electric potentials (millions of volts).

13.2.1 Combining magnetic and electric sails

Perakis and Hein (2016) investigate braking a spacecraft using a combination of a magnetic sail and an electric sail. The magnetic sail is more efficient at decelerating from large velocities, so this is used initially, then ejected once the velocity is small. The remaining deceleration at lower velocities is done with a lighter electric sail. The study authors adopt a similar approach as in section 13.1 to compute the deceleration from the magnetic sail, but adopt a larger value of the maximum current density, $J_s = 2 \times 10^{10}\,\text{A m}^{-2}$, a larger loop radius, $a = 50\,\text{km}$, and a slightly lower ISM plasma density, $n = 0.05 \times 10^6\,\text{m}^{-3}$. They find that a spacecraft of mass 8250 kg can be decelerated from $0.05\,c$ to $35\,\text{km s}^{-1}$ in 29 years. Of this mass, 7500 kg is taken up by the deceleration system (10 times the payload), with just over two-thirds of this taken by the magnetic sail part. The magnetic sail is ejected at a velocity of $0.03\,c$ after about 14 years. Just using the magnetic or electric sail alone of the same mass (7500 kg) would have required 40 years and 35 years respectively to brake by the same amount. If, however, a final speed above a few hundred kilometres per second is acceptable, then they find that the magnetic sail alone is optimal.

13.3 Summary

Charged particles in the solar wind or ISM can be used to accelerate a spacecraft. A magnetic sail works by using an electric current in a superconducting loop to establish an artificial magnetosphere around a spacecraft. Moving ions are deflected by this magnetic field via the Lorentz force and exchange momentum with the sail. An electric sail works by positively charging long wires. These repel ions via the Coulomb force, thereby accelerating the sail.

Both magnetic and electric sails could be used to accelerate a spacecraft away from the Sun using the solar wind. Spacecraft velocities of 50–$100\,\mathrm{km\,s^{-1}}$ could probably be achieved in practice, but these sails are ultimately limited by the speed of the solar wind, because there is no force when the spacecraft is stationary relative to the wind.

An alternative and more promising use of these sails is to decelerate a high-velocity spacecraft using the drag on the ions in the ISM. A deceleration from 0.04–$0.1\,c$ almost to rest is theoretically possible over timescales of a few to a few tens of years using an electric sail. This requires a total wire mass of many tonnes, plus significantly more mass for the power supply and for shielding the wires against particle impacts. Magnetic and electric sails are therefore only feasible for massive spacecraft. But they are one of the few mechanisms available for significant deceleration at a target star.

Not knowing accurately the actual ISM ion density is a challenge for using such sails in practice. If the density is lower than expected, the deceleration will be lower. In the case of a magnetic sail, provision could be made to increase the current in the loop to compensate for this. But this implies that the nominal current is below the maximum possible, and so the loop is larger than necessary for the nominal mission. Such a margin comes at a high cost in performance if it is not required.

13.4 Exercises

1. Derive equations 13.5 and 13.6.
2. Investigate how the force per unit length on an electric sail varies with the plasma number density and the electron temperature in equation 13.14. Note that both variables appear in the expression for r_0. Do this for the nominal values listed in the text, for velocities of $0.01\,c$ and $0.05\,c$. Explain the results physically.

To fulfil its mission, a spacecraft needs to know where it is and how fast it is moving. This is the goal of navigation. Here we look first at how navigation is done for spacecraft in the solar system, using internal measurements as well as artificial signals sent from the Earth acting as navigation beacons. The latter has some relation to the concept of global navigation satellite systems such as GPS, which we also take a brief look at. At interstellar distances, signals from the Earth would be too weak and take too long to arrive, so interstellar spacecraft will have to navigate autonomously. We will see how this can be done using measurements of the positions or radial velocities of stars, by exploiting aberration and parallax, or by timing pulses from neutron stars.

14.1 The need for navigation

An interstellar spacecraft must arrive close enough to its target star to perform the intended science. Suppose that for a flyby mission we want to pass within 1 au of the star. For Proxima Cen at a distance of 1.3 pc, this means that the spacecraft's velocity vector, when launched ballistically from the solar system, needs to be accurate to better than $0.8''$. If our spacecraft is travelling at $0.085\,c$ (50 years to Proxima Cen), this means the velocity transverse to the line of sight to the star has to be less than $95\,\mathrm{m\,s^{-1}}$. This could be a real challenge for rocketless propulsion.

It is actually more complicated than this because stars are moving relative to the Sun. Proxima Cen, for example, has an angular velocity, or proper motion, of about $3.8''\mathrm{yr^{-1}}$, meaning that in 50 years it will have moved $193''$. So we have to target the spacecraft this much ahead of the current star's position. However, the current uncertainty in this proper motion is $0.074\,\mathrm{mas\,yr^{-1}}$, which accumulates to $3.7''$ in 50 years (1 mas = $0.001''$), much more than our targetting requirement. On top of this comes the fact that Proxima Cen is approaching the Sun at $22\,\mathrm{km\,s^{-1}}$, meaning that in 50 years it will be about 0.001 pc closer. Some targets are also in binary stars systems with relatively short periods (e.g. Alpha Cen A and B with a period of 80 years), and this too has to be taken into account. The upshot is that at the current level of accuracy, we cannot target a spacecraft to better than a few au at the target star if we just send it off ballistically with no means of changing its course.

Although the position and motion of the target system would be characterized better by the time an interstellar mission starts, there remain many factors that can alter the trajectory of the spacecraft, by altering the magnitude or direction of its velocity. One factor is the means of propulsion. A rocket may not perform quite as expected, or the laser power arriving at a sail may be slightly different from that predicted, or small-scale instabilities may cause the sail to veer slightly off course. If the Δv provided by the propulsion system differs from expectations by just 0.1%, then after 50 years of travel the spacecraft will arrive at the expected interception point 18 days too late or too early. As Proxima Cen is moving at $32\,\mathrm{km\,s}^{-1}$ relative to the Sun, it would then be $0.34\,\mathrm{au}$ away from its expected position. Interstellar magnetic fields acting on the spacecraft, the cumulative effect of particle impacts (drag), and of course the gravity of the target star will change the spacecraft's course and velocity. Finally, during the long travel time, we on Earth are likely to learn a lot more about the target system, such as the presence and orbits of planets, so we might want to steer the spacecraft to a different flyby or rendezvous point within the target system.

All this means that we need to have some control over the velocity of the spacecraft at multiple points during the mission. Aside from making velocity changes, knowledge of the spacecraft's position and velocity are essential for interpreting its science results (e.g. the variation of particle flux as a function of stellar distance), for determining when to switch on instruments, and for knowing in what direction the Earth is for transmitting data.

Navigation is the process of determining the three-dimensional (3D) position and 3D velocity of the spacecraft in some reference frame. This frame is typically the International Celestial Reference Frame (ICRF), an inertial reference frame centred on the solar system barycentre (SSB). Generally speaking, navigation also needs some measure of time. As relativistic time dilation is likely to be significant, this is typically the time in the ICRF (at the SSB, properly known as the barycentric coordinate time, TCB), but it could be the time in the spacecraft's rest frame. The spacecraft will also need a means of determining its orientation in the reference frame, so as to point instruments, antennae, or engines in the right direction.

14.2 Inertial navigation

The idea of inertial navigation is to determine the change in position, velocity, and orientation of a spacecraft without using external references.

When a spacecraft moves away from a reference point with known constant velocity in a known direction, determining the future position is a simple matter of measuring the time of travel with an onboard clock. This is known as *dead reckoning*. The error in the determined position comes directly from the error in the initial velocity and orientation and in the measured time. If moving at a relativistic speed, this calculation needs to accommodate the time dilation (or length contraction) to determine the distance travelled in the reference system's rest frame.

14.2.1 Accelerometers

When a spacecraft is accelerating, this can be measured without external reference using an accelerometer. This is essentially a mass on a spring that can move freely within a rigid case and works via the principle of inertia (figure 14.1). In a non-rotating rocket in free space that is not firing its engines, the mass and spring hang freely and there is no acceleration. When the rocket fires its engines, the spring will compress or expand by a length Δx. Knowing the

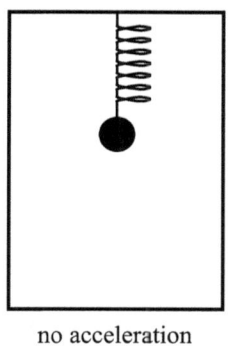
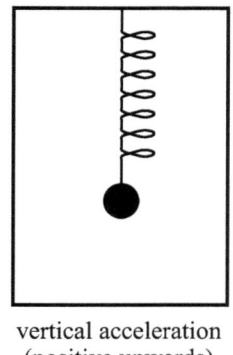

no acceleration
(free fall)

vertical acceleration
(positive upwards)

The principle of an accelerometer.   Fig. 14.1

spring constant k and the mass m, we compute the acceleration as $(\Delta x)k/m$ from Hooke's law. By orienting three such accelerometers perpendicular to one another, the acceleration in the three spatial directions can be measured. Integrating the acceleration over time we can determine the velocity over time, and by integrating again we establish the position as a function of time.

Accelerometers are insensitive to uniform gravity because gravity acts on all parts of the device equally. Thus an accelerometer in free fall, for example in a Keplerian orbit about a point mass, will read zero acceleration ($\Delta x = 0$). Likewise, the accelerometer on a skydiver who has just jumped out of a plane will read zero (neglecting air drag). Note that in both cases there is a net force acting on the accelerometer. The orbiting accelerometer is accelerating in the sense that its velocity relative to the central mass is changing with time. But this does not register on the accelerometer. In contrast, an accelerometer at rest on the surface of the Earth reads $9.8 \, \mathrm{m \, s^{-2}}$ in the upward direction, because there is now a reaction force from the ground, which is transmitted through all the rigid parts of the accelerometer case. In this situation we tend to think of gravity as pulling the mass down to stretch the spring, but of course gravity acts on every other part of the accelerometer too. But in contrast to the rigid parts, there is no reaction force from the ground on the mass, which is free to move and stretch the spring. The same happens in the situation with the accelerometer in the accelerating rocket, and the equivalence of these two is in fact one statement of the equivalence principle. It is the reaction force that makes us feel the effects of gravity when we are standing on the surface of the Earth. Whether we are skydiving in a gravitational field or are in a non-accelerating rocket in deep space, there is no reaction force and so we feel no gravity.

In general relativity, gravity is not considered as a force, but is rather a result of the curvature of spacetime. If there are no (other) forces acting, then objects are unaccelerated and move along so-called geodesics, an example of which is a Keplerian orbit. The apparent force of gravity only becomes measurable when something prevents motion along a geodesic, such as the reaction force of the surface of the Earth when we stand on it. Because they work on the principle of inertia, accelerometers measure the proper acceleration rather than the coordinate acceleration. As discussed in section 7.4, coordinate acceleration is a relative (frame-dependent) concept, like velocity (see section 7.2). But proper acceleration, in the sense of being the acceleration relative to a geodesic, is in some sense absolute.

14.2.2 Gyroscopes and orientation

Knowing and controlling the orientation, or attitude, of a spacecraft is important for many reasons, such as pointing instruments, antennae, and engines in the right direction. It is also needed for dead reckoning, so that the integration of the acceleration and/or velocity is done in a known direction.

Gyroscopes are used in some aircraft to determine orientation. A mechanical gyroscope is a spinning wheel mounted such that its spin axis is free to move in all three dimensions without any torque acting on it (figure 14.2). The angular momentum of the wheel is therefore conserved, so even if the vehicle in which the wheel is mounted rotates, the axis of the wheel remains pointing in a constant direction. By measuring the change in orientation of the vehicle relative to the axis, we can determine the change in the vehicle's orientation.

Mechanical gyroscopes inevitably spin down over time due to friction, so need to be spun up from time to time. An alternative to a mechanical gyroscope is the fibre optic gyroscope, shown schematically on the right of figure 14.2. This comprises a piece of fibre optic wrapped into a circle in a plane. Light from a laser beam is injected into the coil at a point, split into two using a half-silvered mirror, and each part sent in opposite directions through the fibre optic. If the coil is not rotating about the axis perpendicular to the plane, the distance travelled by each beam is the same: once around the circle. Hence the beams arrive back at the injection point still in phase. If the coil rotates clockwise by some angle during the travel, then because the photons are not dragged along with the medium, the beam travelling in the clockwise direction must travel further before it arrives back at the injection point. The anticlockwise beam, in contrast, travels less than the full circle. The beams will now be out of phase when they arrive back at the injection point. The phase difference, measured by interferometry, is a measure of the rotation angle in a given time period. By monitoring the change in the angle continuously over time, we can not only track the orientation but also determine the angular velocity and acceleration.

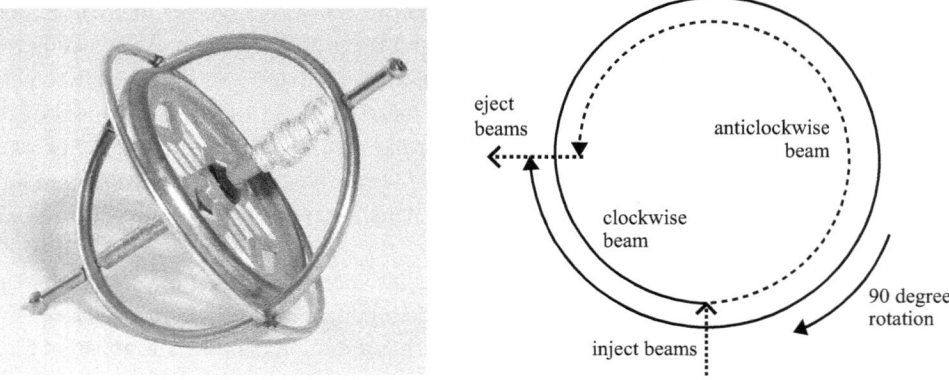

A mechanical gyroscope (left; licensed under CC0-1.0, from Wikimedia Commons) and the principle of a fibre optic gyroscope (right).

This principle underlying the fibre optic gyroscope is called the Sagnac effect and works by exploiting the constancy of the speed of light in a given medium. In order to build up a more easily measurable phase shift, a fibre optic gyroscope in practice comprises a long (kilometres) loop of fibre optic coiled up many times to increase the difference between the path lengths. Despite this, fibre optic gyroscopes are typically compact (centimetres in size) and, by having no moving parts, are more robust than mechanical gyroscopes.

Often a set of three mutually perpendicular accelerometers and three gyroscopes are combined together into a single unit called an *inertial measurement unit* (IMU). IMUs are used extensively for navigation in cars, ships, aircraft, and missiles, often as a complement to satellite navigation (see section 14.6), in particular when satellite signals are not available, such as in tunnels or from fear of interference. Many smartphones and digital wearables for fitness also have IMUs.

14.2.3 Clocks

Onboard clocks are vital for many purposes. They are an essential part of inertial navigation for integrating the linear accelerations measured by an accelerometer. As we will see below, accurate clocks are also required for pulsar navigation, and in satellite navigation systems.

Wrist watches that use a quartz oscillator to maintain time are accurate to about 10 s per year, or one part in 3×10^6. A watch or phone connected to the internet appears to be more accurate than this because it can be regularly synchronized to a reference clock.

The most accurate manufactured clocks are atomic clocks, which exploit the universality of the energy levels in atoms. Electrons transitioning between energy levels release photons of very precise frequency. The constancy of this is even used in the SI definition of the second. Atomic clocks use the hyperfine transitions between energy levels that are split due to electromagnetic interactions between the nucleus and the electrons. These transitions are in the microwave region. In the case of the definition of the second, it is a 9 GHz transition in caesium-133.

A NASA deep space atomic clock based on the hyperfine transition of mercury. Credit: NASA.

Fig. 14.3

An atomic clock commonly used in spacecraft is based on a transition in rubidium-87, which results in an accuracy of 1 part in 10^{12}, or $30\,\mu s\,yr^{-1}$. Another technology is hydrogen maser clocks that use the 1.4 GHz (21 cm) transition in hydrogen. These are used in the Galileo satellites for Earth-based navigation, where they have an accuracy of 0.45 ns in 12 hours, or 1 part in 10^{14}. A more recent development is based on mercury–iron with an accuracy of 1 part in 10^{15}, or $0.03\,\mu s\,yr^{-1}$. Perhaps the best accuracy achieved so far in an experiment – not yet in a deployed clock – is based on strontium that has achieved an accuracy of 1 part in 10^{18}.

14.2.4 Use in space travel

The main disadvantage of dead reckoning for navigation is the steady accumulation of errors with time. Commercial accelerometers for navigation have accuracies of order $\delta a = 10^{-4}\,m\,s^{-2}$. Over a period of time t at constant acceleration, the errors in velocity and distance from dead reckoning are $\delta a t$ and $(1/2)\delta a t^2$, respectively. These are $0.4\,m\,s^{-1}$ and 650 m, respectively, after one hour, and $3.2\,km\,s^{-1}$ and 0.3 au after one year. These are too high for many applications.

Nonetheless, if we can get periodic external determinations of position and velocity, then inertial navigation can be reset and used to provide updates to the position and velocity over the shorter timescales between updates. Inertial navigation was used on the Apollo missions to take humans to the Moon, for example, in particular for the periods when the spacecraft had no contact with the Earth. (The astronauts were also provided with sextants to enable a manual determination of the orientation of the spacecraft.) IMUs are frequently used even in modern spacecraft, for example in the New Horizons spacecraft that encountered Pluto in 2015. Inertial methods are particularly useful during impulsive manoeuvres, when the ranging techniques that we'll discuss in the next section would take too long to provide a new determination.

14.3 Navigation in the solar system by ranging

Navigation of spacecraft in the solar system is routinely assisted by tracking them from the Earth. By sending a signal to a spacecraft, which it then returns, the spacecraft's distance from the transmitter can be computed from the time delay. This is called *two-way ranging*. The radial velocity is determined from the Doppler effect. One-way ranging – in which the time delay over just one leg of the journey is measured – can also be used, but this suffers from any discrepancies between the clocks on the spacecraft and ground station, so is generally less accurate. It may be necessary to use this for distances more than several light hours, however, because the ground station may lose line-of-sight contact to the spacecraft as the Earth rotates. Alternatively a second ground station could be used, to give three-way ranging.

For sending ranging signals, ESA and NASA generally use X-band radio at 7–8 GHz. Sometimes the higher-frequency Ka-band (32–34 GHz) is used as this is dispersed less by the ionosphere and solar wind (recall the plasma frequency discussed in section 12.3.1), but it is more attenuated by dust and water in the troposphere. Using a 35 m radio telescope with a minute of exposure time to collect many signals, the accuracy of these ranging systems is about 1 m in distance and 0.1 mm s^{-1} in radial velocity. We need to know the radial velocity v to get an accurate range, because if the spacecraft is at a distance d, the light travel time is $t = 2d/c$, during which time the spacecraft moves $vt = 2dv/c$. With $v = 1$ km s^{-1} and $d = 1$ au, for example, this is a displacement of 1000 km.

The accuracy of the angular position of the spacecraft determined from the radio telescope during ranging is relatively poor. The spacecraft's transverse position can be determined more accurately by measuring the radial distance and velocity as a function of time, and then fitting a model for the motion of the spacecraft. This model will typically be a Keplerian orbit (if the spacecraft's engines are not firing), although deviations from this due to the non-sphericity of the Earth or planetary perturbations, for example, may also have to be fitted. During critical parts of a mission, such as orbit insertion or target encounter, this modelling may not be accurate, especially if the position and velocity are changing rapidly.

Angular coordinates can be determined better for distant spacecraft using *differential one-way ranging* (DOR). Two ground stations A and B with known positions receive a signal from the spacecraft (figure 14.4, left). The difference in the arrival times Δt gives the direction towards the satellite in the plane containing the spacecraft and the two ground stations through simple trigonometry:

$$\cos \theta = \frac{c\Delta t}{b} . \tag{14.1}$$

This assumes that the wavefronts arriving at the two ground stations are parallel, which is a good approximation if the distance to the spacecraft is much larger than the separation between the ground stations.[1] The second angular coordinate (out of the plane) can generally be found by waiting for the baseline to rotate as the Earth rotates, or by using a third ground station to define a second plane. This method is called one-way ranging because it does not

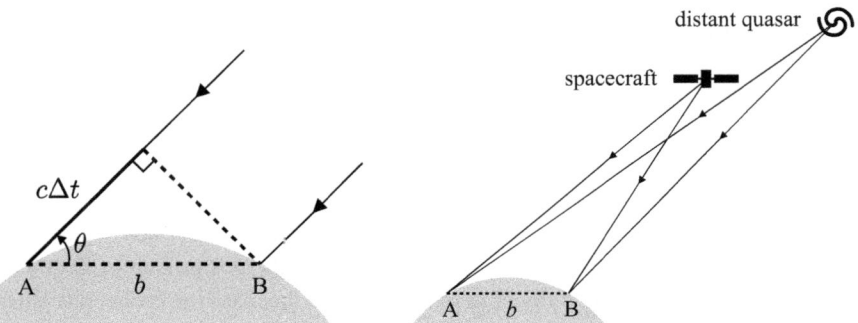

An illustration of DOR (left) and delta-DOR (right) for finding the angular coordinates of a spacecraft. The right diagram is not to scale. The distances to the spacecraft and quasar are much larger, such that the incoming wavefronts from the spacecraft to A and B are virtually parallel, and likewise for the quasar, such that the geometry in the left panel still holds to a good approximation. The angular separation between the spacecraft and quasar is of the order of a few degrees.

Fig. 14.4

---

[1] If we had no idea of the distance or direction to the sources, then a measured difference in the arrival times would only allow us to place the spacecraft on a hyperbola with the semi-major axis equal to half the ground station separation. This is the basis of *hyperbolic navigation*, which has been used historically for terrestrial navigation over small enough areas such that the Earth's curvature can be neglected. A third 'ground station' in the plane constrains the position to the interception of two hyperbolae, which is just two points, additional information then normally being used to choose between them.

rely on the signal being sent back, and it is differential because it only uses the difference in the path lengths. As the baseline is very large, often transcontinental, this is also called very long baseline inferferometry (VLBI), which is commonly used in astronomy to get high-resolution radio maps rather than for ranging (see section 11.2.2).

The accuracy of DOR is limited by a number of factors. When the spacecraft is at a finite distance, the incoming wavefronts are not parallel and the time delay does not strictly follow from the simple geometry in the left panel of figure 14.4. The error in the angular position from making this assumption is very small, however (see exercises). When the ground stations are separated by 1 Earth radius, and the spacecraft is 0.1 au away (corresponding to a parallax between the ground stations of $88''$), then when the spacecraft is positioned at $\theta = 45°$ from the baseline (which maximizes the error), the angle θ inferred from equation 14.1 is wrong by only 2.3 mas. With the spacecraft at 0.5 au, this drops to 0.1 mas.

More significant factors limiting the accuracy of DOR include clock timing errors and instrument noise, as well as phase delays introduced by the fact that the waves take different paths through the solar plasma and the Earth's atmosphere to the two ground stations. These problems can be reduced by repeating the measurement on a source with well-known angular coordinates that lies a small angular distance (order of degrees) from the spacecraft as seen from the Earth. This is illustrated in the right panel of figure 14.4. Typically quasars with positions known to a fraction of a milliarcsecond are used. Quasars don't send pre-arranged signals, of course. But their intensities are not constant, and although their variations are essentially white noise over short timescales, it is nonetheless the same signal that is received at the two ground stations, just delayed in time. So by correlating the two signals, the difference in arrival times can be determined. By taking a DOR measurement on the spacecraft and then on the quasar, then taking the difference, many of the systematic error sources cancel. This technique is known as *delta-DOR*.

Delta-DOR has been used in various interplanetary missions, for example NASA's Mars Science Laboratory (the Curiosity rover), which arrived at Mars in 2012. This achieved a timing uncertainty of 0.06 ns over an 8000 km baseline between ground stations, which corresponds to an angular uncertainty of 2.2 nrad, or 0.5 mas (using the error propagation of equation 14.1). This in turn corresponds to a transverse position accuracy of 350 m for a spacecraft 1 au from the Earth (the distance is found by two-way ranging).

Beyond determining the position and velocity of the spacecraft relative to the ICRF, there will be times when we also need to accurately determine the position and velocity relative to the target, for example during flyby or at rendezvous. In that case the spacecraft must observe the target directly. As an example, New Horizons used its onboard cameras to determine the direction to Pluto (and its largest moon Charon), during its flyby in 2015. Together with onboard IMU data and Doppler ranging information from the Earth, this led to an improved determination of the spacecraft velocity and position relative to Pluto. The velocity in three dimensions was then established to within 1 mm s^{-1}, and the two-dimensional position transverse to the line of sight to Pluto was established to an accuracy of $1'' \times d$ in each direction, where d is the Pluto–spacecraft separation, $1''$ (5 μrad) being the resolution of the camera. This corresponds to 5 km when $d = 10^6$ km, for example. The sixth component – the distance to Pluto – was less well determined. Of particular importance was the remaining time to encounter, and here the acceleration of the spacecraft due to Pluto's gravity had to be taken into account. The imaging data taken from the spacecraft allowed the uncertainty in this to be reduced to about 20 s, one day before the encounter.

The Pluto flyby was a success, passing just 12 500 km above the surface of Pluto at a relative velocity of 14 km s^{-1} when it was 34 au from the Sun. Four years later in 2019 this feat was bettered, when New Horizons steered to within 3500 km of the Kuiper belt object 486958 Arrokoth.

14.4 Stellar astrometric navigation

Interstellar spacecraft will be too far from the Earth for the use of ranging methods. They instead have to navigate autonomously. Here we will see how, by measuring only the angular positions of stars, a spacecraft can establish its 3D position and 3D velocity.

14.4.1 Principle

To determine our position in a coordinate system, we need to determine our location relative to one or more objects of known position in that coordinate system. Consider two such objects, or beacons, on the surface of the Earth, shown as A and B in figure 14.5. By measuring the angle to these relative to a reference direction (e.g. north, perhaps found using a compass), we can determine our position S uniquely. If we do not know the reference direction, then a third beacon – C in the figure – can effectively take its place (it does not have to lie in the reference direction). In three dimensions, the number of beacons required depends on the type of measurement we make. If we can only measure the angles between pairs of beacons, then provided we do not lie in a common plane with all three beacons, we can measure three independent angles and from this determine our 3D position. If we instead can measure the two angular coordinates (like right ascension and declination) to a beacon, then from two beacons we can acquire four measurements, which suffice for us to determine our position.

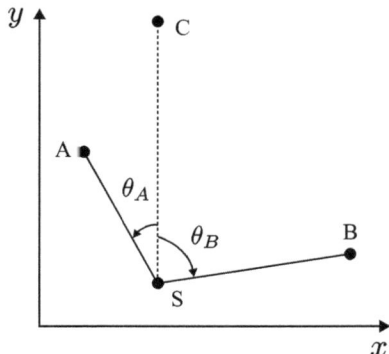

Navigation by beacons in two dimensions to find the position of S. If we know the positions of two beacons, A and B, and measure the angles θ_A and θ_B between them and a reference direction (dotted line), the location of S can be found.

Fig. 14.5

We can use these ideas to determine our position in interstellar space. Specifically, by measuring just the angular positions of stars with known 3D positions in the ICRF, we can determine our position in the ICRF. For this to work we need accurate 3D positions of those stars.

For over two thousand years, astronomers have catalogued the two angular coordinates – right ascension and declination – of stars in the sky. Only much more recently could the third dimension – the distance – be determined with any precision. The fundamental way to do this is with parallax, which exploits the annual motion of the observer. As the Earth moves around the Sun, the apparent position of a nearby star shifts with respect to very distant stars (or even better, extragalactic quasars). As we know the size of the Earth's orbit very accurately, the measured angular shift – the parallax – gives the distance to the star via simple trigonometry.

The nearest stars have parallaxes below $1''$, so their shifts were unobservable to early astronomers. It was not until the 1830s that the first statistically significant measurements were reported, first by Friedrich Bessel in 1838 with a measurement of 314 ± 20 mas for the binary star 61 Cygni. Parallaxes of more stars with ever-increasing accuracy were collected during the twentieth century, culminating in the first mission dedicated to astrometry from space. The satellite Hipparcos measured parallaxes for 120 000 stars to an accuracy of around 1 mas. This was followed by Gaia, which between 2014 and 2025 observed around two billion stars, achieving parallax uncertainties as low as 10μas, and effectively extended the horizon of good parallaxes from around 100 pc to a few kiloparsecs.[2]

As a result of these surveys, we now know the distances of all but the faintest stars within 100 pc to within 0.3% (median accuracy), with a higher accuracy for the nearer and brighter stars. A 0.1% distance accuracy for a star at 10 pc is about 2000 au, for example. The angular positions of the stars are typically known to within a few microarcseconds, so their transverse positions are known much better than their distances: 10μas at 10 pc corresponds to 10^{-4} au, or just 15 000 km. By the time interstellar spacecraft are built, we can assume that the distances to potential navigation stars will be determined even more accurately.

These nearby stars can be used as beacons by an interstellar spacecraft to determine its position in space. This is complicated, however, by the fact that the spacecraft is moving with respect to the stars. As we saw in section 9.1, even a small non-relativistic velocity introduces a significant aberration. As we would not know the velocity of the spacecraft with respect to each star, it seems we could not account for this. However, if we know the velocities of the stars relative to the ICRF, then we can turn the problem around and use the measured aberration to determine the velocity of the spacecraft (if we also know the position of the spacecraft, that is; we'll get to this circular problem shortly).

Fortunately, we do know the 3D velocities for many stars. As part of its astrometric observations, Gaia also determines the proper motions – angular velocities – of the stars. It has to do this, in fact, in order to account for the motion of the stars over the several years of its observations. The median accuracy for each proper motion component for stars within 20 pc is about 30μas yr^{-1}, which for a star at 20 pc corresponds to a transverse velocity accuracy of just 3 m s^{-1}. Gaia also measures the radial velocities for many stars to an accuracy of a few kilometres per second, although these can also be obtained from ground-based observations, which can achieve an accuracy of a few tens of metres per second.

---

[2] In its third data release (DR3), Gaia determined the parallaxes of the two components of 61 Cygni to be 285.995 ± 0.060 mas and 286.005 ± 0.029 mas. Bessel was off by less than two sigma.

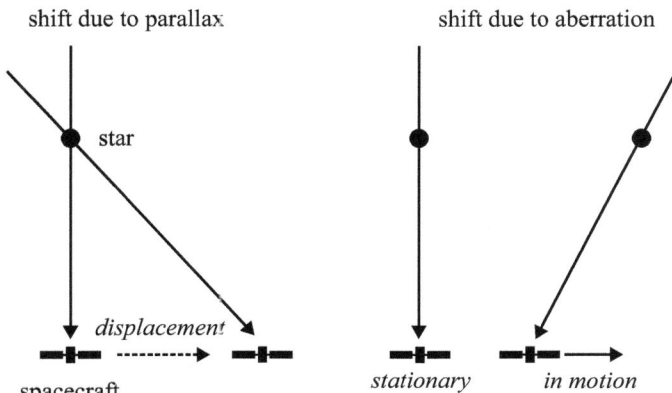

shift due to parallax

shift due to aberration

star

displacement

spacecraft

stationary *in motion*

Parallax and aberration induce different shifts in the observed star location. In both cases the star initially appears straight up from the stationary spacecraft. Left: When the spacecraft has moved to the right (and is again at rest), the star appears to move to the left, by an amount that depends on the size of the displacement and the distance to the star. Right: If the spacecraft acquires a velocity, then before it has even moved, the star appears to shift to the right due to aberration (see figure 9.1) by an amount that depends on the velocity but is independent of distance.

Fig. 14.6

The overall problem, then, is to determine both the 3D position and 3D velocity of the spacecraft in the ICRF from the known 3D positions and 3D velocities of a number of stars in the ICRF. When a spacecraft is stationary relative to the ICRF, but displaced some distance away from its origin (the SSB), the observed positions of the stars shift because of the parallax effect. But the finite velocity of the spacecraft also creates shifts due to aberration. To interpret the shifts as due to aberration, and thus velocity, we would need to know the spacecraft's position. And to interpret the shifts as due to parallax, and thus position, we would need to know the spacecraft's velocity. However, the effects of aberration and parallax are not the same. The size of the parallax depends on the distance to the source, whereas the size of the aberration does not. Furthermore, a displacement of the spacecraft in a given direction, and a velocity of the spacecraft in the same direction, give rise to parallax and aberration shifts with opposite signs, as illustrated in figure 14.6. By measuring the apparent shifts in the positions of several stars, we can disentangle the effects of parallax and aberration and determine both the 3D position and 3D velocity of the spacecraft simultaneously.

14.4.2 Implementation and simulation results

The technique of navigation outlined in the previous section has of course never been applied to interstellar travel. We therefore look at the results of a simulation.

Because of the interdependence of the parallax and aberration, and the inherent nonlinear relationship between the measured angles and the spacecraft position and velocity, an iterative approach is used, as summarized in figure 14.7. It works as follows. We start off by assuming a 3D position ($\mathbf{x}$) and 3D velocity ($\dot{\mathbf{x}}$) of the spacecraft. Then, from the catalogued ICRF coordinates of the beacon stars, we calculate the effects of parallax and aberration to predict the angular positions of the stars that we expect to observe from the spacecraft. We then compare these predicted angular positions to the measured angular positions (which are

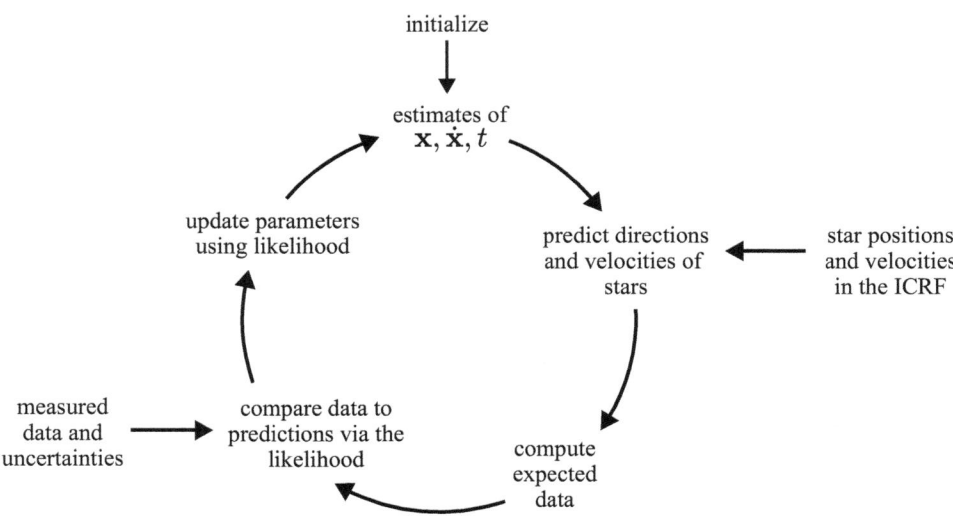

Fig. 14.7 Stellar astrometric navigation. An iterative approach is used to estimate the 3D position (**x**) and 3D velocity of the spacecraft ($\dot{\mathbf{x}}$) from measurements of the angular separations of pairs of stars (and/or their radial velocities).

simulated here from their true positions with measurement noise added). Then we adjust the spacecraft position and velocity in such a way that aims to minimize the difference between the predictions and measurements. The whole process is iterated until the predictions converge on the measurements, which occurs when the true spacecraft position and velocity has been found (to within the noise).

As there is no angular coordinate grid printed on the sky, we cannot directly measure angular positions of stars, in the sense of right ascension and declination in a coordinate system. In practice we can only measure with a high degree of accuracy the angular separation θ between a pair of stars, as illustrated in figure 14.8. We therefore convert our predicted angular coordinates into predicted angular separations, and compare these to measurements. The angular separations can be measured using a pair of telescopes, or better, using a single camera with two different fields of view. This is in fact the principle of a sextant, traditionally used to determine the angle of the Sun or stars above the horizon (figure 14.9).

There are various ways to implement the iterative scheme, but a good approach is to compare the predictions and the data via a probabilistic likelihood, as this takes into account the measurement uncertainties. In the simplest case, the likelihood is just the product of one-dimensional Gaussian distributions, each one describing the probability of the measurement given its true but unknown value (the one we are trying to infer). The space of the parameters – **x** and $\dot{\mathbf{x}}$ – can then be explored with a Monte Carlo method to determine the relative probabilities of different solutions. This produces a probability distribution over the parameters, which we can summarize with a point estimate and uncertainties. An alternative approach is to use an optimization algorithm to search the space to locate just the maximum of the likelihood.

To determine **x** and $\dot{\mathbf{x}}$ we need to measure at least six independent angular separations. We can do this by measuring the angular separation between six stars and a seventh reference star. The reference star is arbitrary, but the Sun is a natural choice because for communication

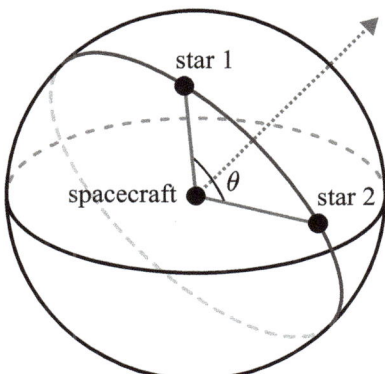

Stellar astrometric navigation relies on measurements of the angular separation θ between pairs of stars. It does not need the separate angular coordinates (right ascension and declination) of the stars.

Fig. 14.8

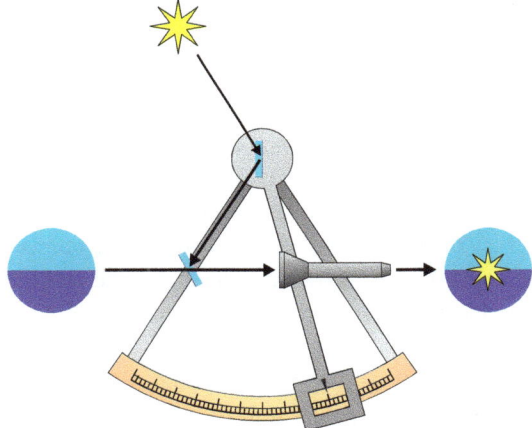

A sextant uses a half-silvered mirror to superimpose the images of two objects – here the horizon and the Sun – onto a common focal plane in order to measure the angular separation between them. Credit: Public domain, from Wikimedia Commons, modified by the author.

Fig. 14.9

purposes, at least, we always need to know where it is. But there is no lack of suitable – bright, nearby – stars, and having more stars spread out evenly over the sky is beneficial.

Note that in this method we do not need to determine either the distances to, or the velocities of, the stars from the spacecraft. These cannot anyway be determined instantaneously from onboard the spacecraft. Only the radial velocity component of the stars can be measured quasi-instantly, using a spectrograph to measure the Doppler shift of the stellar spectral lines. This could, in fact, be used as an alternative measurement to the angular separations (or in addition to them), as it too encodes the position and velocity of the spacecraft via parallax and aberration. But we won't consider it here.

How well does this method perform? Bailer-Jones (2021a) performed various simulations to determine this. As expected, the performance depends on both the number of stars used and

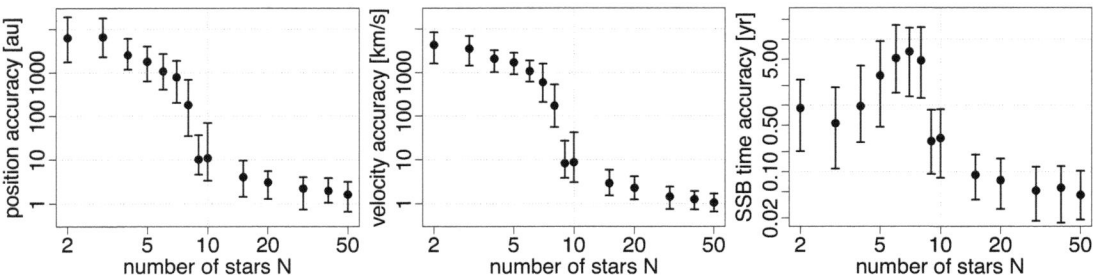

Astrometric stellar navigation. This shows how the spacecraft navigation accuracy (the error in the estimation) varies with the number of stars used, for the overall spacecraft position (left), the magnitude of the 3D velocity (centre), and the SSB (= ICRF) time (right). This assumes angular separations between pairs of stars are measured at a very modest accuracy of 1″. The navigation accuracies scale linearly with this measurement accuracy, so for example they would all improve by a factor of 1000 with measurements accurate to 1 mas. Credit: Adapted from figure 11 of Bailer-Jones (2021a) published under a CC-BY-3.0 licence.

the accuracy of the onboard measurements. These simulations first assume that the angular separations between the N nearest stars to the Sun are measured to a very modest accuracy of 1″. This can easily be achieved with a very small telescope aperture ($\sim$ 1 cm). For a given N, many simulations were made with the spacecraft at different positions up to 10 ly away, and moving at velocities up to $0.5\,c$. For each simulation, the iterative method was then used to infer the position and velocity of the spacecraft and this compared with the true values, from which the error in the position and velocity was computed.

Figure 14.10 shows how the performance varies with N. The points show the median error over a set of simulations, and the bars show the spread over the set (16th and 84th quantiles). We see that with $N = 10$, the position of the spacecraft can be determined to within 10 au (left panel) and the magnitude of the velocity to within $10\,\mathrm{km\,s}^{-1}$ (centre panel). With more stars the performance improves roughly as $N^{-1/2}$. For smaller N the behaviour is more complex. As there are seven unknowns to determine – three positions, three velocites, and the time (as I will explain shortly) – we need at least seven measurements, that is, $N = 8$ stars. With fewer stars we see a big increase in the errors, as we formally cannot solve for all the parameters (although there is still some information, and as the problem is not linear there is not a complete degeneracy).

A positional knowledge of 10 au may be adequate for some interstellar navigation purposes (Proxima Cen is 269 000 au away) but is too poor for injecting a spacecraft into a star system, where we may want to pass within 0.1 au or less of a planet. One solution is to use more stars, as figure 14.10 shows. However, further simulations show that the position and velocity improve directly in proportion to the onboard angular measurement accuracy. So if angular separations can be measured to 1 mas between 10 stars, then we could determine the position and velocity to within 0.01 au and $10\,\mathrm{m\,s}^{-1}$, respectively.

The above explanation of the method glosses over one aspect. The stars are moving relative to the ICRF, so their expected positions relative to the spacecraft depend on the time elapsed since their ICRF positions were recorded in the catalogue. Over the course of 50 years, a nearby star moving at $30\,\mathrm{km\,s}^{-1}$ relative to the ICRF moves 0.0015 pc. If the star is at 20 pc and its movement is entirely in the transverse direction, then even if the spacecraft is still at the SSB (the origin of the ICRF), the star is displaced by 16″, which is much more than our measurement accuracy. To compensate for this we need to know the time of the spacecraft's observations.

Time can be measured precisely using an onboard clock (section 14.2.3). But if our spacecraft has been moving at relativistic speeds, the onboard clock will no longer show ICRF time due to time dilation. So if we cannot keep track of all the velocity changes, it may not be possible to know ICRF time onboard our spacecraft with sufficient accuracy. It turns out, however, that if we observe one additional star, we can solve for the ICRF time as part of the iterative method. The right panel of figure 14.10 shows that the time is not actually determined very precisely in this way: to the order of just a few weeks once we have more than about 10 stars with $1''$ measurement accuracy. This is approximately the time it takes the stars to move the distance of the angular measurement uncertainty, so we cannot – and need not – resolve the time to better than this. When we have onboard measurement accuracy of 1 mas, then we can determine the ICRF time to within about an hour. This will prove useful for tracking the position of the Earth for communication purposes (see section 15.3).

It turns out that this astrometric navigation method is very insensitive to the initially assumed position, velocity, and time required to initialize the calculations. The performance is also largely independent of the true positions or velocity of the spacecraft. It should, therefore, be quite reliable.

This method of navigation works by using the known positions and velocities of stars to determine the position and velocity of the observer (spacecraft). The Gaia astronomical survey spacecraft mentioned earlier does the opposite. It uses the known position and velocity of the spacecraft – its orbit in the ICRF – to determine the positions and velocities of the stars. It does this too by measuring angular separations between the stars.

One practical aspect we have not yet considered is the need to identify the correct stars so we can match them with their catalogue data. This is a standard task for satellites, and is solved by the use of star trackers. These are wide-field cameras coupled with software that can identify patterns of the brightest stars (asterisms, typically as triangles). If our spacecraft travels far compared to the typical distances to the beacon stars, the asterisms will change significantly and eventually become unrecognizable (and it is in fact this change that we are exploiting in the navigation). This could be addressed by keeping track of the directions to the relevant stars during the course of the mission. If there is a spectrograph on the spacecraft, the spectral signature can also help identify the stars from their characteristic line features and continuum shapes.

14.5 Pulsar navigation

We now turn to another way of using astronomical objects to navigate autonomously. This makes use of the fact that some objects send out very regular pulses, although any highly regular predictable signal would do. Observations of several such objects in different directions can be used to constrain the position of the spacecraft.

14.5.1 Pulsars

Once nuclear fusion ceases in stars, they evolve in to one of several end states. Stars with masses between $8\,M_\odot$ and $30\,M_\odot$ explode as supernovae, shedding much of their mass. With insufficient radiation pressure to counter the pull of gravity from the remaining mass, they collapse. The gravity is strong enough to overcome the electromagnetic forces that otherwise prevent less massive objects such as planets from collapsing. The electrons create a large degeneracy pressure that, if the stellar mass were low enough, would balance gravity to leave what is known as a white dwarf. But in higher-mass stars, the gravity is large enough even to overcome the electron degeneracy pressure, and the temperature increases enough to split the nuclei via photodisintegration, forcing protons to interact with electrons to produce neutrons. Further collapse is finally stopped by the repulsive strong nuclear force and the neutron degeneracy pressure to create a neutron star. If the mass were even larger, then even this balance would eventually be overcome, and the end state would be the singularity of a black hole.

Neutron stars typically have masses between $1.2\,M_\odot$ and $2.0\,M_\odot$ and radii of $10\,\mathrm{km}$, which gives them an enormous average density of the order of $10^{18}\,\mathrm{kg\,m^{-3}}$. Each cubic centimetre has a mass of nearly a billion tonnes. All stars rotate, and now that the star is orders of magnitude smaller than before it collapsed, conversation of angular momentum produces a very rapid rotation. Typical rotation periods range from milliseconds to seconds. Some neutron stars also possess strong magnetic fields – 10^4 to $10^8\,\mathrm{T}$, up to $10^{11}\,\mathrm{T}$ for magnetars – and emit radiation at a wide range of wavelengths (gamma to radio) along their magnetic poles, with a spreading angle of a few degrees. If the magnetic axis and spin axis are misaligned, the magnetic pole will sweep out a circle over the sky, and any observer positioned in its path will see regular pulses of radiation from the neutron star (figure 14.11). This is a pulsar. The emitted electromagnetic energy is derived from the rotation of the neutron star, so pulsars are losing rotational energy and slowing down, albeit very slowly. The Crab pulsar (PSR B0531+21) with a period of $33\,\mathrm{ms}$, for example, is spinning down at a rate of 4.2×10^{-13} (a dimensionless number, e.g. seconds per second). This is a very young pulsar, formed from the supernova that occurred in the year 1054 CE (for which historical observing reports exist), and is the central star in the Crab nebula that was produced by this supernova.

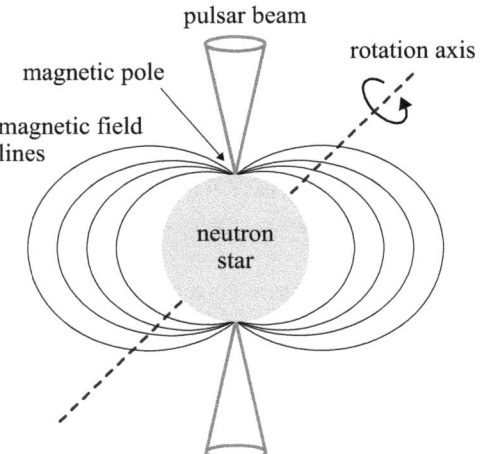

Fig. 14.11 The concept of a pulsar. A neutron star emits X-rays and radio waves in a narrow cone along the magnetic axis. When this axis is inclined to the rotation axis, the electromagnetic radiation is received in pulses, once per rotation, by a distant observer.

Ninety per cent of rotation-powered pulsars have periods above 20 ms. The rest with shorter periods are called millisecond pulsars. These are older, with weaker magnetic fields, and so smaller spin-down rates, of order 10^{-18} to 10^{-21}. This rate is so small that the constancy of their periodic pulses is even better than the stability of atomic clocks (see section 14.2.3). Just over 2000 rotation-powered pulsars are known, of which 150 are detected in X-rays, and 50 of those are millisecond pulsars. The periods of pulsars can be measured to an accuracy of about 10^{-15}. This is crucial for being able to use their pulses in navigation, as we will see shortly.

The are others classes of neutron stars that emit pulses and so appear as pulsars. Accretion-powered neutron stars are neutron stars in binary systems in which mass, and therefore angular momentum, is accreted by the neutron star. This causes strong X-ray emission (but not radio emission), the pulses being caused by the rotation of the accretion hotspot. However, the accretion and the pulses are irregular, so they are not suitable for navigation. Another type of neutron star, a magnetar, is a single star with a very strong magnetic field. Their pulses vary in ways that are not understood well enough for them to be reliable beacons for navigation at this time.

14.5.2 Navigation principle

The rough idea of pulsar navigation is as follows. If we can precisely measure the arrival times of pulses from known pulsars, this constrains the position of the spacecraft in the radial direction to the pulsar, with one important caveat that we'll get to. Repeating this for at least two more pulsars in different known directions, we can locate the spacecraft in three dimensions.

Pulsar navigation requires us to know the period and phase of the pulses, such that we can predict when pulses will arrive at the origin of our coordinate system, which we take to be the SSB. We then compare the measured arrival time of a pulse at the spacecraft with the expected arrival time at the SSB. Suppose at first that the pulse period is very long, so we only have to deal with one pulse, and that the difference in arrival times at the spacecraft and the SSB is Δt. In figure 14.12, the known direction to the pulsar is $\hat{\mathbf{x}}_p$ (a unit vector) and the unknown position vector of the spacecraft S relative to the SSB is $\mathbf{x}_s$. Assuming the pulsar is infinitely far away, the pulse will arrive simultaneously at points S and A. The extra time needed for the pulse to travel from A to the SSB is $a = c\Delta t$. From the scalar product we see that

$$\hat{\mathbf{x}}_p \cdot \mathbf{x}_s = |\hat{\mathbf{x}}_p||\mathbf{x}_s| \cos \theta = a, \qquad (14.2)$$

where θ is the angular separation between the spacecraft and pulsar as seen from the SSB. If we have three pulsars in different directions, then we can write down the three corresponding equations together in matrix form as

$$\mathbf{X}_p \mathbf{x}_s = \mathbf{a} \qquad (14.3)$$

where $\mathbf{X}_p$ is the 3×3 matrix containing $\hat{\mathbf{x}}_p$ for each pulsar in each row, and $\mathbf{a}$ is the vector of distances computed from the three time delays. Provided the three pulsars and the spacecraft do not lie in a common plane, the matrix is invertible and we get a unique solution for the spacecraft position. If we have more than three pulsars, we can solve equation 14.3 by least squares to also find a unique solution.

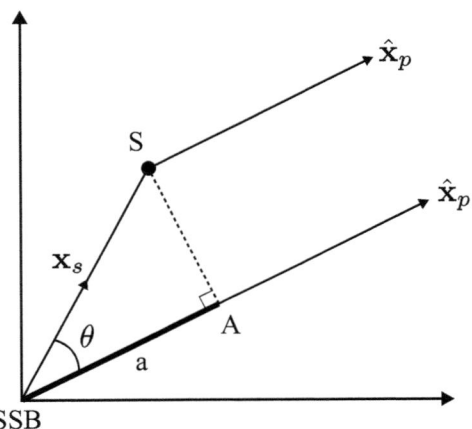

Fig. 14.12   The geometry of pulsar navigation when assuming the pulsar is infinitely distant in direction $\hat{\mathbf{x}}_p$ (unit vector). The spacecraft S is at position $\mathbf{x}_s$ relative to the SSB.

Now to the caveat. Pulsars emit pulses continuously. These pulses are indistinguishable, so we cannot know when a particular pulse is received at the SSB and compare that to the corresponding pulse at the spacecraft. All we know is that pulses are received at the SSB every P seconds, where P is the period, relative to some known reference time t_0. That is, pulses arrive at the SSB at times $t_0 + nP$ where n is an integer. When the spacecraft receives a pulse, it does not know the value of n. Thus we cannot measure the actual time delay Δt, but instead only the delay with respect to the nearest cycle, expressed as a phase shift f (fraction of a cycle; see top of figure 14.13). Thus in figure 14.12 we have $a = cP(f+n)$ for unknown n.

The consequence of this phase ambiguity is that measuring the first pulsar only places the spacecraft on one of an infinite number of parallel planes with separation cP, as shown in the lower left of figure 14.13. If we know the initial position of the spacecraft to within a distance $cP/2$, then we know which cycle we are in – the value of n – and thus which plane the spacecraft is on. A second pulsar in another direction constrains the position to a second set of infinite planes, inclined to the first, and the intercept of these planes with the first set of planes constrains the position to a set of lines (pointing in/out of the page). A third pulsar constrains the set of possible spacecraft positions further (bottom right of figure 14.13), and if this pulsar lies out of the plane of the page, it limits the set of points in the third dimension too. The distance between these solutions will be smaller or larger depending on the orientation and periodicity of the planes. With sufficiently accurate knowledge of the initial position of the spacecraft (from astrometric navigation, or dead reckoning), a single most-probable solution can be found. On the other hand, if the initialization is less precise than the separation between the solutions, a phase ambiguity, and therefore a spacecraft position ambiguity, will remain.

Reality is of course more complicated. The pulses are not perfectly narrow spikes, and the signals are weak, so the spacecraft has to measure more than one pulse in order to get a sufficient signal-to-noise ratio to determine the phase shift. Consequently the planes in figure 14.13 have finite width and their interceptions are finite-sized blobs. There are also uncertainties in the pulsar periods and directions. All these propagate into uncertainties in

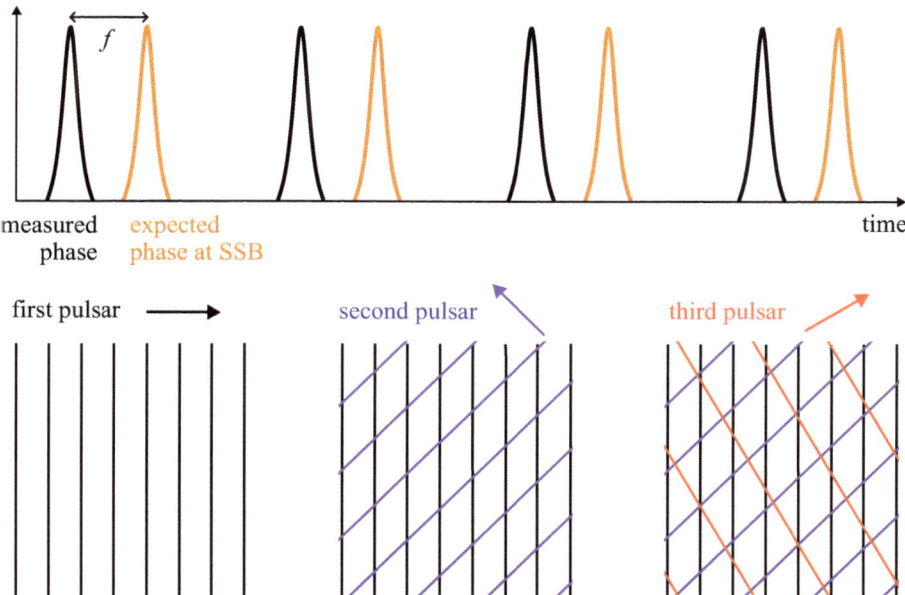

measured phase   expected phase at SSB   time

first pulsar ⟶   second pulsar   third pulsar

Pulsar navigation. Top: the times of arrival of the pulses at the spacecraft ('measured phase') are offset in phase from their arrival times at the SSB by a large number of integer periods plus a fraction of a period, but only that fraction – the phase difference f – can be measured. Bottom: Measuring the arrival times from one pulsar would therefore place the spacecraft on any one of an infinite number of parallel planes (bottom left). Adding a second pulsar (middle) and then a third pulsar (right) helps to solve resolve the ambiguity by reducing the volume of plausible solutions where the planes intercept.

Fig. 14.13

the spacecraft position. Because the spacecraft has to measure multiple pulses, it will move during the course of its measurements, and this too has to be taken into account. Some knowledge of the spacecraft's velocity is therefore necessary. Note that the pulses have a different period at the spacecraft than in the SSB due to the Doppler effect, and the pulsars will also appear to lie in different directions due to aberration, so the velocity of the spacecraft must be known or determined also to account for these.

In practice an iterative procedure is required. As with astrometric navigation (section 14.4), this is initialized with an assumption of the spacecraft's position and velocity relative to the SSB. These are used to predict the phase offset, and from the difference with respect to the measured offset we can identify the best solution. This solution involves a trade-off between reproducing the measured phases as well as possible to within the uncertainties, but also being as close to the initialization as possible (within its uncertainties). As with astrometric navigation, we can regard this as a probabilistic inference problem, with a likelihood defined by the phase measurements, and the prior probability corresponding to the initialization.

This navigation method relies on measuring the time of arrival of pulses on a reference clock that is keeping track of SSB time. An accurate onboard clock is therefore essential. However, as the spacecraft has been moving and accelerating, we have variable time dilation, and it will be almost impossible to compute precisely enough all the relativistic corrections required to keep the clock synchronized with the SSB. However, by observing a fourth pulsar we can also solve for the SSB time, or rather a correction to the onboard clock that resynchronizes it with SSB time.

An important part of the determination of the pulse arrival times is measuring and fitting the profiles of the received pulses. Pulsars give out periodic signals at both radio and X-ray wavelengths, and we could measure them at either wavelength. The pulses are not always simple shapes or even single profiles, as can be seen in figure 14.14, and they can also change with time. They must therefore be fitted against a reference pulse profile for that particular pulsar. X-ray pulsars are relatively faint, so the determination of the pulse profile is limited by the photon statistics (note the error bars in figure 14.14). This is less of an issue in the radio, although even here a large telescope (10 metres or more in diameter) is needed, which may not be compatible with low-mass interstellar spacecraft. Instead of a single dish one could use a phased array antenna (see section 15.8.3). This too would be large and also requires a lot of computer power for the processing. But phased arrays have the advantage of being able to see half the sky – to observe multiple pulsars – without having to be physically steered. Another aspect that needs to be taken into account is the dispersion of electromagnetic waves by charged particles in the ISM, which cause the pulses to spread in time, making it harder to determine an accurate time of arrival. This affects the radio more than the X-ray, which is another reason to prefer the use of X-ray wavelengths.

14.5.3 Application within the solar system

Several groups have considered how pulsar navigation could work in practice, and using simulations based on real pulsars have estimated how accurately the spacecraft position could be determined. This accuracy depends on the size of the instrument that records the pulses, the brightness of the pulsars, how accurately their positions are known, and also on their distribution across the sky (all three lying in the same direction gives little information on two of the dimensions, for example). Shemar et al. (2016) assume an X-ray receiver with an effective area of $50\,cm^2$ operating at 1 keV that observes each pulsar for 10 hours. They identify a set of 10 suitable pulsars with periods between 2 ms and 150 ms. Using three pulsars and an accurate atomic clock, they conclude that a 3D positional accuracy of between 30 km and 100 km can be attained depending on which pulsars are used. A velocity accuracy of the order of $0.5\,m\,s^{-1}$ for the spacecraft can also be obtained. These results assume that the spacecraft is no more than 30 au from the Sun. At larger distances, the parallax of the pulsars becomes an issue (see next section). For comparison, the transverse (plane of sky) position errors for a spacecraft at 30 au obtained with a 70 m antenna on the Earth via ranging is around 4.5 km.

The instrument must have a good time resolution in order to record the pulses; here it is assumed to be 1 μs. The X-ray telescope could either be a photon counter with a collimator tube to select the source, or an imaging instrument. The latter potentially offers a better signal-to-noise ratio by allowing the sky around the target to be removed. But these are quite bulky, because a long focal length is generally needed to focus the X-rays. For example, the MIXS instrument on the BepiColombo mission to Mercury, which also has a collecting area of $50\,cm^2$, is about 1 m long and 0.5 m in diameter. Shemar et al. (2016) outline an analogous instrument optimized for pulsar timing of similar size with a mass of around 12 kg and power consumption of 16 W.

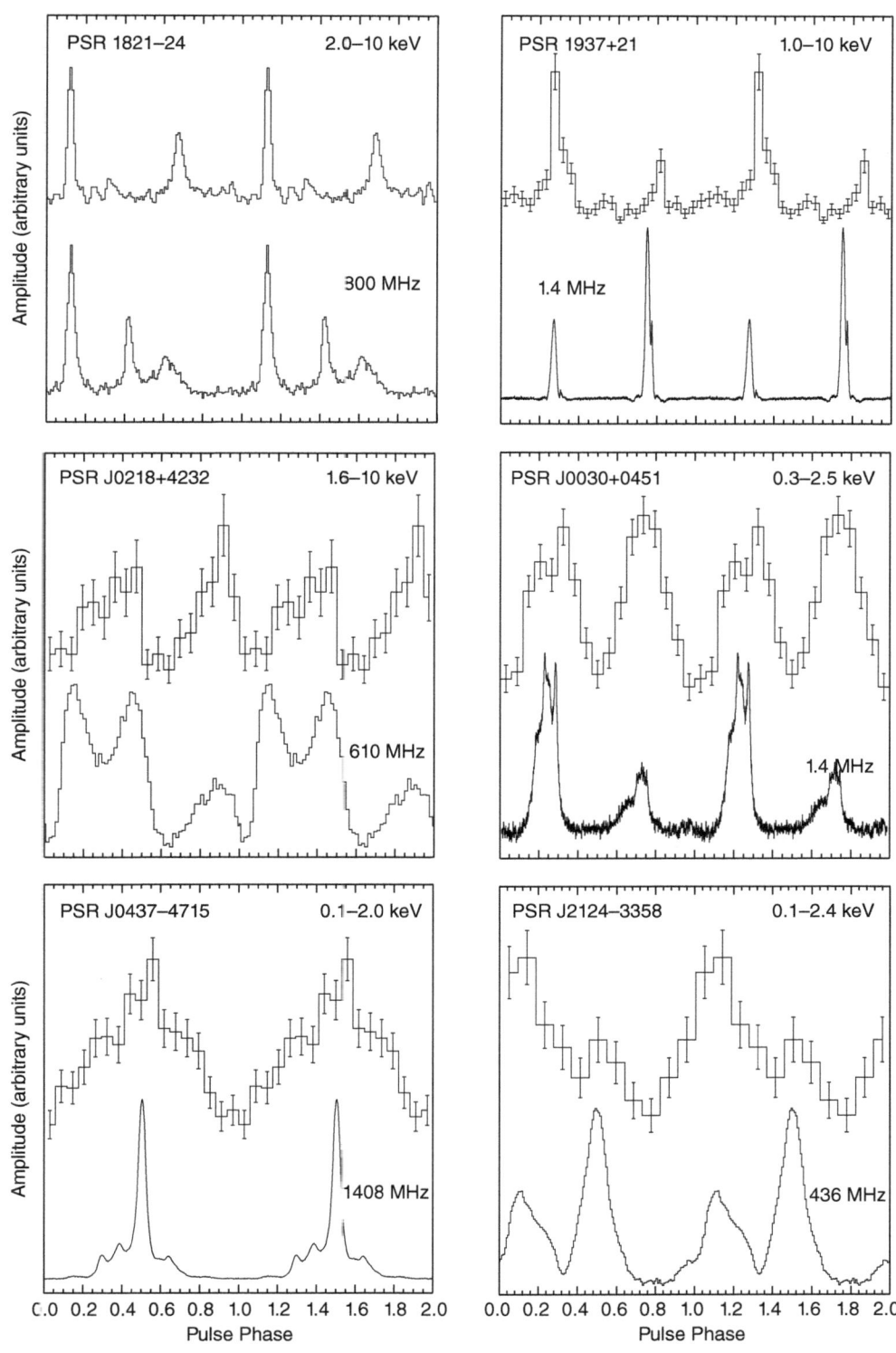

Pulsar profiles in the X-ray (upper curve in each panel) and radio (lower curve) for the six brightest millisecond pulsars. The pulse amplitude in arbitrary units vs phase is plotted. Two identical phases are shown. Credit: Used with permission of Springer Nature BV, from Becker (2009), conveyed through Copyright Clearance Center, Inc.

Fig. 14.14

Pulsar navigation has been tested in the real world from Earth's orbit. NASA attached an X-ray experiment called NICER to the International Space Station in 2017. Part of this is the SEXTANT (Station Explorer for X-ray Timing and Navigation Technology) experiment. Using several millisecond pulsars, it could determine its position to within 10 km (Yu et al. 2020). China launched an experimental X-ray navigation satellite called XPNAV-1 into an Earth orbit in 2016. From observations of the Crab nebula it was able to determine its position at certain control points in its orbit to within an average of 38 km over an 85 day period (Huang et al. 2019).

So far we've been assuming that pulsars have a constant period, or at least that the period is changing in a known way. This is generally true for millisecond pulsars, which is why they are selected for navigation. Other pulsars, and at least one millisecond pulsar, have exhibited 'glitches', which are sudden changes in the rotation period by a factor of around 10^{-6}, although sometimes much smaller. These are believed to be a result of a redistribution of mass, and thus moment of inertia, within the star, at constant angular momentum. Such glitches could be catastrophic for navigation if they were not noticed.

14.5.4 Pulsars for interstellar navigation

Up until now we have been assuming that the pulsar is so distant that the direction to the pulsar as seen from the spacecraft is the same direction to the pulsar as seen from the SSB (as in figure 14.12). This approximation no longer holds once the spacecraft is far from the SSB, as we can see in figure 14.15. The difference in the spacecraft–pulsar and SSB–pulsar distance is, for small angle α,

$$\Delta x = \frac{x_p}{\cos \alpha} - x_p = x_p(\sec \alpha - 1) \simeq \frac{x_p \alpha^2}{2} \simeq \frac{x_s^2}{2x_p} . \tag{14.4}$$

The Crab pulsar, for example, is about $x_p = 2$ kpc away. When the spacecraft is $x_s = 30$ au from the SSB, then $\Delta x = 160$ km (corresponding to $\alpha = 15$ mas). Once Δx is more than the uncertainty in the spacecraft position derived from the navigation, the assumption of a common direction no longer holds. As we expect navigation uncertainties of tens of kilometres, the assumption breaks down once the spacecraft is more than a few tens of astronomical units from the SSB.

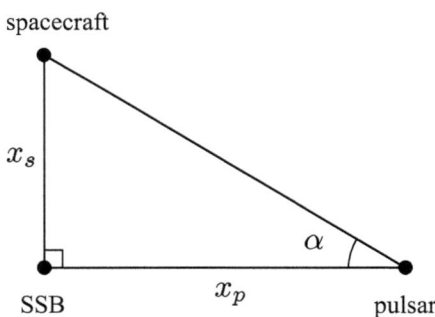

Fig. 14.15 Navigation using pulsars at finite distances.

To use pulsars for interstellar navigation, we therefore need to have some idea of their distances, although not necessarily as accurately as we did when measuring the angle between stars for navigation in section 14.4. Knowing the distances, we modify the expected phase difference when solving for the spacecraft position (and we would use the modified directions to the pulsars too). Unfortunately, other than the Crab pulsar, all other known rotation-powered pulsars are too faint in the optical to have had their parallaxes (and thus distances) measured by Gaia. Distances to some pulsars have been determined to a greater or lesser degree of precision in the radio, however. As the spatial resolution of a telescope aperture is of the order of λ/d, where λ is the observing wavelength and d is the diameter of the telescope, then the parallax precision from a single dish ($d \leq 100\,\text{m}$) is very low in the radio. However, by linking many dishes over thousands of kilometres via VLBI, it is possible to obtain parallax precisions at the milliarcsecond level.

There are other complicating factors when the spacecraft is at large distances from the SSB. The path through interstellar space to the spacecraft is now different from the path to the SSB, so the pulses will be distorted by charged particles in the ISM in different ways, potentially leading to systematic errors in the measurement of the arrival times. A further issue is that a pulsar is only seen as a pulsar because the Earth happens to lie in the rotating beam of the neutron star (figure 14.11). Once the spacecraft is sufficiently far from the Earth, it will no longer fall in the beam. However, as the opening angle of pulsars appears to be on the scale of degrees, this is unlikely to be a problem for early interstellar travel.

Finally, the finite proper motion of a pulsar will change its observed direction over time. Some pulsars have been ejected from binary systems at velocities of hundreds of kilometers per second, and for pulsars 1 kpc away, $500\,\text{km s}^{-1}$ correspond to a proper motion of $100\,\text{mas yr}^{-1}$. Thus within a year the position will change by more than the positional uncertainty. Of course, if the proper motion has been measured, it too can be incorporated into a time-dependent model of the pulsar.

14.6 Relation to global navigation satellite systems

Pulsar navigation bears some relation to global navigation satellite systems (GNSS), such as the Global Positioning System (GPS), for determining locations on the Earth. Yet there are important differences.

GNSS works by a navigator receiving signals from several satellites. These satellites are in low-Earth orbits that are known to high accuracy. The signal from a given satellite encodes both the satellite's position at the time the signal was sent, as well as this time, which is provided by an accurate clock onboard the spacecraft. If the navigator had an equally accurate clock synchronized to the satellite's clock, then the difference between the transmission and reception times gives the distance to the satellite on account of the constancy of the speed of light (neglecting atmospheric effects). This information allows the navigator to place themselves somewhere on the surface of a sphere centred on the satellite. The signal from a second satellite places the navigator on the surface of a second sphere. The intercept of two spheres is a circle (see figure 14.16). A third satellite gives another sphere, and its intersection with the circle gives two points. Only one of these will be near to the Earth's surface, thus giving a unique 3D position of the navigator. This has been achieved without measuring the directions to the satellites, just the signal arrival times.

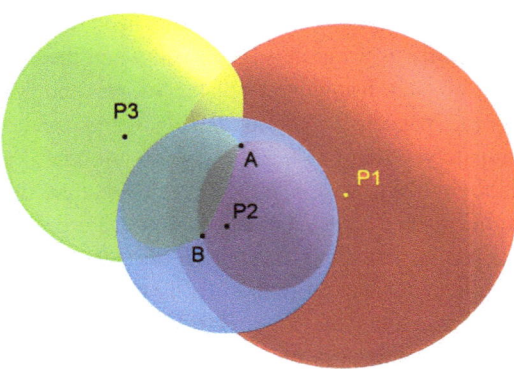

For the above to work as stated, the navigator would need to have a clock that is accurately synchronized to the satellite clocks. With a synchronization error of Δt, the positional error from one satellite is $c\,\Delta t$. Thus to achieve a positional accuracy of 3 m (typical for GNSS), the navigator's clock would have to be desynchronized by less than $\Delta t = 10$ ns over long durations. This is an unrealistic expectation for most clocks (see section 14.2.3) and also ignores relativistic time dilation. Therefore, a signal from a fourth satellite is used to solve for the time t at which the signals are simultaneously received at the navigator's position $\mathbf{x}$. If $\mathbf{x}_i$ is the position of the ith satellite when it sends its signal at time t_i, navigation involves solving the set of equations

$$|\mathbf{x} - \mathbf{x}_i| = c(t - t_i) \quad \text{for} \quad i = 1, 2, 3, 4 \tag{14.5}$$

for the unknowns $\mathbf{x}$ and t. As the receiver usually has a clock of low accuracy (e.g. a GPS wristwatch), then instead of solving for the absolute time, the receiver solves for an offset from the common satellite time (the different satellite clocks must be synchronized to one another to high accuracy). Furthermore, the signals from the satellites will not be received simultaneously, so the receiver has to measure the differences between the arrival times. A watch with an accuracy of 10 s per year accumulates the 10 ns error after just 30 ms, implying it needs to receive the different GPS signals within 30 ms in order to achieve the nominal 3 m position accuracy. Thus while the receiver's clock does need a high time resolution (better than Δt), and must be accurate over the duration of the observations, it does not need long-term accuracy.

To determine our position with GNSS on the Earth, we therefore need signals from at least four satellites, or three if we know our altitude. More satellites will give higher accuracy. One example of GNSS is the Galileo system, which comprises nominally 24 satellites orbiting in three different planes at an altitude of 23 000 km above the Earth's surface. This ensures that more than four satellites are visible from any point on the Earth at any time, to allow for possible satellite outages. The system can deliver a positional accuracy of around 20 cm.

To maintain the accuracy of the satellite's clocks, they must be regularly synchronized with a reference clock on the Earth's surface, and they must be sufficiently accurate to maintain this until the next synchronization. If they are synchronized once per orbital period (just over half a day), then adopting 1 ns as the target accuracy, we see that the clock must be accurate to about one part in 10^{13}. This can only be achieved with atomic clocks.

Because the navigation satellites are moving relative to the reference clock (at around $3.5 \, \text{km s}^{-1}$), there is a significant special relativistic time dilation (section 7.2). This makes the clocks on the satellites appear to run slow by about 7 μs per day. There is an additional general relativistic time dilation arising from the fact that the satellites are moving through a lower gravitational potential than the reference clock on the Earth's surface. This correction leads to the satellite clock advancing by 45 μs per day. The net effect is that clocks at the GNSS altitude advance by about 38 μs per day.[3] These relativistic effects are large ($38 \, \mu\text{s} \times c = 11 \, \text{km}$) and orders of magnitude larger than the required timing accuracy, so must be corrected for when solving for the navigator's position. There are additional relativistic effects, such as the Sagnac effect arising from motion in a rotating coordinate system. These are smaller in magnitude, but must also be corrected for.

Deep space pulsar navigation is similar to GNSS in that both use beacons of known positions, transmitting unique identification signals. With GNSS, the satellites transmit their continuously changing 3D locations, whereas with pulsars we assume that the direction to the pulsar is fixed but the distance is unknown. (Once we get to interstellar space then we do need to know the distance to some degree of accuracy.) The pulsar does not transmit its position, but we can identify it from the shape of its pulse and so look up its direction in a catalogue. In both cases we use the timing of the received signal to determine our position accurately in the radial direction to the beacon, but without constraining the tangential position much, or at all. Multiple beacons lying in different direction then allow us to determine our 3D position.

Another major difference between GNSS and pulsar navigation is that GNSS broadcasts a universal timescale (e.g. Galileo System Time, GST) whereas pulsars of course do not. The pulses from pulsars are instead markers on a universal timescale (SSB time), but with a phase ambiguity that is solved to within some distance range by observing multiple pulsars.

Stellar navigation as outlined in section 14.4 is very different from both pulsar navigation and GNSS, because it observes the directions to the beacons by measuring the angular separations between them. It is therefore using the parallax displacement, which is not used in either GNSS or pulsar navigation. The variation on stellar navigation mentioned earlier using the stellar radial velocities, on the other hand, does not measure angular positions. It relies instead on the displacement of the stars to change their relative directions of travel, and thus observed radial velocities. So this too is a radial, rather than tangential measurement, and again multiple sources will reveal the 3D position.

---

[3] The net timing correction from these relativistic effects is a function of orbital radius. For satellites in low-Earth orbits (up to 2000 km) such as the International Space Station, the time dilation from the faster orbital speed dominates over the smaller gravitational effect, and the clocks lose time. For higher orbits the gravitational effects dominate, and so the orbiting clocks gain time.

14.7 Spacecraft control

Knowledge of the position, velocity, and orientation of a spacecraft is required for two distinct tasks. First, it is needed to tag observations. For example, as the spacecraft crosses from the solar system into the ISM, mapping the change in plasma density, we need to know where the spacecraft is. Similarly, when arriving at the target system, the spacecraft will need to know where it is in order to point its instruments in the right directions.

The second use of navigation is to change course by changing the velocity of the spacecraft. Interplanetary spacecraft typically achieve this with rockets. Changes in spacecraft orientation are often achieved using cold gas thrusters (diametrically opposed so that they generate a torque but not a lateral acceleration), although for space observatories in particular this is also done with gyroscopes.

What options a spacecraft has for controlling its velocity or orientation depends very much on the overall propulsion design and mass budget. A spacecraft with a fusion drive (chapter 8) could be designed to reuse this during the mission to adjust its course. Even if that has been used up, or jettisoned, it is likely that the spacecraft has enough mass to host thrusters for small corrections. A massive laser sailcraft (chapter 11) may likewise have enough mass budget for microthrusters. Another option for steering might be the use of magnetic and electric sails (chapter 13), by changing the orientation of the loop or grid with respect to the direction of travel.

The situation is very different for ultralight (gramme-scale) sailcraft, which would not be able to accommodate any kind of matter rocket. One option here is to vary the reflectivity of different parts of the sail. When illuminated by the star in the target star system, this will induce a torque that can be used to change the orientation and thus direction of the travel of the sailcraft (as tested by IKAROS; see section 10.6.1). The same could be done using the Sun when the spacecraft is still near the beginning of its journey (after the laser has been turned off). The sail could also be equipped with semiconductor light-emitting diodes on the sail rim to reorient it via their emitted photon momentum.

14.8 Summary

Navigation addresses the need to determine the position and velocity of a spacecraft within a coordinate system. When an interstellar spacecraft is within a few tens of astronomical units of the Earth, it can use Earth-based beacons to help it navigate, just as interplanetary spacecraft do today. This is done with two-way ranging and the Doppler effect, in which signals are sent from the Earth to the spacecraft and back again. This can achieve an accuracy in the radial distance and velocity of 1 m and 1 mm s^{-1}, respectively (but worse for large distances). DOR (differential one-way ranging) and delta-DOR, which rely on two widely spaced ground stations receiving signals from the spacecraft, may be used to determine the angular coordinates of the spacecraft (figure 14.4). Accuracies of a few tenths of a milliarcsecond have been achieved.

Inertial techniques use gyroscopes, accelerometers, and clocks to monitor the change in velocity, and thus position, over short time periods. They provide a differential navigation between receiving external updates of the position and velocity of the spacecraft.

At larger distances from the Earth, autonomous methods of external navigation are required that do not rely on receiving signals from the Earth. Stellar astrometric navigation uses a catalogue of stars with accurately known 3D positions and 3D velocities relative to the Sun. A spacecraft at another position with another velocity will see these stars displaced by the effects of parallax and aberration (figure 14.6). By measuring the angular separation between multiple pairs of stars, it is possible to untangle these effects to determine the 3D position and 3D velocity of the spacecraft. By measuring the angular separations of 10 star pairs to an accuracy of 10 mas, we could achieve accuracies of 0.01 au and $10 \, \mathrm{m \, s^{-1}}$ in position and velocity, respectively, even at interstellar distances and relativistic speeds.

Another autonomous method is the use of highly regular pulses from pulsars to determine the position of the spacecraft along the line of sight to the pulsar, albeit with a phase ambiguity (figure 14.13). Using multiple pulsars, the 3D position can be determined. With sufficiently accurate initial knowledge of the spacecraft position and velocity, a final accuracy of 10–100 km should be achievable. For navigation near to the Sun, only the directions to the pulsars need to be known accurately; beyond about 30 au, their approximate distances also need to be known.

14.9 Exercises

1. Convince yourself – with a sketch, and mathematically – that the statement in the footnote on page 255 about hyperbolic navigation is true.
2. The DOR technique in section 14.3 assumes that the spacecraft is infinitely far away when interpreting the time delay at the ground stations as a position angle (equation 14.1). Suppose instead that the spacecraft is at a large but finite distance qb from the midpoint of the ground stations, where b is the distance between the ground stations, and q is positive real number. Denote the angular position of the spacecraft from this midpoint as ϕ. What angular position would equation 14.1 predict? That is, find an expression for θ as a function of ϕ. Plot $\theta - \phi$ vs ϕ for a range of values of q for $b = R_\oplus$. Below what distance is the difference between θ and ϕ significant, such that equation 14.1 is no longer a good approximation? (Note that for $q \gg 1$, $1/q$ is the parallax in radians of the spacecraft between the ground stations.)
3. Show that neglecting the special relativistic time dilation between a receiver on the Earth's surface and a GNSS satellite orbiting at 22 300 km above the Earth's surface introduces a timing error of around 7 μs per day.

Communication

In this chapter we look at the issue of sending messages over very large distances using electromagnetic radiation. An interstellar spacecraft is likely to have a small, low-power transmitter. Diffraction will reduce the intensity received at the Earth to tiny amounts, making the signal hard to detect even with large receivers. This limits the achievable data rates significantly. We will see how to compute the power of the received signal, how to modulate messages into electromagnetic radiation, and how to use signal redundancy – error-correction coding – to reduce information loss due to noise. The received power may be so low that reception occurs in the regime of individual photon counting; we will examine how many bits of information photons can carry. We will also examine the choice of wavelength – in particular optical vs radio – and the relative merits of placing receivers on the Earth compared to in space.

15.1 Requirements and issues

An interstellar spacecraft must return data back to Earth. Assuming the spacecraft itself does not return, this means it has to transmit the data electromagnetically. The very large distances, combined with the limited transmission power, makes this a major challenge.

One of the first questions we face is the choice of communication wavelength. As the technology used to generate, transmit, and receive electromagnetic radiation varies considerably over wavelength, this choice influences all aspects of the system design. Broadly speaking we have two choices: optical, meaning visible plus ultraviolet and infrared wavelengths, and radio. To date, interplanetary spacecraft have all used well-developed radio communication techniques, usually operating in the X-band at 7–8 GHz (4 cm) or Ka-band at 32–34 GHz (0.9 cm). Shorter optical wavelengths have the significant advantage of much less diffraction, and also larger bandwidths. Although optical communication is used extensively for short ranges over fibre optic cables, its use over long distances in free space has been limited. A notable exception is the Deep Space Optical Communications (DSOC) experiment on the Psyche spacecraft, which in 2023 sent signals to the Earth from a distance of 3 au.

Much shorter wavelength X-rays could in principle be used for communication, as they have even less diffraction than optical wavelengths. But X-rays require more energy to generate, are harder to collimate, and the technology to encode/decode signals with them is much less developed. The Earth's atmosphere is also opaque to X-rays. We'll examine the issue of wavelength choice in more detail in section 15.8.

As we will learn, the data rates from an interstellar spacecraft will be low, so transmission times will have to be long. This is already the case for some interplanetary missions. NASA's New Horizons spacecraft, launched in 2006, returned data to the Earth at a rate of 1–2 kb s^{-1} after its 2019 encounter with Pluto, when it was 32 au from the Sun. The 6.25 GB of data it gathered took around 15 months to transmit. For comparison, a single 1024×1024 uncompressed image with $2^8 = 256$ grey scales requires $1024 \times 1024 \times 8\,b = 8.4\,Mb \simeq 1\,MB$. It is likely that the data transfer from an interstellar spacecraft would take many years.

A defining characteristic of interstellar spacecraft is their very large distances from the Earth, with light travel times of years. Other than some potential uplink from the Earth early in the mission to provide software or data updates, the communication will be entirely one way. The spacecraft will therefore have to operate autonomously, which also means it cannot coordinate with the Earth when it will send data, and Earth cannot request that corrupt data be resent. This is a major challenge if data are to be received over a duration of years. Even if we arrange for continuous contact with the spacecraft, perhaps with multiple receivers, there will be inevitable outages. An appropriate degree and timescale of redundancy therefore needs to be built into the communications protocol to accommodate not only noise but also longer-term signal loss.

Finally, the spacecraft may gather more data than it can ever transmit. In that case, the spacecraft's software needs to be intelligent enough to prioritize the data it sends, perhaps based on the quality of the collected data, or based on what it finds. We will consider autonomous control briefly in the next chapter.

15.2 Power received: the Friis transmission equation

One of the major problems of sending electromagnetic waves over large distances is diffraction, as this reduces the received intensity in proportion to the distance squared. Consider a circular transmitting aperture, such as a telescope mirror, of diameter d_t transmitting waves of power P_t. The telescope is focused to infinity, but due to diffraction the waves will spread out. This issue is the same as illuminating a sail with a laser, and as discussed in section 11.2.1, the beam pattern is no longer a point but an extended region of characteristic angular size $\alpha = a\lambda/d_t$, where λ is the wavelength. (We assume for now that the transmitter and receiver are stationary.) For a circular aperture, $a = 1.22$ marks the first minimum of the Airy disk, the region that contains 84% of the power (see figure 11.10). Thus if a 1 m transmitter emits radiation at 1 μm from the distance of Proxima Cen (4.25 ly), the Airy disk at the Earth has a radius of $ra\lambda/d_t \simeq 49$ million kilometres. Any realistically sized receiver can therefore collect only a tiny proportion of the transmitted power.

To calculate how much power is received, let us first assume for simplicity that the power from the transmitter in the Airy disk is distributed uniformly. The physical radius of the Airy disk at distance r is $s = \alpha r$. Assuming there is no loss of signal along the path, then the received intensity – the power per unit area – in the Airy disk is

$$I \simeq \frac{0.84P_t}{\pi s^2} = \frac{0.84P_t}{\pi \alpha^2 r^2} = \frac{0.84P_t d_t^2}{\pi a^2 \lambda^2 r^2} . \tag{15.1}$$

For a receiver of area A_r that is much smaller than the Airy disk and contained entirely within it, the power received is $P_r = IA_r$. Hence

$$P_r \simeq 0.23 \frac{P_t A_t A_r}{\lambda^2 r^2} \quad \text{(uniform disk)} . \tag{15.2}$$

Of course, the intensity is not uniform over the Airy disk, so in general we instead have to integrate the intensity variation of the diffraction pattern (equation 11.28) over the area of the receiver. In practice, though, the Airy disk is much larger than the receiver (tens of millions of kilometres compared to perhaps a few kilometres at most). Provided the receiver is aligned on the centre of the Airy disk, the received intensity is more or less constant at $I_0 = P_t A_t / \lambda^2 r^2$, as given by equation 11.29. In that case the received power is[1]

$$P_r = \frac{P_t A_t A_r}{\lambda^2 r^2} . \tag{15.3}$$

Equation 15.3 is the *Friis transmission equation*. The equation holds for non-circular transmitters and receivers too. It sometimes appears with a numerical factor in front, the value of which depends on the geometry and nature of the receivers. Sometimes the numerical factor is absorbed into the areas of the transmitter and/or receiver to give effective areas. This is often done in the context of radio communication where the concept of antenna gains is used. Other factors may be included to account for signal losses along the path, finite efficiencies of receivers, and relativistic effects (see section 15.2.1). Although equation 15.3 has been derived assuming monochromatic radiation, it also holds for signals with a finite narrow bandwidth.

The Friis transmission equation shows how the received power depends on the transmitted power, the wavelength, the areas of the transmitter and receiver, and the distance between them. The $1/r^2$ dependence comes from the power being spread over a spherical cap. This term is very significant for interstellar distances. Due to mass limitations on the spacecraft, both the power and area of the transmitter are likely to be small. The Friis transmission equation shows that we can compensate for these to some extent by using a large receiver on the Earth (or in Earth's orbit). The $1/\lambda^2$ term also has a big impact: changing the wavelength from 1 cm to 1 μm increases the received power by a factor of 10^8, all other things being equal.

Table 15.1 shows some example values for the various terms in the Friis transmission equation. In addition to the received power, it also shows the received signal photon rate,

$$\dot{n}_s = \frac{P_r \lambda}{hc} . \tag{15.4}$$

The first line of the table corresponds to receiving a signal from a modestly powered satellite in low Earth orbit with a 10 cm diameter receiver in the X-band. The second line corresponds approximately to New Horizons at Pluto. The subsequent lines are for radio and optical communications from Proxima Cen for different transmitter and receiver sizes at different wavelengths. The number of receivers, N_r, is also varied. These are combined incoherently, so this is equivalent to, but more practical than, using a single larger receiver.

Using a 1 m diameter radio transmitter and a 70 m receiver, we see that the power received from the interstellar spacecraft is many orders of magnitude lower than what could be received from New Horizons. But by using a larger transmitter and 1000 receivers, we can attain a received power that is only about a thousand times lower than New Horizons. This is a lot of receivers, but perhaps is not the hardest or most expensive part of interstellar travel.

---

[1] This assumes our receiver measures all of the light over area A_r. The arriving wavefront is plane. Thus when the receiver is an imaging system (like a telescope), the transmitter can be considered as a point source, and this is imaged with some diffraction pattern on the detector (perhaps another Airy pattern). In the derivations, I am assuming that all of the light in this diffraction pattern is measured to obtain P_r. It may be the case, however, that in order to suppress light from the target star next to the transmitter, we only measure the signal in the central part of the diffraction pattern, as shown later in figure 15.3. In that case P_r would be reduced.

Table 15.1 Received power P_r and photon counts $\dot{n}_s$ for different transmitter–receiver configurations, as computed from the Friis transmission equation (equation 15.3). This adopts a circular transmitter of diameter d_t and N_r circular receivers, each of diameter d_r and combined incoherently. In all cases the receiver is much smaller than the size of the transmitted beam's Airy disk at the receiver.

P_t [W]	d_t [m]	d_r [m]	N_r	λ	r	P_r [W]	$\dot{n}_s$ [s^{-1}]
10	1	0.1	1	3.6 cm	500 km	1.9×10^{-10}	3.4×10^{13}
24	2.1	70	1	3.6 cm	32 au	1.1×10^{-17}	1.9×10^{6}
10	1	70	1	3.6 cm	4.25 ly	1.4×10^{-26}	2.6×10^{-3}
10	1	70	1000	3.6 cm	4.25 ly	1.4×10^{-23}	2.6×10^{0}
10	10	70	1000	3.6 cm	4.25 ly	1.4×10^{-21}	2.6×10^{2}
10	10	70	1000	1 cm	4.25 ly	1.9×10^{-20}	9.4×10^{2}
10	1	10	1	1 μm	4.25 ly	3.8×10^{-19}	1.9×10^{0}
10	4	10	1	1 μm	4.25 ly	6.1×10^{-18}	3.1×10^{1}
10	1	30	1	1 μm	4.25 ly	3.4×10^{-18}	1.7×10^{1}
10	10	30	1	1 μm	4.25 ly	3.4×10^{-16}	1.7×10^{3}
10	1	30	1	0.25 μm	4.25 ly	5.5×10^{-17}	6.9×10^{1}
10	1	10	100	1 μm	4.25 ly	3.8×10^{-17}	1.9×10^{2}

Operating in the optical (1 μm) at the same power (10 W) with the same size transmitter (1 m) and using a 30 m telescope as a receiver, we receive much more power: just a third that received from New Horizons when it was 8000 times nearer to the Earth. Note, however, that the photon rate is much lower in the optical because, for a given power, each individual photon carries more energy. This is particularly relevant if reception is done in the photon-counting regime, which we discuss later.

In all of the interstellar examples listed in the table, the power at the receiver is reduced relative to the power at the transmitter by 17 to 27 orders of magnitude.[2] The received signal photon rates are only of order 1–1000 per second, unless much larger transmitters, receivers, or transmission powers can be used. How these photon rates corresponds to data rates (bits per second) depends on a number of other factors, in particular the amount of noise and how messages can be encoded into electromagnetic waves, which we will look at later in this chapter.

15.2.1 A relativistic correction

So far we have been assuming that the transmitter and receiver are stationary with respect to one another. This is not the case for a flyby mission, in which relativistic effects need to be considered on account of the large relative velocities.

---

[2] In physics we tend to think of increases or decreases in terms of factors, or orders of magnitude. In engineering one often uses decibels dB, which are tenths of orders of magnitude. A factor f corresponds to $10 \log_{10}(f)$ dB. Thus 20 dB corresponds to a factor of 100, -10 dB to a factor of 0.1, and so on. Use of dB is convenient when there are many loss and gain terms that need to be combined, as it replaces multiplicative terms with additive ones. The loss of the signal described by equation 15.3 is $f = P_r/P_t$, which for the bottom line of Table 15.1 is -174 dB.

If we go back over the derivation of the Friis transmission equation, we see that relativistic effects must be taken into account in two places. These are the same considerations as for a moving laser sail in chapter 11, here for a receiver receding from the transmitter at velocity β. First, following equation 11.27, the power received in the receiver's rest frame is reduced relative to the transmitted power by a factor of $(1-\beta)/(1+\beta)$, the square of the 'Doppler factor'. Second, the diffraction angle in the receiver's rest frame is increased by the Doppler factor due to aberration, as given by equation 9.5. This angle enters as the square in equation 15.1. Taking both of these effects into account, the relativistic version of equation 15.3 is

$$P_r = \frac{P_t A_t A_r}{\lambda^2 r^2} \left(\frac{1-\beta}{1+\beta} \right)^2 . \tag{15.5}$$

Relativistic effects therefore reduce the power on the receiver moving at $\beta = 0.1$ to 67% of that for a stationary receiver. For $\beta = 0.2$, the power is reduced to 44%. These factors are significant, but not dramatic for our purposes, given that we are often ignoring numerical factors of similar size, such as those arising from aperture geometry or path losses.

The wavelength in equation 15.5 is the wavelength measured in the transmitter's rest frame (as is the distance r). (I considered the diffraction to occur in the transmitter's rest frame and then applied the Lorentz transformation to the received angle.) The wavelength of the signals in the receiver's rest frame is redshifted by a factor of $\sqrt{(1+\beta)/(1-\beta)}$ due to the Doppler effect (equation 9.12).

We derived a very similar result to the one above in equations 9.16 and 9.17, where we saw that the radiance of a moving source is modified by a factor of ψ^4. This equals $[(1-\beta)/(1+\beta)]^2$ for a radially moving source (ψ is defined in equation 9.7).

15.3 Spacecraft pointing accuracy

In deriving equation 15.3, we assumed that the receiver was pointed exactly at the spacecraft, so that the receiver is centred on the Airy disk. When transmitting at $\lambda = 1\mu m$ with a $d_t = 1$ m aperture, the diffraction angle is $1.22\lambda/d_t = 0.25''$. The spacecraft needs to be able to point to the Earth with an accuracy much better than this. If it cannot, the receiver will only see the edges of the diffraction pattern, and the received power will be greatly reduced (see figure 11.10).

Numerous forces act on a spacecraft that could change its orientation, including stellar radiation pressure and incident particles, as well as the anisotropic radiation of onboard heat. The spacecraft must, therefore, actively track the receiver. Today's space telescopes can track bright sources much more accurately than this fiducial value of $0.25''$ for extended periods. The James Webb Space Telescope, for example, can track stationary targets to an accuracy of $0.001''$ and targets moving at $0.003''$ s^{-1} to an accuracy better than $0.01''$. Thus an interstellar spacecraft could probably track the bright Sun to sufficient accuracy, provided it also has some means to manoeuvre.

Tracking the Sun is a start, but the Earth, and therefore the receiver, orbits the Sun. As seen from Proxima Cen, the Earth therefore changes its angular position relative to the Sun by $\pm 0.77''$ over the course of the year. This is larger than the example diffraction angle of $0.25''$ mentioned above. Thus to remain centred on the Airy disk, the spacecraft will need to know where the Earth is in its orbit at any time. As it cannot see the faint Earth – and even if it could, that would provide its position 4.25 years before signals arrive – the spacecraft instead needs to predict its position. For this the spacecraft needs to know the time on Earth, more specifically the barycentric coordinate time (TCB). The maximum angular velocity of the Earth relative to the Sun as seen from Proxima Cen is $0.013''$ per day, so the TCB needs to be known to an accuracy of about a day or two. An accurate atomic clock on board the spacecraft, synchronized to the TCB at launch, can in principle maintain this time to one part in 10^{12} (see section 14.2.3). This corresponds to an error of 2 ms after a journey of 50 years. However, by travelling at relativistic speeds, this clock would suffer from time dilation and so no longer track the TCB. If the spacecraft can keep track of all its velocity changes, then it might be able to correct the clock to the TCB. Another option is to uplink the TCB to the spacecraft from the Earth just before it starts transmitting. This would have to accommodate the long light travel time, of course, for which an accurate position, and thus good navigation, is also required (chapter 14). As we saw in section 14.4, stellar astrometric navigation could instead provide the TCB to the required accuracy on the spacecraft without needing any contact to Earth.

If the spacecraft cannot achieve a high pointing accuracy for extended periods, there will be an unpredictable reduction in received power and thus a potential loss of transmitted information. One solution to this is to relax the pointing requirements by using a larger diffraction pattern, for example by using a longer wavelength or a smaller transmitter. This would also lower the received power (and data rate), but the variation in received power would be reduced and thus unexpected signal losses mitigated. We could then compensate for the larger diffraction angle by using a larger transmitted power or a larger receiver area, as per the Friis transmission equation.

15.4 Modulation

For an electromagnetic wave to carry information, it must vary in some way. This could be a variation in amplitude, frequency, phase, or polarization. How we vary these to imprint a message on the signal is the topic of modulation. For now we assume that we have enough power to form continuous waves, so we can ignore the quantization of radiation.

To illustrate the concept of modulation, we start with an electromagnetic wave at a single frequency, the so-called carrier wave. By varying the amplitude of the wave, with different amplitudes corresponding to different letters in an alphabet for example, we can send a message. This is the basis of amplitude modulation (AM). We can instead vary the frequency by small amounts compared to the carrier wave frequency. This is frequency modulation (FM). They are illustrated in figure 15.1. These analogue modulation schemes were used extensively in terrestrial radio broadcasts in the past, where AM carrier waves typically had frequencies between 0.5 MHz and 2 MHz (medium frequency bands), although carrier waves over a broader range of 0.1–20 MHz were also used. FM radio used higher frequencies, typically 90–110 MHz.

An FM signal of course requires a finite bandwidth for modulation (frequency changes) to occur. In principle an AM carrier wave could be monochromatic, but only if it had an infinite duration. From Fourier transforms we know that a signal of finite duration must occupy a finite frequency range. Moreover, in order to send a message, each amplitude level can only last a finite amount of time, and therefore also occupies a range of frequencies.

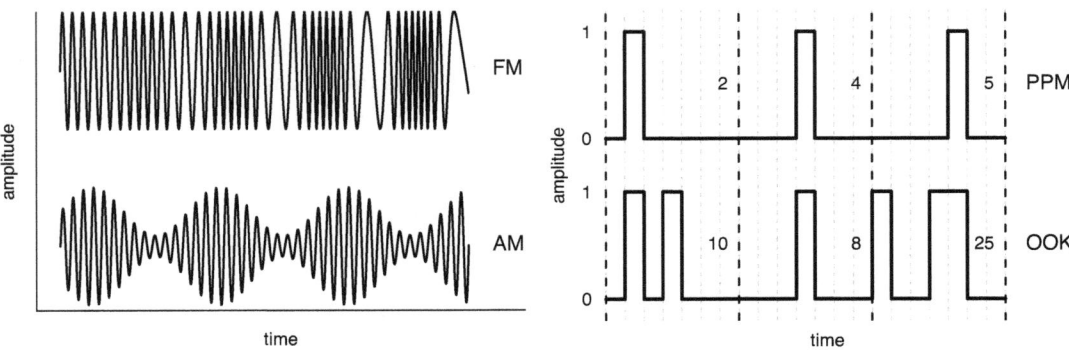

Fig. 15.1 Modulation schemes: analogue schemes AM and FM (left) and digital schemes OOK and PPM (right). In both digital schemes shown, $m = 7$ slots make up a frame. With PPM, only one slot is activated, so the three frames shown could represent the numbers 2, 4 and 5. With OOK, multiple slots can be activated and are interpreted as bits of a binary number. The three frames represent $2^1 + 2^3 = 10$, $2^3 = 8$ and $2^0 + 2^3 + 2^4 = 25$.

No matter what the modulation scheme, to send a signal we require a finite bandwidth. All other things being equal, the larger the bandwidth, the more variation of the carrier wave that can be distinguished, and so the larger the rate at which we can send data. An important result from signal processing quantifies this statement. Suppose we are trying to modulate a number of pulses in a signal that has bandwidth B (in Hz). The smaller the bandwidth, the broader each pulse in time. The largest number of independent pulses that can be resolved (per second) is $2B$, which is known as the *Nyquist rate*. If we send pulses at a higher rate, we can no longer distinguish between them. Another way of thinking about this is, if we want to send more data per second, the signal must vary more rapidly. To properly record this, we need to measure the signal more frequently, which requires a larger bandwidth.

AM and FM are examples of analogue modulation schemes, whereby a carrier wave is modulated by a continuously varying signal. In digital modulation, the carrier wave is modulated by a digital signal. The digital equivalent of AM is amplitude shift keying (ASK), in which the amplitude changes are confined to a number of discrete levels. Similarly, frequency shift keying (FSK) encodes a signal as discrete changes in the frequency. Digital modulation has now superseded analogue modulation for most purposes. Digital modulation is also suitable when intensities are so low that we count individual photons.

To compute the data rate for a particular modulation scheme, we need to know how many different levels m it uses. The smallest number of levels is a binary, $m = 2$, scheme. This means that we send 0s or 1s encoded, for example, as two different amplitudes or two different frequencies. Each pulse then conveys one bit of information. If we instead have $m = 4$ different levels, then we can send the numbers 0, 1, 2, 3, which we can interpret in binary terms as 00, 01, 10, and 11, which is two bits of information. In general, the number of bits we can send per pulse for m levels is $k = \log_2 m$. If pulses are sent at rate f_p, the data rate in bits per second is $R = f_p k = f_p \log_2 m$. The maximum pulse rate for a channel of bandwidth B is the Nyquist rate, $f_p = 2B$. Hence the maximum data rate, or *channel capacity*, is

$$C\,[\text{b s}^{-1}] = 2B \log_2 m \quad \text{(noise-free)}. \tag{15.6}$$

This suggests that, even with a finite bandwidth, the channel capacity can be made arbitrarily large if m is made arbitrarily large. For example, when using FSK modulation, then in theory we could separate the levels by vanishingly small frequencies and so fit arbitrarily many of them into a channel of any finite bandwidth. In practice, noise limits the number of levels we can distinguish, and this places a limit on the channel capacity if we are not to incur errors. We shall look at this in section 15.7.2.

There exist many digital modulation schemes. One of the simplest is on–off keying (OOK), show in figure 15.1. This transmits a binary message as a series of pulses in slots of fixed time width T_s. A pulse indicates a 1 and no pulse indicates a 0. A series of m slots makes up a frame, which can be interpreted as an m-bit binary number. This in turn can be interpreted as a codeword via a codebook. For example, to send ASCII (American Standard Code for Information Interchange) characters, a scheme of 128 different characters (codewords), we require $\log_2 128 = 7$ bits and thus $m = 7$ slots in a frame. A series of frames then makes up a message. In the absence of noise, OOK conveys m bits of information per frame. Each pulse, and thus OOK as a whole, requires a bandwidth of order $1/T_s$.

Another digital modulation scheme is pulse position modulation (PPM), also shown in figure 15.1). Like OOK, a message consists of a series of frames, and each frame is again divided up into m equal-duration time slots. But in PPM, only one slot in a given frame is activated (pulse sent). This allows us to convey one of m different codewords. For example, with $m = 128$ we can convey the ASCII characters. Assuming each slot has an equal probability of activation, this conveys $\log_2 m$ bits of information per frame on average.

When the duration of a single slot in either scheme is T_s, then in OOK a frame has a duration of $T_s \log_2 m$, whereas in PPM a frame has a duration of $T_s m$. Thus with a common slot (pulse) duration, OOK has a larger data rate by a factor of $m/\log_2 m$ (in the absence of noise). Alternatively, to achieve the same data rate in both schemes, PPM requires shorter pulses, and thus requires a larger bandwidth by a factor of $m/\log_2 m$.

One reason we may nonetheless choose PPM over OOK is that PPM usually requires less average power, and is usually more energy efficient. To see this, consider both schemes transmitting a signal at the same bit rate, so PPM has shorter slots. PPM requires energy to activate just one slot, whereas OOK must activate any number of slots. The number depends on the message sent, but on average it is more than one. Assuming we send the same number of photons per slot in the two schemes, PPM requires fewer photons, and thus energy, per frame, so lower average power. However, as its slots are shorter, it will have to send these photons more quickly, so needs more peak power to send a pulse. These considerations are very relevant for interstellar communications, where spacecraft are likely to be power limited.

15.5 Noise

When discussing the Friis transmission equation in section 15.2, the implication was that a larger received power – equivalently a larger received photon rate at a given wavelength – permits a higher data rate. This is only part of the story. The data rate also depends on the noise, because this affects our ability to distinguish between the levels in the modulation scheme.

The effect of noise is shown schematically figure 15.2. The signal is created by modulating the intensity of light between two levels that represent the two binary digits. But there are many things that could affect this signal by the time it arrives at the receiver. The transmitter's power may not be stable, there may be a contaminating light source, the atmosphere

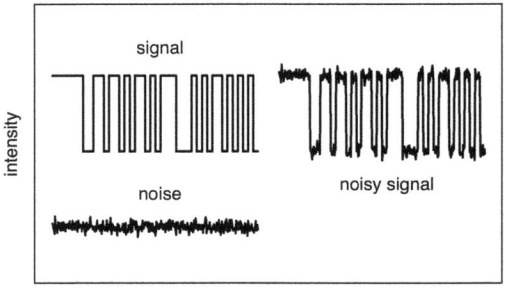
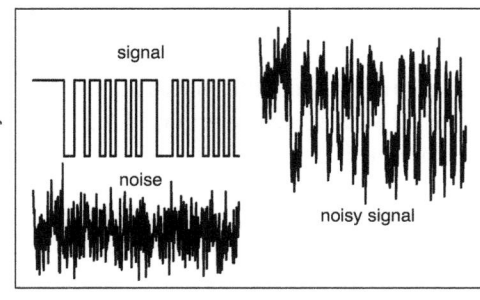

time   time

Fig. 15.2   A demonstration of noise. In the left panel, the message is encoded as a signal by using two intensity levels, show at the top. The additive/subtractive noise is below, and their sum – the received noisy signal – is shown to the right. The right panel shows the same signal, but with a higher noise level.

may scatter some of the signal, or electronics may emit radiation, to mention just a few examples. These phenomena can be considered as random additions to or subtractions from the signal. This is noise and, as we see in figure 15.2, it corrupts the signal, making it harder to distinguish the two levels. If the noise is large enough, we may not be able to reconstruct the original signal correctly.

Noise is characterized not by its overall intensity, but by its degree of fluctuation. This is because if a contaminating source is bright, but constant, we still see the same variations in our signal, just at a larger overall intensity. We could subtract that constant – by measuring it when sending no signal – and so retrieve the original signal. Thus when we refer to the degree of noise, we do not mean the average power, but rather its standard deviation.

Our ability to identify the original signal depends, therefore, not so much on the power of the signal – the difference between the levels in figure 15.2 – but on how large this power is compared to the noise. This is the signal-to-noise ratio (SNR).

For many noise sources, the average power of the noise and its standard deviation are not independent. This is the case for photon counting, because photons are emitted by a random quantum process that can be described by Poisson statistics. In our two-level signal example, suppose that a contaminating source (e.g. the star seen near the spacecraft) contributes on average n_b photons to each pulse. (Such contaminants are conventionally referred to as the 'background', motivating the 'b' subscript.) Due to the random nature of light emission, not every pulse contains exactly n_b photons. Instead, the number of photons per pulse follows a Poisson distribution with mean n_b. The variance of a Poisson distribution equals its mean. The standard deviation, which is the noise associated with the pulse, is therefore $\sqrt{n_b}$.

The source of the original signal is subject to the same statistics, so even in the absence of any contaminating sources there is noise in the signal itself. Denoting the average number of photons in the signal pulse with n_s, this 'self-noise' is $\sqrt{n_s}$. In the absence of any background ($n_b = 0$) the SNR is therefore $n_s/\sqrt{n_s} = \sqrt{n_s}$. This is an important result, because it shows that although the signal itself contains noise, the SNR is increased by using a larger intensity. This is essentially why we want to maximize the received power from our interstellar spacecraft.

When there is a background, we will measure $n_m = n_s + n_b$ photons (over a time slot). To extract the signal, we need to subtract the background. As we cannot distinguish background photons from signal photons, we cannot do this directly. What we typically do is measure the background in the absence of the signal over many time slots in order to determine its average $\overline{n_b}$. Subtracting this from the measurement, our estimate of the signal is $n_s + n_b - \overline{n_b}$. This is the sum of three independent Poisson-distributed variables, which is itself a Poisson-distributed variable with the variance equal to the sum of the individual variances. When $\overline{n_b}$ is measured so accurately as to have a negligible variance, the variance in our estimate of the signal is $n_s + n_b$. The SNR of our signal estimate is

$$\text{SNR} = \frac{n_s}{\sqrt{n_s + n_b}} . \tag{15.7}$$

This tells is that if we have an intense background, then even if its average can be perfectly subtracted in a statistical sense, its fluctuations – which scale as $\sqrt{n_b}$ – still contribute to reducing the SNR of the estimated signal. However, because the signal in the numerator grows linearly with n_s, whereas the noise in the denominator grows only with the square root of the number of photons, gathering more photons still increases the SNR.

The actual calculation of noise and SNR in real situations is more complicated than has been described here. The main point is that we want to have as many signal photons and as few background photons as possible in order to maximize the SNR. The situation is also more complicated when it comes to single photon counting, something we will examine separately later.

15.6 Contaminating light from the target star

When a spacecraft passes Proxima Cen at a distance of 0.1 au, their angular separation as seen by an Earth-based receiver is just 75 mas. Both the star and signal transmitted by the spacecraft deliver plane wavefronts of uniform intensity to the distant receiver. The star is much brighter than the spacecraft's signal (quantified below), so following the discussion from the previous section, the noise contaminating the signal will be very large.

One way to reduce the contamination is to use a wavelength filter. A star emits light over a very broad range of wavelengths. If we can confine our signal to a narrow bandwidth, we can use a narrow filter centred on the signal to remove most of the starlight. If the transmitter is an optical or infrared laser, then a bandwidth of less than 1 nm is plausible.

In principle we could receive the signal using a bare detector, but this would receive light from the entire sky, which includes a lot of more contaminating sources beside the contaminating target star. We could use a baffle to narrow this down. But in practice we would use an imaging telescope as a receiver to spatially separate the spacecraft's signal from the star. Let us calculate what improvement this provides.

Both star and spacecraft can be considered as point sources. When the telescope is diffraction-limited, each point source will form an Airy diffraction pattern on the focal plane of the telescope. For a telescope with an aperture of 3 m diameter receiving at 1 μm, the diffraction angle is about 70 mas, which is similar to the geometric angular separation between the star and the spacecraft mentioned above. This configuration is shown in figure 15.3, where the star is only five times brighter than the spacecraft's signal (we'll see below how we may achieve this). We could only get close to the telescope's diffraction limit when observing above the atmosphere or when using adaptive optics (see section 11.2.3). Otherwise the telescope is seeing-limited, and the images on the focal plane will be broader, and the overlap larger.

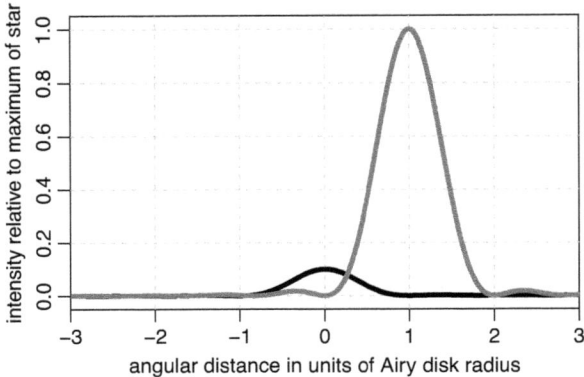

Fig. 15.3 Overlapping Airy diffraction patterns. When the receiver is an imaging telescope, the diffraction pattern of the on-axis faint signal from the spacecraft's signal will probably be significantly overlapped by the diffraction pattern of the off-axis brighter star. In the case shown the two sources are separated by one Airy disk radius and the star is five times brighter than than the signal.

We already saw in section 15.2 how to compute the power of the signal at the receiver. We can compute the power from the star at the receiver as follows. Let I_λ be the spectral intensity (units $\mathrm{W\,m^{-2}\,m^{-1}}$) at the star's surface. For an isotropically radiating star of radius $R_\star$ observed from distance r, the spectral intensity at the receiver is $I_\lambda (R_\star/r)^2$, assuming no losses along the path. Integrating this over the area A_r of the telescope's aperture gives the spectral flux (units $\mathrm{W\,m^{-1}}$) as $I_\lambda A_r (R_\star/r)^2$. To block out as much of this light as possible, the receiver only observes over the narrow bandwidth $\Delta\lambda$. The power (units W) received at the telescope from the star is therefore

$$P_{r,\star} = I_\lambda \Delta\lambda A_r \frac{R_\star^2}{r^2} \,. \tag{15.8}$$

This power is distributed according to the diffraction pattern of the intensity given by equation 11.28.

If we approximate the star as a black body of effective temperature $T_\star$, then the star's spectral radiance – power emitted per unit surface area per unit solid angle per unit wavelength – is given by the Planck function,

$$B(\lambda, T_\star) = \frac{2hc^2}{\lambda^5} \left[\exp\left(\frac{hc}{\lambda k T_\star} \right) - 1 \right]^{-1} \,. \tag{15.9}$$

Each surface element of the star emits isotropically over the half sphere, so integrating the Planck function over this gives its spectral intensity,

$$I_\lambda = \int B(\lambda, T_\star) \cos\theta \, d\Omega = B(\lambda, T_\star) \int_0^{2\pi} d\phi \int_0^{\pi/2} \cos\theta \sin\theta \, d\theta = \pi B(\lambda, T_\star) \,. \tag{15.10}$$

For a black body we can use the Stefan–Boltzmann law to express the stellar radius in terms of $T_\star$ and the stellar luminosity $L_\star$:

$$L_\star = 4\pi \sigma_{\mathrm{SB}} R_\star^2 T_\star^4 \,, \tag{15.11}$$

where σ_{SB} is the Stefan–Boltzmann constant. Thus for a black body we can write equation 15.8 as

$$P_{r,\star} = \frac{E(\lambda, T_\star)\Delta\lambda A_r L_\star}{4\sigma_{SB}T_\star^4 r^2} \quad \text{(black body).} \tag{15.12}$$

Let us compute the received power per unit area of receiver for our nearest star. Proxima Cen has $L_\star = 0.0015\,L_\odot$, $T_\star = 3000\,\text{K}$, and $r = 4.25\,\text{ly}$ (section 2.5.2). At $\lambda = 1\,\mu\text{m}$ we then have $B(\lambda, T_\star) = 990\,\text{W m}^{-2}\,\text{nm}^{-1}\,\text{str}^{-1}$. Adopting a bandwidth of $\Delta\lambda = 1\,\text{nm}$, we get $P_{r,\star}/A_r = 1.9 \times 10^{-14}\,\text{W m}^{-2}$, which is 97 000 photons $\text{s}^{-1}\,\text{m}^{-2}$.

When our spacecraft at Proxima Cen has a 1 m diameter telescope transmitting at 10 W, then using the Friis transmission equation (equation 15.3), this delivers an intensity at the receiver of $P_r/A_r = 4.9 \times 10^{-21}\,\text{W m}^{-2}$, or 0.024 photons $\text{s}^{-1}\text{m}^{-2}$. The power from the star is four million times larger than the power from the signal. Without further mitigation, it would be very hard to extract the signal from the (noisy) glare of the star.

This is where the spatial separation provided by the imaging system comes in. The ratio of four million assumes that, to extract the signal, we have to integrate the light over the entire diffraction pattern of both signal and star in the image plane. But if they are spatially separated, as in figure 15.3, we can integrate instead in just a narrow region centred on the Airy disk of the signal. Of course, if the Airy pattern of the star is four million times higher than the one of the signal, and not five times as shown in the figure, the small spatial separation shown won't help much on its own. But we can now introduce an aperture mask, or coronagraph, centred on the star to suppress its light before it gets onto the focal plane. This is routinely done in astronomy to look for the faint light of exoplanets orbiting stars. Suppression factors of 10^6 have been achieved for separations of a few tenths of an arcsecond, for example using the SPHERE instrument on the VLT. With this suppression factor, the light from Proxima Cen would be reduced to a level similar to that of the signal in the above example. This increases the SNR of the extracted signal significantly and, therefore, the channel capacity (data rate), as we shall see in the next section.

Higher suppression factors can probably be achieved for telescope receivers in space, which will improve things further. A factor of 10^8 to 10^9 at an angular distance of several λ/d_r is predicted for the upcoming Roman Space Telescope. Furthermore, in the calculation above I assumed a low-power transmitter, 10 W, which is what we might expect to have available on a low- or intermediate-mass sailcraft. If our spacecraft has instead used nuclear propulsion to get to the target star, there is likely orders of magnitude more power available, some of which could be used for the communication system, boosting the SNR further.

Bringing everything together, the ratio of the power received from the star to that received in the signal is

$$\frac{P_{r,\star}}{P_r} = \frac{L_\star B(\lambda, T_\star)\lambda^2\Delta\lambda}{4\sigma_{SB}T_\star^4 P_t A_t C} \quad \text{(black body),} \tag{15.13}$$

where C is the suppression factor of the coronagraph. The dependence of this power ratio on wavelength is complicated by the presence of the Planck function. But for $T_\star = 3000$ K, the ratio decreases with decreasing wavelength for wavelengths below about $1\,\mu\text{m}$. A narrower bandwidth reduces the power ratio, because the star emits over a broad spectrum, whereas the signal only emits over the specified bandwidth.[3]

The dependence of the power ratio on the properties of the star – $T_\star$ and $L_\star$ – is not obvious, because both the Planck function and $L_\star$ are strong functions of temperature. But on the main sequence, cooler stars have much lower luminosities than hotter stars so, all other things being equal, we achieve a higher data rate from a spacecraft at Proxima Cen than from a spacecraft at Alpha Cen.

Finally, note that both the receiver area A_r and distance r cancel in the power ratio. This does not imply we do not want a large receiver or a nearby target. Collecting more photons from the signal, even if that increases the number of photons from the star by the same factor, still increases the SNR, as discussed in section 15.5.

15.7 Communication in the presence of noise

Interstellar communication with photons is an example of sending signals over a noisy channel. Whatever the message, and however it is encoded into a signal, this needs to be transmitted, received, and then decoded, as outlined in figure 15.4. The channel is everything between the transmitter and receiver. This can include the transmitters and receivers themselves – both their electronics and optics – as well as the medium between them, in our case interstellar space, but in terrestrial communications this could be cables or optical fibres. Due to noise in the channel, the received signal is not identical to the transmitted signal, as discussed in section 15.5.

In its simplest form, *coding* is the technique of converting a message into a transmittable signal, or code.[4] Suppose we want to send text over a binary channel, which is one that can only transmit two different signals, for example 0 and 1. We need to define a codebook that

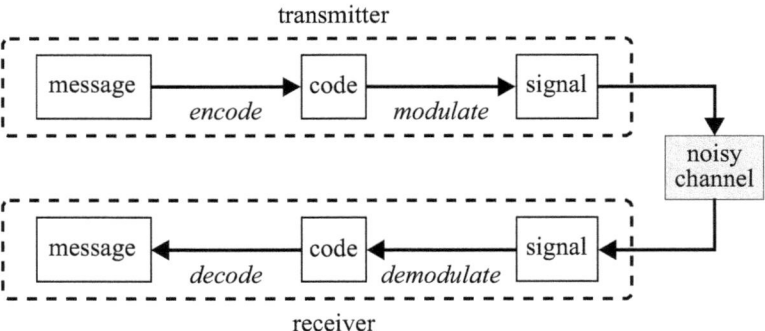

Fig. 15.4   The concept of a noisy channel. A message is encoded and modulated before being sent over a noisy channel that may corrupt it. The received signal is then demodulated and decoded in an attempt to reconstruct the original message. The coding/decoding is designed to make the message robust to noise.

---

[3] This makes the situation somewhat different to the otherwise similar problem of imaging a faint planet next to a bright star. A planet emits or reflects light over a broad wavelength range, so using a narrow filter would reduce the light from the planet too, leading to a net decrease in the SNR.

[4] 'Encoding' refers to converting the message into the code and 'decoding' as converting it back again. Modulation, discussed in section 15.4, describes mapping the code to an electromagnetic signal that can be transmitted.

maps our text to binary digits. This may map individual letters, as in ASCII or Morse code, or it may map entire words. Coding can also involve compressing a message, for example by removing redundant data, such as areas of an image with the same brightness, or it can involve encrypting the message to prevent it being read by eavesdroppers. Here we are interested specifically in channel coding, the goal of which is to make the message robust, to some level, to errors induced by noise. A channel code, or *error-correcting code*, is one that can both detect the presence of errors and correct them to some degree. This is achieved by introducing redundancy into the signal. We are further concerned here only with forward error-correcting codes, which are codes that enable the receiver to detect and correct errors without having to ask the transmitter to re-transmit parts of the message. This is essential for the multi-year light travel times of interstellar communication.

15.7.1 Example error-correcting code: repetition coding

Consider a signal digitally encoded into a string of L binary characters, 0 and 1. Due to noise in the channel, each character has a probability p that it can be flipped. This occurs independently for each character. Table 15.2 shows an example with $L = 10$ and $p = 0.1$. The top row is the original message; beneath this are various possible received messages. As we would expect with these parameters, most received messages contain at least one error. How can we reliably send a message in such a situation?

A simple but inefficient approach is to resend the message multiple times. For each character the receiver takes a majority vote to decide which value it should have. Suppose we send the message R times ('repetitions'). For any one character in the message, the probability that r of these repetitions are errors is given by the binomial probability distribution,

$$P(r|p, R) = \frac{R!}{r!(R-r)!} p^r (1-p)^{R-r}. \tag{15.14}$$

Majority voting means that we choose the character that appears in at least $r_{min} = \lceil R/2 \rceil$ of the repetitions. The ceiling function $\lceil \rceil$ gives the smallest integer larger than or equal to its

Table 15.2 Repetition coding. The original message is in the top line. The following lines show examples of randomly generated received messages when the bit flip probability is $p = 0.1$.

1	1	1	0	1	0	1	1	0	1
1	1	1	0	1	0	1	1	0	0
1	1	1	0	1	0	1	1	0	1
1	1	0	0	1	0	1	1	0	0
1	1	1	0	1	0	1	1	0	1
1	0	1	0	0	1	0	1	0	0
1	1	0	0	1	0	1	1	0	1
1	1	1	1	1	0	1	1	0	1
1	1	1	0	1	0	1	0	0	1

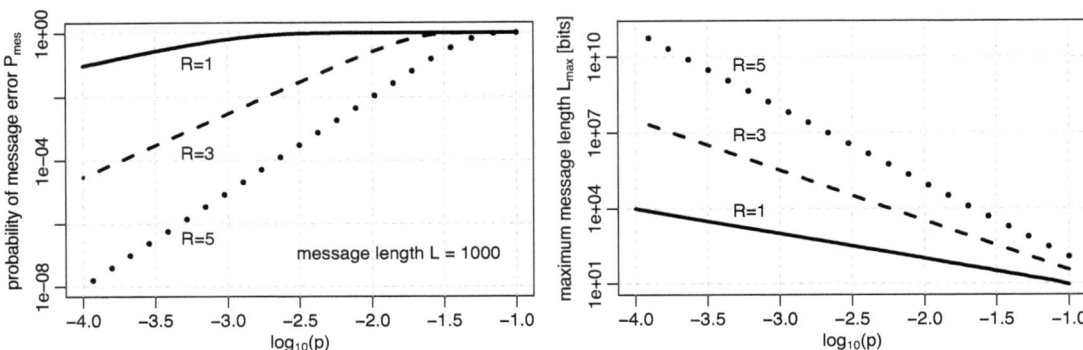

Fig. 15.5 Error suppression by repetition coding. Left: Variation of the probability P_{mes} (equation 15.16) that there is at least one error in a message of length $L = 1000$, as a function of the probability p that a single bit can be flipped, for different numbers of message repetitions R. Right: Variation of the maximum message length L_{max} that can be sent before we expect one bit to be an error.

argument (for example when $R = 5$, $\lceil R/2 \rceil = 3$). To ensure we have no tie-breaks, we select R to be odd.

The probability that our majority vote for this one character is wrong is the probability that r_{min} or more of the repetitions were flipped, which is

$$P_{char} = \sum_{r_{min}}^{R} \frac{R!}{r!(R-r)!} p^r (1-p)^{R-r} . \qquad (15.15)$$

As the character flipping takes place independently for each character, the probability that a message of length L contains at least one error is

$$P_{mes} = 1 - (1 - P_{char})^L . \qquad (15.16)$$

The left panel of figure 15.5 shows how this probability varies with the noise level p, for three values of R, for a message length of $L = 1000$. As we would expect, the probability decreases for smaller noise or more repetitions. But note how nonlinear the variation is: with $p = 10^{-3}$, the probability of a message error for $R = 3$ is 3.0×10^{-3}, whereas for $R = 5$ it is only 1.0×10^{-5}. That is, we increase the message length by a factor of $5/3 = 1.7$ but lower the probability of getting an error by a factor of 300. The plot also shows that for a given noise level p, we can decrease the probability of an error in the message to arbitrarily low values by increasing the number of repetitions.

Another way to look at this is to ask what is the expected number of characters in error. This is LP_{char}. Hence the longest message we can send before we expect to get a single error somewhere in the message is $L_{max} = 1/P_{char}$. The right panel of figure 15.5 shows how L_{max} varies with p for different values of R. For $p = 10^{-3}$, when sending the message just once ($R = 1$), $L_{max} = 999$, but if we increase the number of repetitions to $R = 3$ we get $L_{max} = 3.3 \times 10^5$, and with $R = 5$ we get $L_{max} = 1.0 \times 10^8$. Thus at the cost of reducing the information rate by a factor of five (for $R = 5$), we can increase the reliable message length by a factor of 10^5.

This simple method of repetition coding is inefficient. Yet it illustrates the fundamental concept of adding redundancy to a message in order to increase the probability of recovering the original message in the presence of noise. Moreover, it shows that it is possible to achieve an arbitrarily low message error probability for a given message length by increasing the total amount of data sent. This is a key result that we will explore formally in the next section.

Given the huge amounts of data sent over noisy channels every day (think of the internet), error-correcting codes have become critical to human society. There are many sophisticated error-correcting codes available, the general goal of which is to minimize the amount of data that has to be sent to recover a message at some specified error rate. Examples include block codes (such as Reed–Solomon), convolution codes, and low-density parity-check codes. They work in different ways, but the general idea is to spread out the redundant signal over time so that the effects of noise can be averaged out.

One particular challenge of interstellar communication is that in addition to noise, which corrupts signals on short timescales, there are inevitably longer periods of receiver outage. The transmitting satellite will not know these have occurred. So the coding scheme needs to spread out signals over a sufficiently long timescale in order to reduce message loss to an acceptable level.

15.7.2 Channel capacity from the Shannon–Hartley theorem

For communication, we are ultimately interested in how many bits of information per second we can receive. In section 15.4 we saw that for continuous waves the maximum data rate, or channel capacity, is given by $C = 2B \log_2 m$ (units b s^{-1}), where B is the bandwidth of the channel, and m is the number of distinguishable levels we can achieve with the waves. This no longer holds in the presence of noise, however, because the levels will overlap and become indistinguishable, thus causing the received message to differ from the transmitted message. It can be proven that for an analogue channel with additive white[5] Gaussian noise, the maximum achievable data rate, the channel capacity, is given by

$$C = B \log_2(1 + \text{SNR}), \tag{15.17}$$

where SNR is the signal-to-noise ratio. This is known as the *Shannon–Hartley theorem*. When the bandwidth is 1 MHz and the SNR is 10, the channel capacity is $3.5 \, \text{Mb s}^{-1}$. With an SNR of 1, $C = B$. By comparing this to the no-noise case of equation 15.6, we see that the SNR can be related to an effective number of distinguishable levels as

$$m = \sqrt{1 + \text{SNR}}. \tag{15.18}$$

The Shannon–Hartley theorem is a special case of the more general *noisy channel coding theorem*, which says it is theoretically possible to communicate over a noisy channel up to some maximum rate with an arbitrarily low probability of error. This is quite remarkable: no matter how large the noise, we can, in principle, still pass information reliably at a non-zero rate. How low the error rate will be depends on the actual error-correcting code used to encode and decode the message. If we can define a good enough algorithm, we can achieve an arbitrarily low error rate at the channel capacity. In practice, the actual data rate achievable at some error rate is lower than the channel capacity. However, error-correcting codes have been invented that get quite close to the theoretical limit.

It is possible to transmit at a rate above the channel capacity, but only at a non-zero error rate. In that case, the larger the data rate, the larger the minimum error rate.

---

[5] White noise is noise that has a uniform spectral power density across the bandwidth.

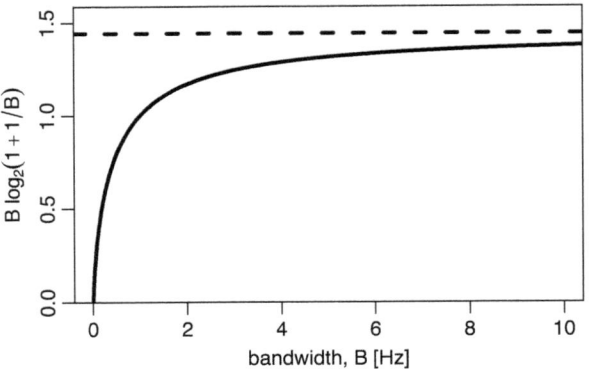

Fig. 15.6 The variation of the channel capacity with bandwidth for a channel with Gaussian white noise and the ratio of the signal power to the noise power per bandwidth of $P/N_B = 1$ (equation 15.19). For all P/N_B, the capacity increases with B and asymptotes for infinite bandwidth to a value of $1.44(P/N_B)$ (dashed line).

The linear dependence of the channel capacity on the bandwidth of the signal in equation 15.17 suggests that to raise the channel capacity, we should set the bandwidth as high as possible. In contrast, given the logarithmic dependence of channel capacity on the SNR, once the SNR is of order 1, increasing it further only increases the channel capacity slowly. Thus to achieve a high channel capacity, it appears we should focus on achieving a high bandwidth. However, a larger signal bandwidth will often include more noise too, because many noise sources have a broad power spectrum. For white Gaussian noise, the noise per unit bandwidth N_B (units $W\,Hz^{-1}$) is constant, so the total noise is BN_B. If P is the received power in the signal, then the SNR is P/BN_B and the Shannon–Hartley theorem becomes

$$C = B \log_2 \left(1 + \frac{P}{BN_B} \right) . \tag{15.19}$$

Figure 15.6 shows that in this situation the channel capacity increases monotonically with increasing B (the curve has a similar shape for all P/N_B). Even though the SNR decreases with increasing bandwidth, which implies we want a smaller bandwidth, the extra carrying capacity of the larger bandwidth more than compensates for this. Thus the larger the bandwidth, the higher the maximum data rate (for given P/N_B). There are diminishing returns for ever larger bandwidth, however: at large B the channel capacity asymptotes to a value of $(1/\ln 2)(P/N_B) = 1.44(P/N_B)$. This is nonetheless relevant to interstellar travel. While constraints onboard the spacecraft may significantly limit how large P can be, and there is a limit to how small we can make N_B, we can compensate for this by using a large bandwidth.

Another way to see this dependence on bandwidth is to consider the energy E required to send a bit of information (units $J\,b^{-1}$). If R is the actual data rate ($R < C$, in $b\,s^{-1}$) then $E = P/R$. Writing the bandwidth per unit data rate as $\gamma = B/R$ – a dimensionless quantity with units b^{-1} – then equation 15.19 can be rewritten as

$$\frac{E}{N_B} > \gamma(2^{1/\gamma} - 1) . \tag{15.20}$$

The quantity E/N_B, with units b^{-1}, can be considered as the energy required to send a bit for a given noise per unit bandwidth. It depends only on γ, which measures how many times larger

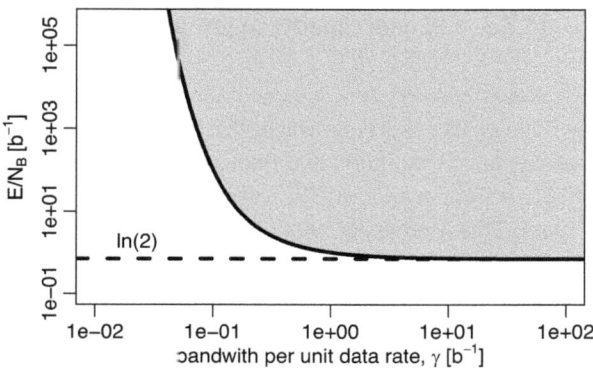

The variation of the minimum energy per bit per noise level, E/N_B from equation 15.20, as a function of the bandwidth per unit data rate, γ. The grey region is that accessible according to the Shannon–Hartley theorem.

Fig. 15.7

the bandwidth is than the data rate. This function is shown in figure 15.7. Consider sending a signal at some fixed data rate R. When we use a relatively narrow bandwidth ($\gamma \ll 1$) this plot shows that we require a large amount of energy for a given noise per unit bandwidth (large E/N_B). This situation is typical of communication on Earth. Many people want to send different messages simultaneously, and so that these don't interfere (overlap), we have to use narrow bandwidths. To achieve this for a given noise level we must use a large amount of energy. To increase the data rate, γ must become smaller, and the figure shows that the energy received must increase. On Earth we can build stationary transmitters fed with large amounts of power, and as the distances over which signals must be sent is relatively small, this is achievable.

On a spacecraft, in contrast, the amount of power available is often limited, so for a given noise per unit bandwidth the term E/N_B is small. The only way we can send a signal – to be in the accessible region in figure 15.7 – is to use a larger bandwidth. Even then we see that there is a minimum energy per bit (for each unit power of noise per unit bandwidth) of $\ln(2) = 0.693\,\text{b}^{-1}$ required to receive any signal, even with an arbitrarily large bandwidth.

Hence communication in the terrestrial and in the interstellar domains in the presence of noise are quite different: terrestrial is low bandwidth, high power; interstellar is high bandwidth, low power. In theory we can use arbitrarily small bandwidths if we have enough power. In contrast, even with an arbitrarily large bandwidth we still need a minimum power.

The above discussion assumes that both signal and noise occupy the full bandwidth B. In practice, the signal is bandwidth-limited. A laser, for example, might have a bandwidth of 1 nm. There is then no point in using a receiver with a broader bandwidth, as this simply adds noise but no signal. So in practice we would use a filter to limit the bandwidth at the receiver, as we did in section 15.6, but then spread the transmitted signal over the entire bandwidth to maximize the energy efficiency.

To now compute the channel capacity we would need to compute the SNR at the receiver. This involves specific details of the modulation method, which we won't delve into here. We would then need to specify an error-correcting code in order to determine the error rate and the actual data rate, which may be well below the theoretical limit. We will look at some examples in section 15.9.

15.7.3 Channel capacity at low photon rates

When the received photon rate is very low, we can no longer consider communication being done with continuous waves like those shown in the left of Figure 15.1, but must instead consider the information carried by individual photons. In this limit, the Shannon–Hartley theorem of section 15.7.2 no longer applies. The upper limit on the channel capacity is instead given by *Holevo's bound*. This is an upper bound to the amount of classical information in a quantum state that is accessible to a receiver.[6]

An important concept in this approach is the amount of information that a single photon can carry, or more specifically, the number of modes a photon can take. A mode can be the photon's frequency, polarization, or time of arrival. In the context of pulse position modulation, discussed in section 15.4, the time of arrival is the position of a pulse in the frame. Each slot corresponds to one mode. If we can count individual photons and place them in one of seven slots (enough to encode ASCII for example), there are $m = 7$ modes. We can also combine types of modes. If we additionally had two polarizations (e.g. left and right circular) and 10 different frequencies, we would have $7 \times 2 \times 10 = 140$ modes per photon. The inverse of this number we write as M, the number of photons per mode.

Without diving into the rather complicated theory, let's just take a brief look at some of the concepts involved. In the absence of noise, the capacity of the channel in bits per mode is, from the Holevo bound,

$$C_m = g(\eta M) \quad \text{(noise-free)} \tag{15.21}$$

where η is the efficiency of the detector (fraction of photons detected), and g is the function

$$g(x) = (1 + x) \log_2(1 + x) - x \log_2 x . \tag{15.22}$$

When there are N noise photons per mode on average, the number of bits per mode is instead

$$C_m = g(\eta M + [1 - \eta]N) - g([1 - \eta]N) \quad \text{(noisy)}. \tag{15.23}$$

This expression is for a specific type of noise, known as the thermal channel. The channel capacity in bits per photon, in both the noise-free and noisy cases, is

$$C_p = \frac{C_m}{\eta M}, \tag{15.24}$$

the factor η being here because this is the number of bits per *detected* photon.

Figure 15.8 shows how the channel capacity C_p varies with the number of photons per mode M. The dashed line is for the noise-free case. As we would expect, the channel capacity increases as M decreases, because there are more information-carrying modes for each photon.

When there is noise, we see from figure 15.8 that the higher the noise per mode N, that is, the lower the SNR in a mode, the lower the channel capacity C_p for any M. In the limit of sufficiently large M, the capacity of a noisy channel tends toward that of a noise-free channel (the lines converging on the right of the figure). This is intuitive, because with a fixed number of noise photons per mode, the more signal photons there are, the closer this is to the noise-free case. As with the noise-free case, increasing the modes per photon (decreasing M) at

---

[6] Although this considers quantum effects, it is not the same as quantum communication, which is a different topic that will not be considered here.

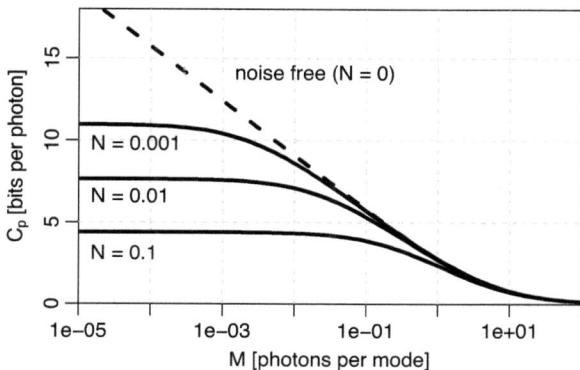

The variation of the channel capacity in bits per photon (equation 15.24), as a function of the number of photons per mode in a photon-counting communication channel, for a detection efficiency $\eta = 0.5$. The dashed line shows the noise-free relation. The solid lines show three examples of noisy channels, where N is the average number of noise photons per mode. As the variation of C_m with M is fairly linear for most of the range shown, the plots of C_p and C_m with M look qualitatively similar.

Fig. 15.8

constant N increases the channel capacity. That is, even though the SNR in a given mode is decreasing to the left in the figure (at fixed N), the corresponding increase in the number of modes for each photon to carry the information more than offsets this. The result is an overall increase in the channel capacity. This has a limit, however, as we see that for a given N, the channel capacity stops increasing once M drops below some level. For example, if we have $M = 0.01$ signal photons in each mode, yet $N = 0.1$ noise photons in each mode, then at $\eta = 0.5$ we get $C_p = 4.3$. Decreasing M further – packing more modes into each photon – barely increases the channel capacity.

The dependency of the channel capacity on the noise is more intuitive when we plot these quantities for a given value of M. This is shown in figure 15.9 for several values of M, again at $n = 0.5$. As the noise per node decreases, the SNR for a given M (and thus per mode) increases, and the channel capacity increases. Again, this increases continues only up to some asymptotic limit.

Note that the channel capacity in bits per mode, equation 15.23, depends on η, the detector efficiency. We find that as this efficiency increases to 1.0, all of the noisy lines in figure 15.8 converge on the noise-free case (see exercises). The variation of C_p with η is counter-intuitive, however, because C_p is the channel capacity per *detected* photon. For the noise-free case, C_p actually increases with decreasing η (at given M). But the rate at which photons are detected has decreased, so the bit rate has also increased. If we plot the noise-free channel capacity per *incident* photon (prior to inefficient detection), C_m/M, then this is indeed lower for lower efficiency.

The channel capacity in bits per second is

$$C = \dot{n}_s \eta C_p, \tag{15.25}$$

where $\dot{n}_s$ is the incident signal photon rate (prior to detection), and thus $\dot{n}_s \eta$ is the detected signal photon rate. Note that ηC_p is the number of bits per incident photon, so the channel

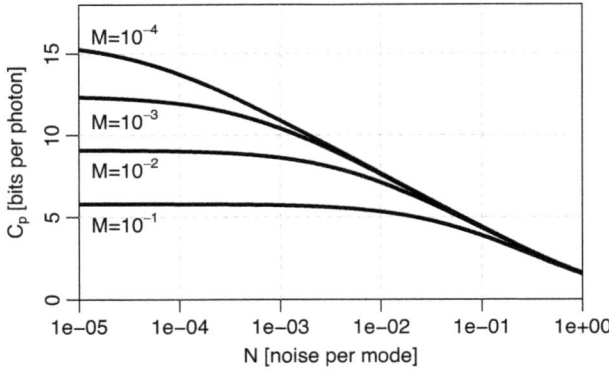

Fig. 15.9 The variation of the channel capacity in bits per photon (equation 15.24), as a function of the number of noise photons per mode in a photon-counting communication channel, for four different values of M (photons per mode), with $\eta = 0.5$.

capacity can also be seen as the product of the number of bits per incident photon and the incident photon rate.

Comparing equation 15.23 to the classical expression in equation 15.17, we see that in the former, the bandwidth and SNR have been replaced by the number of photons per mode and the noise per mode (as well as an efficiency factor). Photons per mode can be thought of as a generalization of the concept of bandwidth.

It is important to appreciate that the ideas sketched in this section give a fundamental theoretical limit for the maximum data rate. Achieving this will be difficult in practice. The value of N, the number of noise photons per mode, as well as the actual noise model, depends on the physical implementation, which will need low-noise single photon-counting detectors. We also need an efficient error-correcting code (section 15.7.1) that can approach the theoretical limit. Studies that assume this method can be used for interstellar communication often compute the incident photon rate $\dot{n}_s$ using the Friis transmission equation, but then adopt a value for ηC_p, the number of bits per incident photon, without specifying the implementation. From these two numbers we can compute an expected data rate using equation 15.25.

To get some idea of what the limit is in the ideal, noise-free case with $\eta = 1$, consider the case of communicating from Proxima Cen in the bottom row of table 15.1, which has $\dot{n}_s = 190$. Adopting $M = 10^{-2}$, equations 15.21 and 15.24 give $C_m = 0.081$ bits per mode and $C_p = 8.1$ bits per detected photon. This gives a theoretical maximum data rate of $C = 1540\,\mathrm{b\,s^{-1}}$.

Theoretical studies aside, there have been practical implementations of communication channels using individual photons. Farr et al. (2013) demonstrate an optical communication channel using PPM that achieves $C_p = 10\text{–}15$ bits per photon. In one example, using $m = 2^{19}$ slots (19 bits per slot, which ideally is $M = 2^{-19} = 1.6 \times 10^{-6}$ modes per photon), they achieved 15.4 bits per detected photon, or 13.2 bits per incident photon ($\eta = 0.86$). This compares to a noise-free theoretical limit of 20.7 bits per detected photon, as computed from equations 15.21 and 15.24. The data rate was $3.2\,\mathrm{kb\,s^{-1}}$. Although the goal of Farr et al. (2013) was to demonstrate a large number of bits per photon rather than per second, they could achieve a higher data rate of $142\,\mathrm{kb\,s^{-1}}$ at the only slightly lower value of 10.4 bits per detected photon.

The experiments of Farr et al. (2013) were done in the lab, not in space, let alone at interstellar distances. However, this approach has been employed in space communication. Villoresi et al. (2008) demonstrated transmission and receipt of single photons between the Earth's surface and a satellite 1500 km away, a distance over which the signal was attenuated by nearly 16 orders of magnitude. Interplanetary-scale optical communication has also been done with large intensities. The DSOC experiment on the Psyche spacecraft has demonstrated optical communications with the Earth from a distance of up to 3 au. This uses a 4 W laser operating at 1.55 μm with a 22 cm transmitting telescope together with a 5 m receiving telescope on the Earth, to achieve data rates of up to 270 Mb s^{-1}. When at 2.7 au, it measured a data rate of 8.3 Mb s^{-1}.

15.8 Choice of wavelength and location of receiver

Earlier in this chapter we saw how the wavelength of a signal influences its diffraction and bandwidth and thus the maximum data rate achievable. Another important factor affecting both data rate and choice of wavelength is where we place the receiver. It could be located on the surface of the Earth, in space, or on the surface of another solar system body such as the Moon. The interdependent choice of receiver location and communication wavelength involves several factors, including: the maturity of technology for building transmitters and receivers; sources of absorption and emission, and thus noise; and the cost of installing and maintaining the receivers. We must also remember that the received wavelength is not the same as the transmitted wavelength due to the Doppler effect. To receive signals from the spacecraft when it is travelling at different velocities, we need to be able to tune to the received wavelength (and variable bandwidth) over the course of the mission.

In this section we look at the impact of the Earth's atmosphere on communication, the merits of putting receivers in space, and some technological issues concerning the choice of wavelength.

15.8.1 Absorption and emission from the Earth's atmosphere

When receivers are placed on the Earth, the spacecraft's signals must pass through the atmosphere. This generates two problems. The first is that the atmosphere is opaque to electromagnetic radiation at certain wavelengths, as seen in figure 15.10. Blueward of about 0.3 μm the atmosphere is almost entirely opaque due to absorption by oxygen and ozone (including the ozone layer that protects life from solar ultraviolet radiation). Between 0.3 μm and 1.0 μm there is little absorption. It is of course no coincidence that animals can see and that plants photosynthesize using light in this region. In the range 1–8 μm there is again a lot of absorption, mostly from water and carbon dioxide, although there are a few gaps (exploited by astronomers to make infrared observations from the ground). From 8 μm to 14 μm there is a low-absorption window (at least in low-humidity regions), apart from an absorption band of ozone in the middle. Redward of 14 μm out to radio wavelengths of a few millimetres, the atmosphere is opaque due initially to carbon dioxide and then to water. Astronomy at these wavelengths must be done from space. The opacity at the long wavelength end of this region is dependent on the amount of water vapour, which is why (sub)millimetre telescopes tend to be placed on very high altitude dry sites, like the Atacama Large Millimeter/submillimeter Array (ALMA) on the Chajnantor plateau in northern Chile at an altitude of 5000 m. Between wavelengths of a few millimetres and a few tens of metres, the atmosphere remains largely transparent. At longer wavelengths the atmosphere becomes opaque again due to scattering by free electrons in the ionosphere.

Which of these atmospheric windows could we use for interstellar communication from the ground? Most communication with satellites is done in the range 0.7–7 cm (4–40 GHz). Shorter wavelengths down to 3 mm (frequencies up to 100 GHz) can be used, but absorption by oxygen and water in the troposphere become increasingly problematic. Shorter wavelengths are preferable on account of the lower diffraction at the transmitting spacecraft (section 15.2) and the larger bandwidths obtainable. This leaves either the infrared windows between 1 μm and 14 μm or the visible window at 0.3–1 μm.

This brings us to the second problem with the Earth's atmosphere: emission, and more specifically thermal emission. Black bodies at room temperature have their peak spectral radiance at about 10 μm. This can be seen from Wien's displacement law parametrized by wavelength, which is $\lambda_{max} [\mu m] = 2898/T$. This means that telescopes, instruments, and detectors may all contribute significant background noise. Detectors and parts of instruments can be cooled, and this is routinely done in astronomy when observing at these wavelengths. But neither the telescope nor the sky can be cooled, so the background remains significant.

The black body distribution is broad, so to avoid thermal emission we should use signals at much shorter wavelengths than the thermal peak. This points to the visible window. The problem here is that the sky is much brighter during the day than the night, which either

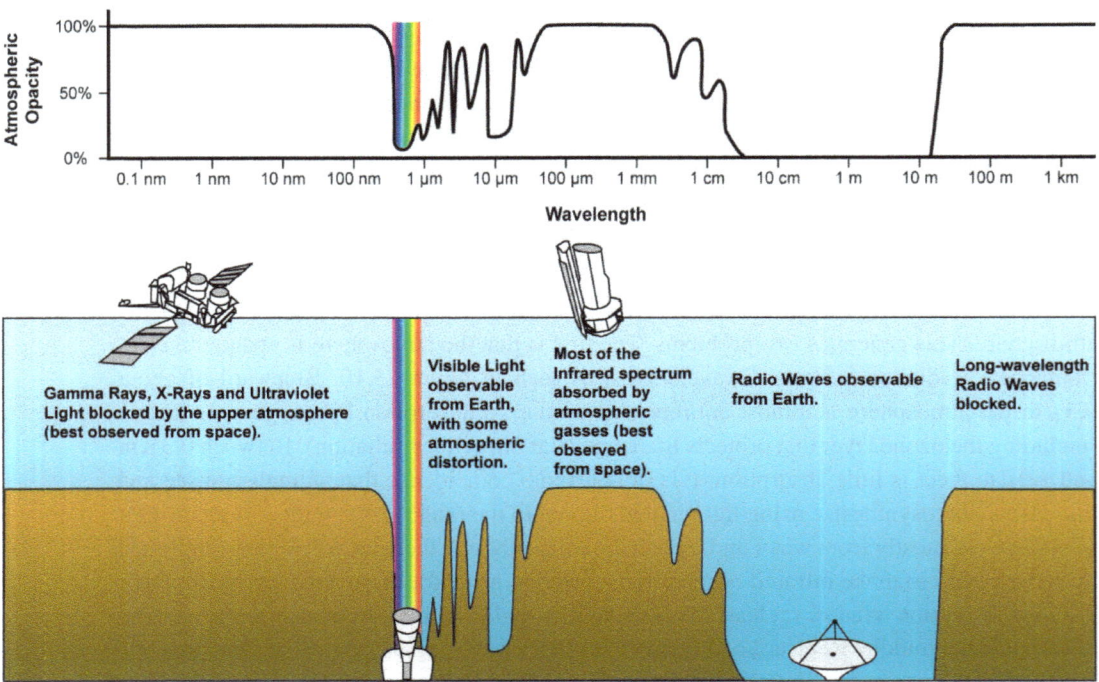

Fig. 15.10   Absorption by the Earth's atmosphere as a function of wavelength. Credit: NASA, public domain, from Wikimedia Commons.

significantly lowers the SNR or inhibits detection entirely, thus reducing the time available for reception by a single receiver. This is not such a limitation for much of the radio spectrum.

Even at night there are additional emission sources that can contaminate an optical or infrared signal. The most obvious is the Moon. Just as it does for the Sun, the Earth's atmosphere scatters light from the Moon, so the night sky brightness depends on the angular distance of the Moon from the direction we observe. The Moon is bright enough that observations of faint astronomical sources are only done around the time of the new Moon. Yet even on moonless nights there is emission from the Earth's atmosphere itself, known collectively as airglow. One source of airglow is the recombination of ions in the upper atmosphere that were ionized by sunlight during the day. The decay of molecules excited into metastable states by sunlight also emits light. One metastable state of molecular oxygen has a radiative lifetime of about 75 minutes, for example, and so this emission is strongest at dusk. Another source of airglow is photochemical luminescence (chemiluminescence), the process whereby light is released following a chemical reaction that has left a molecule in an excited state. One example of this is the combination of hydrogen and ozone that produces oxygen and excited hydroxyl radicals (OH) near the mesopause at an altitude of 85 km. Another is the reaction of sodium and ozone that results in sodium D-line emission. Contributions of different natural emission sources seen from the surface of the Earth are shown in figure 15.11.

15.8.2 Receivers in space

In order to receive signals from our spacecraft, the receiver must have a direct line of sight to it. Depending on the declination of the target star, the spacecraft may not be visible for all of the day/night on account of the Earth's rotation. Alpha Cen has a high declination, just 27° from the south pole, so is continuously visible from a site sufficiently far south. But daytime reception is anyway not possible for signals at visible/infrared wavelengths, so multiple ground stations at different longitudes would be necessary to ensure continuous contact.

One way around these problems is to place the receiver in space or on the surface of the Moon. This is expensive due to the extra costs of launching and maintaining (or replacing) structures in space. If the receiver is in Earth's orbit, the orbit has to be designed to avoid the target star being eclipsed by the Earth or Sun.

Observing from space also avoids the problems with the Earth's atmosphere outlined in the previous section, but it does not remove all sources of absorption and emission. If a spacecraft travelled to a star-forming region or far into the Galactic plane, optical signals in particular would be attenuated by absorption from gas and dust in the ISM. However, as we saw in section 12.3, the Sun is located in a low-density region of the ISM, so this absorption is barely relevant for nearby stars.

Electrons in the ISM significantly disperse electromagnetic waves with frequencies below several kilohertz, so we could not use these very low frequencies. But as such low frequencies suffer from significant diffraction and only offer limited bandwidth, they are not really of interest for communication.

Celestial emission might be a problem no matter where the receiver is placed. Unresolved stars generate a diffuse background, which is strongest at low Galactic latitudes. Zodiacal light, the scattering of sunlight by dust in the solar system, is strongest in the ecliptic plane, and is easily visible with the naked eye from the Earth's surface on a dark night. These two

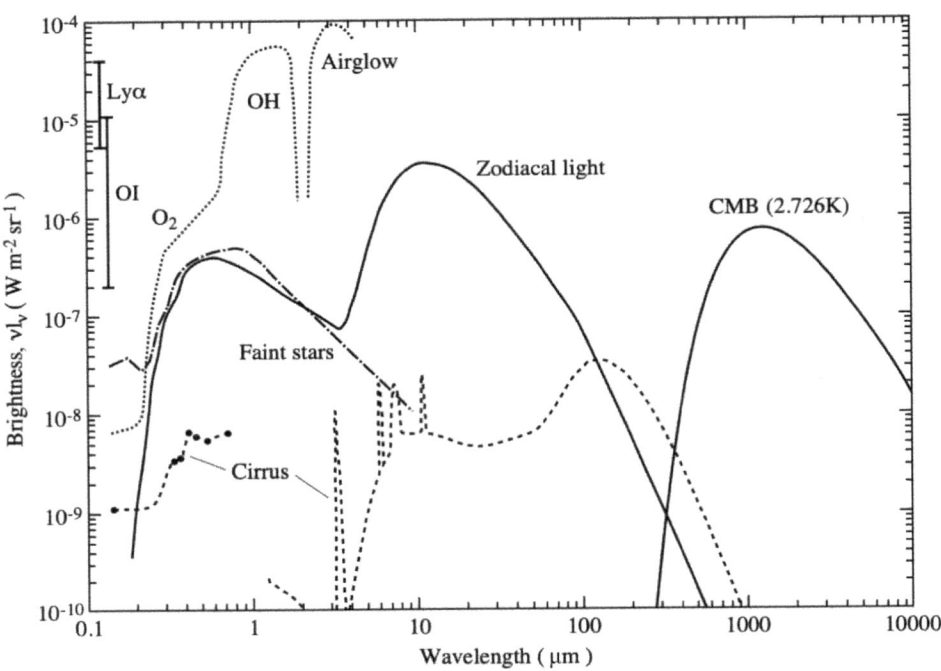

Fig. 15.11 Sources of sky emission from a high-altitude site on the Earth's surface looking away from the Galactic and ecliptic planes. 'Cirrus' refers to emission from interstellar dust, not Earth clouds. OI is neutral oxygen. Credit: Reproduced with permission of EDP, from Leinert et al. (1998).

sources mostly affect the optical and near-infrared. At longer wavelengths there is emission from warm interstellar dust particles. At the longest wavelengths we see emission from the cosmic microwave background (CMB) with a peak flux density at around 1 mm. The relative contributions of these different sources of emission are shown in figure 15.11.

A receiver placed on the surface of the Moon, like one in space, would also benefit from the lack of an atmosphere. It may also be easier to construct or maintain than a receiver in an Earth or solar orbit. There may still be a problem of maintaining a line of sight to the target, however, as the Moon rotates once on its axis every 27 days, so multiple receiving locations may again be required.

15.8.3 Technological issues affecting the choice of wavelength

As already discussed in section 15.2, minimizing the diffraction angle of the transmitted signal is a major factor in maximizing the signal intensity received at the Earth. The diffraction angle is of order λ/d_t. Whereas the ratio of radio to optical wavelengths is of the order of 1000, the ratio of radio to optical telescope diameters that could plausibly be installed on interstellar spacecraft is likely to be much lower: an optical telescope diameter of 1 m is plausible, a radio telescope diameter of 1 km less so. This suggests that, all other things being equal, the limited mass and size of an interstellar spacecraft favours the use of optical rather than radio wavelengths for communication.

In order to achieve a diffraction-limited beam, the transmitting telescope's surface needs to have the correct curvature to within a fraction of a wavelength. In the near-infrared this means of the order of 0.1 μm. This has usually been achieved by casting and polishing glass mirrors and then covering them with a thin coating of reflective aluminium. At radio wavelengths the surface accuracy requirements are significantly more relaxed, more like 1 mm. As high-energy particles will damage the leading surface of a spacecraft (see section 12.4), we may want to deploy the telescope only near the destination (although it would be pointing against the direction of travel). Deploying structures to achieve a very specific and exact shape is challenging.

One idea suited for laser sailcraft in particular is to repurpose the sail as a transmitting telescope. This would save mass. However, as we saw in section 11.4.1, the optimal material for a laser-powered sail may not be highly reflective (e.g. graphene), making it unsuitable as a telescope. The sail may also not have the right shape, as the optimal shape from a stability perspective is probably not ideal for a telescope (see section 11.4.2). The sail could be reshaped, and perhaps even recoated, in situ, but both of these add considerable complexity and risk to the mission.

The properties of the transmitter are only part of the equation, literally in the Friis transmission equation (section 15.2). To increase the data rate we can also increase the total area of the receiver. There are practical limits to the size of individual telescopes. Currently, the largest steerable optical and radio telescopes have diameters of order 40 m and 100 m, respectively. Larger areas need much more massive structures to support them on the Earth or Moon (area $\sim d^2$, but supporting mass $\sim d^3$). So to achieve a larger area, we would probably use arrays of small telescopes.

So that the receiver can spatially separate the spacecraft signal from the bright target star, the receiver would probably be an imaging system (see section 15.6). To achieve the required angular separation, each telescope in the receiver array needs a minimum diameter due to diffraction, and must be equipped with a coronagraph. If situated on the Earth's surface, each telescope must also be fitted with adaptive optics (see section 11.2.3). If the spacecraft and star are physically separated by 0.1 au, then to separate them in the focal plane by at least the diffraction angle at 1 μm, we need individual telescopes at least 3 m in diameter (and in practice larger, as it's better to separate the star and spacecraft by more than one diffraction angle). In that case the individual telescopes can be combined incoherently, meaning each just acts as a separate light bucket. If the individual telescopes are smaller, then to achieve the required angular separation, we need to combine at least subsets of the telescopes coherently to synthesize a larger aperture (see section 11.2.2).

The idea of coherent light combination could be used to produce the transmitted beam on the spacecraft too. That is, instead of using a single large mirror, we could retain just parts of the telescope's surface (see figure 11.11). This changes the diffraction pattern at the receiver, in particular by putting more power into the side lobes (see figure 11.12), but it would make it easier to synthesize a large transmitter on the spacecraft.

Transmitting a coherent signal using a synthesized aperture requires that the individual mirror segments be positioned precisely (sub-wavelength) on the equivalent aperture. If this cannot be achieved, then instead of mirror segments at the correct locations, we can instead use a set of small individual transmitters at arbitrary locations. Each transmits the same signal, but at different phases that account for them not being in the correct locations. That is, we emulate a single telescope (with large gaps) by introducing the appropriate phase shifts into individual small transmitters. This is known as a phased array antenna, which we also mentioned in the context of laser sails in section 11.3.2. This gives us the freedom to place the individual transmitters essentially anywhere, which gives more flexibility onboard a spacecraft. The direction in which the signal is sent is then determined by the phase offsets. This is also referred to as an electronically steerable antenna. Such transmitters are already used on astronomical satellites where the physical movement of an antenna would be too disruptive, for example on Gaia.

Current electronics technology only allows the measurement of phases for frequencies below about 1 THz, that is, wavelengths above 300 μm, so phased array antennae are mostly used at radio wavelengths. However, optical phased arrays have also been made, in which optical path differences are introduced by changing the refractive index of materials electronically. This avoids the need for cumbersome delay lines that are otherwise used in optical aperture synthesis (e.g. figure 11.11).

In summary, there are many factors that influence the optimal wavelength for communication. The choice for a specific mission will be a trade-off between all factors, and will be strongly influenced by the spacecraft mass budget, the required data rate, risk, and cost. However, it seems likely that interstellar spacecraft will use visible or near-infrared wavelengths, and a receiver that is located in orbit or on the Moon that has continuous contact with the spacecraft.

15.9 Case studies

We now look at some specific concepts for sending signals back from interstellar spacecraft. To put this in some context, the New Horizons mission to Pluto (and later Arrokoth) transmitted its data from over 30 au back to Earth at radio frequencies (X-band, 3.6 cm) using a 2.1 m diameter antenna. The signal was received by a 70 m antenna. The data rate was around 1–$2\,\mathrm{kb\,s^{-1}}$, and it took around 15 months to transmit the 6.25 GB dataset obtained at Pluto.

Milne et al. (2016) outline a very large, high-power antenna to achieve a very high data rate from Alpha Cen as part of the Icarus project (see section 8.6). The radio transmitter is a parabolic reflector of 1 km diameter and 40 t mass that is fabricated and deployed after arrival. It transmits 1 MW of power at 32 GHz (1 cm) with a 10% (3 GHz) bandwidth. With this they aim for a data rate of $10\,\mathrm{Gb\,s^{-1}}$. For comparison, high-definition television is $1.5\,\mathrm{Gb\,s^{-1}}$, which is already much higher than is achieved from current space telescopes. Using equation 15.17, the data rate and bandwidth imply an SNR in the received signal of 9.1. Heat dissipation is an issue at these high powers, especially at high frequencies because the transmitting electronics are smaller. To achieve this high data rate, a 15 km diameter receiver is required which, given its size, is placed in space at the Earth/Sun L4 or L5 Lagrange point.

Milne et al. (2016) also consider an optical downlink with a 0.532 μm laser with an average power of 1 kW transmitted by a 10 m diameter mirror. A 200 m diameter telescope receives the signal on the Earth. Using the Friis transmission equation (equation 15.3) we can compute that this would deliver an average power at the receiver of $5 \times 10^{-12}\,\mathrm{W}$, or about 14 million photons per second. They actually report values a factor of 10 lower, because they more realistically take into account optical losses in the transmitting and receiving telescopes, absorption by the Earth's atmosphere, and imperfect detectors. The downlink has a very narrow bandwidth of 3 GHz, or 0.003 nm. They use a PPM modulation scheme with 1024 slots in the frame, which with optimal encoding corresponds to 10 bits per frame (see section 15.4). On average they expect this optical link to achieve a data rate of $1 \, \mathrm{kb \, s}^{-1}$.

Parkin (2020) describes an optical communication channel for a laser sailcraft at Alpha Cen as part of the Breakthrough Starshot initiative. It uses a laser of 100 W power operating at 1.02 μm with a bandwidth of 0.1 nm. This is transmitted via the sail, which has been reconfigured as a reflecting telescope of 4.1 m diameter. The receiver is a 30 m diameter telescope. The spacecraft is moving at a velocity of $0.2 \, c$ away from the Earth, so the received signal is redshifted to 1.25 μm. The 'relativistically moving' Friis transmission equation (equation 15.5) predicts that this would deliver photons at a rate of $1210 \, \mathrm{s}^{-1}$ at the receiver. This assumes perfect efficiency of the optics and electronics and no signal losses. Overall the communication system is stated to deliver 288 signal photons per second at the receiver, indicating a system efficiency of about 25%.

The study of Parkin (2020) uses a telescope to collect and focus the optical signal from the spacecraft. As we saw in section 15.6, there can be considerable contamination from the light of the target star, which here is Alpha Cen, and this is taken to be the dominant source of noise. Adopting a PPM modulation of the signal, the study concludes that the received 288 signal photons per second can carry data at a rate of $260 \, \mathrm{b \, s}^{-1}$ with an error rate of 10^{-3}. This is quite a large error rate but the expectation is that it could be reduced to 10^{-6} with a better coding scheme.

A larger data rate and/or lower error rate is possible if we have larger optics. Lubin (2016) describes a system with a 30 m diameter telescope transmitting a 10 W signal at 1.06 μm from Alpha Cen. Employing an enormous array of telescopes on or near the Earth, with an equivalent single-telescope diameter of 10 km, signal photons are received at a rate of $1.6 \times 10^{10} \, \mathrm{s}^{-1}$ (from equation 15.1). The author then assumes that each photon can carry 0.025 bits on average, a value that reflects some SNR (channel capacity calculations are not provided). Combined with the quoted photon rate, this delivers a data rate of $390 \, \mathrm{Mb \, s}^{-1}$. This is enough to stream a 640×480 video at 30 frames per second, which requires $221 \, \mathrm{Mb \, s}^{-1}$ when adopting 24 bits per pixel and no compression or channel coding.

Hippke (2019) considers an optical system operating at visual wavelengths (0.43 μm; bandwidth 1 nm) that uses a relatively large power (1000 W) but a small transmitter (1 m diameter). The receiver on the Earth's surface has a diameter of 39 m, the diameter of the ELT in Chile. The photon rate from the spacecraft at Alpha Cen on the receiver (prior to detection) is given to be $1370 \, \mathrm{s}^{-1}$. This compares to $6500 \, \mathrm{s}^{-1}$ from equation 15.3, which is prior to accounting for losses of any relativistic velocity. They assume the light from Alpha Cen can be perfectly suppressed so contributes no noise. They instead assume that the dominant noise source is emission from the Earth's atmosphere, estimated to be $0.6 \, \mathrm{photons \, s}^{-1} \, \mathrm{nm}^{-1} \, \mathrm{m}^{-2} \, \mathrm{arcsec}^{-1}$. Taking the collecting area of the telescope as $1170 \, \mathrm{m}^2$, this corresponds to 702 sky photons per second. They then use the approach from section 15.7.3 to compute the data rate. Specifically, they adopt a very low value of the number of photons per mode of $M = 10^{-5}$ and give the number of bits per incident photon to be 1.81, which for the stated detector efficiency of $\eta = 0.5$ corresponds to $C_p = 1.81/0.5 = 3.62$ bits per detected photon. Using equation 15.23 together with equation 15.24, we see that this has assumed about $N = 0.18$ noise photons per mode. As the incident signal photon rate is $1370 \, \mathrm{s}^{-1}$, we get a data rate of $1.81 \times 1370 = 2480 \, \mathrm{b \, s}^{-1}$.

15.10 Summary

Getting data back to the Earth from a distant spacecraft is a major challenge. As the transmitter will inevitably be limited in size, diffraction of the transmitted beam results in an extremely low intensity at the Earth. Limitations of the transmitter can be compensated to some degree by using a large area receiver, as shown by the Friis transmission equation (equation 15.3). Shorter wavelengths reduce diffraction, and for this reason in particular we probably want to use optical rather than radio communications.

To give some example numbers, a $10\,\text{W}$ transmission with a $1\,\mu\text{m}$ laser from a $4\,\text{m}$ diameter telescope at Proxima Cen delivers an intensity at the Earth of $6 \times 10^{-18}\,\text{W}\,\text{m}^{-2}$ at the centre of an Airy diffraction pattern, the central disc of which is 24 million kilometres in diameter. To maximally block out the far brighter target star, which typically has a small angular separation from the spacecraft, the receiver should be an imaging telescope equipped with a coronagraph and narrow band filter. This telescope needs a minimum diameter of a few metres to separate the signal and star on its focal plane, as well as adaptive optics correction if the telescope is observing through the Earth's atmosphere. With a single $10\,\text{m}$ receiving telescope, the above signal intensity corresponds to just 31 photons per second, less when considering losses in the telescope and detector. With an array of 100 such telescopes, combined incoherently, this increases to 3100 photons per second.

The amount of information individual photons can carry depends on how a message is encoded and modulated. At low photon counts, pulse position modulation (PPM) is convenient. By modulating in time, an individual photon can convey more than one bit of information. Up to 15 bits has been demonstrated experimentally with PPM.

The received data rate depends also on the noise, of which there are many sources, including the target star, absorption and emission in the Earth's atmosphere, and detector electronics. Noise introduces errors in the messages. To overcome this, redundancy must be built into the signal by means of an error-correcting code. Remarkably, even in the presence of large noise, it is theoretically possible to communicate at non-zero data rates with an arbitrarily low probability of message error. Codes have been developed that in some situations can almost reach the theoretical data rate limit, or channel capacity.

Assuming we can extract 10 bits per photon from the signal, then 3100 received photons per second gives a data rate of $31\,\text{kbs}^{-1}$, or nearly $1000\,\text{Gb}$ per year. This corresponds to about 30 000 uncompressed 2048×2048 images of 8 bits per pixel. Compression would reduce the raw data size, but the error-correcting code required to overcome noise would increase it. Losses and inefficiencies would also lower the final data rate, perhaps by several factors.

A spacecraft would probably transmit its preciously won data over a period of several years (whether a flyby or a rendezvous). As the transmitter and receiver cannot coordinate, the spacecraft cannot retransmit data in the case of inevitable signal loss. To cope with this, the coding algorithm needs to provide some long timescale redundancy too.

15.11 Exercises

1. I outlined in section 15.2.1 how the Friis transmission equation needs to be modified for relative motion between transmitter and receiver. Verify the resulting equation by re-deriving it from first principles.
2. Compute the power and photon rate on the Earth-based receiving telescope for the DSOC experiment mentioned at the end of section 15.7.3, assuming an end-to-end efficiency of 0.1. How many bits per photon does this correspond to?
3. Show that for the Holevo bound, the channel capacity in bits per mode from equation 15.23 tends towards the noise-free case of equation 15.21 as the detector efficiency $\eta \to 1$.
4. List the relative merits of doing interstellar communication in the optical and in the radio. Under what circumstances might one be clearly preferred over the other?

Payloads

The payload can be considered as everything the spacecraft carries that is not related to propulsion. We examined navigation and communication in previous chapters. Here we look at the scientific instruments that an interstellar spacecraft may carry and what they would be used for, consider possible sources of onboard power, and outline the need for autonomous control of the spacecraft. We also touch briefly on the idea of spreading a mission's payload across multiple spacecraft in the form of swarms or relays.

16.1 Scientific instruments

The main purpose of an interstellar spacecraft is to carry instruments to the target star system, to enable these instruments to obtain data, and to transmit the resulting data back to the Earth. The main targets of scientific interest are the star itself and planets around the star. Further objects of interest include other planetary bodies in the star system, such as comets and moons, any stellar disk, the star's heliosphere, and the space between our Sun and the star, as outlined in chapter 2.

To best study these targets, a range of instruments are required. Just as is done for missions in the solar system, the instruments would be optimized for the specific target, for example whether the planet is known to be a small rocky planet, a gas giant, or an intermediate-mass icy world. Instruments nonetheless need to be sufficiently general to make discoveries and to allow them to adapt in response to new information that will inevitably be discovered about the target system during the spacecraft's long journey from the Earth.

The types of instrument would be broadly similar to those carried by interplanetary spacecraft to explore our solar system. A few example missions are Mars Express, Europa Clipper and JUICE that will explore Jupiter and its moons, and New Horizons that visited Pluto and the Kuiper belt. An overview of the JUICE instruments is given in figure 16.1.

Instruments that fly on interplanetary spacecraft often have masses of tens of kilogrammes and require powers of tens of watts. The New Horizon instruments were relatively light and low power, of the order of a few kilogrammes and watts, as this was a low-mass, fast mission that flew to Pluto in 9.5 years, with just one gravity assist at Jupiter. For an interstellar spacecraft there will be a much stronger need to minimize the instruments' masses. In the case of laser sailcraft weighing just a few tens of grammes or less, this would call for extreme measures, such as having entire sets of instruments on a single computer chip. This greatly limits what any one spacecraft can do, although it could be partly compensated by sending a swarm of spacecraft, as we will consider in section 16.3.

Inspired by the instruments on interplanetary spacecraft, let us now look at the classes of instruments an interstellar spacecraft might carry and what they can be used for.

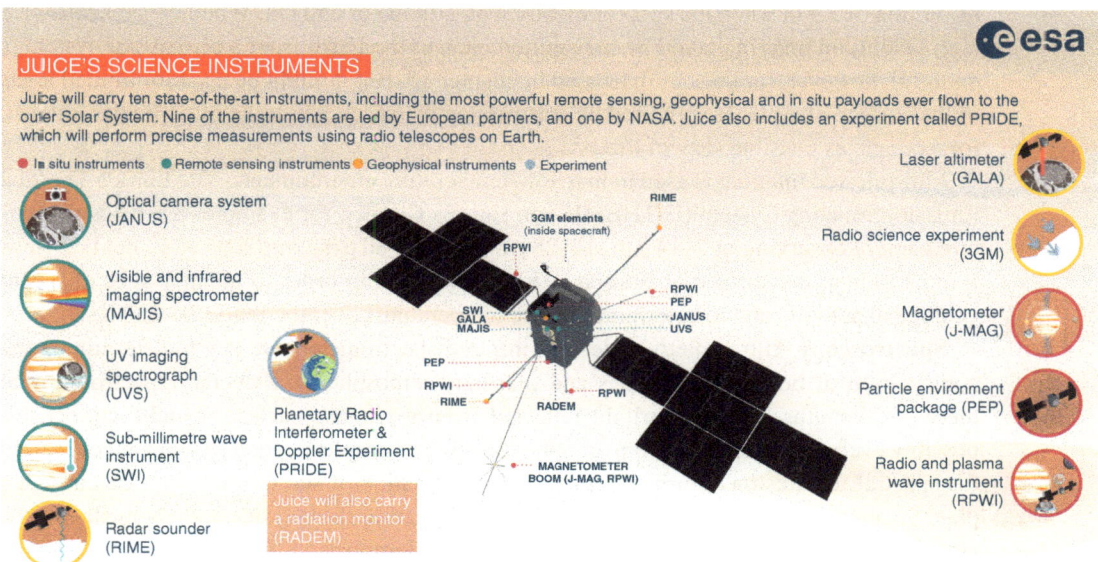

JUICE'S SCIENCE INSTRUMENTS

Juice will carry ten state-of-the-art instruments, including the most powerful remote sensing, geophysical and in situ payloads ever flown to the outer Solar System. Nine of the instruments are led by European partners, and one by NASA. Juice also includes an experiment called PRIDE, which will perform precise measurements using radio telescopes on Earth.

● In situ instruments   ● Remote sensing instruments   ● Geophysical instruments   ● Experiment

Optical camera system (JANUS)

Visible and infrared imaging spectrometer (MAJIS)

UV imaging spectrograph (UVS)

Sub-millimetre wave instrument (SWI)

Planetary Radio Interferometer & Doppler Experiment (PRIDE)

Radar sounder (RIME)

Juice will also carry a radiation monitor (RADEM)

Laser altimeter (GALA)

Radio science experiment (3GM)

Magnetometer (J-MAG)

Particle environment package (PEP)

Radio and plasma wave instrument (RPWI)

The scientific instruments on the Jupiter Icy Moons Explorer (JUICE). Credit: European Space Agency - ESA (acknowledgement: work performed by ATG under contract to ESA), ESA Standard License.

Fig. 16.1

Imaging. Ultaviolet/visible/near-infrared imaging is essential. Narrow-field imaging can be used to map the surface of planetary bodies in reflected starlight, with wide-field optical imaging used to look for new planets as well as moons and rings around planets. Both types of camera would also be used for navigation. It is common for planetary missions to use separate cameras for these two types of imaging because it allows each mode to be optimized separately. Mobile phones often have multiple cameras for the same reason, and because a single camera with optical zoom capability takes up more space. Cameras equipped with multiple filters enable broad-band spectral imaging, which helps to determine the composition and surface temperature of rocky bodies. A spacecraft making a rapid flyby may not want to waste time changing filters, so multiple cameras with different fixed filters might be used instead.

An example camera is Ralph on New Horizons. This is an off-axis three-mirror telescope with a 7.5 cm aperture, which uses a dichroic beamsplitter to send visible light and infrared light to different cameras. One of these is the Multispectral Visible Imaging Camera (MVIC) with a field of view of $5.7° \times 5.7°$ and four optical filters (400–975 nm), delivering $4''$ per pixel on the detector. The infrared channel is the Linear Etalon Imaging Spectral Array (LEISSA), which provides spectral maps in the range 1.25–2.50 μm at a resolution of $\delta\lambda/\lambda = 240$, and 550 in the range 2.10–2.25 μm, over a $0.9° \times 0.9°$ field of view at $12.8''$ per pixel. New Horizons also has a separate narrow-field camera called LORRI (the Long Range Reconnaissance Imager), a 20.8 cm camera with a $0.29° \times 0.20°$ field of view delivering $1''$ per pixel.

Planetary cameras often use methods from Earth remote sensing to construct larger images. A common approach is the 'push broom scanner', in which an image is built up in strips. This is often combined with a spectroscopic capability such that a spectrum is obtained in each pixel along the strip. This is how LEISSA is operated, for example, as does the High Resolution Stereo Camera (HRSC) onboard Mars Express, which achieved a ground resolution of 2.3 m when the spacecraft was at an altitude of 250 km. Whether such extensive high-resolution imaging could be transmitted back to the Earth from a distant star system is doubtful, however, for reasons discussed in chapter 15. But it might be used for limited areas of the surface of a planet identified by software on the spacecraft to be of particular interest, for example as possible sites of life.

Mid-infrared imaging is used to map low-temperature environments. The Europa Thermal Emission Imaging System (E-THEMIS) on Europa Clipper, for example, is a 6.45 cm aperture camera observing at 7–14 μm, 14–28 μm, and 28–80 μm, with a 6° × 1° field of view. It aims to map the surface of Europa at 8 km resolution in order to identify sites of recent geological activity, and to determine surface temperatures over the range 70–230 K.

Spectroscopy. Ultraviolet/visible/near-infrared spectrometers are key to determining the composition of both solid surfaces and planetary atmospheres. Planetary spectroscopy at these wavelengths can use starlight reflected from a planet's surface – rock, ice, oceans, possibly even vegetation – or from an optically thick atmosphere, such as when investigating the spectral features on Saturn or Jupiter. The observation of starlight transmitted through a planet's atmosphere is a powerful means of investigating the physics and chemistry of the atmosphere. This is the same technique as used by telescopes on the Earth to investigate the atmosphere of an exoplanet as it transits across the surface of its host star (see section 2.3.1). However, that only works for the small fraction of systems in which the planet's orbit is seen almost exactly edge-on from the Earth. A telescope that flies to a non-edge-on planetary system can be manoeuvred to place the planet between it and the star, thereby inducing a transit, and so providing transmission spectroscopy we would otherwise never get from the Earth. Such a spacecraft could also spatially resolve different parts of the planet's limb during transit and – if it spent enough time in the star system – obtain transmission spectra at different times of the planet's year. Such observations could reveal a wealth of information about the planet's atmospheric composition and structure.

Different wavelengths reveal a range of information on different parts of an atmosphere or surface. The Visible and Infrared Mineralogical Mapping Spectrometer on Mars Express, for example, uses multiple gratings to observe in 352 spectral channels from 0.36 μm to 5.2 μm. The ultraviolet spectrograph Alice on New Horizons operates over 50–190 nm at 0.2 nm per pixel to investigate the temperature and compositional structure of Pluto's upper atmosphere, focusing on lines/bands of hydrogen, carbon monoxide, argon, neon, nitrogen, and methane. The JUICE spacecraft is equipped with a submillimetre spectrometer operating over the ranges 1080–1275 GHz and 530–600 GHz with a resolving power of 10^7 fed by a 30 cm dish. This will investigate the structure, composition, temperature, and dynamics of Jupiter's stratosphere and troposphere, as well as the surfaces and exospheres of the Galilean moons.

Particles. Imaging and spectroscopic instruments are photon detectors, and photons are essentially the only means astronomers currently have for investigating distant star systems.[1] One of the big advantages of in situ observations by a spacecraft is that massive particles can be directly detected.

The workhorse of particle investigations is the mass spectrometer. The name notwithstanding, this device often measures the mass-to-charge ratio of individual particles rather

---

[1] Astronomy is also done through the detection of high-energy particles (sometimes indirectly), and more recently gravitational waves, but the former cannot be used to tell us about specific individual stars or exoplanets, and the latter are currently limited to observing massive interacting bodies.

than their masses. A typical spectrometer works by using one electric field to ionize neutral particles and another to accelerate them. A magnetic field is then used to deflect the ions through an angle via the Lorentz force. For known electric and magnetic field strengths, the angle of deflection is determined by the particle's mass-to-charge ratio m/q, which is inferred from where the particle lands on a detector. Observing a population of particles with different m/q ratios gives a spectrum showing the relative numbers of each type of particle. Another way of achieving this measurement is with a 'time-of-flight' mass spectrometer: ions are accelerated through a known potential U and so achieve an (unknown) kinetic energy $qU = (1/2)mv^2$. By measuring the time it takes the ion to travel a known distance (often by performing multiple bounces in a linear detector), we can determine the velocity v of the ion, and therefore m/q. One example of such an instrument is MASPEX (MAss Spectrometer for Planetary EXploration) on Europa Clipper. This achieves a mass resolution in excess of $17\,000$ by using thousands of bounces within the spectrometer, since the mass resolution equals the time resolution, which is proportional to the travel time. The goal of this spectrometer is to identify volatile compounds including water and organics in Europa's exosphere (and maybe also in plumes), and to associate them with geological surface features.

In some detectors, the kinetic energy of an ion can be determined from its impact in a solid state detector. As the incident ion passes through the silicon lattice of the detector, it is decelerated by ejecting electrons from the lattice. The current produced by these electrons is a measure of the ion's initial kinetic energy, $(1/2)mv^2$. Combining this with a time-of-flight measurement to determine v, the mass m and not just the mass-to-charge ratio is estimated. An example of such an instrument is the Pluto Energetic Particle Spectrometer Science Investigation (PEPSSI) on New Horizons, which is sensitive to ions with energies in the range $1\,\text{keV}$ to $1\,\text{MeV}$ and achieves a mass resolution of 2–$15\,m_\text{u}$. Another example is the Particle Environment Package (PEP) on JUICE, a suite of instruments that will measure the ions, electrons, and neutral atoms in Jupiter's magnetosphere and near its moons, with a broad range of energies from less than $0.001\,\text{eV}$ to over $1000\,\text{MeV}$.

Particle detectors are not only useful in the vicinity of an exoplanet. They are also crucial for mapping out the interaction of the ISM with the stellar wind, the star's equivalents of our Sun's heliosphere (see section 12.2), and the interaction of this wind with a planet's atmosphere. The SWAP instrument on New Horizons was specifically designed to measure the interaction between the solar wind and ions created by atmospheric loss from Pluto. This provides an estimate of the rate of atmospheric loss from Pluto. It measures the energies and mass-to-charge ratio of charged particles in the solar wind with energies from $25\,\text{eV}$ to $7.5\,\text{keV}$ using a retarding potential analyzer and electrostatic analyzer.

Dust. Larger than individual atoms and molecules, dust particles reveal much about the surface composition of rocky planets. The SUrface Dust Analyzer (SUDA) on Europa Clipper analyses dust via a number of techniques. The flight time of incoming particles measured between a grid and a detector gives their velocities. A time-of-flight impact ionization mass spectrometer then ionizes a fraction of a dust particle through its high-velocity impact in the instrument, from which the mass-to-charge spectrum can be measured to determine the composition of the dust. Separately, this instrument measures the charge of dust particles as a way to estimate their size. New Horizons carries a dust counter to count the rate of dust particle impacts, and measured how this varied with distance to Pluto, thus mapping out the space density of the dust.

How or whether these particle and dust instruments could work in a flyby mission when the spacecraft is moving at, say, $0.1\,c$ relative to the target star or planet is unclear. They would be most effective for an orbiting mission, also because that can get closer to the target and spend more time there.

Magnetic fields. The Earth and all of the gas giant planets in the solar system have strong magnetic fields. The magnetic field at Jupiter is so strong that interplanetary spacecraft need to take special measures to shield their electronics from high-energy particles accelerated by the field. Mercury has a very weak magnetic field, and Mars preserves remnants of past magnetic fields through magnetization of some of its surface.

J-MAG on JUICE is a magnetometer that will characterize Jupiter's magnetic field. This uses a fluxgate magnetometer, which works as follows. Two coils of wire are wrapped around a magnetically susceptible core. An alternating current is applied to one of these coils, which induces an alternating current in the other coil. In the absence of a magnetic field, these two currents are in phase. But when an external field magnetizes the core, this introduces a phase offset. The size of this offset is a measure of the external field strength and, by using multiple sensors with different orientations, the direction of the field can be determined.

JUICE's magnetometer will also be used to investigate the suspected subsurface water oceans on the Galilean satellites Europa, Ganymede, and Callisto. While Ganymede has an intrinsic magnetic field, Europa and Callisto appear not to. However, time variations in Jupiter's field can induce currents in the oceans, which in turn can produce a measurable magnetic field. The strength of this field depends on the depth and salinity of the ocean and how this varies over the surface. Such experiments will be carried out by both JUICE and Europa Clipper. Magnetometers were flown on all the other missions to the outer planets too (Pioneer, Voyager, Galileo, Cassini, Juno). In all cases the instruments are mounted on booms to keep them away from the rest of the spacecraft to minimize magnetic interference with the instrument (figure 16.2).

The Radio & Plasma Wave Investigation (RPWI) instrument on JUICE can also measure magnetic fields, but uniquely it also measures electric fields. For this it uses Langmuir probes. These devices run an electric current along an electrode that is influenced by the temperature, density, and potential of the plasma that surrounds it. These devices may be sensitive enough to detect electric currents within Jupiter's moons.

Magnetic fields can also be explored indirectly using energetic neutral atoms (ENAs). Charged particles inside stellar or planetary magnetospheres are constrained to move on curved paths within the field, so cannot be collected by an external observer. However, if an ion collides with a neutral atom, it is possible for the ion to strip an electron off the neutral atom (a charge exchange collision). The ion retains most of its kinetic energy, but is now neutral, an ENA. As a result, this particle is no longer deflected by the magnetic field and escapes from the magnetosphere where it can be imaged. The Cassini spacecraft used an ENA imager to investigate a thermal ring current in Saturn, and JUICE is also be equipped with such an imager.

Radar. All of the instruments mentioned so far have been passive in the sense that they wait to measure whatever photon, particles, or fields they encounter. Radar ranging, in contrast, bounces electromagnetic pulses generated by the spacecraft off an object in order to measure the distance to it via the returned pulse's travel time. JUICE is equipped with two such instruments. GALA, the GAnymede Laser Altimeter, is a laser radar (a lidar), which will be used to study the shape of Ganymede at high spatial resolution. When JUICE passes

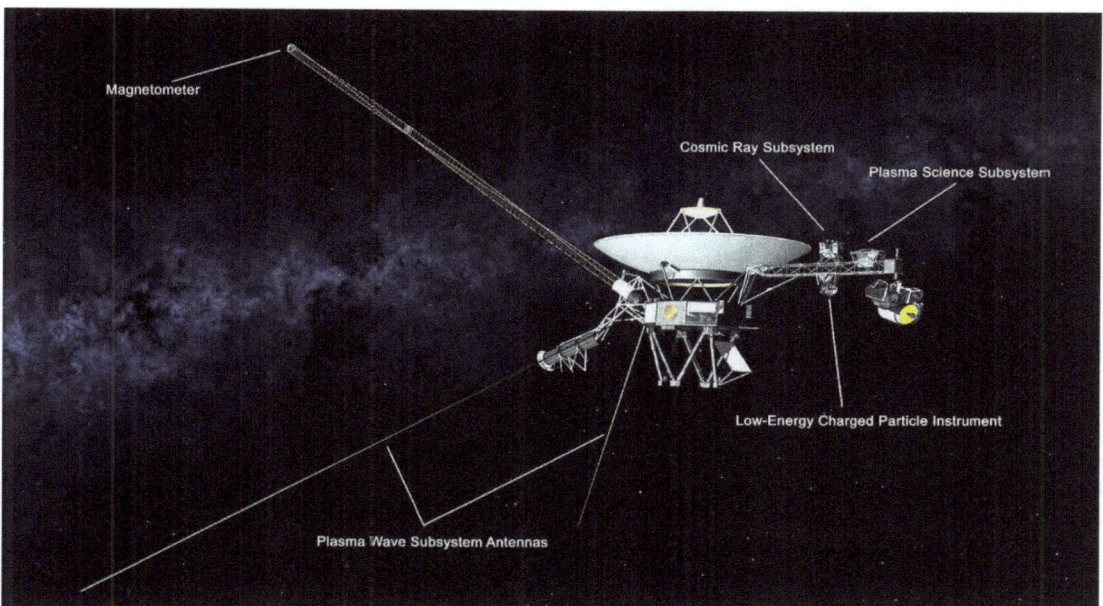

The Voyager spacecraft indicating the location of some of the instruments. Credit: Courtesy NASA/JPL-Caltech.

Fig. 16.2

within 200 km of the moon's surface, GALA will produce a spot size of 20 m on the surface, enabling it to map the surface with a vertical resolution of just 10 cm. In this way it is possible to map in detail surface features that might be caused by impacts, volcanism, or tidal forces. A similar instrument is in use on BepiColombo at Mercury. The brightness of the reflected beam can also be used to estimate the albedo and roughness of the surface.

A second instrument on JUICE called RIME – Radar for Icy Moon Exploration – is an ice-penetrating radar to study the subsurface structure of the moons. Using a 16 m antenna and operating at 9 MHz, this should be able to penetrate to a depth of 9 km and give a vertical resolution of 16 m. This will map the distribution of water in any subsurface ocean, as well as the variation in the ice thickness. A similar instrument is deployed on Europa Clipper and on the Mars Reconnaissance Orbiter.

Radar ranging requires the spacecraft to pass very close to the planet or moon, at altitudes of just hundreds of kilometres. This could be achievable for an interstellar spacecraft that goes into orbit around an exoplanet or exomoon, but is too risky, and of limited value, for a fast flyby.

Gravity. A perfectly spherical planet of uniform density has a gravitational field that varies only with altitude. Variations in the shape or density of a real planet, including from surface features such as mountains or deep oceans, means the gravitational field experienced by a low orbiting spacecraft varies as it passes over different parts of the planet. This in turn causes the position and velocity of the spacecraft to deviate from a Keplerian orbit.

Several interplanetary missions have made use of their radio communication link with the Earth to map the gravitational field of a planet in this way. By accurately measuring the line-of-slight velocity of the spacecraft via the Doppler shift of signals sent from an Earth-based transmitter, as well as variations in the signal travel time, perturbations of the orbit caused by variations in the planet's gravitational field can be measured. This enables some reconstruction of the mass distribution of the planet. Unfortunately, this kind of measurement will not be available to a single interstellar spacecraft. But one could imagine having two interstellar spacecraft, one remaining far from the planet and acting as the analogy of the Earth-based transmitter.

Not long after the first interplanetary missions in the early 1960s, landers were successfully deployed on the surface of planets, initially the Moon, Venus, Mars, and later Saturn's moon Titan. The first interstellar mission may just be a flyby, at a push perhaps an orbiting rendezvous mission. But one day orbiting spacecraft may be able to drop probes into the atmospheres of exoplanets, and eventually perhaps deploy landers or rovers (or submarines or aircraft). This would permit a much more detailed analysis of the atmosphere, geology, and oceans of exoplanets. I'll leave concepts for doing that to the imagination of the reader.

16.2 Power

Power is required onboard a spacecraft in the form of electricity for many functions, including operating scientific instruments, navigation, and communication. Manoeuvres to adjust the velocity or orientation of the spacecraft also require an energy source, even if they are not done with reaction thrusters but, for example, by changing the reflectivity of parts of a sail.

A spacecraft that is a rocket or Bussard ramjet could use some of its fuel as a source of electrical energy, for example to drive a generator in a thermal cycle. Some reactor-based fusion drives like those mentioned in section 8.7 are specifically intended to also provide electricity in this way. A nuclear reactor could even be added specifically for the purpose of generating electricity. Fission reactors have been launched into space, for example the experimental reactor SNAP-10a by the USA, which is defunct but still in orbit, or reactors on the 30 RORSAT spy satellites launched by the Soviet Union. But as nuclear reactors are massive, they are only an option for power generation on an otherwise massive interstellar spacecraft.

Photovoltaics for electricity generation from starlight are frequency used for satellites, increasingly also for deep space missions (e.g. JUICE and Europa Clipper). For a laser-propelled sail, one could even consider using the laser to beam energy. However, due to the enormous diffraction losses at large distances, this could only realistically be used when the spacecraft is still relatively close to the laser. When the spacecraft is far from either the Sun or the destination star, starlight is also far too weak to generate a useful amount of energy: 2 ly from the Sun – half way to Proxima Cen – the solar intensity is 10^{10} times lower than at the Earth, providing only 10^{-7} W m^{-2}. If the mission is designed to require very few operations during the cruise phase – perhaps with the spacecraft in hibernation, as is sometimes done for interstellar missions – then it might be able to rely on some other energy source in this phase, and only make use of starlight when it comes within a few astronomical units of the target star.

Other than photons, the only other source of external energy a spacecraft could rely on far from a star is the kinetic energy of interstellar particles. For a spacecraft travelling at a velocity v relative to the ISM, the power of particles stopped by the spacecraft is $P = (1/2)\dot{m}v^2$, where $\dot{m} = \rho A v$ is the mass flow rate, ρ is the space density of the particles, and A is the cross-sectional area of the spacecraft. Thus $P = (1/2)\rho A v^3$. If we consider just protons (of mass m_p) and adopt the local interstellar cloud space density of $n = 0.3 \times 10^6\,\mathrm{m}^{-3}$ ($= \rho/m_p$), then for a spacecraft travelling at $v = 0.1\,c$ the power per unit area is $P/A = (1/2)nm_p v^3 = 6.7\,\mathrm{W\,m}^{-2}$. This power would have to be converted to electricity somehow, and this would be much less than 100% efficient. As the colliding particles heat the front of the spacecraft, it might be possible to exploit the temperature gradient along the spacecraft to extract this energy as electricity via the thermoelectric effect (discussed below). But to extract several watts of power in this way we would need a significant cross-sectional area. Sails have large areas, but they would also need devices to extract the power, which could add significant mass.

If we cannot use external sources of power, the spacecraft will need an internal one. Options include batteries, capacitors, fuel cells, and radioisotope thermoelectric generators, each of which we will now consider.

Batteries. Current commercial lithium-ion batteries are able to store about $1\,\mathrm{MJ\,kg}^{-1}$ or $300\,\mathrm{Wh\,kg}^{-1}$ of energy. The best achieved so far have a capacity of around $2.5\,\mathrm{MJ\,kg}^{-1}$. Even with a low payload power of 1 W on average, one kilogramme of such a battery could provide power for only one month. Larger masses of battery could be used, but as we shall see, other options have higher energy densities. Batteries also drain with time, lowering their stored energy. As the sole power source for a multi-decade mission, even if the payload is switched off most of the time, batteries do not look like a good option.

Capacitors. A capacitor also stores electrical energy. Electrolytic capacitors have very low energy densities, however, not much more than about $0.001\,\mathrm{MJ\,kg}^{-1}$ and with very short drain times (days at room temperature). More recent electrostatic double-layer capacitors can increase this to up to $0.7\,\mathrm{MJ\,kg}^{-1}$, but this is still less than batteries. The one advantage of batteries and capacitors, however, is that they can be very small, with little mass overhead for the structure (unlike for fuel cells, heat engines, etc.). So if only very small energies are required, batteries and capacitors may have a niche. They may also be required to regulate power and serve as short-term backups.

Fuel cells. A fuel cell extracts an electric current from the exothermic reaction of combining hydrogen with oxygen. After molecular hydrogen is ionized at an anode, the protons flow through an electrolyte to combine with the electrons (which flow via a separate circuit) and oxygen at the cathode, forming water. The energy density of hydrogen, $120\,\mathrm{MJ\,kg}^{-1}$, is much higher than that of a battery, but a spacecraft must of course also carry the oxygen, which increases the mass by a factor of nine, reducing the energy density to $13\,\mathrm{MJ\,kg}^{-1}$. Including the mass of the cell itself reduces this further. This is still superior to batteries for all but the smallest amounts of energy (when the mass of the empty cell would dominate). Fuel cells were used as one of the onboard power sources on Apollo and the Space Shuttle, with average powers of $1.5\,\mathrm{kW}$ and $7\,\mathrm{kW}$, respectively.

Radioisotope thermoelectric generators (RTGs). One of the main sources of power for deep space missions has been the radioisotope thermoelectric generator (figure 16.3). This extracts electrical energy from the heat produced by the radioactive decay of an isotope, usually plutonium.

A single atom of plutonium-238 decays to uranium-234 by alpha decay, releasing $5.6\,\mathrm{MeV}$ energy, most of which goes into the alpha particle. Together with the rate of decay of plutonium-238 (the half-life is 87.8 years), we can compute that this corresponds to a power of $570\,\mathrm{W\,kg}^{-1}$. This heat source, when connected to a heat sink such as a radiator in space, creates a temperature gradient. Electricity can be extracted from this gradient via the Seebeck

An RTG on the New Horizons spacecraft at Kennedy Space Center shortly before launch. The RTG is the black cylinder on the left. It is 1.14 m long and covered with cooling fins. Credit: NASA.

effect: when two ends of a conductor are maintained at different temperatures, a potential difference between the two ends is generated (due to differences in the energy level occupation in the conductor, together with electron diffusion). The size of the potential difference, known as the Seebeck coefficient, depends on the material, and varies from around $-100\,\mu\mathrm{V\,K^{-1}}$ to $+1000\,\mu\mathrm{V\,K^{-1}}$. Connecting two different materials with different Seebeck coefficients in a circuit allows a current to flow, from which we can extract electrical energy. The energy comes from the heat source used to maintain the temperature difference. The Seebeck effect is widely used, for example in power stations, wrist watches, and oven fans, and is the method conventionally used in RTGs to generate electrical energy.

RTGs were used in the Pioneer and Voyager spacecraft, New Horizons, as well as the Mars landers and rovers. The Cassini mission to Saturn used 33 kg of plutonium that produced 885 W of electrical power at the beginning of its 20-year mission. This corresponds to an electrical power of $27\,\mathrm{W\,kg^{-1}}$, a factor of 20 lower than the heat power. This reflects the inefficiency of the energy conversion. Once we also take into account the mass of the entire device, the initial power rating is only about $5\,\mathrm{W\,kg^{-1}}$. This power decreases with time t as $\exp(-t/\tau)$, where the half-life is $\tau \ln 2$. After 50 years of operation, the power from the RTG drops to 67% of its initial value. In this time it has delivered $6.5\,\mathrm{GJ\,kg^{-1}}$ of energy, which is orders of magnitudes more than batteries or fuel cells.

Given its long half-life, plutonium-238 is suited well for interplanetary missions, and would also be suitable for interstellar missions with durations less than about 100 years. Plutonium-238 also has a very low gamma-ray and neutron flux (from spontaneous fission), so the shielding requirements are minimal, much less than for other candidate isotopes. It is typically supplied in the form of plutonium dioxide (PuO_2), which has a high melting temperature. Plutonium-238 itself is usually generated in fission reactors by irradiating neptunium-237 with neutrons, neptunium-237 itself being generated through a series of neutron reactions and beta decay from uranium-235.

A recent problem with RTGs is the limited availability of plutonium-238, as production has decreased significantly since the 1980s. Partly for this reason, other radioactive sources are being investigated for use in RTGs. Polonium-210 alpha decays to lead-206, releasing 5.4 MeV. This is similar to the energy released in a plutonium-238 alpha decay. But with a much shorter half-life of 138 days, polonium-210 delivers a much higher power of 144 kW kg^{-1}. However, this short half-life renders it useless for an interstellar mission. Strontium-90, which beta decays via yttrium-90 to zirconium-90, also has a similar power output to plutonium-238, but a half-life of 28.9 years. For long missions, americium-241, which alpha decays to neptunium-237 with a half-life of 433 years, is an attractive option, although its power density is four times less than that of plutonium-238.

There are ways other than the Seebeck effect to extract electrical energy from an RTG. By using the thermodynamic Stirling cycle, about four times as much electrical power can be extracted per kilogramme of plutonium than has been achieved so far in space missions. A disadvantage of this, however, is that unlike a traditional RTG, which has no moving parts, the heat engine involves continuous movement and a working fluid. This generates vibrations and increases the risk of mechanical failure. One prototype developed by NASA is the Advanced Stirling Radioisotope Generator. This uses 1.2 kg of plutonium-238 to generate 130 W of electrical power (initially) in a unit that weighs 32 kg.

16.3 Relays and swarms

Developing the first interstellar spacecraft will be expensive. There will be significant costs in research, development, production, and operations. How these costs are distributed depends on which technologies are developed, not least which propulsion method is used. As a single fusion engine is likely to be massive and expensive, the spacecraft itself will be massive and the cost of producing one spacecraft relatively high. With a laser sail, in contrast, much of the construction cost will be in the laser array, which will remain at the Earth and can be reused many times. The production cost of an individual sailcraft is probably small in comparison.

For laser sailcraft at least, it therefore makes economic sense to build and send many spacecraft. Moreover, given that the sailcraft may be very light (grammes) and therefore only able to carry a very limited set of instruments, the complete set of instruments required to fulfil the scientific objectives could be spread across multiple sailcraft. This makes even more sense if the spacecraft is to make a fast flyby. At 10% the speed of light, the spacecraft traverses 1 au in just over an hour, with the result that one camera on one spacecraft could not gather much data. An individual spacecraft would also fly past just one side of a star or planet, whereas multiple spacecraft could be navigated to view the target from multiple perspectives.

Single point failures would destroy a spacecraft. Examples include a large dust grain, or critical hardware damage caused by cosmic rays. Sending multiple spacecraft offers more security against such catastrophic failure, to the extent that one could even anticipate losing some fraction of the spacecraft. We could therefore build in across-spacecraft redundancy, just as now there is usually within-spacecraft redundancy for critical systems.

The idea of sending multiple spacecraft is not necessarily limited to laser sails or to flyby missions, however. As soon as the marginal cost of producing and operating a spacecraft is small compared to the research, development, and other set-up costs, it makes sense to send multiple spacecraft.

Given the unknowns of interstellar travel, there is some advantage in sending multiple spacecraft consecutively rather than simultaneously. This way, information gathered by the first spacecraft could be used to inform the later ones. For example, if an early spacecraft identifies a previously undetected planet or moon, one of the later spacecraft could modify its course to fly closer to that. Or if one of the first spacecraft fails to make a high-priority observation for some reason, such as an instrument failure or course error, a later spacecraft could make it instead. To do this, the spacecraft need to communicate with one another. This complicates their communication systems, and also implies that they cannot be too far from one another (think Friis transmission equation), which may limit the extent of possible course corrections. It also assumes that the spacecraft are equipped with intelligent software, but this, at least, is one of the areas of technological development we can be fairly confident to attain (see section 16.4).

The frequency and timing of launches of multiple spacecraft from the Earth (more likely, Earth's orbit) could depend on a number of factors. Laser-propelled sails require use of the laser for an extended period. As we saw in chapter 11, some very low-mass sail missions have beam times of just minutes. But even those require enormous amounts of energy, which will probably have to be generated over much longer periods and stored. A launch rate of one spacecraft per day may be a reasonable highest frequency estimate. For a cruising speed of $0.1\,c$, this corresponds to a distance of 17 au between them. Of course, the spacecraft may not all receive exactly the same amount of energy and so not travel at the same speed, giving rise to a possibly uncontrollable spread in their arrival times. But if the laser could be operated for a year, 365 spacecraft could be sent, leading to high levels of redundancy and/or payload distribution.

Sending multiple spacecraft has implications for the ground segment, because ground-based telescopes are needed to receive the data from all of them ('ground' here includes Earth's orbit or the Moon). Given that flyby spacecraft will store their data and send it back over a duration that is much longer than their encounter durations, this means receiving data from multiple spacecraft simultaneously in the same telescopes. The signal coding and modulation discussed in chapter 15 will have to be designed to prevent interference between the signals, and allow us to distinguish between the spacecraft.

If the spacecraft are launched over a time span that is similar to their journey time, they could act as a relay of the data obtained by the currently encountering spacecraft. This could result in a larger data rate for a given transmitter power and area. Suppose that the ratio of the area of an Earth-based receiver to the spacecraft receiver is f ($f \gg 1$), and that each spacecraft receiver has the same area as its transmitter. All other things being equal, the data rate for spacecraft-to-spacecraft communication is then f times smaller than for spacecraft-to-Earth communication (from the Friis transmission equation in section 15.2). But in a relay, the individual spacecraft are much closer to one another than the encountering spacecraft is to the Earth, let's say by a factor of n, which is also the number of equally spaced spacecraft between the Earth and the target star. Considering again the Friis transmission equation, and neglecting other losses, it follows that when $n^2 > f$, a relay system delivers a higher data rate than a direct transmission. If the Earth-based receiver is an array of 100 telescopes each of 10 m diameter, and the spacecraft have 1 m diameter mirrors, then $f = 10^4$. In that case we would need more than $n = 100$ spacecraft for the relay to be advantageous. In reality we would need closer spacings, and so more spacecraft, because every reception and transmission in the relay adds instrument noise that we only have once in direct transmission. Furthermore, the benefit of the relay diminishes with time when the later spacecraft arrive at the target and lack the full set of relay spacecraft behind them. On the other hand, if the last n spacecraft are only used for relaying signals, then we no longer need a large receiver on the Earth, but just one equal to the size of the spacecraft receiver, because that is already the limiting factor in the relay.

The alternative to a relay of multiple spacecraft is to send them in a swarm, such that they arrive over a time period that is not much more than their encounter duration. Even when spacecraft are launched consecutively, for example by a single laser array, this could be achieved by giving them different velocities. There are several benefits of a swarm. First, it allows simultaneous observations of the same object from different viewing angles; this could be particularly interesting for transient phenomena. Second, it allows more active observations, in which one spacecraft transmits a signal that is received by another, for example in radar ranging or atmospheric transmission spectroscopy (see section 16.1). Third, the spacecraft could even together synthesize a large telescope aperture array for transmitting information back to the Earth. This gives much smaller diffraction losses than any single spacecraft transmitter acting alone. This could be achieved – in the radio at least – with a phased array antenna, mentioned in section 15.8.3.

16.4 Autonomous control

The multi-year round-trip communication time to the nearest stars implies that interstellar spacecraft have to operate autonomously for most of their mission. Intelligent software is therefore essential for many tasks that require obtaining and interpreting data, making a decision, and taking a course of action. These tasks include:

- navigation measurements and corresponding adjustments in spacecraft course and orientation;
- identifying targets, in particular planets and moons, in the target star system;
- switching instruments on and off, not only at the target star system but also at critical phases during the mission, such as crossing the Sun's heliopause;
- operating instruments, including tracking targets during a rapid flyby;
- prioritizing observations during critical mission phases in the face of limited observation time and power;
- reconfiguring computer hardware and modifying software as the result of damage, for example if some processors or memory are damaged by cosmic rays;
- prioritizing the storage and transmission of scientific data. The spacecraft may collect more data than it has the capacity to store or transmit back to the Earth (this is already the case in some astronomy missions). It will therefore need to assess the quality and the importance of the data. The channel capacity of the downlink may also be different from expectations (e.g. due to a damaged antenna or reduced power), which would also require a modification of the nominal data plan;

- communicating and coordinating with other spacecraft in a relay or swarm. Earlier-arriving spacecraft may fail to achieve specific goals, or may identify new objects worthy of investigation, requiring later spacecraft to reassess their goals and capabilities. If spacecraft are operating in concert, such as for radar or transmission spectroscopy measurements, then coordination is critical and complicated.

Interplanetary spacecraft have achieved increasing degrees of autonomy in recent years. Comet and asteroid intercept missions like Deep Impact have operated autonomously for hours at a time during their short mission-critical encounters. Other missions such as Rosetta, OSIRIS-Rex, and Hayabusa, which have spent extended periods of time at a comet or asteroid, also operated autonomously during their investigations, in each case bringing the spacecraft or a separate lander into contact with the body. Mars and Moon landers are also able to control their descent autonomously in response to real-time data on velocity and altitude, as well as the identification of unsuitable landing locations (rocks, craters). Generations of Mars rovers have used increasingly sophisticated algorithms to navigate autonomously over the surface of Mars, avoiding obstacles without intervention from Earth-based controllers.

As it will be difficult to pre-program software to allow for all possibilities, and as software updates will be difficult to perform, spacecraft will probably need to rely on artificial intelligence (AI) to some degree. Although we have seen huge increases in the performance of AI systems over the past few years, these require vast data sets to train. Such data do not exist for interstellar travel, so they would have to be simulated in some way. Trained AI systems often have large storage demands, and although computer memory is compact and light, the chips used on spacecraft must include sufficient shielding against ionizing radiation. Sophisticated computing systems may also require considerable power, and therefore cooling, both of which add mass to the spacecraft.

16.5 Summary

Interstellar spacecraft should carry a number of different scientific instruments to explore their target star systems. Many of these will be similar in concept to those used on planetary exploration missions in our solar system. This includes imagers, spectrometers, magnetometers, and particle and dust analyzers. As mass is even more critical for an interstellar mission, instruments will have to be optimized and miniaturized. An orbiting mission will allow much more science to be done with more sophisticated instruments than a fast flyby.

When the marginal cost of launching a single spacecraft is small compared to the total mission cost, it makes economic sense to launch multiple spacecraft. This allows the scientific instruments to be spread across multiple spacecraft or provides redundancy against the failure of some spacecraft. Multiple spacecraft launched near-simultaneously in a swarm can act together to increase the science return. When launched consecutively, they could instead adapt to the achievements of earlier spacecraft, and even act as a relay for communication back to Earth.

One of the biggest challenges of interstellar travel is achieving sufficient onboard power. Batteries, capacitors, and fuel cells are too massive or have too low energy densities for them to be a useful option. For a nuclear rocket it would be feasible to carry relatively massive power systems, such as a nuclear reactor or an RTG. Perhaps RTGs can be miniaturized, but currently they and reactors are far too massive to be carried by realistic laser sails. Higher-mass lightsails could be equipped with wafer photovoltaics, but more than a few astronomical units from a star the starlight is too weak for these to be effective. Whether sailcraft could survive most of their journey in hibernation with no power is unclear. In principle, power could be generated from ions in the ISM incident on the high-speed spacecraft, but extracting this requires devices that add mass.

16.6 Exercises

1. Go through the calculations in section 16.2 for the RTG to compute the total energy delivered per unit mass, but now after 200 years.
2. How is the operation of the instruments discussed in this chapter affected when used on a spacecraft passing the target star and planet at a relative speed of $0.1\ c$?

Outlook

This final chapter first takes a brief look at some speculative ideas for propulsion that currently lie at or beyond the edge of known physics. I then return to reality and summarize the main challenges of interstellar travel along with the most promising solutions we have considered in this book. I look again at the big picture of what we want to achieve and its ethical implications, and finish by briefly describing some steps that could be taken to get us on our way to the stars.

17.1 The edge of physics

I started this book by saying I would limit the scope to known physics, even if the engineering or financing could be extreme. Let us nonetheless take a brief look at some ideas that operate at the edge of physics, without necessarily violating accepted physical laws. The ideas focus mostly on moving between two points faster than light can, but without contradicting general relativity. Direct faster-than-light travel is not possible within even special relativity, because it would require an infinite amount of energy to get a finite mass up to the speed of light. If it were possible, it would also violate causality by allowing information – or in principle a person – to travel back in time, thereby creating paradoxes such as what happens if you kill your parents before you were born. Some of the ideas below also violate causality.

There are of course descriptions in science fiction or popular culture for faster-than-light travel that simply ignore known physics entirely. But without a replacement theory that can be tested, these descriptions are free to claim anything, so I don't consider them.

While the following ideas are curious, inspiring, and fun, they are at best impractical, at worst speculative, and cannot (yet) be the basis for any serious work on interstellar travel.

17.1.1 Warp drives

Captain Picard commands 'engage' and the Enterprise zips off at faster than the speed of light. Science fiction aside, things called warp drives have been the subject of serious scientific study. A 1994 paper by Miguel Alcubierre showed that within the framework of general relativity, it is in principle possible to distort spacetime around a small bubble, such that this bubble acquires an arbitrarily large velocity relative to an external observer. This is achieved by contracting spacetime in front of the spacecraft and expanding it behind the spacecraft, making it travel to a distant point in an arbitrarily short time. Within its local spacetime the spacecraft does not travel faster than the speed of light. In fact, it does not move at all, and the passengers experience no acceleration.

The drawback of this idea is that it requires a negative energy density to produce the necessary spacetime distortion. Although tiny amounts of negative energy do occur in the quantum mechanical Casimir effect, this warp drive needs vast amounts of it. An energy equivalent to the mass of the Sun would be needed to create a bubble about a metre in size. Even if such amounts of negative energy were available, there is no known way to manipulate it to create the required distortion. And even if there were, the spacecraft itself could not control the bubble. This would have to be done from the outside.

Alcubierre's paper nonetheless inspired follow-up work. Alternative warp drive models arising from general relativity have since been suggested. Although these do not overcome the significant practical problems, they have provided a deeper understanding of the concept. Interestingly, there are warp drive solutions for subluminal velocities that require only normal positive energies.

17.1.2 Wormholes

Black holes are sometimes used as convenient plot devices in science fiction stories to take a short cut to a distant star system. The problem with black holes is that (i) all but the most massive black holes induce enormous tidal forces that would rip a spacecraft apart, (ii) observed from the Earth, a spacecraft would take an infinite amount of time to cross the event horizon, and (iii) nothing can escape from the event horizon. This final inconvenience is sometimes countered by invoking a white hole exit at the destination, this being the hypothetical opposite of a black hole. But white holes are unstable, and the exit would seal itself before the spacecraft had a chance to emerge.

Rotating black holes, described by the Kerr metric, in principle provide a route back to flat spacetime, but this route is plagued by singularities and is also unstable. Even if these difficulties could miraculously be solved, these so-called Kerr tunnels lead to causality violations of the you-can-kill-your-parents type.

The wormhole is a solution to the general relativistic field equations that either connects two different universes or two arbitrarily distant points of a single universe. The connection, sometimes called an Einstein–Rosen bridge, suffers from many of the same problems as the black/white hole and Kerr tunnel between them, however, so this type of wormhole (Schwarzschild wormhole) is considered to be non-traversable. One particular difficulty is keeping the wormhole open: the throat of the tunnel expands and contracts (to zero, leading to disconnection) too rapidly to be able to pass through even at the speed of light.

There are, however, solutions of Einstein's equations that overcome the above problems, although we have no idea how to construct such traversable wormholes, or whether the laws of physics even allow them to be constructed. In addition to lower tidal forces and finite travel time (for both the spacecraft and external observers), the wormhole has no horizons, so two-way travel is in principle possible. However, the wormhole would have to be stabilized to perturbations, such as passing a spacecraft through it, and it is not clear whether that is possible. A big remaining problem is that to construct the wormhole, exotic matter with a negative energy density is required (as for the warp drive). Maybe there are quantum mechanical solutions (suggested by the Casimir effect), but until we have a quantum mechanical theory of gravity, we can only speculate.

17.1.3 Other ideas

Dark matter and dark energy appear to be prevalent in the universe, as inferred from their effects on visible matter, yet we have very little idea what they are. Sometimes dark matter or dark energy is invoked as a source of propulsion, or of exotic matter, but without having a testable theory of what they are, this remains purely speculative.

Various ideas to use astrophysical objects to provide propulsion have been suggested. These include performing extreme gravity assists around compact binary neutron stars, or deploying light sails or electric sails very close to supernovae or microquasars, where the energy fluxes are extremely large. Although these approaches are physically permitted, we would first have to travel the enormous distances to these exotic objects before we could use them to get anywhere else.

Another idea, familiar from science fiction, involves crossing into additional spatial dimensions beyond the familiar three. The purpose of jumping into this so-called hyperspace is to travel between two points in a time that is less than their light–travel time separation in normal space. Somehow time is not spent, or is reduced, when travelling in the additional dimension(s). String theory motivates the idea of higher dimensions, but without an actual theory of hyperspace we can say nothing about how we would cross into it.

A somewhat different idea for propulsion is that of antigravity. If gravity could be shielded in the way that electromagnetic fields can be, this would make escape from the solar system much easier, requiring much less propellant. An anti-gravitational force would even mean that a spacecraft would be accelerated away from the Sun. But assuming inertial mass remained unaffected, the gravity-free rocket equation would still be a limitation, so antigravity wouldn't be a big help.

Finally, there have been experimental investigations into so-called reactionless drives. These are essentially rockets without propellant, and so defy the law of conservation of momentum. There have been some experimental claims of their existence, but none of these could be reproduced under carefully controlled conditions. Giving up on a concept as fundamental as momentum conservation naturally requires very strong evidence (and might leave us wondering about a lot of phenomena and physical devices that depend on it). But as scientists we should be open to ideas and in particular to experimental evidence that could push the boundaries of known physics. Who knows what may lie beyond the edge?

17.2 Challenges and solutions

At the start of this book, I set down the goal of getting to a nearby star within a human lifetime. One of the biggest challenges is achieving the high velocity of around 10% the speed of light that this implies. It requires phenomenal amounts of energy even for small payloads. To reach this speed we can forget chemical rockets, ion engines, solar sails, and in fact anything else currently in use. Antimatter engines and the Bussard ramjet are physically possible but are so far fetched that we can rule them out for at least the foreseeable future. The two most promising propulsion technologies are fusion rockets and laser-driven sails. Both of these are extremely challenging.

Fusion rockets work either by inducing a series of small fusion explosions and coupling the charged products to the spacecraft via a magnetic field, or by having a sustained fusion reaction that leaks some plasma to act as a high-momentum propellant. Both approaches come with enormous challenges. One of the most promising fusion reactions for propulsion uses helium-3, but this barely exists on the Earth. It would have to be mined from the surface of the Moon or from the gas giant planets. Reactions involving only more-readily available isotopes such as deuterium and boron are more demanding to achieve. So far we have not achieved net energy-producing fusion on the Earth. And even for those serious fusion projects in development, their reactors are far too massive to conceive putting them in a spacecraft. Although there has been some research and development on fusion engines that siphon off plasma as propellant, they all remain far from their goal.

Laser-driven sails require very light materials and extremely powerful lasers focused by large telescope arrays. This offers a very different approach to propulsion than fusion engines. As a fusion engine itself (plus its propellant) would be relatively massive, surely tens of tonnes or more, it could carry a relatively massive payload, perhaps tens or hundreds of kilogrammes. Due to limits in laser power, a laser sail, in contrast, is likely to be just a few tens of grammes or less, with a payload of just a few grammes. This severely limits the scientific instruments that can be carried. However, because much of the cost and effort probably goes into building the laser array needed to propel the sail, and because the laser only operates for minutes to hours, many sailcraft could be launched over the course of many months that would collectively fulfil the scientific mission.

There are more requirements to interstellar travel than just propulsion, of course. One is onboard power for instruments and communication. For a fusion engine this may be relatively straightforward to address, as the nuclear reactions could provide electricity too. For a low-mass laser sail, in contrast, onboard power is a serious problem for which there is no obvious solution, as things like RTGs are still far too heavy. One option might be to hibernate the spacecraft for its journey and then use photovoltaic cells powered from the light of the target star during its flyby.

The third major challenge is communication. This is probably better done in the optical than in the radio (ideally in the ultraviolet, but more likely at visual or near-infrared wavelengths). Amazingly, signals from a telescope transmitter a few metres in diameter operating at several watts at the distance of Proxima Cen could be detected by a 30 m telescope on the Earth (or in orbit). But the data rates are extremely small, just a few tens of bits per second, even assuming we can transmit several bits per photon (which has been demonstrated). To scale this up we can either make the transmitter larger or more powerful – difficult given onboard mass limitations – or make the receiver larger, which is comparatively easier. With a hundred 10 m telescopes we could attain data rates of the order of $10\,\mathrm{kb\,s^{-1}}$, and sending data over many years then yields a reasonable data volume. The very large optical array required to propel laser sails could perhaps be repurposed as a receiver, which gives an extra incentive to use this type of propulsion technology. In the 50 years between launch and data reception, astronomers would be overjoyed to use this enormous array for science. It could also be used as part of the mission to study potential target systems, perhaps discovering new planets and helping optimize the science observations.

The time available for collecting data will be very limited for a flyby mission, so a spacecraft that goes into orbit is greatly preferred. A sailcraft has very limited opportunities to decelerate at the target star (the photons from the star reflecting off the sail won't decelerate it much). If the target is a binary or triple star system (as Alpha Cen is), a larger deceleration could be achieved with a combination of stellar photons and gravity assists. Electric and/or magnetic sails potentially give a much larger deceleration, but these are large, massive structures, incompatible with the lightness of a sailcraft. It's hard to see how we could get a fast laser sail into orbit at the destination. A fusion rocket has a big advantage here. By taking its propellant with it, it can decelerate significantly. The cost, however, is that the total amount of propellant needed is the square of that required to accelerate it to cruising speed (for a given payload mass). An interstellar orbital mission is a much larger undertaking than a flyby mission.

The fourth major challenge is the survivability of the spacecraft. As it flies at $0.1\,c$, particles will impact the front of the spacecraft with very high energies. Atoms and ions can be absorbed by a sufficiently thick (around 1 cm) layer of shielding. It therefore makes sense to minimize the cross-sectional area in the direction of travel by making the spacecraft long and thin. Any deployed sails and antennae may be thin enough that they would be perforated. This reduces their effective area, but provided critical components are not hit, it may not inhibit their function. Dust particles are a bigger problem. The larger ones cannot be shielded against and could destroy the spacecraft. The probability of this is low, however. Multiple spacecraft, as foreseen at least with laser sails, nonetheless provide redundancy against total destruction of individual spacecraft.

In summary, I identify the main challenges of fast interstellar travel as propulsion, communication, survivability, and – for low-mass laser sails – onboard power. There are other challenges too, such as navigation, autonomous control, and miniaturizing scientific instruments. These are by no means trivial, but from the current perspective they appear less demanding on the relative scale of things.

17.3   Broader issues of interstellar travel

In this book we have looked at the *physics* of interstellar travel. There are of course wider considerations that need to be addressed before undertaking such a journey. Here I briefly outline a few of the questions we should be asking, but without offering any definitive answers.

Should we undertake interstellar travel? Fast interstellar travel is physically possible. Achieving it will take time, effort, ingenuity, and money. In section 2.1 I outlined some of the possible motivations for pursuing it, first and foremost scientific, but also technological, economic, educational, and cultural. Some people think that we should not invest in technological projects as long as there is poverty and inequality on the Earth, and that we should instead focus on meeting our basic needs first. But what are our basic needs? Their definition has evolved with the development of society, and no doubt varies over the globe. Could we ever agree on when our basic needs have been met? While most people would agree that improving standards of living for everyone is an important task, is it the only thing we should be doing as a civilization? The average human condition has arguably improved with time, and this despite – or perhaps because – we didn't only focus on what we then thought were our basic needs. Learning, vision, and striving for something beyond our grasp surely has as much a place now as it did in the past, and as it will hopefully still have in the future.

No one pretends that interstellar travel is a basic need. But then again neither is the internet, paved roads, or central heating. Money is ultimately a currency for human effort and limited resources, so whether we want to undertake interstellar travel is primarily a question of opportunity cost. What other things would we *not* then do? It is unlikely that striving for interstellar travel would slow progress on medical development or tackling global warming, for example. But it might detract resources from similar projects, such as sending people to Mars or building an even bigger particle accelerator. So choices do have to be made and opportunities traded off, but not across the entire spectrum of human desires and activities.

Is interstellar travel ethical? Beside engineering, financial, and political aspects, there are also ethical questions regarding sending a spacecraft to the nearest stars.

Suppose the star system that we target with our spacecraft contains an inhabited planet. Our spacecraft could collide with that planet, potentially contaminating it with terrestrial bacteria or viruses that survive the long cold journey, which then out-compete and wipe out the indigenous life. Is that morally acceptable?

I have only discussed sending robotic spacecraft, yet some people see this merely as a precursor to sending humans. While big generational ships from science fiction novels immediately jump to mind, there are also more modest ideas of sending frozen human embryos or even just DNA sequences, along with the machinery to help them mature upon arrival at a suitable host planet. Is human expansion into the universe morally acceptable? What if the target planet turns out to be inhabited? The reader need not be reminded that humans have a pretty poor ethical record when it comes to colonization. But does some particular form of life have an indefinite and exclusive right to occupy a planet? Some people argue that expansion beyond the Earth is even an obligation, on the grounds that it significantly improves the chance of survival of the human race should we be destroyed on Earth, either through our own actions or by an external agent.

Is interstellar travel dangerous? As with many technologies, there are dangers of dual use. A powerful laser array would be very attractive to a malevolent actor for use as a weapon, for example. Because of this, one could argue that civilized societies should not pursue certain technologies. But one could instead argue that open development and control is a better way of preventing technology from being developed in secret and misused. Furthermore, some seemingly harmless technological development, for example in materials research, may ultimately have dual uses that we cannot anticipate. Does this mean we should close off whole fields of development 'just in case'?

If there is not only life, but intelligent life, in the target star system, how might it interpret a swarm of incoming spacecraft, even if none of these accidentally hit anything? As a reconnaissance prior to an invasion? Anything moving at 10% the speed of light can be interpreted as a weapon. Intelligent life that could observe these could presumably also trace their origin back to our solar system. The multi-hundred-gigawatt narrow-band laser used to propel sails is also a pretty good way of broadcasting our presence. How might aliens react? Movies generally like to portray aliens as hostile – it makes for good action – whereas scientists often like to think they would be other scientists. The truth is that we have no idea and no way of predicting. But we still have to ask ourselves what the consequences could be of sending a spacecraft.

As discussed in the previous chapter, the computers on an uncrewed spacecraft need to be smart for it to operate autonomously. They may even qualify as general AI. Is there a risk that such computers in deep space go rogue, for example using resources of an alien planet to reproduce and then colonize the Galaxy? If our spacecraft met and communicated with intelligent life, might it misrepresent humanity to our detriment?

Should we wait? One thing that some large, technological projects have in common is a progress or incentive trap. If we assume that technology will continue to progress, then it might be better to wait to take advantage of this to build a cheaper, better, faster space-craft in the future. Otherwise we might end up sending a spacecraft now that is overtaken by a faster and more capable one in the future. This, the argument goes, becomes a disincentive to develop anything now, or indeed at any point in the future. In some instances it may well be prudent to wait for specific developments. But as often as not, it is the very act of trying that expedites the developments we need. The incentive trap can also be circumvented if we proceed in smaller steps with precursor missions, as we'll look at in the next section.

Do we need to go so fast? I have talked about 'fast' interstellar travel in this book, with speeds of the order of $0.1\,c$, because I set the goal of a travel time less than a human lifetime. My choice of this timescale comes from the assumption that society would not be willing or able to pursue a longer mission. Young adults and children who witness the launch could live to see the results sent back, and so would continue to support it in anticipation of this.

But is this single-generational thinking necessary? Maybe it's a symptom of our times. Many medieval cathedrals took multiple human lifetimes to complete, and presumably their architects, builders, and funders often realized this when they started. Although the cathedral analogy should not be stretched, it shows not only that we can run very long-term projects, but also that sometimes this is the only way to achieve remarkable things. Even now, senior scientists initiate – and governments fund – projects they know they will never see the end of. Perhaps not all are motivated by selfish thoughts of their legacy.[1]

A major challenge of interstellar travel is the energy and power needed to get to high velocities. If the travel time can be increased from 50 to 100 or even 200 years, and thus the average velocity lowered to $0.05\,c$ or $0.025\,c$, some things become disproportionately easier: lasers can have lower power; rockets can use simpler engines or less propellant; damage from interstellar dust particles is reduced; braking with magnetic sails becomes more effective; payload masses can be increased. For travel times of a few hundred years, extremely light solar sails that dive to within a few solar radii of the Sun become an option for propulsion (section 10.7). If we can get used to thinking on longer timescales and being more open to multi-generational projects, this may be a more efficient and effective way to explore nearby stars, despite the longer travel times of individual spacecraft. Each human generation could send out its own spacecraft that will be exploited by a later generation, while simultaneously being able to analyze the results of a mission sent by an earlier generation. Maybe the assumption, embodied in the incentive trap, that we will always want to seek cheaper, better, faster ways of travelling to the stars, is wrong.

17.4   Where do we go from here?

If the technological hurdles to achieving fast interstellar travel appear huge, that's because they are. But to finish something we have to start. Even though there may be some techno-logical breakthroughs which allow us to leap-frog over intermediate steps, we surely need significant amounts of novel research and technology development.

---

[1] How long a cathedral took to complete depends on your definition of 'complete', plus cathedrals were used before they were finished. Some were also built very quickly: Hagia Sophia in Constantinople was completed in 537 CE after less than six years.

In terms of rockets, the focus will be on nuclear fusion. Much necessary research on achieving and sustaining net energy-producing fusion is already underway as part of the goal to replace fossil fuels for electricity production. But more specific research will be needed, not least on optimal fuels and on achieving efficient propulsion. Concerning laser sails, the main areas of research are high-power lasers as well as strong ultralight materials that are either highly reflective or are absorbent with a high melting point. Extremely large phase-controlled optical telescope arrays, complete with adaptive optics (if located on the Earth), must also be developed. This goes hand in hand with desirable developments in astronomy.

Beyond propulsion, more research is required on the effect of high-velocity impacts from atoms and dust particles and how to shield spacecraft against them. Mass is a critical factor on spacecraft, so scientific instruments need to be miniaturized and their power consumption reduced. In the realm of communication, we calculated that data from a small spacecraft at Proxima Cen can in principle be received at the Earth. But there need to be significant advances in optical communications technology at the level of individual photons if we are to get anywhere near to the required data rates with acceptably low errors.

We will also need in-flight tests of all technologies. Rather than jumping straight to an interstellar mission, we should first develop one or more precursor missions. These should have significant science goals in their own right in order to gather the support of a wide swathe of the scientific community.

One type of precursor mission is to the outer solar system, to visit Kuiper belt objects, find Oort cloud objects, and explore the heliopause. Multiple spacecraft sent in different directions would allow the shape of the heliopause to be measured directly. They could continue on to directly sample gas and dust in the ISM in order to characterize the particle impacts an interstellar spacecraft will have to survive. Another destination is the solar gravitational lens, starting at 548 au from the Sun in all directions. All of these missions could test not only fast propulsion but also deep space navigation and long-distance communication.

Another type of mission is to use solar sails to rendezvous with long-period comets or interstellar objects. Comets that orbit the Sun on near-parabolic or hyperbolic orbits move rapidly and are only detectable from the Earth shortly before perihelion, leaving little time to launch a spacecraft to intercept them. Solar sailcraft could instead be launched at any time and put on selected parking orbits, or even made to hover if their lightness numbers are high enough. After a target comet has been observed, the sails could then be tilted to follow a fast trajectory to intercept the target. Such sailcraft have some requirements in common with interstellar sailcraft.

A more direct precursor of laser sails – requiring lower- but still very high-power laser arrays on the Earth or in space – is fast interplanetary sailcraft. One example is transport to Mars in a few days, whereby a second laser on Mars would be used to decelerate incoming spacecraft. The laser array required for this could be scaled up over time to propel faster or deeper space missions, and then ultimately used for interstellar missions.

When such precursor missions could start is hard to say, not least because such projects can stand or fall at the whim of other commitments or short-term thinking. It is plausible, though, that some of them could start development by the middle of this century. When the first mission to another star could start depends very much on progress in the technologies I have outlined – or perhaps the arrival of others at the edge of physics – so a prediction is almost impossible. If I have to go out on a limb, then I like to think that the first mission could be launched by the end of this century.

Achieving interstellar travel will require imagination, dedication, and patience. Whether or when we achieve it depends ultimately on what we value, and what we are willing to commit. Time will tell. In any case, the journey in this book examining interstellar travel has now come to an end. But the real journey to the stars is only just beginning.

Appendix A Further reading

There are some non-technical books on the topic of interstellar travel. Gilster (2004) is particularly recommended, as well as Johnson (2022). Also very interesting, although now a little dated, is the slightly more technical book by Mallove and Matloff (1989). The book edited by Johnson and Roy (2024) contains some relevant chapters on propulsion, on-board power, and communication.

There are a number of relevant websites that at the time of writing are kept up to date. These include Centauri Dreams,[1] the Initiative for Interstellar Studies (in particular their magazine Principium),[2] the Interstellar Research Group,[3] and the British Interplanetary Society.[4] The latter publishes the *Journal of the British Interplanetary Society* (*JBIS*) in which many articles on interstellar travel appear. Another important journal for this field is *Acta Astronautica*,[5] although relevant articles are spread over many other journals in science and engineering. A useful resource for searching for articles and preprints in the astronomical literature is the SAO Astrophysical Data System.[6] In the following I discuss the main sources used in each chapter as well as some suggestions for further reading.

A.1 Introduction

Evans (2022) provides a detailed history of the Voyager project from precursor proposals through to the early 2020s. Bailer-Jones and Farnocchia (2019) computed the future trajectories of the Voyager and Pioneer spacecraft relative to the nearby stars. References to the Daedalus project are given in the relevant sections below.

A.2 Why travel to the stars?

Some motivations for interstellar travel are discussed by Crawford (2009). The exoplanet science cases for an interstellar spacecraft overlap significantly with the science cases for solar system missions, some examples of which are covered by the references listed under the payloads chapter below.

---

[1] https://centauri-dreams.org
[2] https://i4is.org
[3] https://irg.space
[4] https://bis-space.com
[5] https://sciencedirect.com/journal/acta-astronautica
[6] https://ui.adsabs.harvard.edu

A comprehensive overview of exoplanets and observation techniques is given by Perryman (2018). For a review of exoplanet statistics see Zhu and Dong (2021), although this field is developing quickly. Optical and infrared interferometry is reviewed by Eisenhauer et al. (2023). Space-based interferometry to obtain spectra of rocky exoplanets could be done by the proposed LIFE (Large Interferometer For Exoplanets) mission (Quanz et al. 2018). A milestone paper on the concept of the habitable zone is Kasting et al. (1993). An important update to this was made by Kopparapu et al. (2013), from which I take my definition. Biosignatures and their measurement are discussed by Kiang et al. (2007), Kaltenegger (2017), Meadows et al. (2018), and Schwieterman et al. (2018).

To compile the list of the nearest stars in table 2.1, I started with the CNS5 catalogue of nearby stars of Golovin et al. (2023) then cross-checked against other sources. For the list of planets in this table, I started with Reylé et al. (2021) then updated it with more recent publications on exoplanets around nearby stars. Rather than listing the many articles claiming, confirming, or refuting exoplanet detections, I refer the reader to the NASA exoplanet archive,[7] which compiles much of the information on the planets discussed in section 2.5.1.

Data for the Alpha Centauri system in section 2.5.2 are taken preferentially from Akeson et al. (2021). Dumusque et al. (2012) claimed to have detected a planet around Alpha Cen B via the radial velocity method, but the signal was weak, and later reanalyses of the data show it was probably a false detection (Hatzes 2013; Rajpaul et al. 2016). Some limits on possible planets based on non-detections are given by Zhao et al. (2018). The planet Proxima Cen b was discovered by Anglada-Escudé et al. (2016). The additional candidates c and d were reported by Damasso et al. (2020) and Faria et al. (2022), respectively.

The solar gravitational lens has been explored in some detail by Turyshev and Toth (2020). Helvajian et al. (2023) describe a possible mission to exploit it. The Pioneer anomaly is described by Anderson et al. (1998), and support for the anisotropic radiation solution is presented by Turyshev et al. (2012). References on the science of the heliosphere and the ISM can be found under the interstellar medium chapter below.

A.3 Rockets

The classical rocket equation and its application to rocketry is generally attributed to Konstantin Tsiolkovsky (1857–1935) in 1903 (reprinted in Tsiolkovsky 2004), although it was independently derived in the early twentieth century by other early rocket pioneers such as Hermann Oberth (1884–1989), Robert Esnault-Pelterie (1881–1957), and Robert Goddard (1882–1945). As early as 1810, William Moore (dates unknown) had described rockets and arrived at a closely related expression (Moore 1810). There are some variations on the derivation of the rocket equation in the literature. One can debate whether the velocity of dm relative to the rocket should be $V + dV - v_e$ rather than $V - v_e$ in equation 3.1a (Pinheiro 2004 argues for this), in which case $dVdm$ cancels instead of having to be neglected. But as we are dealing with differentials, it makes no difference for our case. The variable v_e 'perfect' rocket is discussed by Gowdy (1995). The instantaneous and discrete rocket equations are analyzed by Blanco (2019). Two general books on rockets are Turner (2009) and Walter (2018).

---

[7] https://exoplanetarchive.ipac.caltech.edu/index.html

A.4 Orbital mechanics

I cover only the basics of orbits along with some specific details used in later chapters. There are many books and pedagogical articles on various aspects of orbits. A good general reference is Curtis (2014).

In discussing the escape velocity as a function of planet size, I have made a number of crude approximations, such as very fast mass expulsion from the rocket and a fixed mass–radius relation for planets. These issues are addressed in more detail in several papers, in particular Gonzalez (2020).

The Hohmann transfer is named after the engineer Walter Hohmann (1880–1945), who published it in Hohmann (1925). The principle of the Oberth effect was published by Hermann Oberth in Oberth (1929). Orbital escape manoeuvres with two and three boosts are analyzed by Blanco and Mungan (2021). In my discussion of the efficiency of the bielliptic transfer orbit vs the simpler Hohmann transfer, I have used the threshold values from Curtis (2014). Technical details of the Voyager 2 gravity assists at the outer planets can be found in Cesarone (1989).

A.5 Thermal rockets

General overviews of thermal rocket engines can found in Turner (2009) and Walter (2018). The former also discusses propellants. Isentropic flows and the convergent–divergent nozzle are discussed in most text books on aerodynamics, such as Anderson (2016). Tajmar (2003) and Frisbee (2003b) discuss advanced propellants such as atomic or metallic hydrogen. Details of the Saturn V and Ariane 6 rockets can be found on the NASA and ESA websites.

A.6 Ion engines

My presentation of the principle of the gridded ion engine follows that given by Turner (2009). The Dawn ion propulsion system using NSTAR is described by Brophy (2011). More details of the Dawn spacecraft can be found in Thomas et al. (2011). Some of the variants on ion engines are discussed by Frisbee (2003b). An early paper for using nuclear electric propulsion for interstellar missions is Aston (1986).

A.7 Relativistic motion

There are many text books on special relativity at a range of levels with diverse approaches. Readers are advised to try out different ones if any one doesn't meet their needs. Some texts do not cover acceleration, which may explain the widespread misconception that special relativity does not apply to accelerating systems. There are also many didactic articles in the literature explaining various aspects or promising new insights. An excellent article that inspired my coverage of the journey to Alpha Cen is Müller et al. (2008).

The relativistic rocket equation goes back at least to Robert Esnault-Pelterie in 1928, but a more general derivation was provided by Ackeret (1946). There are several other derivations, including those of Forward (1995) and Walter (2018) (their section 1.4). The proper speed (rapidity) is discussed by Walter (2006). My discussion of the general matter–photon rocket draws on material from Walter (2006), Westmoreland (2010), and Walter (2018) (their sections 1.4.2 and 1.4.3). An earlier presentation along similar lines can be found in Vulpetti (1985). Note that Westmoreland (2010) identified an error in Walter (2006) in the derivation of the effective exhaust velocity. The latter neglected the factor of γ_m in the momentum of the matter particles in equation 7.36. This is corrected in section 1.4 of Walter (2018). Frisbee (2003a) claims that the relativistic rocket equation itself, and not just the expression for effective exhaust velocity, is different when there are mass losses, but they don't expand on the claim, which is criticized by Niang (2020).

A.8 Nuclear rockets

The data quoted for the uranium-235 reaction is from De Sanctis et al. (2016).

The NERVA programme is described in various technical reports available online, such as Robins and Finger (1991). Its predecessor project, ROVER, is described by Finseth (1991). Liquid and gas core nuclear thermal rockets are discussed by Thomas et al. (2021) and Frisbee (2003b), respectively. My outline of fission fragment rockets is based on Clark and Sheldon (2005) and Werka et al. (2012). An earlier description with a different design was given by Chapline et al. (1988).

Fission pulse rockets are described in various technical reports and articles, such as Platt and Hanner (1965) (which was originally classified), Martin and Bond (1979), and Schmidt et al. (2000) (the source of my quoted v_e values). The latter two also give a historical review of NASA's Project Orion. Platt and Hanner (1965) and Martin and Bond (1979) argue that the internal combustion arrangement – detonating nuclear bombs within your spacecraft – would be infeasible.

Further details of the fusion reactions I discuss can be found in any good textbook or review article. Ball (2019) discusses in greater depth the catalyzed deuterium reaction for use in rockets. Wittenberg et al. (1992) outline the sources of helium-3 from the Earth and Moon and how to extract it.

While various governments are investing in fusion projects, the most notable being ITER,[8] there are numerous commercial ventures. Their approaches and status are reviewed by Meschini et al. (2023)

There were various designs for the fusion pulse rocket Daedalus. My description uses data from Bond and Martin (1978) and Long et al. (2011). Long (2022) reviews concepts for fusion engines that use inertial confinement fusion, including the lower-mass Icarus design mentioned in the chapter, and provides many references on earlier studies. The z-pinch variation of Icarus is described by Freedland and Lamontagne (2015). The direct fusion drive

[8] https://iter.org/

is described in several articles, including Razin et al. (2014) and Cohen et al. (2019). There is little concrete material available on either the helicity drive or the fusion drive rocket, but some details are available from You (2020) and Slough (2024), respectively (both conference proceedings).

My derivation of the effective exhaust velocity for the antimatter (hydrogen–antihydrogen) rocket follows Westmoreland (2010) using data from Frisbee (2003a). Antimatter rockets (including thrust augmentation) were explored in some detail in a series of articles in *JBIS* in 1982, in particular Cassenti (1982), Forward (1982), and Morgan (1982). A more recent reference is Semyonov (2014). An approach to using antimatter-induced fission to decelerate a spacecraft is discussed by Jackson (2022).

A.9   Relativistic optical effects

Derivations of aberration and the Doppler effect can be found in many textbooks. My derivations are inspired by those in Christodoulides (2016). Greber and Blatter (1990), Lagoute and Davoust (1995), and Kraus (2000) discuss several of the optical effects I discuss in this chapter. Kraus (2000) provides a good derivation of the brightness effects (spectral radiance etc.).

Different authors use different conventions for the definition of the direction angle (α in my notation), namely whether it is the direction from the source to the observer or vice versa. The angle can therefore differ by $180°$, which results in a difference in the sign before the cosine of the angle.

The Terrell–Penrose rotation is named after the works of Terrell (1959) and Penrose (1959). Kraus (2000) shows that it is often unobservable in practice. Beig and Heinzle (2008) examine in detail the counter-intuitive effects of aberration on the appearance of an object as an observer accelerates away from it. Greber and Blatter (1990) and Yurtsever and Wilkinson (2018) show the impact of aberration and the Doppler effect on the appearance of the cosmic microwave background.

A.10   Solar sails

There are a lot of books and articles on solar sailing. A good overall reference for many aspects is McInnes (1999). This, as well as McInnes and Brown (1990), discuss the impact of the extended size of the Sun on a sail's dynamics.

The logarithmic spiral orbit dates back at least to 1959 (Bacon 1959, Tsu 1959) and is further covered by Bassetto et al. (2018), McInnes (1999), and other authors. Tsu (1959) also mentions briefly the energy of this non-Keplerian orbit. McInnes (1999) gives the expression linking γ to α and λ. Bailer-Jones (2021b) examines the sundiver combined with the Oberth effect.

Forward (1990) looks at imperfectly reflecting (grey) solar sails in some detail. Sail material properties is a very active area of research, with many publications. Kezerashvili

(2008) looks at the skin depth and the thickness of metallic sails. References discussing non-conducting sail materials are given in the chapter on laser sails below.

The LightSail 2 mission is described by Spencer et al. (2021). A summary of the experience from the DLR Gossamer programme is given by Spietz et al. (2021). The IKAROS mission is described by Tsuda et al. (2013); more information can be found on the eoPortal website.[9] JPL's concept for a large interstellar sail is described by Liewer et al. (2000), and ESA's heliopause probe is presented by Lyngvi et al. (2005). Macdonald (2014) provides a collection of articles on various aspects of the science and technology of sails. Ancona and Kezerashvili (2024) review some more recent developments presented at an international solar sailing conference in 2023, and Turyshev et al. (2023) discuss some of the science that can be done in and near the solar system with solar sailcraft.

A.11 Laser sails

The literature on laser sails goes back to not long after the development of the first lasers in the 1950s. Ever since, there have been debates in the literature about both the exact mathematical expressions and the interpretations of the phenomena. As this can be confusing for someone new to the field, I try now to give some overview.

In 1966, Marx published a paper on the basic principle of a laser-propelled sail and derived an expression for the efficiency of the energy transfer (Marx 1966). A year later, Redding claimed in a half-page response paper that Marx had made an error (Redding 1967). More than two decades after this, Simmons and McInnes (1993) published a terse but detailed derivation of the sail kinematics that leads to the expression for the velocity of the sail in terms of the energy supplied (equation 11.7) and the equation of motion in terms of retarded time (equation 11.25). The authors of this paper denounced Redding's correction of Marx as 'nonsensical', attributing it to Redding having mixed up coordinate time and retarded time. They state that Marx's mathematical solution was right, but for the wrong reasons, and go on to provide a corrected interpretation.

Another two decades later, Kipping (2017) examined the basic physics of laser sail kinematics and claimed that the results in an earlier paper – Kulkarni et al. (2016) – had assumed the incident and reflected photon energies were equal, thus leading to a 10% error in the final velocity of the sailcraft. The authors of the 2016 paper refuted this in Kulkarni et al. (2018) and gave a more detailed analysis of the problem, highlighting the error in the criticism, which was again related to mixing up coordinate and retarded time. Kipping conceded the error in an erratum (Kipping 2018). My derivation in section 11.1.3 does not assume the reflected photon has the same energy as the incident photon (see equation 11.12). Parkin (2018) too published a detailed analysis of laser sail kinematics. Both this paper and Kulkarni et al. (2018) agree on the equation of motion in terms of coordinate time (equation 11.16), and the latter paper also gives its integral explicitly (equation 11.19). Parkin (2018) goes on to use this model to describe the kinematics of a laser-accelerated sail in some detail, some of which I summarize in the chapter. (This model was updated and extended in Parkin 2024.) This paper also makes some economic calculations and sail optimizations.

---

[9] https://eoportal.org/satellite-missions/ikaros

My derivations of the laser sail kinematic equations in chapter 11 are based on – and are a synthesis of – Simmons and McInnes (1993), Kulkarni et al. (2018), and Parkin (2018). Kulkarni et al. (2018) extend my results in section 11.1.4 to take into account diffraction losses in a simplified way. Diffraction losses, also for synthesized apertures, are treated in more detail by Haas (2025).

A review of optical and infrared interferometry as used in astronomy is given by Eisenhauer et al. (2023). Aperture synthesis is discussed more specifically by Saha (2011), and Davies and Kasper (2012) review adaptive optics in astronomy. Peretz et al. (2022) discuss the idea of using an orbiting laser beacon to correct the atmospheric distortions when synthesizing a large laser aperture on the Earth's surface.

The properties of various candidate materials for laser sails are discussed by Matloff (2013) (graphene), Hein et al. (2017) (layered dielectrics), Mecklenburg et al. (2012) (aerographite), and Atwater et al. (2018) and Brewer et al. (2022) (2D photonic crystals). The stability and optimal shape of sails is discussed by several authors, for example, Popova et al. (2017), Gao et al. (2024), and Campbell et al. (2022). The last of these looks into an overall optimization of the curvature of a spherical sail together with other characteristics of the sail and laser system.

The mission concepts or sail designs mentioned in the chapter are described in Frisbee (2004), Hein et al. (2017), Parkin (2018), and Häfner et al. (2019). The Project Dragonfly competition workshop is summarized on the Initiative for Interstellar Studies website.

A number of other authors have published analyses of laser sails, either theory or specific mission profiles, including Forward (1984), Perakis et al. (2016), and Lubin (2016) (see also the two-volume mathematical book of Lubin 2022a,b). Some more advanced considerations are contained in Fuzfa et al. (2020). The idea for braking at the target star using a separable outer sail comes from Forward (1984), and the photo-gravitational assists for deceleration in the Alpha Cen system are studied by Heller and Hippke (2017).

A.12 The interstellar medium

The properties of the solar wind are summarized by Kallenrode (2004) and Owens (2020). More background material on the heliosphere and its boundary to the ISM appears in the review article by Zank (2015), and details on the ISM near to the Sun are provided by Crawford (2011) and Frisch et al. (2011). The quoted space density estimate of ISOs of $0.2\,\mathrm{au}^{-3}$ was derived by Do et al. (2018). The Voyager 1 crossing of the heliopause is reported by Gurnett et al. (2013). Articles that describe some of the science studies that an interstellar spacecraft could undertake of the heliosphere, its boundaries, and nearby interstellar space include McNutt et al. (2019), Brandt et al. (2022), and Brandt et al. (2023).

The relativistic stopping formula was derived by Bethe (1930); an earlier non-relativistic version was derived by Niels Bohr (1885–1962). For plotting figure 12.4 I adopt the same dependence of I on Z as used by Lubin et al. (2022). My consideration of the damage done by ions and dust particles to an interstellar spacecraft draws on results presented by Hoang et al. (2017) and London and Early (2018). Early and London (2000), Lubin et al. (2022), and Long (2023) also discuss impacts. Martin (1972) and Hoang (2017) consider in more detail the drag caused by ISM particles.

The Bussard ramjet is named after Robert Bussard (1928–2007), who first introduced the idea in Bussard (1960). My treatment of the ramjet's acceleration follows that of Blatter and Greber (2017), with some minor changes in the derivations. Semay and Silvestre-Brac (2004) go into more mathematical depth for the mass jet case, giving also the analytic expressions for the time and distance. Schattschneider and Jackson (2022) examine using a magnetic scoop to collect ISM particles (referred to as the Fishback ramjet). Whitmire (1975) discusses using the CNO cycle and the electromagnetic recovery of the kinetic energy of the ISM.

A.13 Magnetic and electric sails

The dipole magnetic sail was introduced by Andrews and Zubrin (1990). Its general concept and use are described by Zubrin and Andrews (1991), Funaki and Yamakawa (2009, 2012), Gros (2017), and Djojodihardjo (2018). My derivations follow the more comprehensive mathematical treatment in Freeland (2015), who also comments on the results derived in several of the earlier papers. The numerical values in section 13.1.2 are taken from Zubrin and Andrews (1991). Häfner et al. (2019) looked at using a magnetic sail to decelerate a high-mass laser sail.

The M2P2 concept was introduced by Winglee et al. (2000) and criticized by Khazanov et al. (2005). Other authors also conclude the M2P2 acceleration would be much less than claimed, because much of the momentum from the solar wind is lost by accelerating the plasma instead of the spacecraft.[10] Further development of the M2P2 concept has been carried out by Funaki and Nakayama (2012).

Electric sails were introduced by Janhunen (2004). My presentation of the theory is based on Janhunen and Sandroos (2007); see also Janhunen et al. (2014). A follow-up study by Janhunen (2009) found that the thrust may be five times higher than the model on which I base my results. Wiegmann (2017) outlines a concept for an electric sail accelerated by the solar wind to explore the outer solar system.

Perakis and Hein (2016) compare magnetic and electric sails and using them together for braking with the ISM. The combined effect of photon radiation pressure and magnetic fields for decelerating a charged interstellar spacecraft is analyzed by Forgan et al. (2018).

A.14 Navigation

For some historical background on navigation on the Earth, Alder (2014) gives an excellent account of mapping France by triangulation in the late eighteenth century.

The DOR and delta-DOR techniques are covered by a number of papers, for example James et al. (2009) and Iess et al. (2014). Some more history and technical details are given by Curkendall and Border (2013). The navigation of New Horizon's flyby of Pluto is described in Miller et al. (2005).

---

[10] For example https://space.fmi.fi/~pjanhune/papers/eMPii_final_1.3.pdf

The Gaia satellite has measured accurate parallaxes (as well as proper motions and radial velocities) for hundreds of millions of stars. The content of its third data release is summarized in Gaia Collaboration (2023). The technique of astrometric star navigation to determine the position and velocity of a relativistic spacecraft in interstellar space (section 14.4) from the angular separation of stars (or their radial velocities) is described and simulated by Bailer-Jones (2021a) (see also Yucalan and Peck 2021). A non-relativistic demonstration of this using observations from New Horizons when it was 47 au from the Sun is described by Lauer et al. (2025). Estimating just the velocity from aberration is investigated by Christian (2019) for the case of a spacecraft near to the Sun, and by Calabro' (2011) when using distant quasars.

A lot has been written about pulsar navigation, for example Sheikh et al. (2006), Becker et al. (2013), and Shemar et al. (2016). The pulsar navigation results that I report are from Shemar et al. (2016), Huang et al. (2019), and Yu et al. (2020). Ashby (2002) provides a non-technical overview of GNSS.

A.15 Communication

Various derivations of the Friis transmission equation can be found in the literature. They make a range of assumptions – perhaps implicit – and sometimes end up with different numerical factors, but all agree on the general dependencies. The correction for relative motion detailed in section 15.2 is only mentioned by a few authors, such as Parkin (2020).

Modulation methods are discussed in standard texts on signal processing. A foundational paper on information theory is Shannon (1948) and a good introduction to information theory and error-correcting codes is given by MacKay (2003). Moision and Farr (2014) show that the $1/r^2$ dependence of the channel capacity with distance r, as expected from the Friis transmission equation, can degrade to $1/r^4$ under certain circumstances.

The technique of coronagraphy is described by Rouan (2011) and Kenworthy and Haffert (2025). An example of its use to look for faint sources – in this case exoplanets – near Proxima Cen is reported in Grattor et al. (2020).

Early work on the introduction of quantum effects into information theory was done by Gordon (1962). Giovannetti et al. (2013) and Giovannetti et al. (2014) summarize the Holevo formulation for the channel capacity in the limit of photon counting. Farr et al. (2013) and Kanth Dacha et al. (2025) demonstrate achieving 10–15 bits of information per incident photon. The latter give more background on the quantum information paradigm and on using PPM for low photon rate communication. Several papers discuss the application of the Holevo bound for interstellar communications, including Hippke (2019) and Messerschmitt et al. (2023).

Single photon counting for optical communication between the Earth and an orbiting satellite was demonstrated by Villoresi et al. (2008). Biswas et al. (2024) summarize the optical communication channel using the DSOC on the Psyche satellite, which has since sent optical signals back to the Earth from a distance of 3 au.

Measurements of the brightness of emission sources from and beyond the Earth's atmosphere across the spectrum are presented by Leinert et al. (1998). Sources of airglow are discussed by von Savigny (2017).

Interstellar communications are discussed in numerous papers. Fridman (2011) and Shostak (2011) both discuss the application of the Friis transmission equation and the Shannon–Hartley theorem, as does Messerschmitt (2015), who also looks at the energy requirements and discusses channel coding. Messerschmitt et al. (2020b) provide an overview of the challenges of communication from a low-mass spacecraft.

The communications system of the New Horizons spacecraft is summarized by Fountain et al. (2008). The communication concept for Icarus is presented by Milne et al. (2016). Their optical alternative for Icarus evolves an earlier concept of Lesh et al. (1994) developed for Daedalus. Those concepts more appropriate for lighter sailcraft mentioned in the chapter are summarized by Parkin (2020), Lubin (2016), and Hippke (2019). Bazzani et al. (2023) outline the use of a phased array antenna as an optical transmitter on the spacecraft.

A.16 Payloads

The following articles summarize the science instruments carried by various interplanetary missions: Cardesin-Moinelo et al. (2024) (Mars Express); Pappalardo et al. (2024) (Europa Clipper); Grasset et al. (2013) (JUICE, to Jupiter and its moons); Weaver et al. (2008) (New Horizons to Pluto and the Kuiper belt). The instruments suggested for an interstellar precursor mission to explore the heliosphere and the nearby ISM are described by Brandt et al. (2023).

The idea of relay spacecraft is investigated by Messerschmitt et al. (2020a). One proposal for a swarm of satellites to synthesize a large transmitting aperture is presented in Eubanks et al. (2023). Nesnas et al. (2021) discuss past and future automation in solar system missions.

A.17 Outlook

The Alcubierre warp drive was introduced by Alcubierre (1994). Bobrick and Martire (2021) generalize the concept and discuss some alternative relativistic solutions, including a subluminal warp drive that works with a positive energy density. An accessible account of black holes and wormholes is given by Morris and Thorne (1988); this provided the means of travel in Carl Sagan's novel *Contact*. Lingam and Loeb (2020) investigate deploying electric sails and light sails near to extreme astrophysical objects as a means of achieving large velocities. Tajmar et al. (2022) give an overview of the theory and claimed measurement of one type of propellantless propulsion concept (the EMDrive), and provide an experimental refutation. The final chapter of Tajmar (2003) looks briefly at some breakthrough propulsion ideas.

The progress or incentive trap in the context of interstellar travel is discussed by Kennedy (2006).

Some ideas for interstellar precursor missions include: a 2 m-sized chemical rocket-propelled spacecraft to explore the heliosphere out to 550 au over a period of 50 years (Brandt et al. 2023); a solar sail made of aerographite that travels to Pluto's orbit within three years (Heller et al. 2020); laser sails that travel to Mars within a few days (Lubin and Hettel 2020); and solar and laser sails to chase interstellar objects (Lingam et al. 2023).

References

Ackeret, Jacob. 1946. Zur Theorie der Raketen. *Helvetica Physica Acta*, **19**, 103–112.

Akeson, Rachel, Beichman, Charles, Kervella, Pierre, et al. 2021. Precision millimeter astrometry of the α Centauri AB system. *Astronomical Journal*, **162**(1), 14. https://ui .adsabs.harvard.edu/abs/2021AJ....162...14A.

Alcubierre, Miguel. 1994. The warp drive: Hyper-fast travel within general relativity. *Classical and Quantum Gravity*, **11**(5), L73–L77. https://ui.adsabs.harvard.edu/abs/ 1994CQGra..11L..73A.

Alder, Ken. 2014. *The measure of all things: The seven-year odyssey and hidden error that transformed the world*. New York: Free Press.

Ancona, Elena, and Kezerashvili, Roman Ya. 2024. Recent advances in space sailing missions and technology: Review of the 6th International Symposium on Space Sailing (ISSS 2023). *arXiv e-prints*, Nov., arXiv:2411.12492. https://ui.adsabs.harvard.edu/abs/ 2024arXiv241112492A.

Anderson, John D. 2016. *Fundamentals of aerodynamics*. 6th ed. New York: McGraw-Hill.

Anderson, John D., Laing, Philip A., Lau, Eunice L., et al. 1998. Indication, from Pioneer 10/11, Galileo, and Ulysses data, of an apparent anomalous, weak, long-range acceleration. *Physical Review Letters*, **81**(14), 2858–2861. https://ui.adsabs.harvard.edu/abs/ 1998PhRvL..81.2858A.

Andrews, Dana G., and Zubrin, Robert M. 1990. Magnetic sails and interstellar travel. *Journal of the British Interplanetary Society*, **43**, 265–272. https://api.semanticscholar.org/ CorpusID:55324095.

Anglada-Escudé, Guillem, Amado, Pedro J., Barnes, John, et al. 2016. A terrestrial planet candidate in a temperate orbit around Proxima Centauri. *Nature*, **536**(7617), 437–440. https://ui.adsabs.harvard.edu/abs/2016Natur.536..437A.

Ashby, Neil. 2002. Relativity and the Global Positioning System. *Physics Today*, **55**(5), 41–47. https://doi.org/10.1063/1.1485583.

Aston, G. 1986. Electric propulsion: A far reaching technology. *Journal of the British Interplanetary Society*, **39**(Nov.), 503. https://ui.adsabs.harvard.edu/abs/1986JBIS...39 ..503A.

Atwater, Harry A., Davoyan, Artur R., Ilic, Ognjen, et al. 2018. Materials challenges for the Starshot lightsail. *Nature Materials*, **17**(Oct.), 861. https://doi.org/10.1038/ s41563-018-0075-8.

Bacon, Ralph Hoyt. 1959. Logarithmic spiral: An ideal trajectory for the interplanetary vehicle with engines of low sustained thrust. *American Journal of Physics*, **27**(3), 164–165. https://doi.org/10.1119/1.1934738.

Bailer-Jones, Coryn A. L. 2021a. Lost in space? Relativistic interstellar navigation using an astrometric star catalog. *Publications of the Astronomical Society of the Pacific*, **133**(1025), 074502. https://ui.adsabs.harvard.edu/abs/2021PASP..133g4502B.

Bailer-Jones, Coryn A. L. 2021b. The sundiver: Combining solar sails with the Oberth effect. *American Journal of Physics*, **89**(3), 235–243. https://ui.adsabs.harvard.edu/abs/2021AmJPh..89..235B.

Bailer-Jones, Coryn A. L., and Farnocchia, Davide. 2019. Future stellar flybys of the Voyager and Pioneer spacecraft. *Research Notes of the American Astronomical Society*, **3**(4), 59. https://ui.adsabs.harvard.edu/abs/2019RNAAS...3...59B.

Ball, Justin. 2019. Maximizing specific energy by breeding deuterium. *Nuclear Fusion*, **59**(10), 106043. https://ui.adsabs.harvard.edu/abs/2019NucFu..59j6043B.

Bassetto, Marco, Niccolai, Lorenzo, Quarta, Alessandro A., and Mengali, Giovanni. 2018. Logarithmic spiral trajectories generated by solar sails. *Celestial Mechanics and Dynamical Astronomy*, **130**(2), 18. https://ui.adsabs.harvard.edu/abs/2018CeMDA.130...18B.

Bazzani, Elisa, Guglielmi, Anna Valeria, Corvaja, Roberto, et al. 2023. The tree of light as interstellar optical transmitter system. *arXiv e-prints*, Aug., arXiv:2308.01900. https://ui.adsabs.harvard.edu/abs/2023arXiv230801900B.

Becker, Werner. 2009. X-Ray emission from pulsars and neutron stars. Page 91 of: Becker, Werner (ed.), *Astrophysics and Space Science Library*. Astrophysics and Space Science Library, vol. 357. https://ui.adsabs.harvard.edu/abs/2009ASSL..357...91B.

Becker, Werner, Bernhardt, Mike G., and Jessner, Axel. 2013. Autonomous spacecraft navigation with pulsars. *Acta Futura*, **7**(Nov.), 11–28. https://ui.adsabs.harvard.edu/abs/2013AcFut...7...11B.

Beig, Robert, and Heinzle, J. Mark. 2008. Relativistic aberration for accelerating observers. *American Journal of Physics*, **76**(7), 663–670. https://ui.adsabs.harvard.edu/abs/2008AmJPh..76..663B.

Bethe, H. 1930. Zur Theorie des Durchgangs schneller Korpuskularstrahlen durch Materie. *Annalen der Physik*, **397**(3), 325–400. https://onlinelibrary.wiley.com/doi/abs/10.1002/andp.19303970303.

Biswas, Abhijit, Srinivasan, Meera, Wollman, Emma, et al. 2024. Photon-counting for deep space optical communications. Page 1302508 of: Itzler, Mark A., Bienfang, Joshua C., and McIntosh, K. Alex (eds.), *Advanced Photon Counting Techniques XVIII*, vol. 13025SPIE, for International Society for Optics and Photonics, https://doi.org/10.1117/12.3013703.

Blanco, Philip R. 2019. A discrete, energetic approach to rocket propulsion. *Physics Education*, **54**(6), 065001. https://ui.adsabs.harvard.edu/abs/2019PhyEd..54f5001B.

Blanco, Philip R., and Mungan, Carl E. 2021. High-speed escape from a circular orbit. *American Journal of Physics*, **89**(1), 72–79. https://ui.adsabs.harvard.edu/abs/2021AmJPh..89...72B.

Blatter, Heinz, and Greber, Thomas. 2017. Tau Zero: In the cockpit of a Bussard ramjet. *American Journal of Physics*, **85**(12), 915–920. https://ui.adsabs.harvard.edu/abs/2017AmJPh..85..915B.

Bobrick, Alexey, and Martire, Gianni. 2021. Introducing physical warp drives. *Classical and Quantum Gravity*, **38**(10), 105009. https://ui.adsabs.harvard.edu/abs/2021CQGra..38j5009B.

Bond, A., and Martin, A. R. 1978. Project Daedalus: The mission profile. *Journal of the British Interplanetary Society*, **31**(Jan.), S37–S42. https://ui.adsabs.harvard.edu/abs/1978JBIS...31S..37B.

Brandt, P. C., Provornikova, E., Bale, S. D., et al. 2023. Future exploration of the outer heliosphere and very local interstellar medium by Interstellar Probe. *Space Science Review*, **219**(2), 18. https://ui.adsabs.harvard.edu/abs/2023SSRv..219...18B.

Brandt, Pontus C., Provornikova, E. A., Cocoros, A., et al. 2022. Interstellar Probe: Humanity's exploration of the Galaxy begins. *Acta Astronautica*, **199**(Oct.), 364–373. https://ui .adsabs.harvard.edu/abs/2022AcAau.199..364B.

Brewer, John, Campbell, Matthew F., Kumar, Pawan, et al. 2022. Multiscale photonic emissivity engineering for relativistic lightsail thermal regulation. *Nano Letters*, **22**(2), 594–601. https://doi.org/10.1021/acs.nanolett.1c03273, PMID: 35014534.

Brophy, John. 2011. The Dawn ion propulsion system. *Space Science Review*, **163**(1-4), 251–261. https://ui.adsabs.harvard.edu/abs/2011SSRv..163..251B.

Bussard, Robert W. 1960. Galactic matter and interstellar flight. *Astronautica Acta*, **6**(4), 179–194.

Calabro', E. 2011. Relativistic aberrational interstellar navigation. *Acta Astronautica*, **69**(7), 360–364. https://ui.adsabs.harvard.edu/abs/2011AcAau..69..360C.

Campbell, Matthew F., Brewer, John, Jariwala, Deep, et al. 2022. Relativistic light sails need to billow. *Nano Letters*, **22**(1), 90–96. https://doi.org/10.1021/acs.nanolett.1c03272, PMID: 34939817.

Cardesin-Moinelo, A., Godfrey, J., Grotheer, E., et al. 2024. Mars Express: 20 years of mission, science operations and data archiving. *Space Science Review*, **220**(2), 25. https://ui .adsabs.harvard.edu/abs/2024SSRv..220...25C.

Cassenti, Brice N. 1982. Design considerations for relativistic antimatter rockets. *Journal of the British Interplanetary Society*, **35**(Sept.), 396. https://ui.adsabs.harvard.edu/abs/ 1982JBIS...35..396C.

Cesarone, R. J. 1989. A gravity assist primer. *AIAA Student Journal*, **27**(Jan.), 16–22. https:// ui.adsabs.harvard.edu/abs/1989AIASJ..27...16C.

Chapline, George F., Dickson, Paul W., and Schnitzler, Bruce G. 1988. Fission fragment rockets – a potential breakthrough. In: *Proceedings of the 1988 International reactor physics conference*, vol. 4. USA: American Nuclear Society, https://inis.iaea.org/search/search .aspx?orig_q=RN:22026614.

Christian, John. 2019. StarNAV: Autonomous optical navigation of a spacecraft by the relativistic perturbation of starlight. *Sensors*, **19**(19), 4064. https://ui.adsabs.harvard.edu/abs/ 2019Senso..19.4064C.

Christodoulides, Costas. 2016. *The special theory of relativity*. Cham: Springer, https://doi .org/10.1007/978-3-319-25274-2.

Clark, Rodney, and Sheldon, Robert. 2005. Dusty Plasma Based Fission Fragment Nuclear Reactor. Pages 1–7 of: *41st AIAA/ASME/SAE/ASEE Joint Propulsion Conference and Exhibit*. American Institute of Aeronautics and Astronautics, https://arc.aiaa.org/doi/abs/ 10.2514/6.2005-4460.

Cohen, Samuel A., Swanson, C. P. S., McGreivy, Nick, et al. 2019. Direct fusion drive for interstellar exploration. *Journal of the British Interplanetary Society*, **72**, 37–50. https:// api.semanticscholar.org/CorpusID:229289135.

Crawford, Ian A. 2009. The astronomical, astrobiological and planetary science case for interstellar spaceflight. *Journal of the British Interplanetary Society*, **62**(Jan.), 415–421. https://ui.adsabs.harvard.edu/abs/2009JBIS...62..415C.

Crawford, Ian A. 2011. Project Icarus: A review of local interstellar medium properties of relevance for space missions to the nearest stars. *Acta Astronautica*, **68**(Apr.), 691–699. https://ui.adsabs.harvard.edu/abs/2011AcAau..68..691C.

Curkendall, David W., and Border, James S. 2013. Delta-DOR: The one-nanoradian navigation measurement system of the deep space network – History, architecture, and

componentry. *Interplanetary Network Progress Report*, **42-193**(May), 1–46. https://ui
.adsabs.harvard.edu/abs/2013IPNPR.193D...1C.

Curtis, Howard D. 2014. *Orbital Mechanics for Engineering Students*. 3rd ed. Boston:
Butterworth–Heinemann.

Damasso, Mario, Del Sordo, Fabio, Anglada-Escudé, Guillem, et al. 2020. A low-mass planet
candidate orbiting Proxima Centauri at a distance of 1.5 AU. *Science Advances*, **6**(3),
eaax7467. https://ui.adsabs.harvard.edu/abs/2020SciA....6.7467D.

Davies, Richard, and Kasper, Markus. 2012. Adaptive optics for astronomy. *Annual Review
of Astronomy and Astrophysics*, **50**(Sept.), 305–351. https://ui.adsabs.harvard.edu/abs/
2012ARA&A..50..305D.

De Sanctis, Enzo, Monti, Stefano, and Ripani, Marco. 2016. *Energy from nuclear fission*.
Cham: Springer, https://doi.org/10.1007/978-3-319-30651-3.

Djojodihardjo, Harijono. 2018. Review of solar magnetic sailing configurations for space
travel. *Advances in Astronautics Science and Technology*, **1**(11), 207–219. https://doi.org/
10.1007/s42423-018-0022-4

Do, Aaron, Tucker, Michael A., and Tonry, John. 2018. Interstellar interlopers: Number density and origin of 'Oumuamua-like objects. *Astrophysical Journal*, **855**(1), L10. https://ui
.adsabs.harvard.edu/abs/2018ApJ...855L..10D.

Dumusque, Xavier, Pepe, Francesco, Lovis, Christophe, et al. 2012. An Earth-mass planet
orbiting α Centauri B. *Nature*, **491**(7423), 207–211. https://ui.adsabs.harvard.edu/abs/
2012Natur.491..207D.

Early, James T., and London, Richard A. 2000. Dust grain damage to interstellar laser-
pushed lightsail. *Journal of Spacecraft and Rockets*, **37**(4), 526–531. https://ui.adsabs
.harvard.edu/abs/2000JSpRo..37..526E.

Eisenhauer, Frank, Monnier, John D., and Pfuhl, Oliver. 2023. Advances in optical/infrared
interferometry. *Annual Review of Astronomy and Astrophysics*, **61**(Aug.), 237–285.
https://ui.adsabs.harvard.edu/abs/2023ARA&A..61..237E.

Eubanks, T. Marshall, Blase, W. Paul, Hein, Andreas, et al. 2023. Swarming Proxima Centauri: Optical communication over interstellar distances. *arXiv e-prints*, Sept.,
arXiv:2309.07061. https://ui.adsabs.harvard.edu/abs/2023arXiv230907061E.

Evans, B. 2022. *NASA's Voyager missions*. 2 ed. Cham: Springer Praxis, https://doi.org/10
.1007/978-3-031-07923-8.

Faria, J. P., Suárez Mascareño, A., Figueira, P., et al. 2022. A candidate short-period sub-
Earth orbiting Proxima Centauri. *Astronomy & Astrophysics*, **658**(Feb.), A115. https://ui
.adsabs.harvard.edu/abs/2022A&A...658A.115F.

Farr, William H., Choi, John M., and Moision, Bruce. 2013. 13 bits per incident photon
optical communications demonstration. Page 861006 of: Hemmati, Hamid, and Boroson,
Don M. (eds), *Free-Space Laser Communication and Atmospheric Propagation XXV*, vol.
8610SPIE, for International Society for Optics and Photonics, https://doi.org/10.1117/12
.2007000.

Finseth, J. L. 1991 (February). *Rover nuclear rocket engine program: Overview of
rover engine tests*. Tech. rept. 19920005899. NASA. https://ntrs.nasa.gov/citations/
19920005899

Forgan, Duncan H., Heller, René, and Hippke, Michael. 2018. Photogravimagnetic assists
of light sails: A mixed blessing for Breakthrough Starshot? *Monthly Notices of
the Royal Astronomical Society*, **474**(3), 3212–3220. https://ui.adsabs.harvard.edu/abs/
2018MNRAS.474.3212F.

Forward, R. L. 1982. Antimatter propulsion. *Journal of the British Interplanetary Society*, **35**(Sept.), 391. https://ui.adsabs.harvard.edu/abs/1982JBIS...35..391F.

Forward, R. L. 1984. Roundtrip interstellar travel using laser-pushed lightsails. *Journal of Spacecraft and Rockets*, **21**(2), 187–195. https://ui.adsabs.harvard.edu/abs/1984JSpRo ..21..187F.

Forward, R. L. 1990. Grey solar sails. *The Journal of the Astronautical Sciences*, **38**(2), 161–185.

Forward, R. L. 1995. A transparent derivation of the relativisitic rocket equation. In: *21st Joint Propulsion Conference and Exhibit.* https://arc.aiaa.org/doi/abs/10.2514/6 .1995-3060.

Fountain, Glen H., Kusnierkiewicz, David Y., Hersman, Christopher B., et al. 2008. The New Horizons spacecraft. *Space Science Review*, **140**(1-4), 23–47. https://ui.adsabs.harvard .edu/abs/2008SSRv..140...23F.

Freedland, Robert M., and Lamontagne, Michel. 2015. Firefly Icarus: An unmanned interstellar probe using z-pinch fusion propulsion. *Journal of the British Interplanetary Society*, **68**(Jan.), 68–80. https://ui.adsabs.harvard.edu/abs/2015JBIS...68...68F.

Freeland, Robert M. 2015. Mathematics of magsails. *Journal of the British Interplanetary Society*, **68**(Jan.), 306–323. https://ui.adsabs.harvard.edu/abs/2015JBIS...68..306F.

Fridman, P. A. 2011. SETI: The transmission rate of radio communication and the signal's detection. *Acta Astronautica*, **69**(9-10), 777–787. https://ui.adsabs.harvard.edu/abs/ 2011AcAau..69..777F.

Frisbee, Robert. 2003a. *How to build an antimatter rocket for interstellar missions: Systems level considerations in designing advanced propulsion technology vehicles.* American Institute of Aeronautics and Astronautics, https://arc.aiaa.org/doi/10.2514/6.2003-4676. Pages 1–26.

Frisbee, Robert. 2004. *Beamed-momentum light sails for interstellar missions: Mission applications and technology requirements.* American Institute of Aeronautics and Astronautics, https://arc.aiaa.org/doi/abs/10.2514/6.2004-3567. Pages 1–24.

Frisbee, Robert H. 2003b. Advanced space propulsion for the 21st century. *Journal of Propulsion and Power*, **19**(6), 1129–1154. https://doi.org/10.2514/2.6948.

Frisch, Priscilla C., Redfield, Seth, and Slavin, Jonathan D. 2011. The interstellar medium surrounding the Sun. *Annual Review of Astronomy and Astrophysics*, **49**(1), 237–279. https://ui.adsabs.harvard.edu/abs/2011ARA&A..49..237F.

Funaki, Ikkoh, and Yamakawa, Hiroshi 2009. Research status of sail propulsion using the solar wind. *Journal of Plasma Fusion Research*, **8**, 1580–1584. www.jspf.or.jp/JPFRS/ index_vol8-9.html.

Funaki, Ikkoh, and Yamakawa, Hiroshi. 2012. Solar wind sails. Chap. 19 of: Lazar, Marian (ed.), *Exploring the solar wind.* Rijeka: IntechOpen, https://doi.org/10.5772/35673.

Fuzfa, André, Dhelonga-Biarufu, Williams, and Welcomme, Olivier. 2020. Sailing towards the stars close to the speed of light. *Physical Review Research*, **2**(4), 043186. https://ui .adsabs.harvard.edu/abs/2020PhRvR...2d3186F.

Gaia Collaboration. 2023. Gaia Data Release 3. Summary of the content and survey properties. *Astronomy & Astrophysics*, **674**(June), A1. https://ui.adsabs.harvard.edu/abs/ 2023A&A...674A...1G.

Gao, Ramon, Kelzenberg, Michael D., and Atwater, Harry A. 2024. Dynamically stable radiation pressure propulsion of flexible lightsails for interstellar exploration. *Nature Communications*, **15**(May), 4203. https://ui.adsabs.harvard.edu/abs/2024NatCo..15.4203G.

Gilster, Paul. 2004. *Centauri Dreams: Imagining and planning interstellar exploration*. New York: Springer Science & Business Media, https://doi.org/10.1007/978-1-4757-3894-0.

Giovannetti, Vittorio, Lloyd, Seth, Lorenzo, Maccone,, and Shapiro, Jeffrey H. 2013. Electromagnetic channel capacity for practical purposes. *Nature Photonics*, **7**(Oct.), 834–838. www.nature.com/articles/nphoton.2013.193.

Giovannetti, V., García-Patrón, R., Cerf, N. J., and Holevo, A. S. 2014. Ultimate classical communication rates of quantum optical channels. *Nature Photonics*, **7**(Sept.), 796–800. www.nature.com/doifinder/10.1038/nphoton.2014.216.

Golovin, Alex, Reffert, Sabine, Just, Andreas, et al. 2023. The fifth catalogue of nearby stars (CNS5). *Astronomy & Astrophysics*, **670**(Feb.), A19. https://ui.adsabs.harvard.edu/abs/2023A&A...670A..19G.

Gonzalez, Guillermo. 2020. The solar system: Favored for space travel. *BIO-Complexity*, **2020**(0). https://bio-complexity.org/ojs/index.php/main/article/view/117.

Gordon, James P. 1962. Quantum effects in communications systems. *Proceedings of the IRE*, **50**(9), 1898–1908. https://doi.org/10.1109/JRPROC.1962.288169

Gowdy, Robert H. 1995. The physics of perfect rockets. *American Journal of Physics*, **63**(3), 229–232. https://ui.adsabs.harvard.edu/abs/1995AmJPh..63..229G.

Grasset, O., Dougherty, M. K., Coustenis, A., et al. 2013. JUpiter ICy moons Explorer (JUICE): An ESA mission to orbit Ganymede and to characterise the Jupiter system. *Planetary Space Science*, **78**(Apr.), 1–21. https://ui.adsabs.harvard.edu/abs/2013P&SS...78....1G.

Gratton, R., Zurlo, A., Le Coroller, H., et al. 2020. Searching for the near-infrared counterpart of Proxima c using multi-epoch high-contrast SPHERE data at VLT. *Astronomy & Astrophysics*, **638**(June), A120. https://ui.adsabs.harvard.edu/abs/2020A&A...638A.120G.

Greber, T., and Blatter, H. 1990. Aberration and Doppler shift: The cosmic background radiation and its rest frame. *American Journal of Physics*, **58**(10), 942–945. https://ui.adsabs.harvard.edu/abs/1990AmJPh..58..942G.

Gros, Claudius. 2017. Universal scaling relation for magnetic sails: Momentum braking in the limit of dilute interstellar media. *Journal of Physics Communications*, **1**(4), 045007. https://ui.adsabs.harvard.edu/abs/2017JPhCo...1d5007G.

Gurnett, D. A., Kurth, W. S., Burlaga, L. F., and Ness, N. F. 2013. In situ observations of interstellar plasma with Voyager 1. *Science*, **341**(6153), 1489–1492. https://ui.adsabs.harvard.edu/abs/2013Sci...341.1489G.

Haas, Elisa. 2025. *Laser sail system optimization for in-situ exploration of exoplanets*. MSc thesis, Heidelberg University.

Häfner, Tobias, Kushwaha, Manisha, Celik, Onur, and Bellizzi, Filippo. 2019. Project Dragonfly: Sail to the stars. *Acta Astronautica*, **154**(Jan.), 311–319. https://ui.adsabs.harvard.edu/abs/2019AcAau.154..311H.

Hatzes, Artie P. 2013. The radial velocity detection of Earth-mass planets in the presence of activity noise: The case of α Centauri Bb. *Astrophysical Journal*, **770**(2), 133. https://ui.adsabs.harvard.edu/abs/2013ApJ...770..133H.

Hein, Andreas M., Long, Kelvin F., Fries, Dan, et al. 2017. The Andromeda study: A femtospacecraft mission to Alpha Centauri. *arXiv e-prints*, Aug., arXiv:1708.03556. https://ui.adsabs.harvard.edu/abs/2017arXiv170803556H.

Heller, René, and Hippke, Michael. 2017. Deceleration of high-velocity interstellar photon sails into bound orbits at α Centauri. *Astrophysical Journal*, **835**(2), L32. https://ui.adsabs.harvard.edu/abs/2017ApJ...835L..32H.

Heller, René, Anglada-Escudé, Guillem, Hippke, Michael, and Kervella, Pierre. 2020. Low-cost precursor of an interstellar mission. *Astronomy & Astrophysics*, **641**(Sept.), A45. https://ui.adsabs.harvard.edu/abs/2020A&A...641A..45H.

Helvajian, Henry, Rosenthal, Alan, Poklemba, John, et al. 2023. Mission architecture to reach and operate at the focal region of the solar gravitational lens. *Journal of Spacecraft and Rockets*, **60**(3), 829–847. https://ui.adsabs.harvard.edu/abs/2023JSpRo..60..829H.

Hippke, Michael. 2019. Interstellar communication. I. Maximized data rate for lightweight space-probes. *International Journal of Astrobiology*, **18**(3), 267–279. https://ui.adsabs.harvard.edu/abs/2019IJAsB..18..267H.

Hoang, Thiem. 2017. Relativistic gas drag on dust grains and implications. *Astrophysical Journal*, **847**(1), 77. https://ui.adsabs.harvard.edu/abs/2017ApJ...847...77H.

Hoang, Thiem, Lazarian, A., Burkhart, Blakesley, and Loeb, Abraham. 2017. The interaction of relativistic spacecrafts with the interstellar medium. *Astrophysical Journal*, **837**(1), 5. https://ui.adsabs.harvard.edu/abs/2017ApJ...837....5H.

Hohmann, W. 1925. *Die Erreichbarkeit der Himmelskörper*. Berlin: Oldenbourg Wissenschaftsverlag, https://doi.org/10.1515/9783486751406.

Huang, Liangwei, Shuai, Ping, Zhang, Xinyuan, and Chen, Shaolong. 2019. Pulsar-based navigation results: Data processing of the x-ray pulsar navigation – I. Telescope. *Journal of Astronomical Telescopes, Instruments, and Systems*, **5**(Jan.), 018003. https://ui.adsabs.harvard.edu/abs/2019JATIS...5a8003H.

Iess, Luciano, Di Benedetto, Mauro, James, Nick, et al. 2014. Astra: Interdisciplinary study on enhancement of the end-to-end accuracy for spacecraft tracking techniques. *Acta Astronautica*, **94**(2), 699–707. https://ui.adsabs.harvard.edu/abs/2014AcAau..94..699I.

Jackson, Gerald P. 2022. Deceleration of exoplanet missions utilizing scarce antimatter. *Acta Astronautica*, **197**, 380–386. www.sciencedirect.com/science/article/pii/S0094576522001163.

James, Nick, Abello, Ricard, Lanucara, Marco, et al. 2009. Implementation of an ESA delta-DOR capability. *Acta Astronautica*, **64**(11), 1041–1049. https://ui.adsabs.harvard.edu/abs/2009AcAau..64.1041J.

Janhunen, P. 2009. Increased electric sail thrust through removal of trapped shielding electrons by orbit chaotisation due to spacecraft body. *Annales Geophysicae*, **27**(8), 3089–3100. https://angeo.copernicus.org/articles/27/3089/2009/.

Janhunen, P., and Sandroos, A. 2007. Simulation study of solar wind push on a charged wire: basis of solar wind electric sail propulsion. *Annales Geophysicae*, **25**(3), 755–767. https://ui.adsabs.harvard.edu/abs/2007AnGeo..25..755J.

Janhunen, Pekka. 2004. Electric sail for spacecraft propulsion. *Journal of Propulsion and Power*, **20**, 763–764. https://api.semanticscholar.org/CorpusID:122272677.

Janhunen, Pekka, Toivanen, Petri, Envall, Jouni, et al. 2014. Overview of electric solar wind sail applications. *Proceedings of the Estonian Academy of Sciences*, **63**(3), 267–278. https://doi.org/10.3176/proc.2014.2S.08.

Johnson, Les. 2022. *A traveler's guide to the stars*. Princeton: Princeton University Press, https://press.princeton.edu/books/hardcover/9780691212371/.

Johnson, Les, and Roy, Kenneth (editors). 2024. *Interstellar Travel. Propulsion, Life Support, Communications, and the Long Journey*. Amsterdam: Elsevier, https://doi.org/10.1016/C2021-0-01074-9.

Kallenrode, May-Britt. 2004. *Space physics: An introduction to plasmas and particles in the heliosphere and magnetospheres*. 3rd ed. Heidelberg: Springer, https://doi.org/10.1007/978-3-662-09959-9.

Kaltenegger, Lisa. 2017. How to characterize habitable worlds and signs of life. *Annual Review of Astronomy and Astrophysics*, **55**(1), 433–485. https://ui.adsabs.harvard.edu/abs/2017ARA&A..55..433K.

Kanth Dacha, Sai, Essiambre, Rene-Jean, Ashikhimin, Alexei, et al. 2025. Communicating at a record 14.5 bits per received photon through a photon-starved channel. *arXiv e-prints*, Jan., arXiv:2501.13356. https://ui.adsabs.harvard.edu/abs/2025arXiv250113356K.

Kasting, James F., Whitmire, Daniel P., and Reynolds, Ray T. 1993. Habitable zones around main sequence stars. *Icarus*, **101**(1), 108–128. https://ui.adsabs.harvard.edu/abs/1993Icar..101..108K.

Kennedy, A. 2006. Interstellar travel: The wait calculation and the incentive trap of progress. *Journal of the British Interplanetary Society*, **59**(Jan.), 239–246. https://ui.adsabs.harvard.edu/abs/2006JBIS...59..239K.

Kenworthy, Matthew A., and Haffert, Sebastiaan Y. 2025. High-contrast coronagraphy. *Annual Review of Astronomy and Astrophysics*. www.annualreviews.org/content/journals/10.1146/annurev-astro-021225-022840.

Kezerashvili, Roman Ya. 2008. Solar sail interstellar travel – 1. Thickness of solar sail films. *Journal of the British Interplanetary Society*, **61**(Jan.), 430–439. https://ui.adsabs.harvard.edu/abs/2008JBIS...61..430Y.

Khazanov, G., Delamere, P., Kabin, K., and Linde, T. J. 2005. Fundamentals of the plasma sail concept: magnetohydrodynamic and kinetic studies. *Journal of Propulsion and Power*, **21**(5), 853–861. https://doi.org/10.2514/1.3737.

Kiang, Nancy Y., Siefert, Janet, Govindjee, and Blankenship, Robert E. 2007. Spectral signatures of photosynthesis. I. Review of Earth organisms. *Astrobiology*, **7**(1), 222–251. https://ui.adsabs.harvard.edu/abs/2007AsBio...7..222K.

Kipping, David. 2017. Relativistic light sails. *Astronomical Journal*, **153**(6), 277. https://ui.adsabs.harvard.edu/abs/2017AJ....153..277K.

Kipping, David. 2018. Erratum: Relativistic light sails. *Astronomical Journal*, **155**(2), 103. https://ui.adsabs.harvard.edu/abs/2018AJ....155..103K.

Kopparapu, Ravi Kumar, Ramirez, Ramses, Kasting, James F., et al. 2013. Habitable zones around main-sequence stars: New estimates. *Astrophysical Journal*, **765**(2), 131. https://ui.adsabs.harvard.edu/abs/2013ApJ...765..131K.

Kraus, U. 2000. Brightness and color of rapidly moving objects: The visual appearance of a large sphere revisited. *American Journal of Physics*, **68**(1), 56–60. https://ui.adsabs.harvard.edu/abs/2000AmJPh..68...56K.

Kulkarni, Neeraj, Lubin, Philip M., and Zhang, Qicheng. 2016 (Sept.). Relativistic solutions to directed energy. Page 998106 of: Hughes, Gary B. (ed.), *Planetary defense and space environment applications*. Society of Photo-Optical Instrumentation Engineers (SPIE) Conference Series, vol. 9981. https://ui.adsabs.harvard.edu/abs/2016SPIE.9981E..06K.

Kulkarni, Neeraj, Lubin, Philip, and Zhang, Qicheng. 2018. Relativistic spacecraft propelled by directed energy. *Astronomical Journal*, **155**(4), 155. https://ui.adsabs.harvard.edu/abs/2018AJ....155..155K.

Lagoute, C., and Davoust, E. 1995. The interstellar traveler. *American Journal of Physics*, **63**(3), 221–227. https://ui.adsabs.harvard.edu/abs/1995AmJPh..63..221L.

Lauer, Tod R., Munro, David H., Spencer, John R., et al. 2025. A demonstration of interstellar navigation using New Horizons. *Astronomical Journal*, **170**(1), 22. https://ui.adsabs.harvard.edu/abs/2025AJ....170...22L.

Leinert, Ch., Bowyer, S., Haikala, L. K., et al. 1998. The 1997 reference of diffuse night sky brightness. *Astronomy & Astrophysics Supplement*, **127**(Jan.), 1–99. https://ui.adsabs.harvard.edu/abs/1998A&AS..127....1L.

Lesh, J., Ruggier, C., and Cesarone, R. 1994 (August). *Space communications technologies for interstellar missions*. Tech. rept. 20060037973. NASA. https://ntrs.nasa.gov/citations/20060037973

Liewer, P. C., Mewaldt, R. A., Ayon, J. A., and Wallace, R. A. 2000. NASA's interstellar probe mission. *AIP Conference Proceedings*, **504**(1), 911–916. https://aip.scitation.org/doi/abs/10.1063/1.1302594.

Lingam, Manasvi, and Loeb, Abraham. 2020. Propulsion of spacecraft to relativistic speeds using natural astrophysical sources. *Astrophysical Journal*, **894**(1), 36. https://ui.adsabs.harvard.edu/abs/2020ApJ...894...36L.

Lingam, Manasvi, Hein, Andreas M., and Eubanks, T. Marshall. 2023. Chasing nomadic worlds: A new class of deep space missions. *Acta Astronautica*, **212**(Nov.), 517–533. https://ui.adsabs.harvard.edu/abs/2023AcAau.212..517L.

London, R. A., and Early, J. T. 2018. Evaluation of the hazard of dust impacts on interstellar spacecraft. *Journal of the British Interplanetary Society*, **71**(Jan.), 133–139. https://ui.adsabs.harvard.edu/abs/2018JBIS...71..133L.

Long, K. 2023. Calculations of particle bombardment due to dust and charged particles in the ISM on the project Starshot gram-scale interstellar probe. *Journal of the British Interplanetary Society*, **76**(Aug.), 262–272. https://ui.adsabs.harvard.edu/abs/2023JBIS...76..262L.

Long, K. F., Obousy, R. K., and Hein, A. 2011. Project Icarus: Optimisation of nuclear fusion propulsion for interstellar missions. *Acta Astronautica*, **68**(June), 1820–1829. https://ui.adsabs.harvard.edu/abs/2011AcAau..68.1820L.

Long, Kelvin F. 2022. Interstellar propulsion using laser-driven inertial confinement fusion physics. *Universe*, **8**(8), 421. https://ui.adsabs.harvard.edu/abs/2022Univ....8..421L.

Lubin, P. 2016. A roadmap to interstellar flight. *Journal of the British Interplanetary Society*, **69**(Jan.), 40–72. https://ui.adsabs.harvard.edu/abs/2016JBIS...69...40L.

Lubin, P., and Hettel, W. 2020. The path to interstellar flight. *Acta Futura*, **12**(Apr.). https://doi.org/10.5281/zenodo.3874099.

Lubin, Philip. 2022a. *The path to transformational space exploration. Volume 1: Fundamentals of Directed Energy*. World Scientific Publishing Company, https://ui.adsabs.harvard.edu/abs/2022pts1.book.....L.

Lubin, Philip. 2022b. *The path to transformational space exploration. Volume 2: Applications of directed energy*. World Scientific Publishing Company, https://ui.adsabs.harvard.edu/abs/2022pts2.book.....L.

Lubin, Philip, Cohen, Alexander N., and Erlikhman, Jacob. 2022. Radiation effects from the interstellar medium and cosmic ray particle impacts on relativistic spacecraft. *Astrophysical Journal*, **932**(2), 134. https://ui.adsabs.harvard.edu/abs/2022ApJ...932..134L.

Lyngvi, A., Falkner, P., and Peacock, A. 2005. The interstellar heliopause probe technology reference study. *Advances in Space Research*, **35**(12), 2073–2077. https://ui.adsabs.harvard.edu/abs/2005AdSpR..35.2073L.

Macdonald, Malcolm (ed.). 2014. *Advances in solar sailing*. Cham: Springer Praxis, https://doi.org/10.1007/978-3-642-34907-2.

MacKay, David J. C. 2003. *Information theory, inference and learning algorithms*. Cambridge: Cambridge University Press, www.inference.org.uk/mackay/itila/book.html.

Mallove, Eugene, and Matloff, Gregory 1989. *The starflight handbook*. John Wiley & Sons, www.wiley.com/en-us/The+Starflight+Handbook3A+A+Pioneer's+Guide+to+%Interstellar+Travel-p-9780471619123.

Martin, A. R. 1972. The effects of drag on relativistic spaceflight. *Journal of the British Interplanetary Society*, **25**(Jan.), 643–653. https://ui.adsabs.harvard.edu/abs/1972JBIS...25..643M.

Martin, Anthony R., and Bond, Alan. 1979. Nuclear pulse propulsion: A historical review of an advanced propulsion concept. *Journal of the British Interplanetary Society*, **32**(Aug.), 283. https://ui.adsabs.harvard.edu/abs/1979JBIS...32..283M.

Marx, G. 1966. Interstellar vehicle propelled by terrestrial laser beam. *Nature*, **211**(5044), 22–23. https://ui.adsabs.harvard.edu/abs/1966Natur.211...22M.

Matloff, G. L. 2013. The speed limit for graphene interstellar photon sails. *Journal of the British Interplanetary Society*, **66**(Jan.), 377–380. https://ui.adsabs.harvard.edu/abs/2013JBIS...66..377M.

McInnes, Colin R. 1999. *Solar sailing*. Cham: Springer Praxis, https://doi.org/10.1007/978-1-4471-3992-8.

McInnes, Colin R., and Brown, John C. 1990. The dynamics of solar sails with a non-point source of radiation pressure. *Celestial Mechanics and Dynamical Astronomy*, **49**(3), 249–264. https://ui.adsabs.harvard.edu/abs/1990CeMDA..49..249M.

McNutt, Ralph L., Wimmer-Schweingruber, Robert F., Gruntman, Mike, et al. 2019. Near-term interstellar probe: First step. *Acta Astronautica*, **162**(Sept.), 284–299. https://ui.adsabs.harvard.edu/abs/2019AcAau.162..284M.

Meadows, Victoria S., Reinhard, Christopher T., Arney, Giada N., et al. 2018. Exoplanet biosignatures: Understanding oxygen as a biosignature in the context of its environment. *Astrobiology*, **18**(6), 630–662. https://ui.adsabs.harvard.edu/abs/2018AsBio..18..630M.

Mecklenburg, Matthias, Schuchardt, Arnim, Mishra, Yogendra Kumar, et al. 2012. Aerographite: Ultra lightweight, flexible nanowall, carbon microtube material with outstanding mechanical performance. *Advanced Materials*, **24**(26), 3486–3490. https://onlinelibrary.wiley.com/doi/abs/10.1002/adma.201200491.

Meschini, Samuele, Laviano, Francesco, Ledda, Federico, et al. 2023. Review of commercial nuclear fusion projects. *Frontiers in Energy Research*, **11**. www.frontiersin.org/journals/energy-research/articles/10.3389/fenrg.2023.1157394.

Messerschmitt, David, Lubin, Philip, and Morrison, Ian. 2020a. Relaying swarms of low-mass interstellar probes. *arXiv e-prints*, July, arXiv:2007.11554. https://ui.adsabs.harvard.edu/abs/2020arXiv200711554M.

Messerschmitt, David, Lubin, Philip, and Morrison, Ian. 2023. Interstellar flyby scientific data downlink design. *arXiv e-prints*, June, arXiv:2306.13550. https://ui.adsabs.harvard.edu/abs/2023arXiv230613550M.

Messerschmitt, David G. 2015. Design for minimum energy in interstellar communication. *Acta Astronautica*, **107**, 20–39. www.sciencedirect.com/science/article/pii/S0094576514004391.

Messerschmitt, David G., Lubin, Philip, and Morrison, Ian. 2020b. Challenges in scientific data communication from low-mass interstellar probes. *Astrophysical Journal Supplement*, **249**(2), 36. https://ui.adsabs.harvard.edu/abs/2020ApJS..249...36M.

Miller, James, Stanbridge, Dale, and Williams, Bobby. 2005. New Horizons Pluto approach navigation. *Advances in the Astronautical Sciences*, **119**(01). https://www2.boulder.swri.edu/~tcase/nhnav-Miller-DaleAASpaper.pdf

Milne, P., Lamontage, M., and Freeland II, R.M. 2016. Project Icarus: Communications data link designs between Icarus and Earth and between Icarus spacecraft. *Journal of the British Interplanetary Society*, **69**(Jan.), 278–288. https://ui.adsabs.harvard.edu/abs/2016JBIS...69..278M.

Moision, Bruce, and Farr, William 2014. Range dependence of the optical communications channel. *Interplanetary Network Progress Report*, **42-199**(Nov.), 1–10. https://ui.adsabs.harvard.edu/abs/2014IPNPR.199B...1M.

Moore, W. 1810. Theory of the motion of rockets. *Journal of Natural Philosophy, Chemistry and the Arts*, **27**(Aug.), 276–285.

Morgan, David L. 1982. Concepts for the design of an antimatter annihilation rocket. *Journal of the British Interplanetary Society*, **35**(Sept.), 405. https://ui.adsabs.harvard.edu/abs/1982JBIS...35..405M.

Morris, Michael S., and Thorne, Kip S. 1988. Wormholes in spacetime and their use for interstellar travel: A tool for teaching general relativity. *American Journal of Physics*, **56**(5), 395–412. https://ui.adsabs.harvard.edu/abs/1988AmJPh..56..395M.

Müller, Thomas, King, Andreas, and Adis, Daria. 2008. A trip to the end of the universe and the twin "paradox". *American Journal of Physics*, **76**(4), 360–373. https://doi.org/10.1119/1.2830528.

Nesnas, Issa, Fesq, Lorraine, and Volpe, Richard. 2021. Autonomy for space robots: Past, present, and future. *Current Robotics Reports*, **2**(09). https://link.springer.com/article/10.1007/s43154-021-00057-2.

Niang, Samuel. 2020 (Dec.). *Optimisation of positron accumulation in the GBAR experiment and study of space propulsion based on antimatter*. Theses, Université Paris-Saclay, https://theses.hal.science/tel-03163848.

Oberth, Hermann. 1929. *Wege zur Raumschiffahrt*. Düsseldorf: VDI-Verlag, https://d-nb.info/860779556.

Owens, Mathew J. 2020. Solar-wind structure. In: *Oxford Research Encyclopedia of Physics*. Oxford University Press, https://doi.org/10.1093/acrefore/9780190871994.013.19.

Pappalardo, Robert T., Buratti, Bonnie J., Korth, Haje, et al. 2024. Science overview of the Europa Clipper mission. *Space Science Review*, **220**(4), 40. https://ui.adsabs.harvard.edu/abs/2024SSRv..220...40P.

Parkin, Kevin L. G. 2018. The Breakthrough Starshot system model. *Acta Astronautica*, **152**(Nov.), 370–384. https://ui.adsabs.harvard.edu/abs/2018AcAau.152..370P.

Parkin, Kevin L. G. 2020. A Starshot communication downlink. *arXiv e-prints*, May, arXiv:2005.08940. https://ui.adsabs.harvard.edu/abs/2020arXiv200508940P.

Parkin, Kevin L. G. 2024. Cost-optimal laser-accelerated lightsails. *arXiv e-prints*, May, arXiv:2205.13138. https://arxiv.org/abs/2205.13138v6.

Penrose, R. 1959. The apparent shape of a relativistically moving sphere. *Proceedings of the Cambridge Philosophical Society*, **55**(1), 137. https://ui.adsabs.harvard.edu/abs/1959PCPS...55..137P.

Perakis, Nikolaos, and Hein, Andreas M. 2016. Combining magnetic and electric sails for interstellar deceleration. *Acta Astronautica*, **128**(July), 13–20. https://ui.adsabs.harvard.edu/abs/2016AcAau.128...13P.

Perakis, Nikolaos, Schrenk, Lukas E., Gutsmiedl, Johannes, et al. 2016. Project Dragonfly: A feasibility study of interstellar travel using laser-powered light sail propulsion. *Acta Astronautica*, **129**(Dec.), 316–324. https://ui.adsabs.harvard.edu/abs/2016AcAau.129..316P.

Peretz, Eliad, Mather, John C., Hamilton, Christine, et al. 2022. Orbiting laser configuration and sky coverage: Coherent reference for Breakthrough Starshot ground-based laser array. *Journal of Astronomical Telescopes, Instruments, and Systems*, **8**(Jan.), 017004. https://ui.adsabs.harvard.edu/abs/2022JATIS...8a7004P.

Perryman, Michael. 2018. *The exoplanet handbook*. 2nd ed. Cambridge: Cambridge University Press. https://doi.org/10.1017/9781108304160

Pinheiro, Mario J. 2004. Some remarks about variable mass systems. *European Journal of Physics*, **25**(1), L5–L7. https://ui.adsabs.harvard.edu/abs/2004EJPh...25L...5P.

Platt, E. A., and Hanner, D. W. 1965 (5). *Effective specific impulse of a pulsed rocket engine*. Tech. rept. Lawrence Radiation Laboratory Livermore. https://www.osti.gov/biblio/1068247

Popova, Elena, Efendiev, Messoud, and Gabitov, Ildar. 2017. On the stability of a space vehicle riding on an intense laser beam. *Mathematical Methods in the Applied Sciences*, **40**(4), 1346–1354. https://onlinelibrary.wiley.com/doi/abs/10.1002/mma.4282.

Quanz, Sascha P., Kammerer, Jens, Defrère, Denis, et al. 2018 (July). Exoplanet science with a space-based mid-infrared nulling interferometer. Page 107011I of: Creech-Eakman, Michelle J., Tuthill, Peter G., and Mérand, Antoine (eds.), *Optical and Infrared Interferometry and Imaging VI*. Society of Photo-Optical Instrumentation Engineers (SPIE) Conference Series, vol. 10701. https://ui.adsabs.harvard.edu/abs/2018SPIE10701E..1IQ.

Rajpaul, Vinesh, Aigrain, Suzanne, and Roberts, Stephen. 2016. Ghost in the time series: No planet for Alpha Cen B. *Monthly Notices of the Royal Astronomical Society*, **456**(1), L6–L10. https://ui.adsabs.harvard.edu/abs/2016MNRAS.456L...6R.

Razin, Yosef S., Pajer, Gary, Breton, Mary, et al. 2014. A direct fusion drive for rocket propulsion. *Acta Astronautica*, **105**(1), 145–155. https://ui.adsabs.harvard.edu/abs/2014AcAau.105..145R.

Redding, J. L. 1967. Interstellar vehicle propelled by terrestrial laser beam. *Nature*, **213**(5076), 588–589. https://ui.adsabs.harvard.edu/abs/1967Natur.213..588R.

Reylé, Céline, Jardine, Kevin, Fouqué, Pascale, et al. 2021. The 10 parsec sample in the Gaia era. *Astronomy & Astrophysics*, **650**(June), A201. https://ui.adsabs.harvard.edu/abs/2021A&A...650A.201R.

Robins, W. H., and Finger, H. B. 1991 (July). *An historical perspective of the NERVA nuclear rocket engine technology program*. Tech. rept. 19910017902. NASA. https://ntrs.nasa.gov/citations/19910017902

Rouan, Daniel. 2011. *Coronagraphy*. Berlin: Springer, https://doi.org/10.1007/978-3-642-11274-4_356. Pages 363–372.

Saha, Swapan Kumar. 2011. *Aperture synthesis*. New York: Springer, https://doi.org/10.1007/978-1-4419-5710-8.

Schattschneider, Peter, and Jackson, Albert A. 2022. The Fishback ramjet revisited. *Acta Astronautica*, **191**(Feb.), 227–234. https://ui.adsabs.harvard.edu/abs/2022AcAau.191..227S.

Schmidt, G., Bonometti, J., and Morton, P. 2000. Nuclear pulse propulsion: Orion and beyond. In: *36th AIAA/ASME/SAE/ASEE Joint Propulsion Conference and Exhibit*. https://arc.aiaa.org/doi/abs/10.2514/6.2000-3856.

Schwieterman, Edward W., Kiang, Nancy Y., Parenteau, Mary N., et al. 2018. Exoplanet biosignatures: A review of remotely detectable signs of life. *Astrobiology*, **18**(6), 663–708. https://ui.adsabs.harvard.edu/abs/2018AsBio..18..663S.

Semay, Claude, and Silvestre-Brac, Bernard. 2004. The equation of motion of an interstellar Bussard ramjet. *European Journal of Physics*, **26**(1), 75. https://dx.doi.org/10.1088/0143-0807/26/1/009.

Semyonov, Oleg G. 2014. Relativistic rocket: Dream and reality. *Acta Astronautica*, **99**(June), 52–70. https://ui.adsabs.harvard.edu/abs/2014AcAau..99...52S.

Shannon, C. E. 1948. A mathematical theory of communication. *Bell System Technical Journal*, **27**(3), 379–423. https://onlinelibrary.wiley.com/doi/abs/10.1002/j.1538-7305.1948.tb01338.x.

Sheikh, Suneel I., Pines, Darryll J., Ray, Paul S., et al. 2006. Spacecraft navigation using X-ray pulsars. *Journal of Guidance Control Dynamics*, **29**(1), 49–63. https://ui.adsabs.harvard.edu/abs/2006JGCD...29...49S.

Shemar, Setnam, Fraser, George, Heil, Lucy, et al. 2016. Towards practical autonomous deep-space navigation using X-ray pulsar timing. *Experimental Astronomy*, **42**(2), 101–138. https://ui.adsabs.harvard.edu/abs/2016ExA....42..101S.

Shostak, Seth. 2011. Limits on interstellar messages. *Acta Astronautica*, **68**(3), 366–371. https://ui.adsabs.harvard.edu/abs/2011AcAau..68..366S.

Simmons, J. F. L., and McInnes, C. R. 1993. Was Marx right? or How efficient are laser driven interstellar spacecraft? *American Journal of Physics*, **61**(3), 205–207. https://ui.adsabs.harvard.edu/abs/1993AmJPh..61..205S.

Slough, John. 2024. The fusion driven rocket. Pages 1–20 of: *2024 IEEE Aerospace Conference*. https://doi.org/10.1109/AERO58975.2024.10521025

Spencer, David A., Betts, Bruce, Bellardo, John M., et al. 2021. The LightSail 2 solar sailing technology demonstration. *Advances in Space Research*, **67**(9), 2878–2889. https://ui.adsabs.harvard.edu/abs/2021AdSpR..67.2878S.

Spietz, Peter, Spröwitz, Tom, Seefeldt, Patric, et al. 2021. Paths not taken: The Gossamer roadmap's other options. *Advances in Space Research*, **67**(9), 2912–2956. www.sciencedirect.com/science/article/pii/S0273117721000995.

Tajmar, Martin. 2003. *Advanced space propulsion systems*. Vienna: Springer, https://doi.org/10.1007/978-3-7091-0547-4.

Tajmar, Martin, Neunzig, Oliver, and Weikert, Marcel. 2022. High-accuracy thrust measurements of the EMDrive and elimination of false-positive effects. *CEAS Space Journal*, **14**(01), 31–44. https://doi.org/10.1007/s12567-021-00385-1.

Terrell, James. 1959. Invisibility of the Lorentz contraction. *Physical Review*, **116**(4), 1041–1045. https://ui.adsabs.harvard.edu/abs/1959PhRv..116.1041T.

Thomas, Dale, Houts, Michael, Walters, William, et al. 2021 (August). *Establishing the feasibility of the centrifugal nuclear thermal rocket.* Tech. rept. 20210019705. NASA.

Thomas, Valerie C., Makowski, Joseph M., Brown, G. Mark, et al. 2011. The Dawn spacecraft. *Space Science Review*, **163**(1-4), 175–249. https://ui.adsabs.harvard.edu/abs/2011SSRv..163..175T. https://ntrs.nasa.gov/citations/20210019705

Tsiolkovsky, Konstantin E. 2004. *K. E. Tsiolkovsky selected works.* 2nd ed. Moscow: Nauka, www.tsiolkovsky.org/.

Tsu, T. C. 1959. Interplanetary travel by solar sail. *ARS Journal*, **29**(6), 422–427. https://doi.org/10.2514/8.4791.

Tsuda, Yuichi, Mori, Osamu, Funase, Ryu, et al. 2013. Achievement of IKAROS Japanese deep space solar sail demonstration mission. *Acta Astronautica*, **82**(2), 183–188. https://ui.adsabs.harvard.edu/abs/2013AcAau..82..183T.

Turner, Martin J. L. 2009. *Rocket and spacecraft propulsion.* 3rd ed. Cham: Springer Praxis, https://doi.org/10.1007/978-3-540-69203-4.

Turyshev, Slava G., and Toth, Viktor T. 2020. Image formation for extended sources with the solar gravitational lens. *Physical Review D*, **102**(2), 024038. https://ui.adsabs.harvard.edu/abs/2020PhRvD.102b4038T.

Turyshev, Slava G., Toth, Viktor T., Kinsella, Gary, et al. 2012. Support for the thermal origin of the Pioneer anomaly. *Physical Review Letters*, **108**(24), 241101. https://ui.adsabs.harvard.edu/abs/2012PhRvL.108x1101T.

Turyshev, Slava G., Garber, Darren, Friedman, Louis D., et al. 2023. Science opportunities with solar sailing smallsats. *Planetary Space Science*, **235**(Oct.), 105744. https://ui.adsabs.harvard.edu/abs/2023P&SS..23505744T.

Villoresi, P., Jennewein, T., Tamburini, F., et al. 2008. Experimental verification of the feasibility of a quantum channel between space and Earth. *New Journal of Physics*, **10**(3), 033038. https://ui.adsabs.harvard.edu/abs/2008NJPh...10c3038V.

von Savigny, C. 2017. Airglow in the Earth atmosphere: Basic characteristics and excitation mechanisms. *ChemTexts*, **3**(Oct.), 14. https://ui.adsabs.harvard.edu/abs/2017ChTxt...3...14V.

Vulpetti, Giovanni. 1985. Maximum terminal velocity of relativistic rocket. *Acta Astronautica*, **12**(2), 81–90. www.sciencedirect.com/science/article/pii/0094576585900761.

Walter, Ulrich. 2018. *Astronautics.* 3rd ed. Cham: Springer, https://doi.org/10.1007/978-3-319-74373-8.

Walter, Ulrich. 2006. Relativistic rocket and space flight. *Acta Astronautica*, **59**(6), 453–461. https://ui.adsabs.harvard.edu/abs/2006AcAau..59..453W.

Weaver, H. A., Gibson, W. C., Tapley, M. B., et al. 2008. Overview of the New Horizons science payload. *Space Science Review*, **140**(1-4), 75–91. https://ui.adsabs.harvard.edu/abs/2008SSRv..140...75W.

Werka, Robert, Clark, Rod, Sheldon, Rob, and Percy, Tom. 2012 (January). *Concept assessment of a fission fragment rocket engine (FFRE) propelled spacecraft.* Tech. rept. 20160010095. NASA. https://ntrs.nasa.gov/citations/20160010095

Westmoreland, Shawn. 2010. A note on relativistic rocketry. *Acta Astronautica*, **67**(9), 1248–1251. https://ui.adsabs.harvard.edu/abs/2010AcAau..67.1248W.

Whitmire, Daniel P. 1975. Relativistic spaceflight and the catalytic nuclear ramjet. *Acta Astronautica*, **2**(5), 497–509. www.sciencedirect.com/science/article/pii/0094576575900636.

Wiegmann, Bruce M. 2017 (February). *The conceptual design of an electric sail technology demonstration mission spacecraft*. Tech. rept. 20170001821. NASA. https://ntrs.nasa.gov/citations/20170001821

Winglee, R. M., Slough, J., Ziemba, T., and Goodson, A. 2000. Mini-magnetospheric plasma propulsion: Tapping the energy of the solar wind for spacecraft propulsion. *Journal of Geophysics Research*, **105**(A9), 21067–21078. https://ui.adsabs.harvard.edu/abs/2000JGR...10521067W.

Wittenberg, L. J., Cameron, E. N., Kulcinski, G. L., et al. 1992. A review of ^{3}He resources and acquisition for use as fusion fuel. *Fusion Technology*, **21**(4), 2230–2253. https://doi.org/10.13182/FST92-A29718.

You, Setthivoine. 2020. Helicity drive: A novel scalable fusion concept for deep space propulsion. In: *AIAA Propulsion and Energy 2020 Forum*. https://arc.aiaa.org/doi/abs/10.2514/6.2020-3835.

Yu, Wayne H., Semper, Sean R., Mitchell, Jason W., et al. 2020. NASA SEXTANT mission operations architecture. *Acta Astronautica*, **176**, 531–541. www.sciencedirect.com/science/article/pii/S0094576520304070.

Yucalan, Doga, and Peck, Mason. 2021. Autonomous navigation of relativistic spacecraft in interstellar space. *Journal of Guidance Control Dynamics*, **44**(6), 1106–1115. https://ui.adsabs.harvard.edu/abs/2021JGCD...44.1106Y.

Yurtsever, Ulvi, and Wilkinson, Steven. 2018. Limits and signatures of relativistic spaceflight. *Acta Astronautica*, **142**, 37–44. www.sciencedirect.com/science/article/pii/S0094576517305143.

Zank, G. P. 2015. Faltering steps into the Galaxy: The boundary regions of the heliosphere. *Annual Review of Astronomy and Astrophysics*, **53**(Aug.), 449–500. https://ui.adsabs.harvard.edu/abs/2015ARA&A..53..449Z.

Zhao, Lily, Fischer, Debra A., Brewer, John, et al. 2018. Planet detectability in the Alpha Centauri system. *Astronomical Journal*, **155**(1), 24. https://ui.adsabs.harvard.edu/abs/2018AJ....155...24Z.

Zhu, Wei, and Dong, Subo. 2021. Exoplanet statistics and theoretical implications. *Annual Review of Astronomy and Astrophysics*, **59**(Sept.), 291–336. https://ui.adsabs.harvard.edu/abs/2021ARA&A..59..291Z.

Zubrin, Robert M., and Andrews, Dana G. 1991. Magnetic sails and interplanetary travel. *Journal of Spacecraft and Rockets*, **28**(2), 197–203. https://ui.adsabs.harvard.edu/abs/1991JSpRo..28..197Z.

Index

For EU product safety concerns, contact us at Calle de José Abascal, 56–1°, 28003 Madrid, Spain or eugpsr@cambridge.org.

www.ingramcontent.com/pod-product-compliance
Ingram Content Group UK Ltd.
Pitfield, Milton Keynes, MK11 3LW, UK
UKHW052215120526
471007UK00007B/791